Achieving sustainable production of eggs

Volume 1: Safety and quality

It is widely recognised that agriculture is a significant contributor to global warming and climate change. Agriculture needs to reduce its environmental impact and adapt to current climate change whilst still feeding a growing population, i.e. become more 'climate-smart'. Burleigh Dodds Science Publishing is playing its part in achieving this by bringing together key research on making the production of the world's most important crops and livestock products more sustainable.

Based on extensive research, our publications specifically target the challenge of climate-smart agriculture. In this way we are using 'smart publishing' to help achieve climate-smart agriculture.

Burleigh Dodds Science Publishing is an independent and innovative publisher delivering high quality customer-focused agricultural science content in both print and online formats for the academic and research communities. Our aim is to build a foundation of knowledge on which researchers can build to meet the challenge of climate-smart agriculture.

For more information about Burleigh Dodds Science Publishing simply call us on +44 (0) 1223 839365, email info@bdspublishing.com or alternatively please visit our website at www.bdspublishing.com.

Related titles:
Achieving sustainable production of eggs Volume 2: Animal welfare and sustainability
Print (ISBN 978-1-78676-080-7); Online (ISBN 978-1-78676-082-1, 978-1-78676-083-8)

Achieving sustainable production of poultry meat Volume 1: Safety, quality and sustainability
Print (ISBN 978-1-78676-064-7); Online (ISBN 978-1-78676-066-1, 978-1-78676-067-8)

Achieving sustainable production of poultry meat Volume 2: Breeding and nutrition
Print (ISBN 978-1-78676-068-5); Online (ISBN 978-1-78676-070-8, 978-1-78676-071-5)

Achieving sustainable production of poultry meat Volume 3: Health and welfare
Print (ISBN 978-1-78676-072-2); Online (ISBN 978-1-78676-074-6, 978-1-78676-075-3)

Chapters are available individually from our online bookshop: https://shop.bdspublishing.com

climate•SMART•publishing
IN AGRICULTURAL SCIENCE

BURLEIGH DODDS SERIES IN AGRICULTURAL SCIENCE

NUMBER 16

Achieving sustainable production of eggs

Volume 1: Safety and quality

Edited by Professor Julie Roberts,
University of New England, Australia

burleigh dodds
SCIENCE PUBLISHING

Published by Burleigh Dodds Science Publishing Limited
82 High Street, Sawston, Cambridge CB22 3HJ, UK
www.bdspublishing.com

Burleigh Dodds Science Publishing, 1518 Walnut Street, Suite 900, Philadelphia, PA 19102-3406, USA

First published 2017 by Burleigh Dodds Science Publishing Limited

Library of Congress Control Number: 2016961789

British Library Cataloguing in Publication Data
A catalogue record for this book is available from the British Library

ISBN 978-1-78676-076-0 (print)
ISBN 978-1-78676-078-4 (online)
ISBN 978-1-78676-079-1 (online)
ISSN 2059-6936 (print)
ISSN 2059-6944 (online)

Typeset by Deanta Global Publishing Services, Chennai, India
Printed by Lightning Source

Contents

Part 2 Safety

Series list

Title	Series Number
Achieving sustainable cultivation of maize - Vol 1 From improved varieties to local applications Edited by: Dr Dave Watson, CGIAR Maize Research Program Manager, CIMMYT, Mexico	001
Achieving sustainable cultivation of maize - Vol 2 Cultivation techniques, pest and disease control Edited by: Dr Dave Watson, CGIAR Maize Research Program Manager, CIMMYT, Mexico	002
Achieving sustainable cultivation of rice - Vol 1 Breeding for higher yield and quality Edited by: Prof. Takuji Sasaki, Tokyo University of Agriculture, Japan	003
Achieving sustainable cultivation of rice - Vol 2 Cultivation, pest and disease management Edited by: Prof. Takuji Sasaki, Tokyo University of Agriculture, Japan	004
Achieving sustainable cultivation of wheat - Vol 1 Breeding, quality traits, pests and diseases Edited by: Prof. Peter Langridge, The University of Adelaide, Australia	005
Achieving sustainable cultivation of wheat - Vol 2 Cultivation techniques Edited by: Prof. Peter Langridge, The University of Adelaide, Australia	006
Achieving sustainable cultivation of tomatoes Edited by: Dr. Autar Mattoo, USDA-ARS, USA & Prof. Avtar Handa, Purdue University, USA	007
Achieving sustainable production of milk - Vol 1 Milk composition, genetics and breeding Edited by: Dr Nico van Belzen, International Dairy Federation (IDF), Belgium	008
Achieving sustainable production of milk - Vol 2 Safety, quality and sustainability Edited by: Dr Nico van Belzen, International Dairy Federation (IDF), Belgium	009
Achieving sustainable production of milk - Vol 3 Dairy herd management and welfare Edited by: Prof. John Webster, University of Bristol, UK	010
Ensuring safety and quality in the production of beef - Vol 1 Safety Edited by: Prof. Gary Acuff, Texas A&M University, USA & Prof.James Dickson, Iowa State University, USA	011
Ensuring safety and quality in the production of beef - Vol 2 Quality Edited by: Prof. Michael Dikeman, Kansas State University, USA	012
Achieving sustainable production of poultry meat - Vol 1 Safety, quality and sustainability Edited by: Prof. Steven C. Ricke, University of Arkansas, USA	013
Achieving sustainable production of poultry meat - Vol 2 Breeding and nutrition Edited by: Prof. Todd Applegate, University of Georgia, USA	014
Achieving sustainable production of poultry meat - Vol 3 Health and welfare Edited by: Prof. Todd Applegate, University of Georgia, USA	015

Introduction

The quality of the egg encompasses its chemical composition, nutritional quality as a human food source, sensory qualities including appearance and special nutraceutical benefits for human health. Eggs contain all essential nutrients for human health except for dietary fibre and vitamin C. This volume also discusses handling, packaging and storage of eggs, conditions that affect the quality of the product that reaches the consumer. The microbiological safety of eggs is of paramount importance as eggs have been implicated in outbreaks of food-borne illness, mainly caused by *Salmonella*. Fortunately, by virtue of its role in nature, the egg possesses many anti-microbial properties, and these properties can be maintained and enhanced for the benefit of consumers. This volume, *Achieving sustainable production of eggs Volume 1: Safety and quality*, explores the wealth of research addressing these themes.

Part 1 Egg composition and chemistry

Part 1 consists of three chapters on egg composition and chemistry of the eggshell, egg white and egg yolk, respectively. The integrity of the eggshell is of primary importance because it ensures the safe delivery of the egg's contents from the point of production to the consumer, as well as influencing the microbiological safety of the egg's contents, as outlined in **Chapter 1**. The layers of the eggshell perform different functions. The mammillary layer and its bond with the shell membranes is the foundation of the eggshell. The palisade layer forms the bulk of the thickness of the eggshell. The surface cuticle is a protective layer that impedes the ingress of bacteria into the egg's contents. The eggshell consists of both the inorganic calcite and an organic matrix, with the synthesis of the organic components determining the dynamics of eggshell production. At the time of publication, 699 proteins have been identified in the eggshell. Most characteristics of eggshell quality have moderate to high heritability, so genetic selection by breeder companies plays a vital role in maintenance and improvement of eggshell quality. In recent years, the availability of high-throughput technologies in fields such as proteomics and transcriptomics has facilitated the study of the organic components of the eggshell. The goal of the commercial industry is to extend worldwide the life span of layer flocks to around 100 weeks, at the same time maintaining eggshell quality.

Chapter 2 describes the structure and role of the egg white (albumen). As for the eggshell, modern techniques have facilitated the identification of many major proteins in egg white (78 identified to 2007), including ovalbumin, ovotransferrin, ovomucoid, ovoglobulins and ovomucin. In addition, there are many 'minor' proteins. Various proteins influence the properties of egg white, from textural and functional to anti-microbiological properties. The allergens in egg white are ovalbumin and ovomucoid. Ovomucin is largely responsible for the jelly-like structure of egg white and plays a major role in the egg white thinning during storage. Egg white contains many antimicrobial substances including lysozyme, ovotransferrin, ovomucin, ovocystatin and avidin, as well as carbohydrates, minerals and vitamins. In the food industries, egg white is used for its functional properties of gelation and foaming.

Egg yolk composition and chemistry is outlined in **Chapter 3**. Egg yolk consists of 15.7–16.6% proteins (lipoprotein, livetin, phosvitin, riboflavin-binding protein, biotin-binding protein, thiamine-binding protein, egg yolk transferrin, proteins with cholinesterase activity), 32–35% lipids (neutral lipids, phospholipids, cholesterol), minerals and vitamins, antibodies and bioactive substances such as sialic acid. Egg yolk immunoglobulins (IgY) can be selectively produced in eggs for human therapeutic purposes. Egg yolk can be separated into two main fractions by dilution and centrifugation: plasma (supernatant – 85% low-density lipoproteins, 15% livetins) and granules (precipitate – 70% high-density lipoprotein or lipovitellin,16% phosvitin). Egg yolk can be fortified with desirable substances such as omega-3 fatty acids, folic acid and vitamins by enriching the diet of the hens. Continuing development in proteomics of egg yolk proteins has great potential for identifying novel egg proteins and their biological functions for use in human health. Egg yolk lipids are used in human infant formula because they closely mimic the composition of human breast milk. In the food industries, egg yolk is used as an emulsifier. Recent studies have shown that earlier concerns about consumption of egg yolk causing elevation of plasma cholesterol in humans are unfounded as eggs contain low levels of saturated fatty acids which are the main cause of elevated plasma cholesterol.

Part 2 Safety

Part 2 comprises Chapters 4–10 on the food safety of eggs and egg products. **Chapter 4** lists the main pathogenic bacteria associated with eggs. Although a large number of microorganisms (bacteria, yeasts, moulds) may be found on the outside of eggs, the organisms of greatest concern are *Salmonella* spp. and, to a lesser extent, *Campylobacter* spp. In many parts of the world, *Salmonella* Enteritidis is the main cause of egg-associated food-borne illness. *S.* Enteritidis is capable of entering the egg during its forming in the hen's oviduct. It is also able to pass across the shell of a fully-formed egg. To date it appears to be the only type of *Salmonella* capable of such vertical transmission. In Australia, where *S.* Enteritidis is not endemic in commercial laying hen flocks, *S.* Typhimurium is the main cause of egg-associated illness. **Chapter 5** explores the mechanisms by which pathogens of concern to human health are transmitted into the contents of eggs, with emphasis on *S.* Enteritidis. Although *S.* Enteritidis is able to enter an egg via the hen's oviduct (primary or vertical transmission), transmission of other pathogens appears to occur exclusively by horizontal transmission from the outside of the eggshell into the egg contents. Other pathogens associated with eggs are discussed including other serovars of *Salmonella* such as Typhimurium, *Staphylococcus aureus, Bacillus cereus, Clostridium perfringens, Campylobacter jejuni* and *Listeria monocytogenes.* Mechanisms of trans-shell penetration (horizontal transmission) are discussed, including risk factors such as cracks or defects in the shell and/or cuticle, storage conditions and water condensation on the eggs. Chapter 5 also covers the natural antimicrobial properties of all regions of eggs.

Chapter 6 reviews the literature on methods of sampling and detection of *Salmonella* in eggs. This information is critical to effective monitoring of commercial eggs. Emphasis is on *S.* Enteritidis, although other bacteria associated with eggs are mentioned. In commercial production systems, pathogens such as *Salmonella* arise mainly from environmental sources. It is difficult to detect bacteria inside eggs so sampling needs to involve large numbers of eggs. Storage time and temperature influence the ease with which bacteria can reach the egg contents. Bacteria are detected on the outside of eggshells by immersing the whole eggs in a nutrient broth in special bags. Shells may be crushed to release bacteria contained in the pores of the shells. Detection of bacteria in egg contents requires pooling and incubation of samples to increase the chances of detection. Following pre-enrichment, samples are plated onto selective agar, incubated and colonies selected for further analysis. Rapid detection methods have increased the speed at which the process can be conducted. Egg yolk antibody detection provides additional information. **Chapter 7** provides a more detailed description of the antimicrobial properties of egg white and addresses ways of enhancing these natural defences. Detailed information is provided on the properties and role of lysozyme; ovotransferrin; the proteinase inhibitors ovomucoid, ovoinhibitor, ovostatin and cystatin; the vitamin-binding proteins avidin and riboflavin-binding protein; the defensins AvBD11 and Gallin (OvoDA1) and others including Tenp (Transiently Expressed in Neural Precursor) (Ovoglobulin G2); Ovalbumin-related protein X (OVAX) and recently identified proteins VMO-1, pleiotrophin and beta-microseminoprotein-like. Factors influencing bacterial growth and spread in egg white include temperature, pH, viscosity, carbon dioxide levels and ionic composition. This chapter also reports on the effects of physiology, genetics, age and immune status of the hen on egg white antimicrobial properties. The effect of changes occurring during egg storage, on the antimicrobial properties, of the egg white are also discussed in detail. The use of modified atmosphere packaging (MAP) and cryogenic cooling of eggs can improve the properties of stored eggs. Also discussed are the changes that occur during incubation of embryonated eggs and how these affect the antimicrobial properties of the egg white.

The egg industry, all around the world, has experienced a trend of moving away from conventional cages, towards cage-free production systems. Conventional cages are banned in the EU and egg production in Australia has been moving progressively towards free-range production. **Chapter 8** reviews the effect of housing system on egg quality and safety. Social pressures resulting from concerns about hen welfare have resulted in this shift towards cage-free production, although this production system does not result in a product of higher quality. The chapter reviews the types of microorganisms that are found on eggs. Some of these are indicators of general cleanliness and others are potentially pathogenic for humans. Research studies comparing housing systems have produced contradictory results, although non-cage systems are more likely to result in higher levels of bacteria on the outside of eggs and this is particularly the case for floor eggs. Strain of hen, flock management, management of the nest boxes and stocking density have all been shown to influence egg cleanliness. Challenge studies using *Salmonella* indicate that non-cage production systems do not necessarily result in worse outcomes for food safety. Similar results were obtained when commercial farms were surveyed for the presence of *Salmonella* and *Campylobacter* with housing system not having a consistent effect on flock microbiological status. Various studies have investigated the effect of production system on eggshell quality and egg internal quality. Although some studies have reported best-quality eggs from conventional cages, others have found no consistent relationship. It appears that some of the earlier egg quality problems that occurred in non-cage production systems have been overcome by improvements in the design of the production systems and in flock management. Ensuring that hens lay in the nest boxes in cage-free systems is essential for the maintenance of egg quality and safety.

Chapter 9 evaluates the effect on food safety of the egg-washing process. Egg washing is not permitted for first quality eggs in Europe, but is commonly practiced in other countries including the United States and Australia. Whereas properly conducted egg washing can improve the microbiological safety of eggs, it can actually make the situation worse if washing is not carried out correctly. A series of steps need to be carried out and important considerations are temperature; time; type, concentration and application of the cleaning chemical used (including agitation, use of brushes); rinsing; application of sanitizer; and drying of the eggs. This needs to be followed by monitoring, process control and verification. Care needs to be taken to avoid damage to the cuticle of the eggshell and to avoid ingress of the egg washing solutions into the egg contents.

The type of packaging used for transport of eggs from the processing plant to the consumer is important, as reviewed in **Chapter 10**. Most egg packaging used throughout the world is 'low barrier', meaning that it is very permeable to air and water vapour, but has the advantage of minimizing the build-up of water condensation on the eggs. The loss of carbon dioxide and water vapour for eggs housed in low barrier packaging results in changes in egg quality during storage. These losses can be reduced by using high barrier packaging but at the expense of increased condensation on the eggs. MAP enables maintenance of egg quality. Carbon dioxide atmosphere packaging improves egg albumen quality in terms of Haugh unit, coagulation and foaming, whipping and gelling properties. Although studies have shown no adverse effect of the MAP on the microbiological safety of the eggs, it is recommended that an absorbent be included in the packaging to reduce condensation of water vapour on the outside of the eggs.

Part 3 Sensory and nutritional quality

Part 3, Chapters 11–17, covers the sensory and nutritional quality of eggs. Maintenance of egg quality is of fundamental importance to the egg industries as consumers are increasingly discriminating in their choice of food items. **Chapter 11** examines consumer preferences for egg quality and the ways in which quality can be measured. Quality indicators of concern to consumers are cleanliness, egg size, shell colour and shell colour uniformity, shell strength and integrity (absence of cracks and defects), albumen quality, yolk colour, absence of meat and blood spots. Preferences for some of these measures vary among countries. Some consumers are strongly influenced by animal welfare considerations and the type of production system. Others focus on health benefits and are prepared to pay a premium for specialty eggs such as organic, or eggs fortified with nutrients such as omega-3 fatty acids. A range of different equipment types are available for measurement of eggshell quality and egg internal quality in research studies and in quality assurance laboratories associated with commercial facilities. However, in commercial egg-processing operations, equipment for measuring quality needs to be compatible with the in-line processing equipment used. Such methodologies are being extended to detection of microbiological agents, as well as of eggs produced in one production system (e.g. conventional cage), but are being sold as having been produced in a different production system (e.g. free range). Measurement of egg appearance and colour is further explored in **Chapter 12**. Egg consumption per annum varies greatly among countries, with Mexico and Japan having the highest consumption at more than 300 eggs per year per person. The attractiveness of eggs as a food item is influenced by many of the quality measurements described in Chapter 12 and also by packaging and marketing. This chapter discusses the regional differences in consumer preferences and outlines ways in which egg appearance and colour can be achieved.

Shell eggs are regarded as having a limited shelf life and **Chapter 13** investigates ways of improving the shelf life of eggs. Shelf life is important both in terms of the sensory qualities of eggs and also in terms of their microbiological safety. This chapter outlines some of the ways in which shelf life is determined by regulators in different countries and ways in which it may be extended. Of primary importance is the handling and storage of eggs, with refrigeration being one of the most important factors. Rapid cooling has been investigated, although it is important that this process does not induce microcracks in the shells of the eggs. Pulsed visible light, electrolysed water, MAP, ultraviolet radiation, ionizing radiation, whole egg pasteurization and ultrasonic treatment have all been investigated, at least experimentally. A range of egg coating substances have been tested: oils and waxes, carbohydrate coatings such as those based on cellulose or chitosan, emulsions of oils and chitosan, proteins and synthetic polymers. Oils, both mineral and vegetable, have excellent moisture-proofing properties, whereas chitosan coatings have antimicrobial properties, which make the combination of these two substances promising for use in extending the shelf life of eggs. One commercial provider reports use of whole-egg pasteurization followed by coating with food-grade wax as extending the shelf life significantly.

Chapter 14 reviews the nutritional role of eggs for humans. Eggs are one of the most affordable, accessible and nutrient-dense foods available around the world. The proteins, lipids, carbohydrates, vitamins, minerals and other functional nutrients in an egg not only are sufficient to support the transition from fertilized cell to newborn chick, but also provide unique nutrient contributions to the human diet. Eggs contain proteins (12.6%), lipids (9.5%), vitamins and minerals. There have been concerns about the association between egg consumption, coronary heart disease (CHD) and blood lipids. However, recent research and analyses of data from a large number of studies have failed to support any association between egg consumption and CHD. There is also no clear evidence to implicate egg consumption in the development of type II diabetes. Because of their effect on satiety, eggs can play a useful role in weight loss. In addition, eggs are the source of many bioactive compounds which promise to be useful for human health. The uses of eggs are further explored in **Chapter 15**, which reviews the nutraceutical benefits of eggs. Biological functions of eggs, reported in the literature, include anti-inflammatory, anti-hypertensive, anti-microbial, anti-oxidant, blood cholesterol/lipid-lowering, immune modulatory, anti-toxin/allergen and skin/cartilage/bone strengthening. Eggshell powder is a natural source of calcium and microelements (magnesium, fluorine, strontium, copper, manganese and zinc), which have a positive effect on bone density. Components of eggshell membranes can assist in maintenance of healthy joints and as treatment for wounds. The section on egg white complements Chapter 2. Egg white proteins that have nutraceutical benefits are ovalbumin, ovotransferrin, ovomucin, ovomucoid, avidin, lysozyme and cystatin. Egg yolk proteins with nutraceutical benefits are lipoprotein, livetins and phosvitin, and lipids with nutraceutical properties are fatty acids and phospholipids. Efficient methods for large-scale isolation and purification of biologically active constituents of eggs are needed for these products to be commercially viable.

Chapter 16 describes how the nutritional profile of eggs can be enhanced by supplementing the diet of hens with bioactive substances such as omega-3 fatty acids, vitamins, minerals and carotenoids that are then deposited into the eggs. Various sources of omega-3 fatty acids may be used including plant sources such as flax and marine sources such as microalgae and fish. Attention needs to be paid to sensory qualities of the eggs produced. Water-soluble B vitamins are deposited into the egg albumen whereas the lipid-soluble vitamins (A, D, E, K) are deposited into yolk. Carotenoids are deposited into yolk. Egg enrichment with the minerals selenium, iodine, manganese, chromium, zinc, copper, cobalt and boron has been investigated. Combinations of supplements have also been trialled. Enriched eggs are generally viewed as a specialist food item, but have the potential to increase dietary intake in selected groups of humans.

Chapter 17 addresses molecular breeding techniques for improving egg quality. Since the release of the chicken genome sequence in 2004 (Hillier et al., 2004), the range of technologies available for genetic improvement of the multiple traits that define egg quality has increased greatly. The genetic improvement techniques include whole-genome methods, genomic selection, the targeted gene approach and investigations of gene expression and regulation. Genome-wide association studies have identified regions of the genome associated with variation in the different egg quality traits. These regions are called quantitative trait loci as it is presumed that they contain genes that influence the quantitative traits. These studies are able to identify candidate genes, leading to marker-assisted selection. Egg production, along with the various indicators of eggshell quality and egg internal quality, can be considered individually and in combination. Genomic selection relies on using whole-genome information to predict breeding values of animals rather than trying to identify specific genomic regions. In the targeted gene approach, genes with potential to influence the traits of interest are identified. These candidate genes are then investigated in more detail. The thousands of genes that constitute the chicken genome do not work independently. They work together within genetic regulatory networks of interacting genes that influence gene expression. Proteomics is the study of large numbers of proteins all of which are expressed at one time or in one tissue. By combining information from genetic networks with proteomic studies of specific tissues, it is possible to understand when genes are turned on or off, and in which tissues or regions of individual organs this occurs. This knowledge can help in understanding traits of interest for improving egg production. The recently developed genetic manipulation tool CRISPR (Cong et al., 2013) allows rapid and highly specific editing of genes. The genome is altered to produce a very specific and desired change in the genes to improve a particular trait. All these methods complement each other in achieving better understanding of genetic determination of egg-quality traits.

Summary

The chapters in volume 1 outline the structure and function of the main egg components: eggshell, egg white and yolk. The microbiological challenges to the food safety of eggs are discussed, as are ways of detecting pathogens, enhancing the egg's natural defences against pathogens and ways of minimizing food safety risks in the egg industries. The appeal of eggs as a food product is critical to the profitability and sustainability of the egg industries, so ways of maintaining and enhancing the sensory and nutritional quality of eggs are outlined. Finally, the way in which modern molecular breeding techniques improve egg quality is discussed.

Part 1

Egg composition and chemistry

Chapter 1

Composition and properties of eggshell

Maureen Bain, University of Glasgow, UK

1 Introduction

As a food commodity, the eggshell provides us with a convenient and natural packaging for the transportation of the egg contents. However, eggs are fragile and therefore need to be handled, transported and stored with care. Imperfect, cracked and damaged eggshells not only lead to financial loss but constitute a risk in terms of food safety due to the easy ingress of microorganisms and other agents that may cause spoilage and possibly food poisoning (Bain et al. 2006a). The development of efficient production methods and quality control procedures, including the use of online sensors, is therefore fundamental to ensuring the successful marketing of eggs (Mertens et al. 2011).

In an attempt at reducing breakages, breeding companies have paid a lot of attention to eggshell traits, such as eggshell thickness and breaking strength, in their selection programmes (Dunn 2011). The natural deterioration in shell quality over the life of the hen, however, is still one of the main reasons for currently replacing commercial flocks at or around 72 weeks of age. Not all hens in an ageing flock produce eggs with shells of reduced quality, rather the variability in eggshell quality within the flock increases. There is therefore an opportunity to select for improved persistency in lay and to stabilise eggshell quality in older flocks (Bain et al. 2016). To do this requires an understanding of the factors affecting the long-term maintenance of the tissues and organs involved in producing the eggs, and the factors controlling the different stages of the egg-forming process.

http://dx.doi.org/10.19103/AS.2016.0012.01

From a biological point of view, the structure of the avian eggshell has evolved to protect the chick embryo throughout its extra-uterine development. In this respect, the shell acts as a physical barrier, permits gaseous exchange to occur between the embryo and its external environment and provides the chick with calcium for skeletal development (Nys et al. 2004). At the end of the incubation process, the chick must be able to hatch out, so the eggshell cannot be excessively thick. The egg is also equipped with an arsenal of chemical defences to prevent the invasion of potentially harmful pathogens from gaining entry to the egg contents (Rehault-Godbert et al. 2011). This protection begins with the cuticle, which covers the outer surface of the shell (Bain et al. 2013, Munoz et al. 2015), but there are other protective factors within the eggshell structure that need further research including the proteins embedded within the calcified shell and its associated membranes. These multifaceted requirements explain the need for such a complex architecture.

Despite a wealth of research spanning over 60 years, we are still poorly informed about all the processes and mechanisms that are involved in the eggshell formation (Hincke et al. 2011, Nys et al. 1999, Romanoff and Romanoff 1949). What is evident is that the structural integrity of the eggshell is not just a matter of thickness but depends on the complex relationships that exist between the eggshell fabric and how this is then organised into the final structure (Bain 1992, Bain et al. 2006b). However, we have a long way to go in terms of unravelling the nature and functional properties of the myriad organic components that comprise the eggshell fabric.

High-throughput technologies (proteomics and transcriptomics), in combination with the newly available chicken genomic sequence and new bioinformatics tools, have made it much easier to characterise the organic components of the eggshell, which until relatively recently remained elusive (Gautron et al. 2011). The challenge is now how best to explore the long lists of proteins that have been generated in these studies and to clarify the biological relevance of these proteins in terms of eggshell fabrication and/or in designing a protective chamber suitable for embryonic development. Being able to reduce the eggshell to its component parts will improve the precision of how we define eggshell quality and how we then try to influence it through breeding and management.

This chapter describes the structure of the eggshell and introduces to our knowledge some of the advances about the composition and properties of the eggshell, which have resulted from the application of new molecular techniques, including proteomic and transcriptomic approaches. Where possible it suggests how these advances may allow us to realise the major challenges in this topic and contribute to sustainable egg production.

2 Introduction to eggshell formation

2.1 Overview

As a prerequisite to developing a discussion on the composition and properties of the eggshell, a short review of how the egg is formed in the oviduct is provided in this section.

Egg formation has been extensively reviewed (Etches 1996, Gilbert 1990, Nys and Guyot 2011, Sauveur and De Reviers 1988, Solomon 1991). The main events include the shedding of the yolk from the ovary, the addition of the albumen in the magnum, the shell

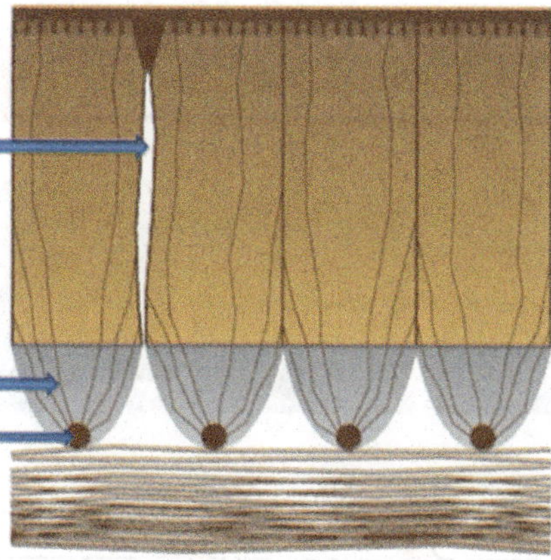

Figure 1 Schematic showing the different layers of the eggshell. The mammillae are the calcified units that form when calcite nucleates onto the mammillary cores during the initial stages of shell formation. The mammillary cores correspond to special sites on the outer shell membrane fibres. The palisade layer forms when the mammillae fuse at their base. The vertical crystal layer is a very thin layer that forms during the final stages of shell formation. The cuticle is an organic layer which forms on the surface of the eggshell.

membranes in the isthmus and the shell in the shell gland or uterus. The eggshell takes about 18 hours to form and consists of the paired shell membranes, the mammillary layer, the palisade layer, the vertical crystal layer and the cuticle (Fig.1).

The eggshell membranes are paired and form two parallel layers of interconnecting fibre networks on the inner surface of the shell. Each fibre consists of a protein core surrounded by a carbohydrate mantle. The inner shell membrane fibres are finer (0.5 nm diameter) than the outer shell membrane fibres (0.8 nm diameter) (Nys and Guyot 2011). Within each layer, the fibres are extensively coupled to each other by irreversible lysine-derived crosslinks (Leach 1982). The inner membrane (thickness ~20 μm) surrounds the albumen and can be easily stripped from the eggshell whereas the outer membrane (thickness ~50 μm) is tightly bound to the eggshell (Nys and Guyot 2011). The first signs of mineralisation appear at specialised sites referred to as the mammillary cores on the outer shell membrane fibres (Solomon 2010). These form on the surface of the outer shell membrane fibres just as the egg passes through the terminal section of the isthmus (red isthmus) and enters the shell gland (Nys et al. 2004) (Fig. 2). The so-called mammillary cores soon become incorporated into the calcified mammillae, which form around the mammillary cores as mineralisation continues.

Within the shell gland, three phases of mineralisation are recognised. The first phase lasts about 4.5 hours and gives rise to the mammillary layer (80–100 μm). During this phase, mineralisation is slow. The second phase lasts about 12 h and experiences rapid growth of the palisade columns that form the bulk of the shell thickness (~300 μm). In the terminal phase (~1.5 h) the vertical crystal layer forms, at which point shell formation ceases. The cuticle (5–10 μm) and the bulk of the shell pigment are also added during this phase (Hincke et al. 2011, Nys et al. 2004, Nys and Guyot 2011). Gaseous exchange pores extend through the calcified shell from the membranes to the cuticle (Fig. 1). The

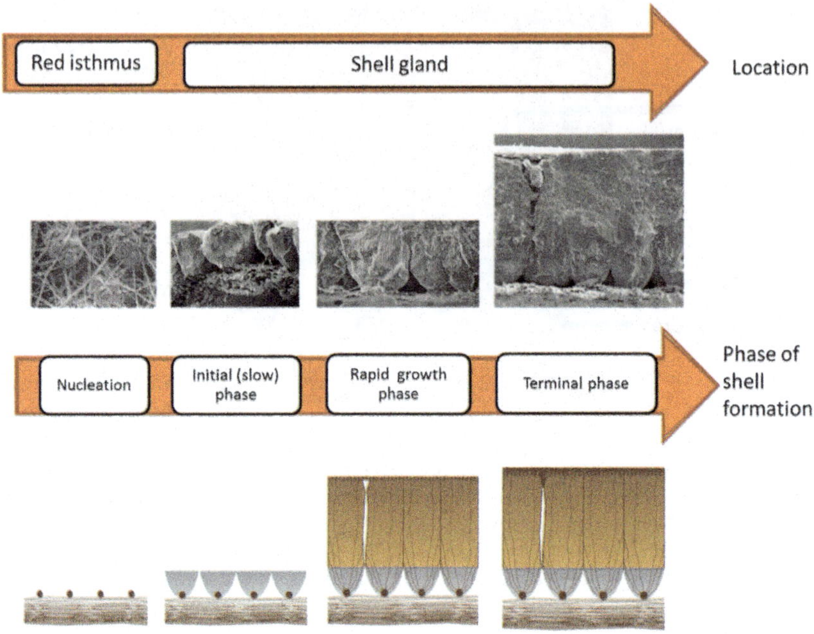

Figure 2 The main phases involved in the formation of the eggshell. Nucleation is initiated in the red isthmus as soon as the mammillary cores form on the outer shell membrane fibres. The egg then passes into the shell gland and the shell forms over three distinct phases of mineralisation of variable duration.

distribution of pores is such that there are more pores towards the blunt end of the egg. The exact mechanism giving rise to the formation of the gaseous exchange pores is not fully understood but seems to involve the distribution and arrangement of the mammillae in the mammillary layer (Tullett 1975).

Solomon (1991) was the first to describe twelve structural variations in the mammillary layer of weak and poor quality eggshells and hypothesised that, if the mammillary layer was structurally imperfect then this would have a 'knock-on effect' on the formation of the rest of the shell and therefore the overall strength of the eggshell. A link between oviduct dysfunction and the incidence of eggshell faults was thus established and this highlighted the importance of the initial stages of shell formation as being critical in terms of shell integrity (Bain 1992, Solomon 2002). Solomon (1991) and Roberts and Brackpool (1994) review the ultrastructural organisation of the eggshell in detail.

Throughout the shell-forming process, the egg is continuously bathed in the uterine fluid, an acellular milieu containing the inorganic and organic precursors that form the shell. It was established fairly early on that the profile of this fluid changes depending on the stage of shell formation (Gautron et al. 1997, Hernández-Hernández et al. 2008). This and later studies imply that there is a temporal release of specific proteins which may either facilitate or modulate the precipitation of calcium carbonate at each stage in the shell-forming process (Hernández-Hernández et al. 2008, Marie et al. 2014).

2.2 The composition and role of uterine fluid

The uterine fluid contains the mineral and organic secretions produced by the cells lining the red isthmus and shell gland as the shell is forming. This fluid remains supersaturated with respect to the solubility product of calcite throughout the mineralisation process (Nys et al. 1991) and contains a large number of proteins some, but not all, of which are directly involved in the fabrication of the eggshell (Gautron et al. 1997). We now have at least two published proteomic inventories of the uterine fluid (Marie et al. 2014, Sun et al. 2013)

Sun et al. (2013) compared the proteome of the uterine fluid bathing the eggshell with the proteome of the calcified shell. This study revealed 577 proteins in the uterine fluid and 466 in the eggshell with an overlap of 244 proteins. On the other hand, Marie et al. (2014) described 308 proteins of which 96 were considered to have a potential functional role in shell formation, the others being considered as 'redundant'. Of the non-redundant proteins, 64 proteins were found to vary depending on the stage of mineralisation (Marie et al. 2014). The temporal appearance of specific proteins in the uterine fluid is therefore of fundamental importance in the fabrication of the eggshell. Obviously, this is an area which needs much further discussion and is returned to in the section on eggshell matrix.

In a complementary transcriptomic study, part of the same team report a repertoire of 302 overexpressed genes in the shell gland when the shell is forming (Brionne et al. 2014). Some of these gene products were found to encode for proteins involved in the transepithelial transfer of Ca_2^+ and CO_3^- ions into the uterine fluid. Further, 57 of the remaining genes encoded for proteins that could be linked to the bio-mineralisation process and/or the chemical defence of the eggshell. Some of these proteins may well prove to be useful biomarkers for the genetic improvement of eggshell quality (Dunn et al. 2009b).

3 The inorganic components of the eggshell

The eggshell consists of 1.6% water, 3.3–3.5% organic material (proteins, proteoglycans and polysaccharides) and 95% inorganic mineral when the membranes are included (Nys and Guyot 2011). In this section, we discuss the inorganic components of the eggshell. Sections 4, 5 and 6 will deal with the organic components.

3.1 Overview

The inorganic or mineral component of the eggshell consists mainly of calcium carbonate in the form of calcite, although low levels of numerous minerals and other trace elements are also found in the shell (Nys et al. 2004). Magnesium, for example, is found in the mammillary layer and both magnesium and phosphorous occur in the outer palisade layer (Cusack et al. 2003). The amount of calcium in the shell is about 37.5% and carbonate 58% (Nys and Guyot 2011).

The supply of calcium required for shell formation is derived mainly from the diet, but as the shell continues to form during the night, calcium is also derived from the mobilisation of bone reserves (Bouvarel et al. 2011). Most of this comes from the medullary bone, a labile type of bone, which is laid down in the shafts of the long bones under the influence of oestrogen at sexual maturity. Medullary bone, unlike structural bone, is capable of undergoing rapid absorption and renewal. Unfortunately, resorption of structural bone also occurs when there is a requirement of calcium for shell formation leading to the symptoms of osteoporosis in older laying flocks (Whitehead 2004). Osteoporosis is caused by a decrease in the amount of fully mineralised structural bone leading to bone fragility and susceptibility to fracture. This is one of the major welfare challenges currently facing the egg industry (Sandilands 2011).

The Ca^{2+} and HCO_3^- ions necessary for shell formation are continuously supplied to the shell gland from the blood vascular system via a transepithelial transport mechanism in the cells lining the shell gland. The classical model describing this process highlighted the importance of calbindin as the main protein involved in intracellular transfer of calcium and a key family of enzymes, the carbonic anhydrases, which catalyse the hydration of plasma CO_2 to HCO_3^- (Jonchere et al. 2012). New molecular approaches (transcriptomics) looking at overexpressed genes in the shell gland have now identified 16 different proteins which could be involved in this process (Brionne et al. 2014). The most up-to-date model provided by these authors describing the possible role of each of these proteins is very complex. It nevertheless contributes towards a better understanding of the mechanisms and regulation of ionic precursors of $CaCO_3$ for egg production and could also be of relevance to other systems where epithelia transport large amounts of calcium or bicarbonate.

3.2 Calcite crystals and crystal growth mechanisms

The eggshell is one of the fastest bio-mineralising systems known in nature; yet, we still do not fully understand the mechanisms involved in its fabrication. The presence of proteins in uterine fluid that surrounds the egg as the shell forms would seem to be critical in terms of directing and modulating the way in which the calcite crystals form and interact with each other, especially at interfacial boundaries. Rodriguez-Navarro and García-Ruiz (2000) used a computer simulation model to demonstrate how calcite grows on a membrane and showed that the size and preferred crystal orientation can be explained by competitive and anisotropic crystal growth. Much of our current thinking is still based on this model.

During the initial stages of calcification, Ca^{2+} and HCO_3^- ions come together to form calcite crystals that nucleate around the mammillary cores. The mammillary cores are rich in proteins and sulphated proteoglycans (Nys et al. 2004). Calcite is the most stable crystalline form of calcium carbonate, aragonite and vaterite being the other forms. Crystal growth initially occurs in all directions but downwards growth is soon inhibited by the presence of the membrane fibres. Further growth continues outwards and upwards to form the mammillae, which eventually fuse to form the base of the palisade layer. At this point, because of the competition for space, sideways crystal growth is inhibited and the palisade columns start to form (Nys et al. 2004, Rodriguez-Navarro and García-Ruiz 2000). The palisade layer consists of columns of rhombohedral calcite crystals that have a honeycomb appearance. The size, shape and crystallographic orientation of the crystals in the palisade layer contribute to the mechanical strength of the eggshell (Rodriguez-Navarro et al. 2002).

Rodriguez-Navarro et al. (2015) have recently demonstrated the presence of a highly soluble transient amorphous calcium carbonate (ACC) phase, the presence of which could be fundamental to the biosynthesis of the eggshell. An ACC phase has previously been reported in the maturation of sea urchin spicules and other shell-forming invertebrates

(Addadi et al. 2003). The new model described by Rodriguez-Navarro et al. (2015) suggests that, during the initial stages of mineralisation, flat disc-shaped ACC particles nucleate preferentially at the mammillary cores but also do so to some degree over the rest of the exposed membrane surface. The ACC deposits at the mammillary cores then transform into the much larger crystal units of calcite. Meanwhile, competition for ions creates under saturation and depletion zones in the surrounding areas, promoting dissolution of the ACC nanoparticles in these areas. More ions then immediately become available to the main nucleation centres where crystal growth continues by the aggregation of more ACC nanoparticles. These progressively crystallise onto the faces of the growing crystals of calcite. According to Rodriguez-Navarro et al. (2015), the ACC phase is also present throughout the rapid phase of mineralisation and provides a highly plausible explanation for how a 340 μm thick eggshell can form in ~18 h (Rodriguez-Navarro et al. 2015). This new model not only helps us to understand the biogenesis and eggshell formation but it also offers a possible explanation for the presence of 'aberrant' crystal forms in the mammillary layer, as described by Solomon (1991) and others. These aberrant crystalline forms of calcite and aragonite typically occur between the normal mammillae and could arise as a result of tissue debris landing on the shell membranes as the APP is formed. This would disrupt the normal dissolution of APP at these sites allowing sufficient time for crystals of calcite or aragonite to form.

4 The organic components of the eggshell: eggshell membranes

Although the eggshell membranes are formed in the isthmus as opposed to the shell gland, they form the base onto which the rest of the shell develops. Understanding of their composition and structure is therefore essential in any discussion relating to the formation of the eggshell.

4.1 Membrane composition

The eggshell membranes consist of glycoproteins, cysteine-rich eggshell membrane proteins (CREMPs) and collagens (types I, V and X) (Arias et al. 1991, Kodali et al. 2011). CREMPs are a fairly recent discovery, and are now considered to be one of the major structural components of the eggshell membranes (Cordeiro and Hincke 2016, Kodali et al. 2011). Discovering the complete inventory of all the proteins associated with the eggshell membranes has proved challenging as the membrane fibres have lysine-derived covalent linkages, which make it very difficult to extract and solubilise them. The most recent and complete listings of all the proteins in the soluble eggshell membrane extract are provided by Cordeiro and Hincke (2016), Makkar et al. (2015) and Rose-Martel and Hincke (2015a).

Du et al. (2015) revealed 135 overexpressed transcripts in the white isthmus corresponding to as many as 107 annotated genes that could be involved in the formation of the eggshell membrane fibres. The genes for both collagen X and CREMPs were overexpressed confirming their importance in membrane synthesis (Du et al. 2015). A detailed annotation of all of the genes and their encoded proteins will be an important next step as this may identify potential useful future applications for eggshell membranes, for example as a biomaterial within industry or in other biomedical applications (Cordeiro and Hincke 2011).

4.2 The mammillary cores, mammillae and mammillary density

As previously discussed, the mammillary cores on the outer membrane fibres are critical to the initiation of the mineralisation process. Keratin sulphate proteoglycan has been shown to be present at these sites when the first signs of nucleation occur (Fernandez et al. 1997). What determines the distribution and spacing of the mammillary cores is not known but it may be genetically predetermined and influenced by egg size (Dunn et al. 2012).

An interesting facet is the strong genetic correlation observed by Dunn et al. (2012) between the size of calcite crystals and the thickness of the mammillary layer. The thickness of the mammillary layer is dependent on the density of mammillae and therefore presumably the mammillary cores. With this in mind, Dunn et al. (2012) proposed that a low mammillary density would result in a thicker mammillary layer and wider calcite crystal columns forming in the palisade layer than where the mammillary density was high and fusion of the mammillae occurred much sooner. The distribution of mammillary cores therefore appears to have a direct effect on the overall structural integrity of the eggshell.

5 The organic components of the eggshell: the eggshell matrix

Proteins, glycoproteins and proteoglycans are the building blocks of the eggshell matrix. These become interwoven with the inorganic constituents during the fabrication of the eggshell and, as a result, have been difficult to extract and analyse.

The first major breakthrough in the identification and characterisation of eggshell matrix proteins was just over 18 years ago and it is interesting that despite all the new discoveries the first few proteins to be characterised still appear to be the main players involved either in eggshell mineralisation or the egg's defence (Hincke et al. 2011).

Among the first identified proteins were three major egg white proteins: ovalbumin, lysozyme and ovotransferrin (Gautron et al. 2001b, Hincke 1995, Hincke et al. 2000). These proteins are well known for their antimicrobial properties and are produced predominately in the magnum region of the oviduct and found in the albumen. Their incorporation into the eggshell matrix may therefore contribute to the egg's antimicrobial defence (Nys et al. 2004). All three proteins, however, are localised in the shell membranes and/or the mammillary layer, which suggests they may also play a critical role in the early stages of shell formation (Marie et al. 2015). Supporting evidence for this comes from *in vitro* and *in vivo* experiments. Ovalbumin has been shown *in vitro* to exhibit calcium-binding properties and to stabilise a series of transient unstable calcium carbonate phases that form prior to the stable crystalline polymorph (Wang et al. 2010). Lysozyme and ovotransferrin have also been shown to alter the morphology of calcium carbonate crystals grown *in vitro* (Gautron et al. 2001b, Hincke et al. 2000). Furthermore, polymorphisms in the genes coding for ovalbumin have also been shown to be associated with eggshell breaking strength and shell thickness in a pedigree population of laying hens (Dunn et al. 2009b).

A number of novel eggshell-specific proteins have also been identified and characterised in the soluble eggshell extract obtained from decalcified eggshells using both conventional and other extraction techniques. These include ovocleidin-17 (Hincke et al. 1995), ovocleidin-116 (Hincke et al. 1999, Mann et al. 2002), ovocalyxin-32 (Gautron et al. 2001a), ovocalyxin-36 (Gautron et al. 2007), ovocalyxin-25 and ovocalyxin-21 (Nys et al. 2004). These proteins are synthesised by the oviducal tissues which are primarily involved in mineralisation (red isthmus or shell gland) (Nys et al. 2004) and have not previously been identified in other tissues or body systems.

The first eggshell-specific protein to be purified and characterised was ovocleidin-17 (OC-17). This C-type lectin protein is found throughout the calcified shell (Hincke et al. 2011). Classical molecular dynamic simulations confirm that OC-17 is capable of binding to calcite (Freeman et al. 2011). OC-17 is therefore likely to play a pivotal role in the shell-forming process (Marie et al. 2015). C-type lectin-like proteins are major components of the calcified eggshell of many avian species. Ansocalcin, for example, is a major matrix protein from goose eggshell (Lakshminarayanan et al. 2003). Purified ovocleidin-17 and ansocalcin both exhibit bactericidal activity so C-type lectin-like proteins may play an important antimicrobial role in defence of the avian embryo (Wellman-Labadie et al. 2008).

Ovocleidin-116 (OC-116) is regarded as one of the most important eggshell matrix protein and is most abundant in the palisade layer of the eggshell (Hincke et al. 1999). The N-terminal sequence of this protein corresponds to the core protein of ovoglycan, an eggshell-specific proteoglycan (Carrino et al. 1997, Fernandez et al. 2001). This protein is overexpressed in the shell gland when a calcified eggshell is forming (Marie et al. 2014). *OC-116* has been reported to be a member of a family of related phosphoprotein genes clustered on avian chromosome 4 (Kawasaki and Weiss 2006). Other members of this family (bone sialoprotein, dental matrix protein-1 and osteopontin) are also present in the eggshell (Horvat-Gordon et al. 2008). These authors confirmed that *OC-116* is also expressed in avian growth plate chondrocytes, chick cortical bone and laying hen medullary bone, as well as in hypertrophic chondrocyte lysates. OC-116 is therefore not strictly an eggshell-specific protein and therefore could be important in both the fabrication of the eggshell and in the mineralisation of the bone (Horvat-Gordon et al. 2008). There is also evidence that variations in genetic markers for OC-116 could be used in selection to improve a number of eggshell quality traits (Dunn et al. 2009b, 2012).

Ovocalyxin-32 (OCX-32) is a member of the latexin family of carboxypeptidase inhibitors and is localised in the mammillary layer and outer palisade/vertical crystal layers of the eggshell (Gautron et al. 2001a, Hincke et al. 2003). *In vivo* studies have demonstrated that *OCX-32* gene polymorphisms are associated with eggshell strength and mammillary layer thickness, supporting the involvement of this protein in the fabrication of the eggshell (Dunn et al. 2009b, 2012, Fulton et al. 2012). Its presence in the uterine fluid during the initial phase of shell mineralisation has also been established (Marie et al. 2014). The overexpression of *OCX-32* in the shell gland at the terminal phase is in accordance with its high concentration in the outer calcified layers and cuticle (Hincke et al. 2003). Recombinant OCX-32 protein exhibits antimicrobial activity against gram-positive bacteria so, within the cuticle, OCX-32 may also contribute to antimicrobial defence (Xing et al. 2007).

Ovocalyxin-36 (OCX-36) is a member of the BPI/LBP/PLUNC family of proteins that are involved with the innate immune response (Gautron et al. 2007, 2011). OCX-36 is mainly found in the mammillary layer of the shell (Gautron et al. 2011). Interestingly, this protein has been shown to be strongly upregulated in the shell gland of sexually mature hens compared to juvenile hens (Dunn et al. 2009a). It has been proposed that OCX-36 contributes to the natural defence mechanisms that keep both the egg and functional oviduct free of pathogens (Rehault-Godbert et al. 2011).

Two other ovocalyxins occasionally mentioned in publications are OCX-25 and OCX-21. OCX-21 contains a BRICHOS domain and has been ascribed the function of a molecular chaperone involved in protein–protein interactions (Gautron et al. 1997). This protein seems to be overabundant in the rapid phase of mineralisation and its presence in the uterine fluid during both the rapid and terminal phases of shell mineralisation has been confirmed (Marie et al. 2014, Sun et al. 2013). *OCX-21* is overexpressed in the shell gland during eggshell calcification (Brionne et al. 2014) and it is consistently found in the cuticle (Bain et al. 2013, Rose-Martel et al. 2012). OCX-25 contains two protease inhibitor domains and so may be involved in the egg's natural defence but, other than being regularly detected in the shell-forming tissues, oviducal fluid or eggshell matrix, little is known about this protein (Rose and Hincke 2009).

Osteopontin (Fernandez et al. 2003, Lavelin et al. 1998), Glypican-4 (Lavelin et al. 2002) and Clusterin (Mann et al. 2003) are considered to be ubiquitous components of the eggshell matrix as these proteins are also found in other tissues. Osteopontin is a phosphoprotein localised at the base of the mammillae and in the outer palisade layer, but it is also found throughout the palisade layer (Fernandez et al. 2003). This protein is considered to be a major protein in the shell mineralisation process as it influences the kinetics of calcium carbonate precipitation and modifies crystal morphology *in vitro* (Chien et al. 2008, 2009, Hincke et al. 2008). This protein is highly overexpressed in the shell gland during shell calcification (Brionne et al. 2014, Pines et al. 1995) and appears to be regulated by the mechanical strain induced on the mucosa when the egg first enters the shell gland (Lavelin et al. 1998). Regional differences in the expression of osteopontin in the surface epithelial cells lining the shell gland has been associated with some common shell abnormalities (Arazi et al. 2009), reinforcing the view that this protein is of prime importance in the control of mineralisation. A candidate gene association analysis showed that osteopontin could be used as a genetic marker for fracture toughness of the eggshell (Dunn et al. 2009b).

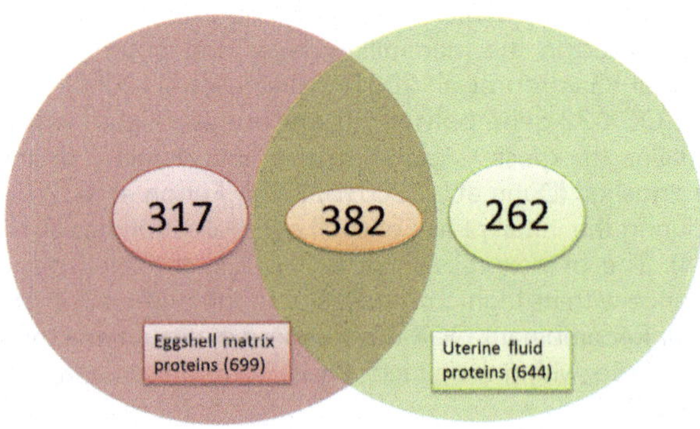

Figure 3 Venn diagram representing the number of proteins identified in the shell matrix and uterine fluid which are unique and common to both.

Glypican-4 is a member of the glypican-related integral membrane proteoglycan family (GRIPS). These proteins have a role in the control of cell division and regulate cell growth. GPC-4 is expressed in the shell gland but it has never been demonstrated in the eggshell matrix (Lavelin et al. 2002). Its functional role remains undetermined.

Clusterin is found in both the mammillary and palisade layers of the shell. This protein is expressed in many other tissues and may function as an extracellular chaperone involved in the stabilisation and prevention of premature aggregation and precipitation of eggshell matrix proteins before and during the assembly of the protein scaffold necessary for ordered mineralisation (Mann et al. 2003, Nys and Guyot 2011).

Having an accurate picture of which proteins play an active role in the different phases of shell formation is a perquisite for developing future quantitative studies (*in vivo* and *in vitro*).

In the last few years, quantitative proteomics has revolutionised our understanding of the eggshell matrix and a number of different approaches have been adopted (Mann et al. 2006, 2007, Marie et al. 2015, Miksík et al. 2007, 2010, Rose-Martel et al. 2015b). In total, 699 major and minor proteins in the eggshell matrix have now been identified, and at least 382 of these are also found in the list of proteins available for the uterine fluid (Fig. 3); but, for most of the newly identified protein constituents of the eggshell matrix, we can only speculate as to their putative functional role.

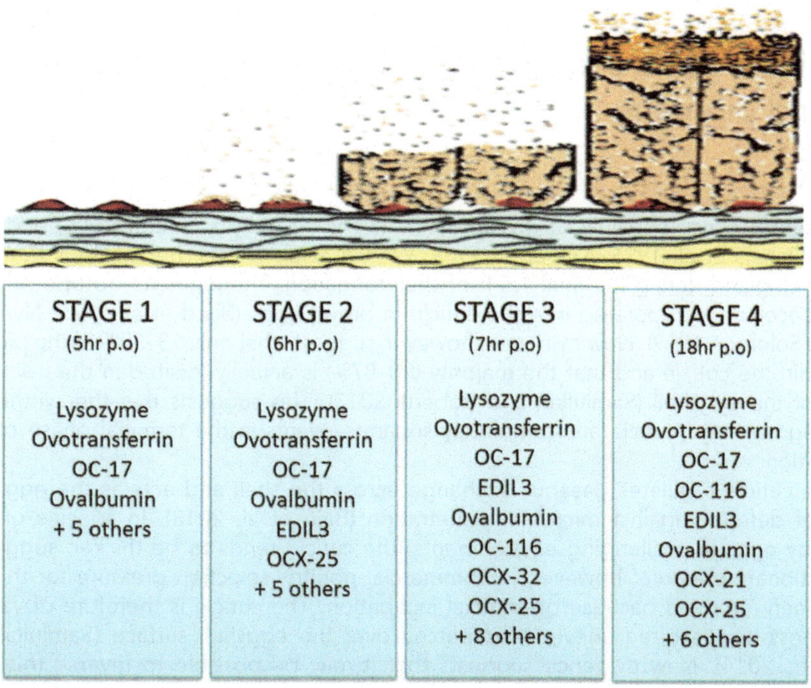

Figure 4 Schematic showing the main molecular actors which are overabundant at 5, 6 and 7 h post ovulation when the shell is just forming (Stages 1–3) and once the rapid phase of mineralisation is well underway (Stage 4). Redrawn from an original schematic drawing by Marie et al., 2015.

A recent study by Marie et al. (2015) provides an inventory of the most abundant proteins present while the eggshell matrix is forming at 5, 6, 7 and 16 hours post ovulation (Fig. 4). These time points correspond to three key events which occur in the initial stages of shell formation (widespread deposition of ACC, ACC transformation into crystalline calcite aggregates and formation of larger calcite crystal units) and the development of the palisade columns in the rapid phase of mineralisation (Rodriguez-Navarro et al. 2015). This study identified 175 non-redundant proteins with different variations in abundance, that have putative temporal functional roles. In addition, 24 of these had not previously been included in the proteomic inventory of whole eggshell matrix. Most importantly, this study provides us with a temporal profile of just 21 proteins, which were assigned a functional role in the control of mineralisation (Fig. 4) including lysozyme, ovotransferrin, ovalbumin, OCX-32 and OC-17. The other proteins of interest include five calcium-binding proteins previously not identified in the eggshell matrix (EDIL3, ALB, MFGE8, HPX and NUCB2), a protein involved in membrane formation (LOXL2), OC-116 and two other proteins associated with the proteoglycan family, three major molecular chaperones including OC-21, three protease inhibitors including OCX-25 and a phosphatase, which could have an important role in the degree of phosphorylation of matrix proteins. Therefore, out of nearly 700 proteins in the eggshell matrix we now have a manageable target list of 21 proteins (including several we already knew about) on which to base future study (Marie et al. 2015).

6 The organic components of the eggshell: cuticle and pigment

6.1 Cuticle

The cuticle consists mainly of glycoproteins (80–90%), polysaccharides (4%) and lipids (3%) (Dennis et al. 1996, Nys et al. 1999, Wedral et al. 1974). The cuticle is formed in the shell gland during the final 1.5 h of shell formation. The pigment Protoporphyrin IX also becomes incorporated into the cuticle in brown eggs (Baird et al. 1975, Nys et al. 1991, Solomon 1999). New evidence, however, suggests that only 13–20% of the pigment is within the cuticle and that the majority (80–87%) is actually located in the calcareous part of the eggshell (Samiullah and Roberts 2013). This suggests that the synthesis of the pigment and cuticle are completely separate events in the terminal phase of shell formation.

The cuticle regulates gaseous exchange across the shell and acts as the egg's first line of defence against microbial penetration (Bain et al. 2013). In species of birds that lay eggs in challenging environments, the cuticle tends to be thicker, suggesting evolutionary pressure; however, in commercial poultry, selection pressure for this trait has been removed because of artificial incubation. The cuticle is therefore of variable thickness and is often uneven and patchy over the eggshell surface (Samiullah and Roberts 2014). New evidence suggests that it may be possible to reverse this trend and improve the cuticle deposition through genetic selection (Bain et al. 2013). Cuticle deposition can be quantified by measuring the per cent reflectance or colour change of eggs pre and post stained with special stains (Bain et al. 2013, De Reu et al. 2006, Leleu et al. 2011).

ATR-FTIR spectrometry provides information about the thickness and chemical composition of the cuticle (Rodriguez-Navarro et al. 2013). This technology suggests that the cuticle consists of two parts: an outer part that is rich in proteins and an inner part that is rich in sulphated polysaccharides and phosphates. A thin layer of crystalline hydroxyapatite layer has also been reported but this seems to be confined to where the cuticle binds to the vertical crystalline layer (Dennis et al. 1996). In commercial layers, the cuticle composition, thickness and degree of coverage of eggs varies depending on bird age and the freshness of the egg (Rodriguez-Navarro et al. 2013). The chemical composition of the mature cuticle has been shown to determine the risk of trans-shell contamination by *Salmonella* (Munoz et al. 2015). This study reported that the cuticle is most effective against bacterial penetration between 6 and 72 h after eggs have being laid when this coating is fully mature and has not dried excessively, and that freshly laid eggs can be easily contaminated as they have an immature cuticle, which is not able to resist bacterial penetration. Eggs with incomplete or absent cuticles are also more susceptible to bacterial contamination (Bain et al. 2013, Wellman-Labadie et al. 2008).

The proteome of the cuticle has been investigated by several researchers (Bain et al. 2013, Miksík et al. 2003, Rose-Martel et al. 2012). Looking across these studies OCX-32, OCX-36 and OC-116 and a Kunitz-like protease inhibitor seem to be the most consistently detected and abundant proteins present in the cuticle. In addition to these, Rose-Martel et al. (2012) identified a number of other proteins known to have antimicrobial activity including lysozyme, ovotransferrin, cystatin, ovoinhibitor and other possible new candidates (OCX-36, myeloperoxidase and members of the SERPIN family).

A significant association has been reported between alleles of OC-116, and cuticle deposition (Bain et al. 2013). A functional explanation for this finding remains to be determined. The characterisation of new cuticle resident antimicrobials may provide new targets for marker-assisted selection in breeding programmes to improve the innate immune defences of the egg for enhanced food safety. Further work in this area is also likely to include measuring the expression of specific genes involved in cuticle protein production and genes known to have roles in glycosylation (Dunn, pers. comm.).

6.2 Pigment

Protoporphyrin IX is the main pigment found in brown-shelled eggs. This pigment is actively synthesised in the shell gland but the mechanisms involved are not yet fully understood. The presence of pigment has been tenuously linked with an improvement in eggshell quality. Ishikawa et al. (2010), however, suggest that pigment may function as a 'light-catalysed microbial tile' against desiccation-resistant gram-positive bacteria. Further research is needed to elucidate the metabolic pathways involved in the synthesis of pigment versus cuticle deposition in the shell gland.

7 Case study: OC-116 as a genetic marker for improved eggshell quality

Throughout this chapter, reference has been made to extensive lists of proteins that comprise the various component of the eggshell. In addition, mention has been made to the 'next steps' in terms of characterising the functional properties of these protein

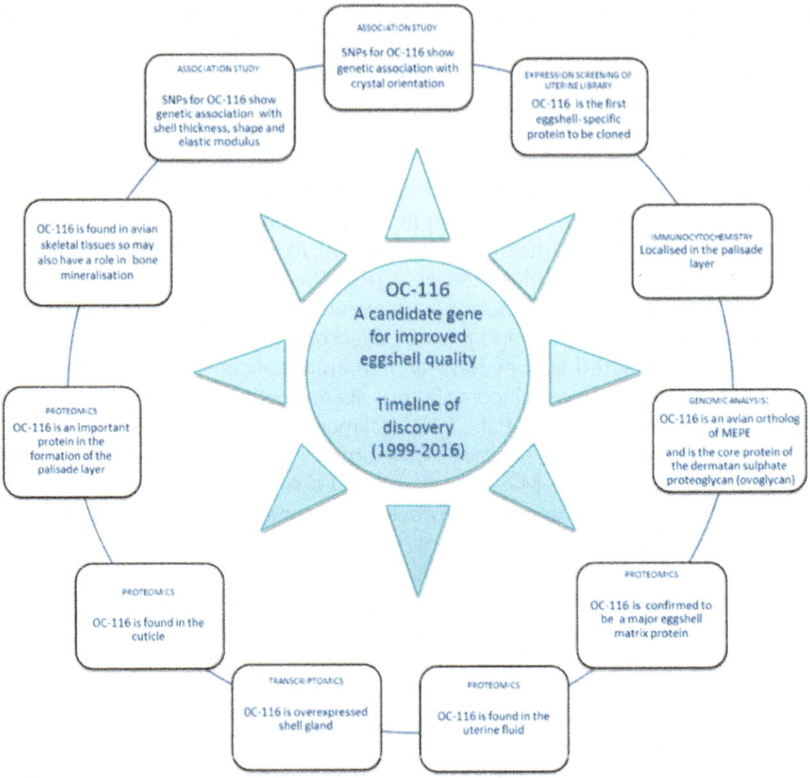

Figure 5 Key steps involved in identifying OC-116 as a potential genetic marker for selection for improvement in eggshell quality.

components of the eggshell. The overall aim for breeding companies is to identify molecular markers that could be applied through marker-assisted selection to improve eggshell quality. In this respect, OC-116 has already shown some promise in terms of achieving this goal (Fig. 5). However, they also offer other insights to biomedicine and possible other industries.

OC-116 was first recognised as an eggshell-specific matrix protein in 1998 and was the first eggshell protein to be cloned. Confirmation that OC-116 was a core protein of the dermatan sulphate proteoglycan highlighted the potential importance of OC-116 in the mineralisation of the eggshell. Localisation of OC-116 in various components of the eggshell confirmed that this protein might be important throughout the shell-forming process but especially during the rapid phase of shell formation. OC-116 was also identified as one of the major proteins in the uterine fluid. In addition, the genes coding for OC-116 are highly overexpressed in the shell gland while the egg is forming. Proteomic investigations of the forming eggshell matrix at different stages of the forming shell confirmed that OC-116 is one of the most abundant proteins present during the rapid phase of mineralisation. Single nucleotide polymorphisms for the OC-116 gene were subsequently described and were shown to be associated with eggshell thickness, egg shape, elastic modulus and the orientation of calcite crystals within the eggshell. The discovery that this protein also plays a significant role in bone mineralisation provides an additional point of interest, not least as it provides a direct link between the two and may offer ways of limiting osteoporosis allowing for longer laying cycles.

8 Summary and future trends

8.1 Summary

With the recent development of high-throughput technologies and the publication of the chicken genome, we now have a protein inventory for the eggshell. In total, these studies have identified a total of 699 proteins and yielded important information on the distribution of some of these proteins in the different compartments of the shell. We also have a better idea of temporal sequence in which some of these proteins appear in the uterine fluid. Proteomic studies have been complimented by transcriptomic studies aiming to identify genes expressed in the uterus with possible importance for eggshell production. These studies have helped to identify the proteins that are actively secreted by the shell gland when the shell is forming and those which are probably present as a by-product of some other process. However, there is still considerable debate as to which of the minor proteins are truly implicated either in the fabrication of the shell or its biological function. Further studies of the proteins of importance in eggshell fabrication in other hard egg-forming species other than the Galliformes may reveal a smaller number of conserved proteins, which will help to focus our attention on the key players in terms of function. We currently have a list of 21 proteins most of which warrant further attention. Some of these proteins may turn out to be useful markers for eggshell quality characteristics that are otherwise difficult to measure and apply in selection programmes. As already mentioned, two of these proteins that are showing considerable promise as genetic markers are OC-116 and OCX-32.

8.2 Future trends

Breeding companies are already selecting for 'long life' layers capable of producing 500 eggs or more in a single laying cycle lasting 100 weeks (O'Sullivan 2009, Van Sambeek 2010). This chapter has described how proteomic and transcriptomic methods have generated a list of novel proteins potentially involved in the structure and function of the eggshell. Sophisticated software that allows large amounts of information in databases to be interrogated simultaneously will offer a broader view of the function and biological significance of these proteins. In conclusion, further research is needed to define the functional properties of these proteins *in vitro* or *in vivo* as some of these proteins may prove to be valuable genetic markers for selection to reinforce the protective systems of the egg (Dunn et al. 2009b). Some may also be of interest to human and veterinary medicine or even other non-food industries.

9 Where to look for further information

A comprehensive description of the myriad of information that exists in the scientific literature relating to egg and egg quality is provided in *Improving the Safety and Quality of Eggs and Egg Products* Volume 1: Egg chemistry, production and consumption Edited by Yves Nys, Maureen Bain and Filip Van Immerseel, Woodhead Publishing Limited. For the most recent publications, the reader is directed to the Web of Science Research Platform.

10 References

Addadi, L., Raz, S. and Weiner, S. (2003). Taking advantage of disorder: Amorphous calcium carbonate and its roles in biomineralization. *Advanced Materials, 15*, pp. 959–70.

Arazi, H., Yoselewitz, I., Malka, Y., Kelner, Y., Genin, O. and Pines, M. (2009). Osteopontin and calbindin gene expression in the eggshell gland as related to eggshell abnormalities. *Poultry Science, 88*(3), pp. 647–53.

Arias, J. L., Fernandez, M. S., Dennis, J. E. and Caplan, A. I. (1991). Collagens of the chicken eggshell membranes. *Connective Tissue Research, 26*(1–2), pp. 37–45.

Bain, M. M. (1992). Eggshell strength: A relationship between the mechanism of failure and the ultrastructural organisation of the mammillary layer. *British Poultry Science, 33*(2), pp. 303–19.

Bain, M. M., Dunn, I. C., Wilson, P. W., Joseph, N., De Ketelaere, B., De Baerdemaeker, J. and Waddington, D. (2006a). Probability of an egg cracking during packing can be predicted using a simple non-destructive acoustic test. *British Poultry Science, 47*(4), pp. 462–9.

Bain, M. M., MacLeod, N., Thomson, R. and Hancock, J. W. (2006b). Microcracks in eggs. *Poultry Science, 85*(11), pp. 2001–8.

Bain, M. M., McDade, K., Burchmore, R., Law, A., Wilson, P. W., Schmutz, M., Preisinger, R. and Dunn, I. C. (2013). Enhancing the egg's natural defence against bacterial penetration by increasing cuticle deposition. *Animal Genetics, 44*(6), pp. 661–8.

Bain, M. M., Nys, Y. and Dunn, I. C. (2016). Increasing persistency in lay and stabilising egg quality in longer laying cycles. What are the challenges? *British Poultry Science, 57*(3), pp. 330–8.

Baird, T., Solomon, S. E. and Tedstone, D. R. (1975). Localisation and characterisation of egg shell porphyrins in several avian species. *British Poultry Science, 16*(2), pp. 201–8.

Bouvarel, I., Nys, Y. and Lescoat, P. (2011). Hen nutrition for sustained egg quality. In Y. Nys, M. Bain and F. Vanimmerseel (Eds), *Improving the Safety and Quality of Eggs and Egg Products, Vol 1: Egg Chemistry, Production and Consumption*, pp. 261–90. Cambridge, Woodhead Publishing Ltd.

Brionne, A., Nys, Y., Hennequet-Antier, C. and Gautron, J. (2014). Hen uterine gene expression profiling during eggshell formation reveals putative proteins involved in the supply of minerals or in the shell mineralization process. *BMC Genomics, 15*, p. 220.

Carrino, D. A., Rodriguez, J. P. and Caplan, A. I. (1997). Dermatan sulfate proteoglycans from the mineralized matrix of the avian eggshell. *Connective Tissue Research, 36*(3), pp. 175–93.

Chien, Y. C., Hincke, M. T. and McKee, M. D. (2009). Avian eggshell structure and osteopontin. *Cells Tissues Organs, 189*(1–4), pp. 38–43.

Chien, Y. C., Hincke, M. T., Vali, H. and McKee, M. D. (2008). Ultrastructural matrix-mineral relationships in avian eggshell, and effects of osteopontin on calcite growth in vitro. *Journal of Structural Biology, 163*(1), pp. 84–99.

Cordeiro, C. M. and Hincke, M. T. (2011). Recent patents on eggshell: shell and membrane applications. *Recent Patents on Food, Nutrition and Agriculture, 3*(1), pp. 1–8.

Cordeiro, C. M. M. and Hincke, M. T. (2016). Quantitative proteomics analysis of eggshell membrane proteins during chick embryonic development. *Journal of Proteomics, 130*, pp. 11–25.

Cusack, M., Fraser, A. C. and Stachel, T. (2003). Magnesium and phosphorus distribution in the avian eggshell. *Comparative Biochemistry and Physiology Part B: Biochemistry and Molecular Biology, 134*(1), pp. 63–9.

De Reu, K., Grijspeerdt, K., Messens, W., Heyndrickx, M., Uyttendaele, M., Debevere, J. and Herman, L. (2006). Eggshell factors influencing eggshell penetration and whole egg contamination by different bacteria, including Salmonella enteritidis. *International Journal of Food Microbiology, 112*(3), pp. 253–60.

Dennis, J. E., Xiao, S.-Q., Agarwal, M., Fink, D. J., Heuer, A. H. and Caplan, A. I. (1996). Microstructure of matrix and mineral components of eggshells from White Leghorn chickens (Gallus gallus). *Journal of Morphology, 228*(3), pp. 287–306.

Du, J. W., Hincke, M. T., Rose-Martel, M., Hennequet-Antier, C., Brionne, A., Cogburn, L. A., Nys, Y. and Gautron, J. (2015). Identifying specific proteins involved in eggshell membrane formation using gene expression analysis and bioinformatics. *Bmc Genomics, 16*, p. 792.

Dunn, I. (2011). Poultry breeding for egg quality: traditional and modern genetic approaches. *Improving the Safety and Quality of Eggs and Egg Products, Vol 1: Egg Chemistry, Production and Consumption*, (213), pp. 245–60.

Dunn, I. C., Ciccone, N. A. and Joseph, N. T. (2009a). Endocrinology and genetics of the hypothalamic-pituitary-gonadal axis. *Biology of Breeding Poultry, 29*, pp. 61–88.

Dunn, I. C., Joseph, N. T., Bain, M., Edmond, A., Wilson, P. W., Milona, P., Nys, Y., Gautron, J., Schmutz, M., Preisinger, R. and Waddington, D. (2009b). Polymorphisms in eggshell organic matrix genes are associated with eggshell quality measurements in pedigree Rhode Island Red hens. *Animal Genetics, 40*(1), pp. 110–4.

Dunn, I. C., Rodriguez-Navarro, A. B., Mcdade, K., Schmutz, M., Preisinger, R., Waddington, D., Wilson, P. W. and Bain, M. M. (2012). Genetic variation in eggshell crystal size and orientation is large and these traits are correlated with shell thickness and are associated with eggshell matrix protein markers. *Animal Genetics, 43*(4), pp. 410–18.

Etches, R. J. (1996). Reproduction in Poultry. Wallingford, CAB International.

Fernandez, M. S., Araya, M. and Arias, J. L. (1997). Eggshells are shaped by a precise spatio-temporal arrangement of sequentially deposited macromolecules. *Matrix Biology, 16*(1), pp. 13–20.

Fernandez, M. S., Escobar, C., Lavelin, I., Pines, M. and Arias, J. L. (2003). Localization of osteopontin in oviduct tissue and eggshell during different stages of the avian egg laying cycle. *Journal of Structural Biology, 143*(3), pp. 171–80.

Fernandez, M. S., Moya, A., Lopez, L. and Arias, J. L. (2001). Secretion pattern, ultrastructural localization and function of extracellular matrix molecules involved in eggshell formation. *Matrix Biology, 19*(8), pp. 793–803.

Freeman, C. L., Harding, J. H., Quigley, D. and Rodger, P. M. (2011). Simulations of ovocleidin-17 binding to calcite surfaces and its implications for eggshell formation. *The Journal of Physical Chemistry C, 115*(16), pp. 8175–83.

Fulton, J. E., Soller, M., Lund, A. R., Arango, J. and Lipkin, E. (2012). Variation in the ovocalyxin-32 gene in commercial egg-laying chickens and its relationship with egg production and egg quality traits. *Animal Genetics, 43*, pp. 102–13.

Gautron, J., Hincke, M. T., Mann, K., Panheleux, M., Bain, M., McKee, M. D., Solomon, S. E. and Nys, Y. (2001a). Ovocalyxin-32, a novel chicken eggshell matrix protein – Isolation, amino acid sequencing, cloning, and immunocytochemical localization. *Journal of Biological Chemistry, 276*(42), pp. 39243–52.

Gautron, J., Hincke, M. T. and Nys, Y. (1997). Precursor matrix proteins in the uterine fluid change with stages of eggshell formation in hens. *Connective Tissue Research, 36*(3), pp. 195–210.

Gautron, J., Hincke, M. T., Panheleux, M., Garcia-Ruiz, J. M., Boldicke, T. and Nys, Y. (2001b). Ovotransferrin is a matrix protein of the hen eggshell membranes and basal calcified layer. *Connective Tissue Research, 42*(4), pp. 255–67.

Gautron, J., Murayama, E., Vignal, A., Morisson, M., McKee, M. D., Réhault, S., Labas, V., Belghazi, M., Vidal, M. L., Nys, Y. and Hincke, M. T. (2007). Cloning of ovocalyxin-36, a novel chicken eggshell protein related to lipopolysaccharide-binding proteins, bactericidal permeability-increasing proteins, and plunc family proteins. *Journal of Biological Chemistry, 282*(8), pp. 5273–86.

Gautron, J., Rehault-Godbert, S., Pascal, G., Nys, Y. and Hincke, M. T. (2011). Ovocalyxin-36 and other LBP/BPI/PLUNC-like proteins as molecular actors of the mechanisms of the avian egg natural defences. *Biochemical Society Transactions, 39*, pp. 971–6.

Gilbert, A. B. (1990). Female genital organs. In A. S. King and J. McLelland (Eds), *Form and Function in Birds*. New York, Academic Press.

Hernández-Hernández, A., Gómez-Morales, J., Rodríguez-Navarro, A. B., Gautron, J., Nys, Y. and García-Ruiz, J. M. (2008). Identification of some active proteins in the process of hen eggshell formation. *Crystal Growth & Design, 8*(12), pp. 4330–9.

Hincke, M., Gautron, J., Nys, Y., Rodriguez-Navarro, A. B. and McKee, M. D. (2011). The eggshell: structure and protective function. *Improving the Safety and Quality of Eggs and Egg Products, Vol 1: Egg Chemistry, Production and Consumption*, (213), pp. 151–82.

Hincke, M. T. (1995). Ovalbumin is a component of the chicken eggshell matrix. *Connective Tissue Research, 31*(3), pp. 227–33.

Hincke, M. T., Chien, Y. C., Gerstenfeld, L. C. and McKee, M. D. (2008). Colloidal-gold immunocytochemical localization of osteopontin in avian eggshell gland and eggshell. *Journal of Histochemistry & Cytochemistry, 56*(5), pp. 467–76.

Hincke, M. T., Gautron, J., Mann, K., Panhéleux, M., McKee, M. D., Bain, M., Solomon, S. E. and Nys, Y. (2003). Purification of ovocalyxin-32, a novel chicken eggshell matrix protein. *Connective Tissue Research, 44*(Suppl. 1), pp. 16–9.

Hincke, M. T., Gautron, J., Panheleux, M., Garcia-Ruiz, J., McKee, M. D. and Nys, Y. (2000). Identification and localization of lysozyme as a component of eggshell membranes and eggshell matrix. *Matrix Biology, 19*(5), pp. 443–53.

Hincke, M. T., Gautron, J., Tsang, C. P., McKee, M. D. and Nys, Y. (1999). Molecular cloning and ultrastructural localization of the core protein of an eggshell matrix proteoglycan, ovocleidin-116. *Journal of Biological Chemistry, 274*(46), pp. 32915–23.

Hincke, M. T., Tsang, C. P. W., Courtney, M., Hill, V. and Narbaitz, R. (1995). Purification and immunochemistry of a soluble matrix protein of the chicken eggshell (Ovocleidin-17). *Calcified Tissue International, 56*(6), pp. 578–83.

Horvat-Gordon, M., Yu, F., Burns, D. and Leach, R. M. (2008). Ovocleidin (OC 116) is present in avian skeletal tissues. *Poultry Science, 87*(8), pp. 1618–23.

Ishikawa, S., Suzuki, K., Fukuda, E., Arihara, K., Yamamoto, Y., Mukai, T. and Itoh, M. (2010). Photodynamic antimicrobial activity of avian eggshell pigments. *FEBS Letters, 584*(4), pp. 770–4.

Jonchere, V., Brionne, A., Gautron, J. and Nys, Y. (2012). Identification of uterine ion transporters for mineralisation precursors of the avian eggshell. *BMC Physiology, 12*, p. 10.

Kawasaki, K. and Weiss, K. M. (2006). Evolutionary genetics of vertebrate tissue mineralization: the origin and evolution of the secretory calcium-binding phosphoprotein family. *Journal of Experimental Zoology Part B: Molecular and Developmental Evolution, 306*(3), pp. 295–316.

Kodali, V. K., Gannon, S. A., Paramasivam, S., Raje, S., Polenova, T. and Thorpe, C. (2011). A novel disulfide-rich protein motif from avian eggshell membranes. *Plos ONE, 6*(3), p. e18187.

Lakshminarayanan, R., Valiyaveettil, S., Rao, V. S. and Kini, R. M. (2003). Purification, characterization, and in vitro mineralization studies of a novel goose eggshell matrix protein, ansocalcin. *Journal of Biological Chemistry, 278*(5), pp. 2928–36.

Lavelin, I., Meiri, N., Einat, M., Genina, O. and Pines, M. (2002). Mechanical strain regulation of the chicken glypican-4 gene expression in the avian eggshell gland. *American Journal of Physiology – Regulatory, Integrative Comparative Physiology, 283*(4), pp. R853–61.

Lavelin, I., Yarden, N., Ben-Bassat, S., Bar, A. and Pines, M. (1998). Regulation of osteopontin gene expression during egg shell formation in the laying hen by mechanical strain. *Matrix Biology, 17*(8–9), pp. 615–23.

Leach, R. M. (1982). Biochemistry of the organic matrix of the eggshell. *Poultry Science, 61*, pp. 2040–7.

Leleu, S., Messens, W., De Reu, K., De Preter, S., Herman, L., Heyndrickx, M., De Baerdemaeker, J., Michiels, C. W. and Bain, M. (2011). Effect of egg washing on the cuticle quality of brown and white table eggs. *Journal of Food Protection, 74*(10), pp. 1649–54.

Makkar, S., Liyanage, R., Kannan, L., Packialakshmi, B., Lay, J. O. and Rath, N. C. (2015). Chicken egg shell membrane associated proteins and peptides. *Journal of Agricultural and Food Chemistry, 63*(44), pp. 9888–98.

Mann, K., Gautron, J., Nys, Y., McKee, M. D., Bajari, T., Schneider, W. J. and Hincke, M. T. (2003). Disulfide-linked heterodimeric clusterin is a component of the chicken eggshell matrix and egg white. *Matrix Biology, 22*(5), pp. 397–407.

Mann, K., Hincke, M. T. and Nys, Y. (2002). Isolation of ovocleidin-116 from chicken eggshells, correction of its amino acid sequence and identification of disulfide bonds and glycosylated Asn. *Matrix Biology, 21*(5), pp. 383–7.

Mann, K., Macek, B. and Olsen, J. V. (2006). Proteomic analysis of the acid-soluble organic matrix of the chicken calcified eggshell layer. *Proteomics, 6*(13), pp. 3801–10.

Mann, K., Olsen, J. V., Macek, B., Gnad, F. and Mann, M. (2007). Phosphoproteins of the chicken eggshell calcified layer. *Proteomics, 7*(1), pp. 106–15.

Marie, P., Labas, V., Brionne, A., Harichaux, G., Hennequet-Antier, C., Nys, Y. and Gautron, J. (2014). Data set for the proteomic inventory and quantitative analysis of chicken uterine fluid during eggshell biomineralization. *Data Brief, 1*, pp. 65–9.

Marie, P., Labas, V., Brionne, A., Harichaux, G., Hennequet-Antier, C., Rodriguez-Navarro, A. B., Nys, Y. and Gautron, J. (2015). Quantitative proteomics provides new insights into chicken eggshell matrix protein functions during the primary events of mineralisation and the active calcification phase. *Journal of Proteomics, 126*, pp. 140–54.

Mertens, K., Kemps, B., Perianu, C., De Baerdemasker, J., Decuypere, E. and DeKetelaere, B. (2011). Advances in egg defect detection, quality assessment and automated sorting and grading. *Improving the Safety and Quality of Eggs and Egg Products, Vol 1: Egg Chemistry, Production and Consumption*, (213), pp. 209–41.

Miksík, I., Charvátová, J., Eckhardt, A. and Deyl, Z. (2003). Insoluble eggshell matrix proteins–their peptide mapping and partial characterization by capillary electrophoresis and high-performance liquid chromatography. *Electrophoresis, 24*(5), pp. 843–52.

Miksík, I., Eckhardt, A., Sedláková, P. and Mikulikova, K. (2007). Proteins of insoluble matrix of avian (gallus gallus) eggshell. *Connective Tissue Research, 48*(1), pp. 1–8.

Miksík, I., Sedláková, P., Lacinová, K., Pataridis, S. and Eckhardt, A. (2010). Determination of insoluble avian eggshell matrix proteins. *Analytical and Bioanalytical Chemistry, 397*(1), pp. 205–14.

Munoz, A., Dominguez-Gasca, N., Jimenez-Lopez, C. and Rodriguez-Navarro, A. B. (2015). Importance of eggshell cuticle composition and maturity for avoiding trans-shell Salmonella contamination in chicken eggs. *Food Control, 55*, pp. 31–8.

Nys, Y., Gautron, J., Garcia-Ruiz, J. M. and Hincke, M. T. (2004). Avian eggshell mineralization: biochemical and functional characterization of matrix proteins. *Comptes Rendus Palevol, 3*(6–7), pp. 549–62.

Nys, Y. and Guyot, N. (2011). Egg formation and chemistry. *Improving the Safety and Quality of Eggs and Egg Products, Vol 1: Egg Chemistry, Production and Consumption*, (213), pp. 83–132.

Nys, Y., Hincke, M. T., Arias, J. L., Garcia-Ruiz, J. M. and Solomon, S. E. (1999). Avian eggshell mineralization. *Poultry and Avian Biology Reviews, 10*, pp. 143–66.

Nys, Y., Zawadzki, J., Gautron, J. and Mills, A. D. (1991). Whitening of brown-shelled eggs: mineral composition of uterine fluid and rate of protoporphyrin deposition. *Poultry Science, 70*(5), pp. 1236–45.

O'Sullivan, N.P. (2009). What are the egg traits that further sustainable production? *International Poultry Production, 17*(6).

Pines, M., Knopov, V. and Bar, A. (1995). Involvement of osteopontin in egg shell formation in the laying chicken. *Matrix Biology, 14*(9), pp. 765–71.

Rehault-Godbert, S., Herve-Grepinet, V., Gautron, J., Cabau, C., Nys, Y. and Hincke, M. (2011). Molecules involved in chemical defence of the chicken egg. *Improving the Safety and Quality of Eggs and Egg Products, Vol 1: Egg Chemistry, Production and Consumption*, (213), pp. 183–208.

Roberts, J. R. and Brackpool, C. E. (1994). The ultrastructure of avian egg shells. *Poultry Science Reviews, 5*(245–72).
Rodriguez-Navarro, A. and García-Ruiz, J. M. (2000). Model of textural development of layered crystal aggregates. *European Journal of Mineralogy, 12*, pp. 609–14.
Rodriguez-Navarro, A., Kalin, O., Nys, Y. and Garcia-Ruiz, J. M. (2002). Influence of the microstructure on the shell strength of eggs laid by hens of different ages. *British Poultry Science, 43*(3), pp. 395–403.
Rodriguez-Navarro, A. B., Dominguez-Gasca, N., Munoz, A. and Ortega-Huertas, M. (2013). Change in the chicken eggshell cuticle with hen age and egg freshness. *Poultry Science, 92*(11), pp. 3026–35.
Rodriguez-Navarro, A. B., Marie, P., Nys, Y., Hincke, M. T. and Gautron, J. (2015). Amorphous calcium carbonate controls avian eggshell mineralization: A new paradigm for understanding rapid eggshell calcification. *Journal of Structural Biology, 190*(3), pp. 291–303.
Romanoff, A. L. and Romanoff, A. J. (1949). *The Avian Egg*. New York, Wiley & Sons.
Rose, M. L. and Hincke, M. T. (2009). Protein constituents of the eggshell: eggshell-specific matrix proteins. *Cellular and Molecular Life Sciences, 66*(16), pp. 2707–19.
Rose-Martel, M., Du, J. W. and Hincke, M. T. (2012). Proteomic analysis provides new insight into the chicken eggshell cuticle. *Journal of Proteomics, 75*(9), pp. 2697–706.
Rose-Martel, M. and Hincke, M. T. (2015a). The eggshell as a barrier to contamination. *Handbook of Eggs in Human Function,* (9), pp. 615–26.
Rose-Martel, M., Smiley, S. and Hincke, M. T. (2015b). Novel identification of matrix proteins involved in calcitic biomineralization. *Journal of Proteomics, 116*, pp. 81–96.
Samiullah, S. and Roberts, J. R. (2013). The location of protoporphyrin in the eggshell of brown-shelled eggs. *Poultry Science, 92*(10), pp. 2783–8.
Samiullah, S. and Roberts, J. R. (2014). The eggshell cuticle of the laying hen. *Worlds Poultry Science Journal, 70*(4), pp. 693–707.
Sandilands, V. (2011). The laying hen and bone fractures. *Veterinary Record, 169*, pp. 411–12.
Sauveur, B. and De Reviers, M. (1988). Reproduction and egg production in poultry. In B. Sauveur and M. De Reviers (Eds), *Reproduction des Volailles et Production d'œufs*. INRA, Paris, pp. 377–436.
Solomon, S. E. (1991). *Egg and Eggshell Quality*. London, Wolfe Publishing.
Solomon, S. E. (1999). Gordon Memorial Lecture. An egg ist ein ei, es un huevo, est un oeuf. *British Poultry Science, 40*(1), pp. 5–11.
Solomon, S. E. (2002). The oviduct in chaos. *World's Poultry Science Journal, 58*, pp. 41–8.
Solomon, S. E. (2010). The eggshell: strength, structure and function. *British Poultry Science, 51*(Suppl. 1), pp. 52–9.
Sun, C. J., Xu, G. and Yang, N. (2013). Differential label-free quantitative proteomic analysis of avian eggshell matrix and uterine fluid proteins associated with eggshell mechanical property. *Proteomics, 13*(23–4), pp. 3523–36.
Tullett, S. G. (1975). Regulation of avian eggshell porosity. *Journal of Zoology, 177*, pp. 339–48.
Van Sambeek, F. (2010). Progress in Layer Genetics 'Longer production cycles from a genetic perspective'. http://www.isapoultry.com/~/media/Files/ISA/Information/Technical%20Bulletins/Breeding/Longer%20production%20cycles%20from%20a%20genetic%20perspective.ashx
Wang, X., Wu, C., Tao, K., Zhao, K., Wang, J., Xu, H., Xia, D., Shan, H. and Lu, J. R. (2010). Influence of ovalbumin on CaCO3 precipitation during in vitro biomineralization. *Journal of Physical Chemistry B, 114*(16), pp. 5301–8.
Wedral, E. M., Vadehra, D. V. and Baker, R. C. (1974). Chemical composition of the cuticle, and the inner and outer shell membranes from eggs of Gallus gallus. *Comparative Biochemistry and Physiology Part B, 47*(3), pp. 631–40.
Wellman-Labadie, O., Picman, J. and Hincke, M. T. (2008). Antimicrobial activity of cuticle and outer eggshell protein extracts from three species of domestic birds. *British Poultry Science, 49*(2), pp. 133–43.

Whitehead, C. C. (2004). Overview of bone biology in the egg-laying hen. *Poultry Science Reviews, 83*, pp. 193–9.

Xing, J., Wellman-Labadie, O., Gautron, J. and Hincke, M. T. (2007). Recombinant eggshell ovocalyxin-32: expression, purification and biological activity of the glutathione S-transferase fusion protein. *Comparative Biochemistry and Physiology Part B: Biochemistry and Molecular Biology, 147*(2), pp. 172–7.

Chapter 2

Composition and properties of egg white

Kaustav Majumder, University of Nebraska, Lincoln, USA and Yoshinori Mine, University of Guelph, Canada

1 Introduction

Since the beginning of human civilization, chicken eggs have been a staple food in the human diet. Egg production and consumption have evolved over time, from collection of eggs from the nest of wild fowls by hunter-gatherers to domesticated chickens, then to modern-day genetically selected birds raised by poultry farmers. As analytical technology developed after the twentieth century, the functional properties and chemical composition of avian eggs have been intensively studied, and most of these studies have focused on the eggs produced by the domestic chicken in commercial poultry farms. Eggs are the most economical and abundant source of high-quality dietary protein. Therefore, it is important to understand the fundamental chemistry and chemical composition of the egg. Eggs are consumed throughout the world among different cultures, irrespective of their race and religion, as one of the prime sources of essential nutrition for human development.

Eggs are well known not only for their nutritional value but also for their functional properties. Functional properties of eggs and the underlying chemical complexities have

http://dx.doi.org/10.19103/AS.2016.0012.02

Table 1 Amino acid composition[1] of egg white

Amino acid	Content (g/100 g egg white)
Tryptophan	0.125
Threonine	0.449
Isoleucine	0.661
Leucine	1.016
Lysine	0.806
Methionine	0.399
Cystine	0.287
Phenylalanine	0.686
Tyrosine	0.457
Valine	0.809
Arginine	0.648
Histidine	0.290
Alanine	0.704
Aspartic acid	1.220
Glutamic acid	1.550
Glycine	0.413
Proline	0.435
Serine	0.798

[1]Values are expressed as g/100 g edible portion of albumen or yolk.
Source: Adapted from the US Department of Agriculture (2006) nutrient database.

been studied by a number of researchers (Li-Chan 1989; Raikos et al. 2007; Abeyrathne et al. 2013). Complete understanding of the functional properties of eggs and the knowledge of chemistry behind the functional properties are critical for the use of eggs in the food industries to develop novel food products or to improve the quality and value of existing products. Apart from the functional qualities, food manufacturers also endeavour to prolong the shelf life of eggs, protect nutritional value of eggs during storage and incorporate egg products into other marketable value-added novel products such as protein shakes, cakes, and ice cream *via* different food processing techniques (Raikos et al. 2007).

Eggs consist of three major components: (1) eggshell – this is the outermost layer of an egg, a polycrystalline structure of calcite that includes a natural outer layer of cuticle; (2) egg white – it is a viscous layer composed mainly of proteins; and (3) egg yolk is the inner light or deep yellow coloured layer of an egg, consisting primarily of lipids.

Among these three major layers of the egg, this chapter will provide detailed information about the compound structure, chemical composition and functionality of egg white.

Table 2 Minerals and vitamins in whole egg white[1]

Constituent (units)	Whole egg	Egg white
Minerals (mg)		
Ca	29.2	3.8
Cl	96.0	66.1
Cu	0.033	0.009
I	0.026	0.001
Fe	1.08	0.053
Mg	6.33	4.15
Mn	0.021	0.002
P	111	8
K	74	57
Na	71	63
S	90	62
Zn	0.72	0.05
Vitamins		
Vitamin A (IU)	264	–
Vitamin D (IU)	27	–
Vitamin E (mg)	0.88	–
Vitamin B12 (μg)	0.48	–
Choline (mg)	11.0	2.58
Folic acid (mg)	237	0.46
Inositol (mg)	0.023	0.006
Niacin (Vitamin B3) (mg)	5.94	1.52
Pantothenic acid (mg)	0.045	0.035
Pyridoxine (mg)	0.83	0.09
Riboflavin (Vitamin B2) (mg)	0.065	0.008
Thiamine (mg)	0.18	0.11
	0.05	0.004

[1] Adapted from Watkins (1995).

Around 60% of the total weight of the egg is composed of egg white (also known as egg albumen). Water is the major constituent of egg white, particularly in the outermost and innermost layers. After water, proteins are the second major component of egg white. An overview of the proteins in egg white is given in Section 3, and then Sections 4–9 provide a detailed review of nine important proteins. Carbohydrates (primarily glucose), lipids and minerals are the minor elements in egg white (Stadelman et al. 1995). Carbohydrates and minerals, along with vitamins, are covered in Section 10. Tables 1 and 2 show the composition of amino acids, minerals and vitamins, respectively, in egg white.

2 The structure of egg white

The egg white is composed of four distinct layers: thin, thick, chalaziferous layer and chalazae cord. About 23.3% of egg white is made up of the thin layer, which is divided into two parts: outer and inner thin layers. The 'outer thin' layer is attached to the inner eggshell membrane and the 'inner thin' layer (16.8% of egg white) is connected with the chalaziferous layer (2.7% of egg white) (Brake et al. 1997; Conrad and Phillips 1938; Li 2006). The viscous or thick layer is the major portion of egg white (57.3%) and separates inner and outer thin layers.

The chalaziferous layer is a thick and gelatinous layer that directly covers and protects the entire egg yolk (Mine 1995). In the long axis of an egg, a portion of the chalaziferous layer is twisted at both sides of the yolk membrane, forming a thick rope-like structure named the chalazae cord (Stadelman et al. 1995). This cord is wrapped clockwise at the sharp (pointed) end of the egg and counterclockwise at the opposite (blunt) end. The chalazae cord stretches into the thick layer of egg white on both sides and creates a pocket to suspend the egg yolk in the egg white (Yamamoto et al. 1996). The chalazae chord is slightly elastic and permit limited rotation of the egg yolk (Mine 1995). In fresh eggs, thick layers of egg white cover the inner thin layer and the chalaziferous layer, holding the egg

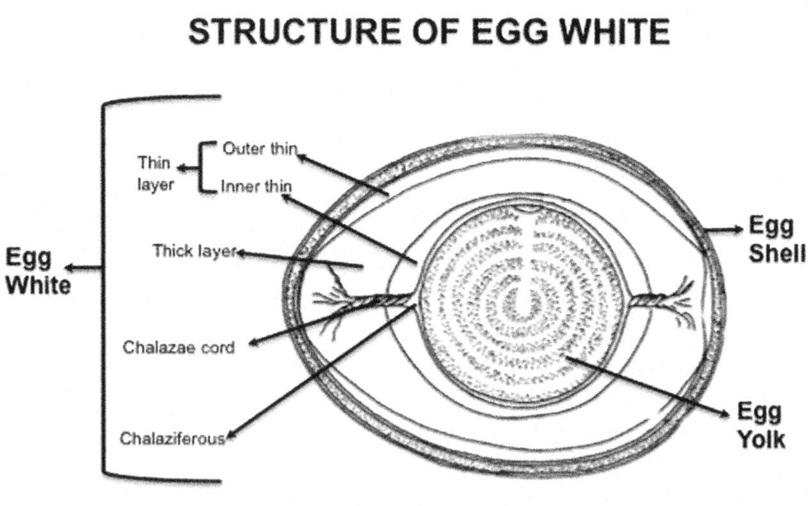

Figure 1 Egg white or albumen consists of four different layers: a thin layer (which consists of an inner thin layer, a layer of egg white between the egg yolk, the rest of the egg white; and an outer thin layer which separates the egg shell membrane and the thick layer or egg white); a thick layer; the chalazae cord; and the chalaziferous layer.

Source: Adapted from American Egg Board, 1981.

yolk in the centre of the egg. Figure 1 explains the complex structure of egg white and identifies the four different layers.

3 Proteins in egg white: an overview

Egg white contains water (87%), proteins (11%), lipids (0.03%), carbohydrates (0.9%), vitamins and minerals (Cotterill and Geiger 1977). Egg proteins are famous for their excellent digestibility and high nutritional quality and contain all the essential amino acids to fulfil human dietary requirements (Friedman 1996).

Egg white contains many different proteins, and these have been identified in studies using modern analytical techniques. However, most of these proteins are not yet characterized (Raikos et al. 2006).

Several researchers identified different protein components in egg white by using high-resolution techniques for proteome analysis. Proteomic analysis of hen egg white using two-dimensional (2D) electrophoresis (2DE) and mass spectrometry identified 16 unique protein spots (Guérin-Dubiard et al. 2006). Subsequently, another study detected 78 egg white proteins using 1-DE and LC-MS/MS and, among those proteins, 54 minor proteins were identified for the first time (Mann 2007).

Table 3 Content, isoelectric point and molecular weight of some proteins found in egg white[1]

Protein	Percentage in egg white	Isoelectric point	Molecular weight, kDa
Ovalbumin	54	4.5	45
Ovotransferrin	12	6.1	76
Ovomucoid	11	4.1	28
Ovomucin	3.5	4.5–5.0	5500–8300
Lysozyme	3.4	10.7	14.3
Ovoglobulin		(6.1–5.3)	
G_2 globulin	4.0	5.5	30–45
G_3 globulin	4.0	4.8	nd[5]
Ovoinhibitor	1.5	5.1	49
Ovoglycoprotein	1.0	3.9	24.4
Ovoflavoprotein	0.8	4.0	32
Ovomacroglobulin	0.5	4.5	769
Cystatin	0.05	5.1	12.7
Avidin	0.05	10	68.3

[1] Data are compiled from Li-Chan et al. (1995) except for those shown in parentheses, which are from Guérin-Dubiard et al. (2006).
[2] Protein with strong homology with bacterial permeability increasing protein.
[3] Quiescence-specific protein or extra fatty acid-binding protein.
[4] Vitelline membrane outer layer protein-1.
[5] nd = not determined.

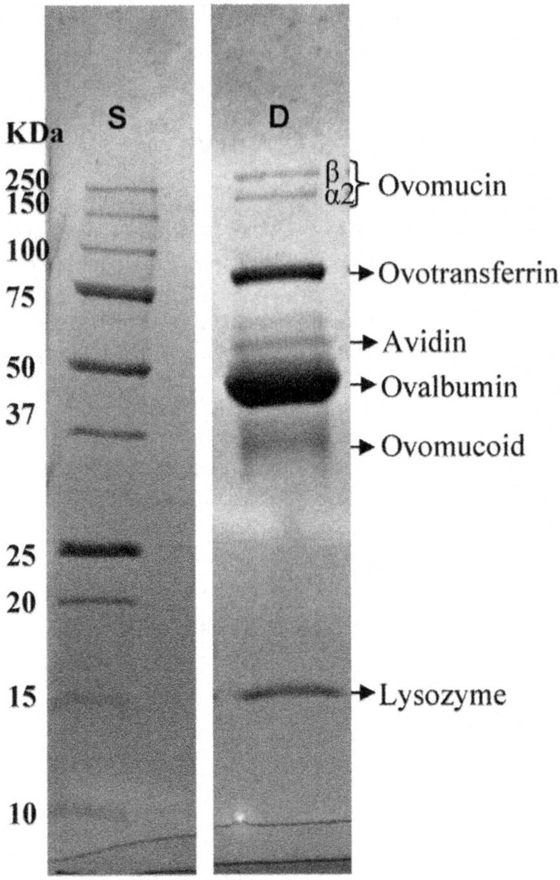

Figure 2 Proteins from egg white, SDS-PAGE pattern of separation of major egg white proteins according to their molecular weight.

Modified from Omana and Wu; *J. Agric. Food Chem.* (2009: 57, 3596–3603).

Sodium dodecyl sulphate–polyacrylamide gel electrophoresis (SDS-PAGE) and 2DE techniques have been used to analyse the protein composition of egg white (Fig. 2) (Omana and Wu 2009; Raikos et al. 2006). SDS-PAGE analysis can clearly identify six major proteins in egg white: ovalbumin (OVA) (54% of the total egg protein), ovotransferrin (12%), ovomucoid (11%), lysozyme (3.5%), ovoglobulins (2%) and ovomucin (3%).

Table 3 lists the content, isoelectric points and molecular weights reported for the major proteins in egg white. The amino acid composition of some of these proteins has been described previously (Li-Chan 1989), and more detailed and updated information is now available through online databases such as the UniProt Knowledgebase (Swiss-Prot and TrEMBL).

Some details of the chemical composition and structure of the egg white proteins are described in the following sections.

4 Ovalbumin

OVA is the most abundant and the major protein in egg white. It is a phosphorylated glycoprotein and belongs to the serpin superfamily, which includes more than 300 homologous proteins with different functions, including the serine protein inhibitors of human plasma (Huntington and Stein 2001). Interestingly, OVA lacks this inhibitory activity, even though it seems to have retained the alleged reactive centre at Als 353-Ser 354 for proteolytic cleavage by elastase. Generally in inhibitory serpins, cleavage leads to a dramatic conformational change thought to be related to the inhibitory activity (McReynolds et al. 1978; Sugimoto et al. 1999).

OVA is composed of 386 amino acids, with genetic polymorphism reported at two residues: 290 (Glu → Gln) and 312 (Asn → Asp). It is one of the major glycoproteins with a single carbohydrate chain linked to Asn293 (McReynolds et al. 1978). OVA is a unique glycoprotein in egg white and has six cysteine residues in total. Only two cysteine residues are involved in a disulphide bond: Cys74 and Cys 12; the other four are composed of sulphydryl (SH) groups, thus creating free SH groups (Fothergill and Fothergill 1970). Three out of four of these SH groups are masked in the native state, whereas the fourth is reactive only in the denatured protein (Fothergill and Fothergill 1970). The three-dimensional confirmation of OVA was established by X-ray crystallographic analysis at 1.95 Å resolution; the study indicates that OVA consists of three beta (β) sheets and nine alpha (α) helices (Huntington et al. 1995). The physiochemical analysis of the amino acid composition revealed that around 50% of the amino acids in OVA are hydrophobic, and 30% of them acidic and charged amino acids (McReynolds et al. 1978; Huntington and Stein 2001). These acidic amino acids presented an acidic isoelectric point (pI) of 4.5 (McReynolds et al. 1978; Huntington and Stein 2001).

The amino acid composition and distribution of OVA are unique as compared to other glycoproteins. OVA does not have an N-terminal ladder sequence but has an acetylated glycine at the N-terminal and proline in the C-terminal, and contains a carbohydrate group attached to the N-terminal amino acids (McReynolds et al. 1978; Huntington and Stein 2001).

Ovalbumin-Y (OVA-Y) is a unique chimeric glycoprotein with high amino acid sequence similarity with native OVA, but it has an adjacent carbohydrate moiety similar to that of ovomucoid (Hirose et al. 2006). OVA-Y protein in egg white was first identified and characterized by 2D PAGE and peptide mass fingerprinting (Nau et al. 2005). OVA-Y protein not only has glycosylated moieties, similar to native OVA, but also has an additional phosphorylated site (Guérin-Dubiard et al. 2006).

Native OVA (N-OVA) is resistant to tryptic digestion but becomes susceptible to trypsin after heat denaturation, pH change or acid treatment. As a result of thermal denaturation, OVA is transformed via an intermediate state into a thermally stable form called OVA-S (Pelegrine and Gasparetto 2006). High pH and temperature are equally capable of converting N-OVA to S-OVA. Moreover, storage temperature has a bigger impact on conversion of N-OVA to S-OVA than storage time which did not affect this conversion at an ambient temperature of 20–25°C (Sugimoto et al. 1999). The conversion of OVA to S-OVA during storage has been attributed to the change in pH, and an increase of pH has a direct effect rather than temperature (Huang et al. 2012). Therefore, at an ambient temperature of 4°C the conversion of N-OVA to S-OVA may reach 81% during a long-term storage period of six months (Huang et al. 2012). During storage, loss of egg mass is associated with the type of OVA present in the egg, N-OVA or S-OVA. The isoelectric point (pI) of OVA also slightly shifted and becomes more acidic during storage (Schäfer and Schwagele 1999; Huang et al. 2012); the change in pI happens due to the formation of S-OVA. Therefore, the percentage of S-OVA could be used an indicator of egg freshness (Huang et al. 2012).

OVA is also the major allergen of egg white and causes IgE-mediated food allergic reactions especially in children (Caubet and Wang 2011). The IgE-binding epitopes of OVA are primarily hydrophobic, polar and charged amino acid residues. These sequences are mainly composed of β-turn and β-sheet structures. Asp_{95}-Ala_{102} is only one allergenic epitope that consists of a single alpha helix. As heat treatment changes the structure of OVA, so IgE binding can also be reduced by heating or irradiation of OVA (Kim 2002), and irradiation has been regarded as one of the suggested methods to reduce allergenicity (Seo et al. 2004).

Along with allergenic epitopes, OVA has some unique sequences that are embedded into its protein sequence and that have exerted beneficial effects upon successful liberation through enzymatic hydrolysis, food processing or fermentation. These protein sequences are called bioactive peptides. A number of bioactive peptides have been reported from OVA, including peptides with antihypertensive activity (angiotensin-converting enzyme: ACE inhibitory) such as ovokinin and ovokinin2-7 (Yu et al. 2012; Matoba et al. 2003) that have been isolated after peptic digest, antimicrobial peptides (Pellegrini et al. 2004; Réhault-Godbert et al. 2013) and antioxidant peptides isolated from various other enzymatic digestion and fermentation (Dávalos et al. 2004; Nimalaratne et al. 2015). Therefore, it is critical to note that, along with its unique physicochemical properties, OVA is also a major allergen and ironically also an excellent source of bioactive peptides.

5 Ovotransferrin

Ovotransferrin (OTf) is a monomeric glycoprotein with structural similarities with serum transferrin-like protein in animals. The transferrin proteins are a family of typical glycoproteins with a specific moiety that tightly binds with ferric iron. OTf has the ability to bind different metal ions in the ratio of two moles of ion per mole of proteins. For the transferrin proteins such as serum transferrin, OTf and lactoferrin, each of their duplicate lobes binds one atom of Fe^{3+} and one carbonate anion (Lambert et al. 2005). Such compound formation of the iron complex in OTf inhibits the growth of microorganisms. A structural study reported that the chicken heat shock protein 108 (HSP108), along with the avian homolog of GRP94, binds with Fe-OTf and was originally isolated from hen oviduct (Weiner et al. 1997).

The homologous N and C lobes in this particular glycoprotein contain a single iron-binding site situated in a deep cleft and consist of 686 amino acids with a molecular weight of 76 kDa and pI of 66.0, with a total of 15 disulphide bridges to maintain the globular structure (Kurokawa et al. 1999). OTf accounts for 12% of the total egg white protein (Desert et al. 2001; Bou Abdallah and el Hage Chahine 1998). OTf was initially called 'conalbumin' but renamed as 'ovotransferrin' after finding that it can bind iron-like transferrin proteins (Wu and Acero-Lopez 2012). The OTf N-lobe and C lobe form two disulphide bonds (Mizutani et al. 2014), and the interaction between both lobes is considered to be necessary for efficient iron acquisition (Alcantara and Schryvers 1996). The anionic binding sites play a critical role in the domain opening and synergistic carbonate anion release, influencing the iron release of the OTf N-lobe. Specifically, the inter-domain disulphide bond has no effect on the iron uptake and release function but determines the conformational stability of the C lobe (Mizutani et al. 2014; Wu and Acero-Lopez 2012).

The mechanism of in vitro Fe^{3+} uptake and release by egg OTf is somewhat similar but not identical when compared to other transferrin proteins like serum transferrin and lactoferrin. Transferrin proteins bind Fe^{3+} very tightly in a closed inter-domain cleft by the surrounding four protein ligands (in the N-lobe of OTf: Tyr_{191}, His_{250}, Asp_{60} and Tyr_{92}) (Abdallah and Chahine 1999). Fe^{3+} can be easily attached with OTf at pH >7.0, but OTf releases the bound iron at pH < 4.5 (Guérin-Dubiard et al. 2006). The two specific iron-binding sites are situated within the inter-domain cleft of the N and C lobes. OTf is folded into two portions and four domains, with each portion composed of two distinct α- and β-domains. These two separate domains, N and C, are linked with antiparallel β strands, which have a hinge joint (Ibrahim et al. 2006; Kurokawa et al. 1999).

The three-dimensional crystal structure of OTf is found in two forms: apo (iron free) and holo (ironbound). The physicochemical properties of these two types are significantly different from each other (Kurokawa et al. 1999). The apo-form does not have any particular colour, but the holo-form has a salmon pink colour, due to the presence of iron. The apo-form is more susceptible to chemical and physical reactions than the holo-form. The apo-OTf or the iron less OTf are refined by simulated annealing and viewed with a 3.0 Å resolution (Kurokawa et al. 1999). In apo-OTf, the iron-binding clefts stay in the open conformation, supporting the theory that Fe^{3+} binding or release in transferrin proceeds via the hinge joint of the antiparallel β strands connecting two lobes through a mechanism that involves domain opening and closure. The areas of each portion contact one another at different sites in the open and closed forms, and area opening or closing produces a seesaw motion between the two alternative close-packed interfaces. An inter-domain disulphide bridge (Cys478–Cys671) is found only in the C lobe and may restrict area opening but does not completely prevent it (Kurokawa et al. 1999; Wu and Acero-Lopez 2012).

Moreover, along with the iron-binding capacity, several recent studies have examined other functional features of the OTf that may be associated with various biological properties. The embryos of eggs from birds and mammals are highly sensitive to oxidative stress. Therefore, maintaining a steady oxygen-reducing environment during embryonic development is a key factor. OTf undergoes autocleavage at specific sites upon reduction with thioredoxin or thiol-reducing system, through the exceptional chemical reactivity of four tri-peptide motifs, which were located upstream and downstream of the two disulphide domains (residues 115–211 and 454–544) of OTf (Ibrahim et al. 2006). The reduction-scissile sequences, His/Cys-X-Thr, are evolutionarily conserved self-cleavage motifs found in many auto-processing proteins. Similarly, a distinct antibacterial peptide domain within the N-lobe of the OTf retains the antibacterial activity independent of its iron-deprivation capacity (Ibrahim et al. 1998). Studies also found various biologically active peptides with OTf. A cationic peptide fragment (OTAP-92) was found within the 109–200 sequence of the N-lobe of OTf (Ibrahim et al. 2000) that possesses bactericidal activity. OTAP-92 is a unique structural motif similar to insect defensins and proficient at destroying Gram-negative bacteria by cleaving the outer membrane by a self-promoted chemical uptake, causing damage to the biological function of the cytoplasmic membrane (Ibrahim et al. 2000). Several epitopes with antiviral activity were also identified from OTf: Asp-Asn-Gln-Lys-Asp-Glu-Tyr-Glu-Leu-Leu (residues 219–227), Lys-Asp-Leu-Leu-Phe-Lys (residues 269–301) and Arg-Phe-Gly-Val-Asn-Gly (residues 633–638). These three peptides work against Marek's disease virus infection of chicken embryo fibroblasts (Giansanti et al.

2005). An antihypertensive peptide with ACE-inhibitory properties, derived from OTf and identified as Lys-Val-Arg-Glu-Gly-Thr-Thr-Tyr, was identified as a pro-drug inhibitor with *in vivo* antihypertensive activity exerted in the form of Lys-Val-Arg-Glu-Gly-Thr (Lee et al. 2006). Another peptide, Iso-Arg-Trp, was later identified as an ACE-inhibitory peptide, and also enhanced the activity of vasodilatory enzyme ACE-2 (angiotensin-converting enzyme-2) (Majumder, Liang, et al. 2015; Majumder, Chakrabarti, et al. 2015; Majumder et al. 2013).

6 Ovomucoid

Ovomucoid (OVM) is another glycoprotein from egg white, which is heat stable and constitutes about 11% of egg white proteins and belongs to the Kazal family of protein inhibitors (Stadelman et al. 1995; Li-Chan E 1989). OVM is composed of three separate domains (domains I, II and III) cross-linked only by intra-domain disulphide bonds, with nine disulphide bonds, has 186 amino acids and the molecular mass is 28ka (Kovacs-Nolan et al. 2000). The domain II of OVM contains the active site that can inactivate trypsin. It contains 20–25% carbohydrates that are attached to the polypeptide chain via N-glycosylation at asparaginyl residues (Li-Chan 1989). Among the carbohydrate-rich portions, around 4% are sialic acids.

Chicken OVM is one of the major egg allergens and plays a critical role in the pathogenesis of IgE-mediated allergic symptoms (Mine and Zhang 2002a; Mine and Zhang 2001; Mine and Rupa 2003a, 2004). The allergenic potential of OVM is related to its susceptibility to gastrointestinal digestion and heat treatment; it was reported that OVM did not lose its immune reactivity after enzymatic hydrolysis by pepsin (Hirose et al. 2005). The pepsin digestion of OVM is pH dependent; the digestibility is good over the pH range from 1.5 to 2.5, but digestibility is poor at pH 3.0 or higher even with prior heat treatment. Additionally, pepsin-digested OVM retains its trypsin (proteinase) inhibitor activities (Yoshino et al. 2004), and disulphide bonds play a crucial role in maintaining the digestive resistance and allergenicity of OVM (Kovacs-Nolan et al. 2000). Simulated gastric fluid cannot digest the allergenic epitopes of OVM; the digested fragments were then tested to IgE in sera from allergic patients, but 21% of the examined patients retained the OVM-specific IgE-binding (Takagi et al. 2005). It was also reported that only epitopes that construct the OVM protein backbone and not the attached carbohydrate residues are responsible for IgE binding (Besler et al. 2008). Studies suggest that eight IgG epitopes, 5–11 amino acids in length and nine IgE epitopes, 5–16 amino acids in length, are embedded within the primary sequence in OVM (Mine and Zhang 2002b).

Furthermore, analyses have indicated that the epitopes with charged (aspartic acid, glutamic acid and lysine), hydrophobic (phenylalanine, leucine and glycine) and polar (serine, threonine, tyrosine and cysteine) amino acids are crucial for antibody binding (Mine and Zhang 2002b). A complete heat stability study that evaluated the effect of heating at 75–110°C, of OVM, showed a time-dependent decrease in trypsin inhibitory activity (Van der Plancken et al. 2004). The heat stability of OVM is also pH dependent, and digestibility is lowest at pH 7.6. The heat stability also depends on the presence of other egg white constituents, particularly the formation of OVM-lysozyme complexes.

Allergenicity is a major issue of OVM, and many studies have been performed to modify the structure and composition of the allergenic epitopes of OVM in an attempt to diminish its allergenicity (Mine and Rupa 2003b, 2006c). Gamma irradiation along with heating (Lee et al. 2002), heating in the presence of wheat flour (Kato et al. 2001; Kovacs-Nolan et al. 2000), genetic modification (Rupa and Mine 2006a), and deglycosylation by endo-beta-*N*-acetylglucosaminidases (Yamamoto et al. 1998) have been attempted, but all these studies showed that allergenic epitopes of OVM are very resistant to any modification. Moreover, these amendments were always associated with loss of functionality.

7 Ovomucin

Ovomucin is another major sulphated glycoprotein from egg white, which accounts for 3.5% of the total egg white protein and confers the jelly-like structure of egg white. Ovomucin is composed of two parts: soluble and insoluble components. The soluble part of ovomucin (8,300 Da) is the main component of the outer and inner albumen, whereas the insoluble component ranges from 220 to 270 kDa and is the major component of the gel-like insoluble fraction of the thick albumen of egg white (Omana and Wu 2009). Both fractions: insoluble and soluble consist of two distinct subunits with different carbohydrate contents: α-ovomucin and β-ovomucin. The insoluble fraction of ovomucin contains 84 α and 20 β-subunits, whereas the soluble fraction of ovomucin is composed of 40 α- and 3 β-subunits (Robinson and Monsey 1971; Omana and Wu 2009). The α-subunits of ovomucin have at least two distinctive subunits α1 and α2 and have comparably fewer carbohydrate groups than β-subunits. The α-subunits are mainly composed of acidic amino acids such as aspartic acid and glutamic acid (Omana and Wu 2009), whereas, along with high carbohydrate content, β-subunits consist mainly of amino acids serine and threonine (Robinson and Monsey, 1971).

OVM is a protein that is well known for its biological and functional properties. OVM plays a critical role in egg white thinning during storage; experimental results have shown that egg white thinning is possible due to two separate processes: a) degradation of Ovomucin by reduction of disulphide bonds and b) dissociation of the lysozyme – Ovomucin complex (Abeyrathne et al. 2014). Apart from this, Ovomucin is famous for its excellent foaming and emulsifying properties and hypocholesterolemic properties (Mann 2007). Several recent studies have characterized the bioactive properties of Ovomucin. Ovomucin has exhibited hemagglutination inhibition activity against bovine rotavirus (Tsuge et al. 1996); this particular activity was dependent on both α- and β-subunits, while the inhibitory activity against hen Newcastle disease virus required the β-subunit only (Tsuge et al. 1997a,b). The P1, P2 and P3 fractions from the β-subunit are composed of *O*-glycoproteins, containing clustered sialic acid moieties and NeuAc residues. The clustered sialic acid moieties play an important role in the binding and growth inhibitory activity of Newcastle disease virus. The peptide-rich P4 and P5 fractions, mainly derived from the α-subunit, had a higher binding activity to the anti-ovomucin antibodies (Watanabe et al. 1998). The glycopeptides from β-subunit of Ovomucin digested with pronase have shown the ability to bind enterohaemorrhagic *Escherichia coli* O157: H7 (Kobayashi et al. 2004). Moreover, another study found that egg Ovomucin inhibited cholesterol absorption in Caco-2 cells and attenuated hypercholesterolaemia in experimental rats (Nagaoka et al. 2006). The attenuation of cholesterol absorption by direct interaction between cholesterol-mixed micelles and Ovomucin in the jejunal epithelial cells of the small intestine is considered to be the underlying mechanism of hypocholesterolemic action of Ovomucin. The study also suggested that, in the ileum region of the small intestine,

Ovomucin might inhibit the reabsorption of bile acids and lower the cholesterol level in blood plasma (Nagaoka et al. 2006).

Ovomucin is a unique protein from egg white with high functional values and vast potential to be developed as functional foods and nutraceutical ingredients. Therefore, simplified extraction technique of Ovomucin with high purity and yield is one of the exciting research areas. Initially, Ovomucin was separated by the principle of isoelectric precipitation (PI). MacDonnell et al. (1955) separated Ovomucin by bringing down the pH of egg white to 4.5, the pI of Ovomucin, and then differentiating Ovomucin by centrifugation. However, the purity of the Ovomucin was hampered due to the co-precipitation of OVA and lysozyme (Omana et al. 2010; Desert et al. 2001). Along with OVA and lysozyme, OVM and OTf also co-precipitate, as their PIs are 4.1 and 6.1, respectively. A dual-column gel-filtration assisted study, to evaluate that the separation of Ovomucin, found it has total eight subunits which resulted in the isolation of eight individual peaks (Hiidenhovi et al. 1999). Various other approaches which include gel electrophoresis (Desert et al. 2001), β-mercaptoethanol treatment (Hiidenhovi et al., 2002), two-step precipitation using sodium chloride (Omana and Wu 2009; Omana et al. 2010; Wang et al. 2012), and NaCl + pH adjustment (Omana and Wu 2009; Omana et al. 2010; Wang et al. 2012) have been used to extract OVM from egg white. Among the methods, the two-step precipitation method was the best way of separating Ovomucin, regarding purity (above 80%), although the yield is not high (Omana and Wu 2009; Omana et al. 2010; Wang et al. 2012). However, still today the structural complexity of Ovomucin is not fully defined due to difficulties in its purification, large size and poor solubility (forming a gel) after isolation.

8 Lysozyme

Lysozyme is a 14.4 kDa protein from egg white; it consists of 129 amino acid residues with an isoelectric point (PI) of 10.7. Naturally, there are many forms of lysozyme abundantly found in different foods, but lysozyme present in the egg is very special as this is the most soluble and stable. In the egg, lysozyme is found as a monomer but is frequently present as a dimer, which leads to its thermal stability. Lysozyme has four unique disulphide bonds that stabilize its tertiary structure (Kato et al. 2006). Because of its primary character, lysozyme actively binds with the negatively charged protein in egg white, more specifically to OVM, OTf and OVA (Abeyrathne et al. 2014). The ovomucin-lysozyme complex is commonly found in the chalaziferous layer and chalaza of egg whites. It makes up to 3.4% of the egg white proteins. Crystallographic analysis with a resolution of 2 Å revealed the three-dimensional structure of lysozyme as had been known since the 1960s (Li-Chan 1989). Further studies delineated that the three-dimensional structure of egg white lysozyme exists as a hexagonal crystal form, as determined by analysis refined at the resolution of 1.46 Å (Blake et al. 1965; Kato et al. 1992). Further studies have identified the changes in lysozyme structure using different techniques, which include aqueous-organic solvent mixtures (Griebenow and Klibanov 1996), re-folding at low concentrations or in the presence of thiol/disulphide reagents (Raman et al. 1996), pH variation (Babu and Bhakuni 1997), sorbitol treatment (Petersen et al. 2004), various guanidine hydrochloride concentrations dependent treatment (Liu et al. 2005), supercritical CO_2 treatment followed by a heat treatment (Liu et al. 2004) and co-crystallization in the presence of various alcohols (Deshpande et al. 2005). Lysozyme was first purified using a high concentration of ammonium sulphate treatment, but the general morphology and characteristics were modified during extraction due to high use of salt and drastic change in pH (Liu et al. 2004; Abeyrathne et al. 2013). The most commonly used separation technique used for lysozyme is based on cation exchange chromatography mainly due to its high pI value (Abeyrathne et al. 2013). However, because of the fine granule sizes of the resin in the cation exchange chromatography, handling is difficult, and it is a time-consuming process due to a low flow rate.

Lysozyme is also characterized as a mucopeptide N-acetylmuramoylhydrolase (E.C. 3.2.1.17), it is an enzyme that catalyses the hydrolysis of the β-1, 4-glycosidic linkage between *N*-acetylmuraminic acid and *N*-acetylglucosamine in polysaccharide components of certain bacterial cell walls, especially in the case of Gram-positive bacteria, but much less against Gram-negative bacteria due to the inability to penetrate the outer lipopolysaccharide membranes in Gram-negative bacteria (Masschalck et al. 2001). It was reported that a helix-loop-helix (HLH) peptide at the upper active site cleft of lysozyme bestows potent antimicrobial activity with membrane permeabilization action. The particular area residues 98–112 of egg white lysozyme, which are a part of an HLH domain, possess broad-spectrum antimicrobial activity *in vitro*, having potent microbicidal activity against Gram-positive and Gram-negative bacteria, and the fungus *Candida albicans* (Ibrahim et al. 2001). Factors such as the age of the hen, egg storage time and conditions, and thermal treatment of egg white can modulate the lysozyme's anti-microbial activity. The lowest antimicrobial activity is observed in eggs from laying hens either younger than 30 weeks or older than 60 weeks, while the highest activity is found in laying hens aged between 40 and 50 weeks when the eggs stored at 4°C (Ibrahim et al. 2001).

Further studies have investigated increasing the antimicrobial activity of lysozyme by modifying physiochemical factors, which include hydrophobicity or emulsifying property, thermal denaturation (Ibrahim, Shinji Higashiguchi, et al. 1996), covalent attachment of saturated fatty acids (Liu et al. 2005), and reduction of disulphide bonds (Touch et al. 2004). Lyophilization of lysozyme, combined with glycosylation, showed enhanced antimicrobial activity towards Gram-negative bacteria with improved yield. It was postulated that the lyophilization leads to increase in bactericidal activity due to the increase in the proportion of the bound short chain fatty acids after the removal of water (Barbiroli et al. 2012). The antibacterial activity of lysozyme is independent of the catalytic function, as enzymatic hydrolysis can enhance its bioactivity (especially anti-microbial) through the production of various small fragments of proteins known as antibacterial peptides including residues 98–108, 98–112 and 15–21 (Mine et al. 2004; Mine and Kovacs-Nolan 2006). Novel and potent antimicrobial peptides have been isolated, purified and characterized from lysozyme hydrolysate (Mine et al. 2004). Partial denaturation of hen egg white lysozyme, with a heat treatment at 80°C for 20 min at a constant pH of 6.0, exhibited strong and enhanced anti-bacterial activity against both Gram-positive and Gram-negative bacteria. The bactericidal effect after heat treatment is probably mediated through a membrane-disrupting mechanism of the dimeric form of the lysozyme-derived peptide, which is independent of the muramidase activity of the native lysozyme (Ibrahim et al. 1996).

Lysozyme also exhibits antimicrobial effects against Gram-negative bacteria upon conformational changes induced by Ca^{2+} by disturbing the normal electrostatic interactions between components of the outer membrane layer of the bacterial cell and divalent cations of lysozyme (Ibrahim et al. 1997). Under high hydrostatic pressure conditions, partial heat denaturation extended the activity spectrum of lysozyme, indicating enhanced uptake of partially denatured lysozyme through the outer membrane (Masschalck et al. 2001). Later on, it was reported that lytic and nonlytic mechanisms of inactivation of Gram-positive bacteria by lysozyme under different pressures (atmospheric and high hydrostatic pressure) showed that the membrane perturbation happens due to the nonlytic mechanism (Masschalck et al. 2002).

Lysozyme also exhibits metal chelating properties, which has received broad interest. Lysozyme binds with 1 to 8 ions of Cu^{2+} and 1 to 6 ions of Zn^{2+} depending on the ratio of ion to lysozyme in the mixture (Moreau et al. 1995). It was reported that presence of some salts such as Ni^{2+}, Mn^{2+}, Co^{2+} and Yb^{2+} chloride increases the solubility of the egg white lysozyme at high concentration, whereas the presence of some other salts such as Mg^{2+} and Ca^{2+} chloride does not have any effect on protein solubility (Li 2006). Similarly, introduction of NaCl induces aggregation of lysozyme; in a 0.5 M NaCl solution total of six chloride anions and at least one sodium cation are bound to preferred sites on the surface of the lysozyme (Fujiwara et al. 2003). From the sensory standpoint, lysozyme has a sweet taste and, like the enzymatic activity, the sensory attribute is dependent on the tertiary structure of the molecule (Masuda et al. 2001). Therefore, the enzymatic activities and sweetness were lost when the lysozyme molecule is reduced and *S*-3-(trimethylated amino) propylated, or by prolonged heating at 95°C for 18 hours (Masuda et al. 2001). Moreover, it was later revealed that the broad surface region formed by five positively charged residues (Lys13, Lys96, Arg14, Arg21 and Arg73) is mainly responsible for the sweetness of native lysozyme (Masuda et al. 2005).

9 Minor proteins in egg white

There are over 50 proteins existing in egg white (Mann 2007). The five most important proteins are described above. Apart from these major proteins, there are some proteins representing a tiny amount of egg white, but which are also an essential part of egg white and called minor proteins. These minor proteins are unique in their sequence, structural and physicochemical properties, but the functional properties of these minor proteins are not yet fully elucidated. The next part of the text will highlight some of the minor egg white proteins and their physiochemical properties and some functional attributes.

9.1 Ovoglobulin

Ovoglobulin is a minor protein found in egg white consisting of two units G2 and G3, although some earlier studies reported three distinct units of ovoglobulin such as G1, G2 and G3. Later on, the G1 unit of ovoglobulin was characterized as a part of lysozyme (Ogawa and Tanabe 1990). Ovoglobulin accounts for about 4% of egg white proteins. The molecular weights of ovoglobulin G2 and G3 are 36 kDa and 45 kDa, respectively. The units of ovoglobulin have similar physicochemical properties and electrophoretic patterns. These proteins showed similar characteristics to those of OVA; they coagulated with heat treatment and were completely soluble in low and high ionic strengths saline solution (Damodaran et al. 1998).

Although the biological function of ovoglobulin is not well known as yet, an earlier study suggested that ovoglobulin is essential for the foaming property of egg white (Stadelman et al. 1995). Further study showed that the high viscosity of ovoglobulin inhibits the loss of liquid after foam formation (Alleoni 2006).

9.2 Cystatin

Cystatin, or ovocystatin (molecular weight 13 kDA), is a cysteine proteinase inhibitor egg white protein, which inhibits the activity of proteases, especially thiol proteases such as ficin and papain. Chicken cystatin contains two distinct disulphide bonds located near the carboxy terminus along with a conserved reactive site (Li-Chan 1989). The PI of cystatin changes in phosphorylated and non-phosphorylated form is 5.6 and 6.5, respectively (Guérin-Dubiard et al. 2006).

Furthermore, in terms of its role as a proteinase inhibitor, several recent reports have characterized the bioactive properties of cystatin. Antimicrobial activity of cystatin against bacterial pathogens has been demonstrated (Wesierska et al. 2005). Cystatin reversibly inhibited bone matrix degradation in the resorption lacunae adjacent to osteoclasts (Brand et al. 2004), suggesting the involvement of cystatin in the modulation of osteoclastic bone degradation.

9.3 Avidin

Avidin is an essential tetrameric glycoprotein from egg white and well known for its biotin-binding properties. It consists of four monomers, and each of the monomers can bind with biotin and establish a strong avidin–biotin interaction, with a dissociation constant of ~10–15 M (Mine and Yang 2010; Stadelman et al. 1995). Avidin consists of around 0.05% of egg white. Although it is found in a very tiny amount in an egg, the protein is renowned for its biotin conjugation activity and could be used as an essential ingredient for the development of diagnostic immune assays. Each avidin chain is composed of 128 amino acid residues, with eight-stranded antiparallel beta-barrel, whose inner region defines the D-biotin-binding site. Fourier-transform infrared spectroscopy (FTIR) has been used to study avidin secondary structure and complexation with biotin and biotin–lipid assemblies (Stadelman et al. 1995). The secondary structure is composed of approximately 66% beta-sheet and extended beta-turns. Conjugation with biotin or with biotinylated lipid did not change the secondary fabric of the protein but increased the hydrogen bond stability of the beta-sheet.

Chicken egg-white avidin and the bacterial streptavidin are in the same family and are popularly used as tools in some affinity-based separations, in diagnostic assays and a variety of other applications. Avidin exhibits insecticide or antimicrobial activity. The insecticidal activity of avidin has been combined with genetically engineered and traditional host plant resistance against Colorado potato beetle (Rupa and Mine 2006b). Avidin was reported to have a bacteriostatic effect against *Salmonella typhimurium*, but this effect is eliminated by addition of biotin (Stadelman et al. 1995). Avidin contains terminal *N*-acetylglucosamine and mannose residues that can bind with lectins. As lectins are important biomarkers for tumour cells and expressed at various levels on the cell surface, the conjugation of cytotoxic agents with avidin can be easily recognized by the cell surface lectins. The phenomenon was explored in the treatment of tumours and avidin served as a promising vehicle for the delivery of drugs, toxins, radioisotopes and therapeutic genes to treat intraperitoneal tumours (Yao et al. 1998), and in therapeutic targeting of different cancer treatments (Hytönen et al. 2003).

9.4 Ovoflavin

Ovoflavin, ovo-flavoprotein or riboflavin-binding protein are phospho-glycoproteins frequently found in nearly equal amounts in the egg white and yolk. It is the protein of egg white that is responsible for the binding of vitamin B2 (riboflavin). Selenium (Se) content in ovo-flavoprotein is the highest (1800 ng/g) among other egg white proteins (Kiliç et al. 2002). But the presence of vitamins or minerals in ovoflavin is directly dependent upon the dietary intake of the hen (Rhodes et al. 1959; Hamazume et al. 1984). Egg white ovoflavin consists of 219 amino acid residues, and the amino terminus contains pyroglutamic acid (Hamazume et al. 1984). Although the function of egg white ovo-flavoprotein is not known entirely, its ability to bind with vitamins and minerals signifies that this protein might be involved in providing vitamins and minerals during embryo development (Hamazume et al. 1984).

In addition to these minor proteins, egg white has several proteins such as ovomacroglobulin, ovoglycoprotein and ovo-inhibitor. These proteins are found in trace amounts in eggs and their physicochemical and functional properties are not fully elucidated yet. The basic information of such as molecular weight and PI are known, but many of their other characteristics remain unknown. As these proteins are present in very tiny amounts in egg white, they are considered not to have significant functional or biological activity.

10 Carbohydrates, minerals and vitamins in egg white

10.1 Carbohydrates

Along with these proteins, some other types of chemical compounds are found in egg white, and carbohydrate is one of them. Carbohydrates are present in egg white mainly in two forms: a) conjugated oligosaccharides, especially with the glycoproteins with N-glycosylation (Yamamoto et al. 1996) and b) glucose. Glucose is the 'free' sugar, comprising ~0.8–1% albumen. The glucose is usually a non-desirable compound in dried egg white due to its ability to triggered browning caused by Maillard reaction. Therefore, excess glucose is always removed by fermentation prior to food processing.

10.2 Minerals

The primary minerals in egg white are sulphur (S), potassium (K), sodium (Na), copper (Cu), phosphorus (P), calcium (Ca) and magnesium (Mg). However, these elements are found in very low quantities, along with various other trace minerals (Table 2) (Stadelman et al. 1995; Mine and Yang 2010). In general, higher levels of minerals such as Cu, Zn, Mg, Ca and K are found in the egg white (Kiliç et al. 2002). Selenium is an essential mineral, mostly associated with proteins such as OVA; flavoprotein has the highest Se content (1800 ng/g). Se content of other proteins ranged from 359 to 1094 ng/g (Kiliç et al. 2002).

10.3 Vitamins

The egg white does not contain any fat-soluble vitamins but does provide significant amounts of water-soluble vitamins in the egg. The most abundant vitamins in egg white are complexes of vitamin B: biotin (vitamin B7), niacin (vitamin B3) and riboflavin (vitamin

B2). The amount of these vitamins is presented in Table 2 (Stadelman et al. 1995). Most of these vitamins are found in a bound form with proteins in egg white. However, egg white is a minor source of some essential vitamins such as vitamins A and C.

11 Summary

Eggs are one of the essential and important food sources; eggs are also regarded as 'nature's perfect food'. Eggs are unique food products and well known for the nutritional and functional properties. Egg proteins, especially from the egg white, have attracted research interest for many years. Along with the academic research interest, food processing industries also have conducted extensive studies of egg proteins, mainly to explore the various functionalities. Recent studies have investigated the various biological activities of egg protein-derived peptides. These bioactive peptides can exert different health beneficial effects, above and beyond their known nutritional benefits. Thus proteins from egg have gained new interest and can be considered as possible ingredients for the development of functional foods. However, egg is a known allergen and using egg as a core element for the development of functional foods is associated challenges. Moreover, the commercial benefits can only be accessed through the development of feasible industrial techniques.

12 Where to look for further information

Egg proteins are well studied and to know more about this subject; the three most important books to study about the functionality of egg proteins and their industrial applications are as follows:

- *Egg Science and Technology*, edited by William J. Stadelman and Owen J. Cotterill.
- *Egg Bioscience and Biotechnology*, edited by Yoshinori Mine.
- *The Amazing Egg Nature's Perfect Functional Food for Health Promotion*, by Jeong Sim; Sunwoo, Hoon.

The bioactivity of egg-derived bioactive compounds is one of the important aspects of modern egg research; it is advisable to read

- *Egg Nutrition and Biotechnology*, edited by Jeong S. Sim, Shuryo Nakai, Wilhelm Guenter
- *Bioactive Egg Compounds*, edited by Rainer Huopalahti, Rosina López-Fandiño, Marc Anton, Rüdiger Schade.

Bioactive proteins and peptides from egg proteins is an emerging field of research; it is recommended to read the following book chapters

- Functional Proteins and Peptides of Hen's Egg Origin, by Adam M. Abdou, Mujo Kim, and Kenji Sato, In: *Bioactive Food Peptides in Health and Disease*, edited by Blanca Hernández-Ledesma and Chia-Chien Hsieh.

- Bioactive Proteins and Peptides from Egg Proteins, by Jianping Wu, Kaustav Majumder and Kristen Gibbons, In: *Bioactive Proteins and Peptides as Functional Foods and Nutraceuticals*, edited by Yoshinori Mine, Eunice Li-Chan, Bo Jiang.

13 References

Abdallah, F. B. and Chahine, J. M., 1999. Transferrins, the mechanism of iron release by ovotransferrin. *European Journal of Biochemistry/FEBS*, 263(3), pp. 912–20.

Abeyrathne, E. D. N. S., Lee, H. Y. and Ahn, D. U., 2013. Egg white proteins and their potential use in food processing or as nutraceutical and pharmaceutical agents – a review. *Poultry Science*, 92(12), pp. 3292–9.

Abeyrathne, E. D. N. S., Lee, H. Y. and Ahn, D. U., 2014. Sequential separation of lysozyme, ovomucin, ovotransferrin, and ovalbumin from egg white. *Poultry Science*, 93(4), pp. 1001–9.

Alcantara, J. and Schryvers, A. B., 1996. Transferrin binding protein two interacts with both the N-lobe and C-lobe of ovotransferrin. *Microbial pathogenesis*, 20(2), pp.73–85.

Alleoni, A. C. C., 2006. Albumen protein and functional properties of gelation and foaming. *Scientia Agricola*, 63(3), pp. 291–8.

Babu, K. R. and Bhakuni, V., 1997. Ionic-strength-dependent transition of hen egg-white lysozyme at low pH to a compact state and its aggregation on thermal denaturation. *European journal of biochemistry/FEBS*, 245(3), pp. 781–9.

Barbiroli, A., Bonomi, F., Capretti, G., Iametti, S., Manzoni, M., Piergiovanni, L. and Rollini, M., 2012. Antimicrobial activity of lysozyme and lactoferrin incorporated in cellulose-based food packaging. *Food Control*, 26(2), pp. 387–92.

Besler, M., Steinhart, H. and Paschke, A., 2008. Allergenicity of hen's egg-white proteins: IgE binding of native and deglycosylated ovomucoid. *Food and Agricultural Immunology*, 9(4), pp. 277–88.

Blake, C. C. F., Koenig, D. F., Mair, G. A., North, A. C. T., Philips, D. C., Sarma and V. R., 1965. Structure of hen egg-white Lysozyme: a three-dimensional fourier synthesis at 2 Å resolution. *Nature*, 206(4986), pp. 757–61.

Bou Abdallah, F. and el Hage Chahine, J. M., 1998. Transferrins. Hen ovo-transferrin, interaction with bicarbonate and iron uptake. *European journal of biochemistry/FEBS*, 258(3), pp. 1022–31.

Brake, J., Walsh, T. J., Benton, C. E. Jr., Petitte, J. N., Meijerhof, R. and Penalva, G. 1997. Egg handling and storage. *Poultry Science*, 76(2), pp. 144–51.

Brand, H. S., Lerner, U. H., Grubb, A., Beertsen, W., Niew-Amerongen, A. V. and Everts, V., 2004. Family 2 cystatins inhibit osteoclast-mediated bone resorption in calvarial bone explants. *Bone*, 35(3), pp.689–96.

Caubet, J.-C. and Wang, J., 2011. Current understanding of egg allergy. *Pediatric Clinics of North America*, 58(2), pp. 427–43, xi.

Conrad, R. M. and Phillips, R. E., 1938. The formation of the chalazae and inner thin white in the hen's egg. *Poultry Science*,. 17(1), pp. 143–6.

Cotterill, O. J. and Geiger, G. S., 1977. Egg product yield trends from shell eggs. *Poultry Science*, 56(3), pp. 1027–31.

Damodaran, S. (University of W.-M. M. W.., Anand, K. and Razumovsky, L., 1998. Competitive absorption of egg white proteins at the air-water interface: direct evidence for electrostatic complex formation between lysozyme and other egg protein at the interface. *Journal of Agricultural and Food Chemistry (USA)*.

Dávalos, A., Miguel, M., Bartloomé, B. and López-Fandiño, R., 2004. Antioxidant activity of peptides derived from egg white proteins by enzymatic hydrolysis. *Journal of Food Protection*, 67(9), pp. 1939–44.

Desert, C., Guérin-Dubiard, C., Nau, F., Jan, G., Val, F. and Mallard, J., 2001. Comparison of Different Electrophoretic Separations of Hen Egg White Proteins. *Journal of Agricultural and Food Chemistry*, 49(10), pp.4553–61.

Deshpande, A., Nimsadkar, S. and Mande, S. C., 2005. Effect of alcohols on protein hydration: crystallographic analysis of hen egg-white lysozyme in the presence of alcohols. *Acta Crystallographica. Section D, Biological Crystallography*, 61(Pt 7), pp. 1005–8.

Fothergill, L. A. and Fothergill, J. E., 1970. Thiol and disulphide contents of hen ovalbumin. C-terminal sequence and location of disulphide bond. *The Biochemical Journal*, 116(4), pp. 555–61.

Friedman, M., 1996. Nutritional value of proteins from different food sources. a review. *Journal of Agricultural and Food Chemistry*, 44(1), pp. 6–29.

Fujiwara, S., Matsumoto, F. and Yonezawa, Y., 2003. Effects of salt concentration on association of the amyloid protofilaments of hen egg white lysozyme studied by time-resolved neutron scattering. *Journal of Molecular Biology*, 331(1), pp. 21–8.

Giansanti, F., Massucci, M. T., Fedrica-Giardi, M. and Antonini, G., 2005. Antiviral activity of ovotransferrin derived peptides. *Biochemical and biophysical research communications*, 331(1), pp.69–73.

Griebenow, K. and Klibanov, A. M., 1996. On protein denaturation in aqueous – organic mixtures but not in pure organic solvents. *Journal of the American Chemical Society*, 118(47), pp. 11695–700.

Guérin-Dubiard, C. et al., 2006. Proteomic analysis of hen egg white. *Journal of Agricultural and Food Chemistry*, 54(11), pp. 3901–10.

Hamazume, Y., Mega, T. and Ikenaka, T., 1984. Characterization of hen egg white- and yolk-riboflavin binding proteins and amino acid sequence of egg white-riboflavin binding protein. *Journal of Biochemistry*, 95(6), pp. 1633–44.

Hiidenhovi, J., Aro, H. S. and Kankare, V., 1999. Separation of ovomucin subunits by gel filtration: enhanced resolution of subunits by using a dual-column system. *Journal of Agricultural and Food Chemistry*, 47(3), pp. 1004–8.

Hirose, J., Murakami-Yamaguchi, Y., Ikeda, M., Kitabatake, N. and Narita, H., 2005. Oligoclonal Enzyme-linked Immunosorbent assay capable of determining the major food allergen, ovomucoid, irrespective of the degree of heat denaturation. *Cytotechnology*, 47(1–3), pp. 145–9.

Hirose, J. et al., 2006. Ovalbumin-related gene Y protein bears carbohydrate chains of the ovomucoid type. *Bioscience, Biotechnology, and Biochemistry*, 70(1), pp. 144–51.

Huang, Q., Qiu, N., Mh, M., Jin, Y. G., Yang, H., Geng, F. and Sun, S. G., 2012. Estimation of egg freshness using S-ovalbumin as an indicator. *Poultry Science*, 91(3), pp. 739–43.

Huntington, J. A., Patston, P. A. and Gettins, P. G., 1995. S-ovalbumin, an ovalbumin conformer with properties analogous to those of loop-inserted serpins. *Protein Science: A Publication of the Protein Society*, 4(4), pp. 613–21.

Huntington, J. A. and Stein, P. E., 2001. Structure and properties of ovalbumin. *Journal of Chromatography B: Biomedical Sciences and Applications*, 756(1–2), pp. 189–98.

Hytönen, V. P., Laitinen, O. H., Grapputo, A., Kettunen, A., Savolainen, J., Kalkkinen, N., Marttila, A. T., Nordlund, H. R., Nyholm, T. K., Paganelli, G. and Kulomaa, M. S., 2003. Characterization of poultry egg-white avidins and their potential as a tool in pretargeting cancer treatment. *The Biochemical Journal*, 372(Pt 1), pp. 219–25.

Ibrahim, H. R., Higashiguchi, S., et al., 1996a. A structural phase of heat-denatured Lysozyme with novel antimicrobial action. *Journal of Agricultural and Food Chemistry*, 44(6), pp. 1416–23.

Ibrahim, H. R., Higashiguchi, S., Koketsu, M., Juneja, L. R., Kim, M., Yamamoto, T., Sugimoto, Y. and Aoki, T., 1996b. Partially unfolded lysozyme at neutral pH agglutinates and kills gram-negative and gram-positive bacteria through mbrane damage mechanism. *Journal of Agricultural and Food Chemistry*, 44(12), pp. 3799–806.

Ibrahim, H. R., Higashiguchi, S., Sugimoto, Y. and Aoki, T., 1997. Role of divalent cations in the novel bactericidal activity of the partially unfolded lysozyme. *Journal of Agricultural and Food Chemistry*, 45(1), pp. 89–94.

Ibrahim, H. R., Iwamori, E., Sugimoto, Y. and Aoki, T., 1998. Identification of a distinct antibacterial domain within the N-lobe of ovotransferrin. *Biochimica et Biophysica Acta*, 1401(3), pp. 289–303.

Ibrahim, H. R., Haraguchi, T. and Aoki, T., 2006. Ovotransferrin is a redox-dependent autoprocessing protein incorporating four consensus self-cleaving motifs flanking the two kringles. *Biochimica et Biophysica Acta*, 1760(3), pp. 347–55.

Ibrahim, H. R., Sugimoto, Y. and Aoki, T., 2000. Ovotransferrin antimicrobial peptide (OTAP-92) kills bacteria through a membrane damage mechanism. *Biochimica et Biophysica Acta*, 1523(2–3), pp. 196–205.

Ibrahim, H. R., Thomas, U. and Pellegrini, A., 2001. A helix-loop-helix peptide at the upper lip of the active site cleft of lysozyme confers potent antimicrobial activity with membrane permeabilization action. *The Journal of Biological Chemistry*, 276(47), pp. 43767–74.

Kato, A. et al., 1992. Structural and functional properties of hen egg-white lysozyme deamidated by protein engineering. *Bioscience, Biotechnology, and Biochemistry*, 56(9), pp.1 424–8.

Kato, A., Tanimoto, S., Muraki, Y., Kobayashi, K. and Kumagai, I., 2006. Engineering hen egg-white lysozyme. *Nutraceutical Proteins and Peptides in Health and Disease*, pp. 583–602.

Kato, Y., Oozawa, E. and Matsuda, T., 2001. Decrease in antigenic and allergenic potentials of ovomucoid by heating in the presence of wheat flour: dependence on wheat variety and intermolecular disulfide bridges. *Journal of Agricultural and Food Chemistry*, 49(8), pp. 3661–5.

Kiliç, Z., Acar, O., Ulasan, M. and Ilim, M., 2002. Determination of lead, copper, zinc, magnesium, calcium and iron in fresh eggs by atomic absorption spectrometry. *Food Chemistry*, 76(1), pp. 107–16.

Kim, E. M., 2002. The effects of supplementation of ascidian tunic shell into laying hen diet on egg quality. *Journal of Animal Science and Technology*, 44(1), pp. 45–54.

Kobayashi, K., Hattori, M., Hara-Kudo, Y., Okubo, T., Yamamoto, S., Takita, T. and Sugita-Konishi, Y., 2004. Glycopeptide derived from hen egg ovomucin has the ability to bind enterohemorrhagic Escherichia coli O157:H7. *Journal of Agricultural and Food Chemistry*, 52(18), pp. 5740–6.

Kovacs-Nolan, J., Zhang, J. W., Hayakawa, S. and Mine, Y., 2000. Immunochemical and structural analysis of pepsin-digested egg white ovomucoid. *Journal of Agricultural and Food Chemistry*, 48(12), pp. 6261–6.

Kurokawa, H. et al., 1999. Crystal structure of hen apo-ovotransferrin. Both lobes adopt an open conformation upon loss of iron. *The Journal of Biological Chemistry*, 274(40), pp. 28445–52.

Lambert, L. A. et al., 2005. Evolution of the transferrin family: conservation of residues associated with iron and anion binding. *Comparative Biochemistry and Physiology. Part B, Biochemistry & Molecular Biology*, 142(2), pp. 129–41.

Lee, N. Y., Cheng, J. T., Enomoto T. N. Y., 2006. One peptide derived from hen ovotransferrin as pro-drug to inhibit angiotensin converting enzyme – proQuest. *Journal of Food Drug and Analysis*, 14(1), pp. 31–5.

Lee, J.-W., Lee, K. Y., Yook, H. S., Lee, S. Y., Kim, H. Y., Jo, C. and Byun, M. W., 2002. Allergenicity of hen's egg ovomucoid gamma irradiated and heated under different pH conditions. *Journal of food protection*, 65(7), pp.1196–9.

Li-Chan, E. N. S., 1989. Biochemical basis for the properties of egg white. *Critical Reviews in Poultry Biology*, 2(1), pp. 51–8.

Li, S. J., 2006. Structural details at active site of hen egg white lysozyme with di- and trivalent metal ions. *Biopolymers*, 81(2), pp. 74–80.

Liu, H.-L., Hsieh, W.-C. and Liu, H.-S., 2004. Molecular dynamics simulations to determine the effect of supercritical carbon dioxide on the structural integrity of hen egg white lysozyme. *Biotechnology Progress*, 20(3), pp. 930–8.

Liu, W., Cellmer, T., Keerl, D., Prausnitz, J. M. and Blanch, H. W., 2005. Interactions of lysozyme in guanidinium chloride solutions from static and dynamic light-scattering measurements. *Biotechnology and Bioengineering*, 90(4), pp. 482–90.

MacDonnell, R. R., Feeney, R. E., Hanson, H. L., Campbell, A. and Sugihara, T. F., 1955. The functional properties of the egg white proteins. *Food Technology*, 9(1), 49–53.

Majumder, K., Chakrabarti, S., Davidge, S. T., Wu, J., 2013. Structure and activity study of egg protein ovotransferrin derived peptides (IRW and IQW) on endothelial inflammatory response and oxidative stress. *Journal of Agricultural and Food Chemistry*, 61(9), pp. 2120–9.

Majumder, K., Liang, G., Chen, Y., Guan, L., Davidge, S. T. and Wu, J., 2015a. Egg ovotransferrin-derived ACE inhibitory peptide IRW increases ACE2 but decreases proinflammatory genes expression in mesenteric artery of spontaneously hypertensive rats. *Molecular Nutrition & Food Research.*

Majumder, K., Chakrabarti, S., et al., 2015b. Egg-derived ACE-inhibitory peptides IQW and LKP reduce blood pressure in spontaneously hypertensive rats. *Journal of Functional Foods*, 13, pp. 50–60.

Mann, K., 2007. The chicken egg white proteome. *Proteomics*, 7(19), pp. 3558–68.

Masschalck, B., Van-Houdt, R., Van-Haver, E. G. and Michiels, C. W., 2001. Inactivation of gram-negative bacteria by lysozyme, denatured lysozyme, and lysozyme-derived peptides under high hydrostatic pressure. *Applied and Environmental Microbiology*, 67(1), pp. 339–44.

Masschalck, B., Deckers, D. and Michiels, C. W., 2002. Lytic and nonlytic mechanism of inactivation of gram-positive bacteria by lysozyme under atmospheric and high hydrostatic pressure. *Journal of Food Protection*, 65(12), pp. 1916–23.

Masuda, T., Ide, N. and Kitabatake, N., 2005. Structure-sweetness relationship in egg white lysozyme: role of lysine and arginine residues on the elicitation of lysozyme sweetness. *Chemical Senses*, 30(8), pp. 667–81.

Masuda, T., Ueno, Y. and Kitabatake, N., 2001. Sweetness and enzymatic activity of lysozyme. *Journal of Agricultural and Food Chemistry*, 49(10), pp. 4937–41.

Matoba, N., Yamada, Y. and Yoshikawa, M., 2003. Design of a genetically modified soybean protein preventing hypertension based on an anti-hypertensive peptide derived from ovalbumin. *Current Medicinal Chemistry. Cardiovascular and Hematological Agents*, 1(2), pp. 197–202.

McReynolds, L., O'Malley, B. W., Nisbet, A. T., FotherGill, J. E., Givol, D., Fields, S., Robertson, M. and Brownlee, G. G., 1978. Sequence of chicken ovalbumin mRNA. *Nature*, 273(5665), pp. 723–8.

Mine, Y., 1995. Recent advances in the understanding of egg white protein functionality. *Trends in Food Science & Technology*, 6(7), pp. 225–32.

Mine, Y. and Kovacs-Nolan, J., 2006. New insights in biologically active proteins and peptides derived from hen egg. *World's Poultry Science Journal*, 62(01), pp. 87–96.

Mine, Y., Ma, F. and Lauriau, S., 2004. Antimicrobial peptides released by enzymatic hydrolysis of hen egg white lysozyme. *Journal of Agricultural and Food Chemistry*, 52(5), pp. 1088–94.

Mine, Y. and Rupa, P., 2003a. Fine mapping and structural analysis of immunodominant IgE allergenic epitopes in chicken egg ovalbumin. *Protein Engineering*, 16(10), pp. 747–52.

Mine, Y. and Rupa, P., 2003b. Genetic attachment of undecane peptides to ovomucoid third domain can suppress the production of specific IgG and IgE antibodies. *Biochemical and Biophysical Research Communications*, 311(1), pp. 223–8.

Mine, Y. and Rupa, P., 2004. Immunological and biochemical properties of egg allergens. *World's Poultry Science Journal*, 60(03), pp. 321–30.

Mine, Y. and Yang, M., 2010. Functional properties of egg components in food systems. *Handbook of Poultry Science and Technology, Volume 1*, pp. 579–630.

Mine, Y. and Zhang, J. W., 2002a. Comparative studies on antigenicity and allergenicity of native and denatured egg white proteins. *Journal of Agricultural and Food Chemistry*, 50(9), pp. 2679–83.

Mine, Y. and Zhang, J. W., 2002b. Identification and fine mapping of IgG and IgE epitopes in ovomucoid. *Biochemical and Biophysical Research Communications*, 292(4), pp. 1070–4.

Mine, Y. and Zhang, J. W., 2001. The allergenicity of ovomucoid and the effect of its elimination from hen's egg white. *Journal of the Science of Food and Agriculture*, 81(15), pp. 1540–6.

Mizutani, K., Yamashita, H., Oe, H. and Hirose, H., 2014. Structural characteristics of the disulfide-reduced ovotransferrin N-Lobe analyzed by protein fragmentation. *Bioscience, Biotechnology and Biochemistry*, 61(4), pp. 641–6.

Moreau, S. et al., 1995. Hen egg white lysozyme-metal ion interactions: investigation by electrospray Ionization Mass Spectrometry. *Journal of Agricultural and Food Chemistry*, 43(4), pp. 883–9.

Nagaoka, S., 2006. Cholesterol-lowering proteins and peptides. *Nutraceutical Proteins and Peptides in Health and Disease*, pp. 41–67.

Nau, F. et al., 2005. Identification and characterization of ovalbumin gene Y in hen egg white. *Journal of Agricultural and Food Chemistry*, 53(6), pp. 2158–63.

Nimalaratne, C., Bandara, N. and Wu, J., 2015. Purification and characterization of antioxidant peptides from enzymatically hydrolyzed chicken egg white. *Food Chemistry*, 188, pp. 467–72.

Ogawa, N. and Tanabe, H., 1990. Effects of washing and oiling on electrophoretic patterns of albumen of the stored chicken eggs. *Japanese Poultry Science*, 27(1), pp. 16–20.

Omana, D. A., Wang, J. and Wu, J., 2010. Ovomucin – a glycoprotein with promising potential. *Trends in Food Science & Technology*, 21(9), pp. 455–63.

Omana, D. A. and Wu, J., 2009. A new method of separating ovomucin from egg white. *Journal of Agricultural and Food Chemistry*, 57(9), pp. 3596–603.

Pelegrine, D. H. G. and Gasparetto, C. A., 2006. A comparative study between ovalbumin and β-lactoglobulin fouling in a tube hot surface, *Journal of Food Engineering*, 73(4), pp. 394–401.

Pellegrini, A. et al., 2004. Proteolytic fragments of ovalbumin display antimicrobial activity. *Biochimica Et Biophysica Acta*, 1672(2), pp. 76–85.

Petersen, S. B. et al., 2004. Sorbitol prevents the self-aggregation of unfolded lysozyme leading to and up to 13 degrees C stabilisation of the folded form. *Journal of Biotechnology*, 114(3), pp. 269–78.

Raikos, V., Hansen, R., Campbell, L., and Euston, S. R. 2006. Separation and identification of hen egg protein isoforms using SDS-PAGE and 2D gel electrophoresis with MALDI-TOF mass spectrometry. *Food Chemistry*, 99(4), pp. 702–10.

Raikos, V., Campbell, L. and Euston, S. R., 2007. Effects of sucrose and sodium chloride on foaming properties of egg white proteins. *Food Research International*, 40(3), pp. 347–55.

Raman, B., Ramakrishna, T. and Rao, C. M., 1996. Refolding of denatured and denatured/reduced Lysozyme at high concentrations. *Journal of Biological Chemistry*, 271(29), pp. 17067–72.

Réhault-Godbert, S. et al., 2013. Ovalbumin-related protein X is a heparin-binding ov-serpin exhibiting antimicrobial activities. *The Journal of Biological Chemistry*, 288(24), pp. 17285–95.

Rhodes, M. B., Bennet, N. and Feeny, R. E., 1959. The flavoprotein-apoprotein system of egg white. *The Journal of Biological Chemistry*, 234(8), pp. 2054–60.

Robinson, D. S. and Monsey, J. B., 1971. Studies on the composition of egg-white ovomucin. *The Biochemical Journal*, 121(3), pp. 537–47.

Rupa, P. and Mine, Y., 2006a. Ablation of ovomucoid-induced allergic response by desensitization with recombinant ovomucoid third domain in a murine model. *Clinical & Experimental Immunology*, 145(3), pp. 493–501.

Rupa, P. and Mine, Y., 2006b. Chapter 22 Egg Proteins. *Nutraceutical Science and Technology*, 4, p. 445.

Rupa, P. and Mine, Y., 2006c. Engineered recombinant ovomucoid third domain can desensitize Balb/c mice of egg allergy. *Allergy*, 61(7), pp. 836–42.

Schäfer, A., Drewes, W. and Schwägele, F., 1999. Effect of storage temperature and time on egg white protein. *Nahrung/Food*, 43(2), pp. 86–9.

Seo, J.-H., Lee, J. W., Lee, Y. S., Lee, S. Y., Kim, M. R., Yook, H. S. and Byun, M. W., 2004. Change of an egg allergen in a white layer cake containing gamma-irradiated egg white. *Journal of Food Protection*, 67(8), pp. 1725–30.

Stadelman, W. J., Newkirk, D. and Newby, L., 1995. *Egg Science and Technology, Fourth Edition*, Boca Raton: CRC Press.

Sugimoto, Y., Sanuki, S., Ohsako, S., Higashimoto, Y., Kondo, M., Kurawaki, J., Ibrahim, H. R., Aoki, T., Kusakabe, T. and Koga, K., 1999. Ovalbumin in developing chicken eggs migrates from egg white to embryonic organs while changing its conformation and thermal stability. *Journal of Biological Chemistry*, 274(16), pp. 11030–7.

Takagi, K. et al., 2005. Kinetic analysis of pepsin digestion of chicken egg white ovomucoid and allergenic potential of pepsin fragments. *International Archives of Allergy and Immunology*, 136(1), pp. 23–32.

Touch, V., Hayakawa, S. and Saitoh, K., 2004. Relationships between conformational changes and antimicrobial activity of lysozyme upon reduction of its disulfide bonds. *Food Chemistry*, 84(3), pp. 421–8.

Tsuge, Y., Shimoyamada, M. and Watanabe, K., 1997a. Bindings of ovomucin to newcastle disease virus and anti-ovomucin antibodies and its heat stability based on binding abilities. *Journal of Agricultural and Food Chemistry*, 45(12), pp. 4629–634.

Tsuge, Y., Shimoyamada, M. and Watanabe, K., 1996. Differences in hemagglutination inhibition activity against bovine rotavirus and hen Newcastle disease virus based on the subunits in hen egg white ovomucin. *Bioscience, Biotechnology, and Biochemistry*, 60(9), pp. 1505–6.

Tsuge, Y., Shimoyamada, M. and Watanabe, K., 1997b. Structural features of newcastle disease virus-and anti-ovomucin antibody-binding glycopeptides from pronase-treated ovomucin. *Journal of Agricultural and Food Chemistry*, 45(7), pp. 2393–8.

Van der Plancken, I. et al., 2004. Trypsin inhibition activity of heat-denatured ovomucoid: a kinetic study. *Biotechnology Progress*, 20(1), pp. 82–6.

Wang, J. et al., 2012. Proteomics analysis of egg white proteins from different egg varieties. *Journal of Agricultural and Food Chemistry*, 60(1), pp. 272–82.

Watanabe, K. et al., 1998. Antitumor effects of pronase-treated fragments, glycopeptides, from ovomucin in hen egg white in a double grafted tumor system. *Journal of Agricultural and Food Chemistry*, 46(8), pp. 3033–8.

Weiner, K. X., Hayes, G. R. and Lucas, J. J., 1997. Binding specificity of avian heat shock protein 108. *Biochemical and Biophysical Research Communications*, 240(3), pp. 673–6.

Wesierska, E., Saleh, Y., Trziszka, T., Kopec, W., Siewinski, M. and Korzekwa, K., 2005. Antimicrobial activity of chicken egg white cystatin. *World Journal of Microbiology and Biotechnology*, 21(1), pp. 59–64.

Wu, J. and Acero-Lopez, A., 2012. Ovotransferrin: structure, bioactivities, and preparation. *Food Research International*, 46(2), pp. 480–7.

Yamamoto, K. et al., 1998. Characterization of Bacillus sp. endo-beta-N-acetylglucosaminidase and its application to deglycosylation of hen ovomucoid. *Biotechnology and Applied Biochemistry*, 28 (Pt 3), pp. 235–42.

Yamamoto, T. et al., 1996. *Hen Eggs: Basic and Applied Science*, Boca Raton: CRC Press.

Yao Z, Zhang M, Sakahara H, Saga T, Arano, Y. and Konishi, J. 1998. Avidin targeting of intraperitoneal tumor xenografts. *Journal of the National Cancer Institute* 7;90(1), pp. 25–9.

Yoshino, K. et al., 2004. Peptic digestibility of raw and heat-coagulated hen's egg white proteins at acidic pH range. *International Journal of Food Sciences and Nutrition*, 55(8), pp. 635–40.

Yu, Z. et al., 2012. Primary and secondary structure of novel ACE-inhibitory peptides from egg white protein. *Food Chemistry*, 133(2), pp. 315–22, 659.

Chapter 3

The nutritional and physiological functions of egg yolk components

Yasumi Horimoto, University of Guelph, Canada and Hajime Hatta, Kyoto Women's University, Japan

1 Introduction

Eggs have been a valuable food source across the world for thousands of years. Almost 2500 years ago, the ancient Greek physician Hippocrates said, 'Let foods be your medicines and medicines be your foods.' Hippocrates had excellent foresight for the relationship between food and disease in modern society; there are now many lifestyle-related diseases such as obesity, diabetics and hypertension. Today, it is recognized that a healthy diet will promote good health and possibly delay the onset of disease.

Egg yolk provides not only essential energy for life but also necessary nutrients – essential amino acids, fatty acids, cholesterol, vitamins and minerals – to maintain and promote health. In fact, one reason that eggs are of interest is that they contain all nutrients necessary for human health, except dietary fibre and vitamin C (Sugino et al., 1997).

Egg yolk consists of 48% water, 34% lipids, 17% proteins and 2.0% minerals, while egg white is composed of 88% water, 11% protein, 0.2% lipids and 0.8% minerals (Sugino et al., 1997) (Fig. 1). Egg proteins are of a quality similar to human breast milk proteins. Egg yolk is also rich in neutral lipids, phospholipids (PLs) and cholesterol (Parkinson, 1966). PLs and cholesterol are important components that promote the formation of somatic cell

http://dx.doi.org/10.19103/AS.2016.0012.03

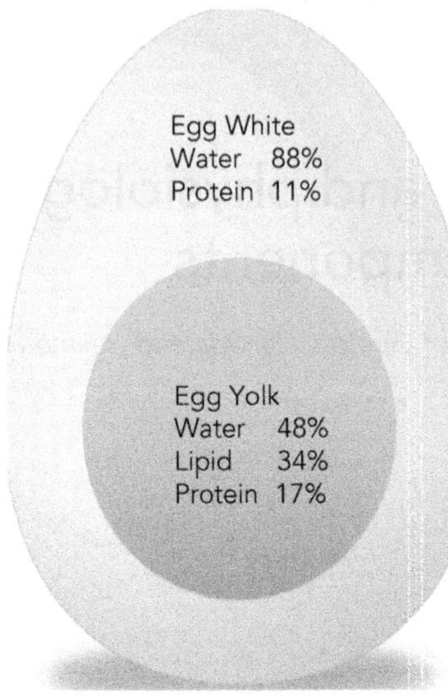

Figure 1 Composition of an egg.

structures and the cell membranes of cranial nerve cells for developing chicks, and may serve a similar function in humans.

To promote health and prevent diseases, much research has explored the nutritional and physiological functions of egg yolk. Key areas of research include bioactive proteins/peptides, egg yolk antibody (immunoglobulin Y (IgY)), PLs and sialic acid (Hatta et al., 2008). Some applications of egg yolk bioactive peptides have reached a market-ready stage (Kobayashi et al., 2015). Hen eggs are presently used as a source of IgY and options to increase production to a larger scale are being considered (Hatta et al., 2008). Useful applications for IgY as an immunological tool have been studied in the fields of diagnosis as well as in pure research (Hatta et al., 2008; Dias da Silva and Tambougi, 2010). The applicability of IgY to prevent infectious diseases has received much attention due to the potential to produce specific antibodies in egg yolk. Considering cholesterol issues, people believe that there is a positive correlation between the ingestion of egg and elevated cholesterol levels in the body. Recently, however, strong scientific evidence has concluded that consumption of one egg per day is not associated with an increased risk of coronary heart disease or stroke in healthy adults (Kritchevsky, 2004; Rong et al., 2013; Robbins et al., 2014; Fuller et al., 2015; Virtanen et al., 2016).

In addition to the research areas mentioned above, recently two subjects have emerged in egg yolk research. First, much research has been conducted on the effects of manipulation of the hen's feed on the nutritional composition of egg yolk. Animal welfare movements, which originated in Europe, have reached North America and influenced animal feed patterns. In 2010, the US National Chicken Council recommended that the chicken industry ensure the humane treatment of animals in order to promote the production of quality products (National Chicken Council, 2010). As a result of this recommendation, studies on chicken husbandry management have increased.

Many researchers are involved in studies concerning the effects of feed formula, the nutrient composition of egg yolk and eggs with special attributes – free-range, omega-3, folic acid and vitamin-fortified – have been successfully commercialized (Sim and Sunwoo, 2002). Some special eggs such as organic eggs and vegetarian eggs have also appeared in the market without scientific evidence of their benefits, but have appealed to health-conscious consumers.

Secondly, the proteomic analysis of egg proteins has emerged as a new area of study, building on previous work on the physiological functions of bioactive peptides. The recent developments of proteomics and transcriptomics have allowed the identification of several hundred new egg proteins (Mann, 2007; Mann et al., 2008; Mann and Mann, 2008; Farinazzo et al., 2009). In future, there will be great potential for identifying novel egg proteins and their biological functions to promote our health.

The aim of this chapter is to discuss current trends in the study of nutritional and physiological functions of egg yolk. In particular, this chapter will focus on the bioactive components derived from egg yolk (focusing on proteins/peptides and lipids), issues related to egg cholesterol and recent applications of IgY. Research on other areas such as sialic acids and lutein are also covered, although in less detail. To place these trends in context, a brief overview of background knowledge about egg yolks is also provided. This chapter updates and expands on the review provided by Hatta et al. (2008).

2 Egg yolk proteins

Egg yolk contains 15.7–16.6% protein (Sugino et al., 1997), including major proteins such as lipoprotein, livetin and phosvitin, as well as other proteins such as riboflavin-binding protein, biotin-binding protein, thiamine-binding protein, egg yolk transferrin and proteins with cholinesterase activity (Muniyappa and Adiga, 1981; Amoresanno et al., 1999; Vasudevan et al., 2001; Niskanen et al., 2005; Hatta et al., 2008) (Table 1). Lipoproteins are classified into three groups: high-density lipoprotein (HDL), low-density lipoproteins (LDL) and very low-density lipoproteins (VLDL).

Egg yolk can be separated into two main fractions by dilution and centrifugation: plasma (supernatant) and granules (precipitate) (Anton, 2013). Plasma is mainly composed of 85% LDL and 15% livetins. On the other hand, granules are mainly composed of 70% HDL, which is also called lipovitellin, and 16% phosvitin. In addition to the major proteins that make up the plasma and granules, several hundred new egg components have been discovered through recent developments in proteomics (Mann, 2007; Mann and Mann, 2008; Mann et al., 2008; Farinazzo et al., 2009).

2.1 Yolk plasma proteins

2.1.1 Low-density lipoprotein

LDL is composed of 11–17% protein and 83–89% lipid, of which 75% is neutral lipids and 26% PLs (Martin et al., 1964). LDL has a lipid core of triglycerides and cholesterol esters in a liquid state surrounded by a layer of PLs and proteins (Anton, 2013).

The LDL particle population is heterogeneous with respect to size, density and composition. LDL could be further separated into two components: LDL-1 (molecular

Table 1 Properties of egg yolk proteins [Reproduced from Hatta et al. 2008. Bioactive components in egg yolk. In Y. Mine (ed.), *Egg Bioscience and Biotechnology*. New Jersey: John Wiley & Sons, p. 187]

Name of proteins	(%)	Localization	Molecular weight	Characteristic
Low-density lipoprotein (LDL)	65	Plasma and granules	10,300 kDa (LDL_1) 3,300 kDa (LDL_2)	Lipid content is about 90%, Apovitellenin I–IV is known as a lipoprotein
Lipovitellin (high-density lipoprotein; HDL)	16	Granules	400 kDa (α, β-lipovitellin, complex)	Lipid content is about 25%, MW of apoprotein are 125, 80, 40, 30 kDa
Livetin	10	Plasma	80 kDa α-livitin	Serum albumin
			40, 42 kDa β-livetin	Fragment of C-terminus in vitellogenin
			180 kDa γ-livetin	IgY (hen's serum IgG)
Phosvitin	4	Granules	33 kDa, 45 kDa	Most phosphorylated protein in nature
Egg yolk riboflavin-binding protein	0.4	Plasma	36 kDa	Similar to flavoprotein in egg white and to serum riboflavin-binding protein

weight (MW) 10×10^6 kDa) and LDL-2 (MW 3×10^6 kDa) (Martin et al., 1964). Furthermore, LDL is subdivided into the super LDL (VLDL), which is equal to a serum VLDL (Cook and Martin, 1969).

The major components of serum VLDL, a precursor of yolk LDL, are apoprotein II (apolipoprotein II; apoVLDL) and apoprotein B (apolipoprotein B; apoB). Figure 2 shows synthesis of yolk proteins and their pathway into the egg yolk. A gene expression of apoVLDL is controlled by oestrogen. ApoB is a common apoprotein and similar to a serum VLDL. When apoB in the serum VLDL passes through the vitelline membrane, it undergoes cleavage into apovitellenins III to VI in egg yolk LDL (Evans and Burley, 1987). Apoprotein of yolk LDL consists of apovitellenin I–IV fractions (Inglis et al., 1982). Among fractions, apovitellenin I (MW 9.4 kDa) is the main fraction and is identical to a serum apoVLDL. Apovitellenin II (MW 20 kDa) is a salt-soluble glycoprotein. Apovitellenin III and IV (MW 65 kDa and 170 kDa, respectively), generated from apoB, are water insoluble and cannot be made soluble using denaturants such as urea in its delipidated state (Evans and Burley, 1987).

Human serum/synthetic LDL particles containing anti-cancer agents have been studied to facilitate the uptake of the drug by cancer cells (Ng et al., 2011). Cancer cells overexpress LDL receptors because of the increased need for cholesterol in rapidly proliferating systems. Anti-cancer drugs encapsulated into the core of LDL interact with LDL receptor and subsequently undergo endocytosis and lysosomal degradation. Lipoproteins as a drug delivery vehicle have advantages and disadvantages (Ng et al., 2011). Compared with liposomes, lipoproteins are approximately 10-fold smaller, which is highly favourable for a drug delivery formulation. In contrast to liposomes, LDL has half-lives of 3 and 4.5 days. A longer circulation time is a highly favourable factor for a drug delivery vehicle. In addition, LDL drug delivery systems involve targeting via chemical modification of the protein or lipid constituents of the LDL complex. As a

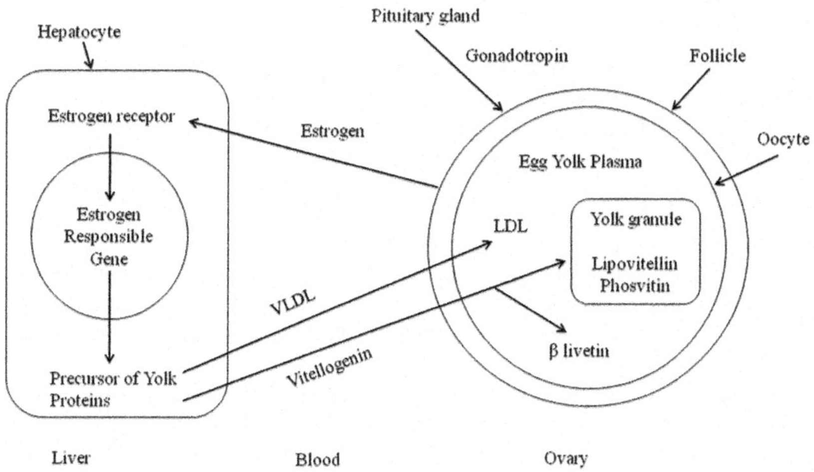

Figure 2 Synthesis of yolk proteins and its transfer into the egg yolk [Reproduced from Matsuda, T. 1998. In R. Nakamura (ed.), Tamago no Kagaku, Asakura Shoten, p. 33].

disadvantage, safety, stability and scale-up are the major concerns for use of blood-derived or synthetic LDL.

Until recently, there have been few reports on egg-derived LDL as a delivery vehicle for cancer drugs. Successful separation of LDL from egg yolk using different gums such as xanthan, gum arabic and lambda-carrageenan has been reported (Hatta et al., 1990; Navidghasemizad et al., 2015). The egg yolk-derived LDL can be further processed for use in the medical and pharmaceutical industries as drug delivery systems for cancer treatment.

2.1.2 Livetins

Livetins are water-soluble proteins and are classified into three groups: α-livetin (MW 80 kDa), β-livetin (MW 40–42 kDa) and γ-livetin (MW 180 kDa) (Williams, 1962). As serum vitellogenin passes through the vitelline membrane, it is degraded and becomes β-livetin. Serum vitellogenin is a known precursor of lipoproteins and phosphoproteins.

Livetins are transferred from blood serum and identified with a serum albumin (α-livetin) and an IgY (γ-livetin) (Williams, 1962). β-livetin, a glycoprotein with a higher cysteine content, could be classified into two components: β-livetin1 (MW 40 kDa) and β-livetin2 (MW 42 kDa). γ-livetin is called IgY due to its difference from mammalian IgG (MW 150 kDa) in terms of structure and nature. IgY as an immunological tool has been studied in the fields of diagnosis as well as in pure research. The applicability of IgY to prevent infectious diseases has been subjected to considerable attention. Details of IgY will be discussed in Section 5.

2.2 Yolk granule proteins

Yolk granules are mainly composed of lipovitellin (also called HDL) (70%) and phosvitin (16%) (Anton, 2013). A limited quantity of LDL also exists in granules. Structurally, granular LDL seems to be identical to plasma LDLs with a similar lipid constitution.

2.2.1 Lipovitellin

Lipovitellin, a globular lipoprotein, is also called HDL. Lipovitellin forms a complex with phosvitin in egg yolk granules; however, the complex can be easily dissociated depending on the ionic strength. Lipovitellin is the smallest and densest among the major lipoproteins in egg yolk because it contains the highest proportion of protein. Lipovitellin does not have a spherical micelle-like structure but rather a pseudo-molecular structure close to that of globular proteins. Lipovitellin forms smaller lipid nanoparticles (5–12 nm) stabilized by the interactions of surface apolipoprotein A-1 (ApoA-1).

Lipovitellin is subdivided into two components: α-lipovitellin and β-lipovitellin (Kurisaki et al., 1981). Apoprotein of α-lipovitellin consists of four subunits having varied MWs: 125, 80, 40 and 30 kDa. Apoprotein of β-lipovitellin consists of two subunits: MW 125 and 30 kDa (Groche et al., 2000).

Lipovitellin and its derived peptides have shown anti-adhesive activities in the intestine. They inhibit colonization of some bacteria including *Salmonella typhimurium*, *Campylobacter jejuni* and *Escherichia coli* O157:H7. Hydrolysates of vitelline, which is one of the apoproteins of lipovitelline, include glycopeptide A and B. Glycopeptide A has a high content of sialic acid, which has been reported to play important roles in biological functions such as cell adhesion, viral infection and toxin neutralization (Schauer, 1982). The details of sialic acid will be discussed in Section 6.

Similar to LDL, the transport behaviour of blood lipovitellin (HDL) has been exploited for enhancing the effectiveness of drug delivery. By interacting with cholesterol receptors on cancer cells, blood serum lipovitellin has been used to deliver drugs in cancer patients (Luthi et al., 2010; Damiano et al., 2013). Besides acting as an efficient carrier of a myriad of biomolecules, lipovitellin also regulates several physiological processes, resulting in anti-inflammatory and anti-atherosclerosis activities (Murphy and Woollard, 2010; Liu et al., 2013). There is limited information about egg-derived lipovitellin (HDL) as a delivery for cancer drugs.

2.2.2 Phosvitin

Phosvitin is a glycoprotein made up of two polypeptides: α-phosvitin (160 kDa) and β-phosvitin (190 kDa) (Itoh et al., 1983). It is classified into five types: B (40 kDa), C (33 kDa), E1 (18 kDa), E2 (15 kDa) and F (13 kDa) (Wallance and Morgan, 1986). Phosvitin is widely studied because of its high phosphorous content (~10%) and is considered to be one of the most phosphorylated proteins in nature (Ito and Fujii, 1962). Thirty to fifty per cent of its amino acids are serine, of which most residues are phosphorylated.

Phosvitin forms an insoluble complex with lipovitellin at low ionic strength, linked by phosphocalcic bridges. At an ionic strength of over 0.3M NaCl, the phosphocalcic bridge begins to be disrupted and at an ionic strength of 1.71M NaCl, a complete disruption of granules occurs. Acidification or alkalization also disrupts the bridge.

The chemical properties of phosvitin have been widely explored for nutritional applications such as metal chelating, antioxidant, emulsifying and antimicrobial activity. Bioactive peptides with metal-binding capabilities have high potential for use in food and cosmetic industries. Phosvitin has a strong metal-binding ability: Fe (II), Fe (III), Co (II), Mn (II), Ca (II) and Mg (II). Ninety-five per cent of an egg's Fe ions exist in the yolk. Due to its strong binding ability with Fe, phosvitin eventually inhibits the use of Fe ions in the human body. However, phosvitin phosphopeptides derived via tryptic hydrolysis followed by partial alkaline phosphorylation enhance calcium-binding capacity and inhibit insoluble calcium phosphate formation (Jiang and Mine, 2000, 2001).

The phosvitin-derived phosphopeptides could be very attractive natural antioxidant agents in food and cosmetic products. Phosvitin and its enzymatic digests provide protection against iron-catalysed hydroxyl radical formation and protect DNA against oxidative damage induced by Fe (II) and peroxide (Ishikawa et al., 2004). These authors suggested that phosvitin may be useful for the prevention of iron-mediated oxidative stress-related diseases, such as colorectal cancer.

2.3 New egg yolk proteins

Recent developments in proteomics have allowed the identification of several hundred new egg components. For example, Mann et al. (2008) discovered 86 previously unidentified proteins in egg yolk, several of which were egg white proteins that had migrated across the vitelline membrane: cystatin, Hep21, lysozyme C, ovalbumin-related proteins Y and X, ovoinhibitor, ovostatin, ovoglycoprotein, ovomucoid, ovotransferrin and ovalbumin. Furthermore, antioxidant enzymes were found in the yolk with sequences similar to extracellular superoxide dismutase, superoxide dismutase, plasma glutathione peroxidase, peroxiredoxin and thioredoxin. These enzymes play an important role in protecting unsaturated fatty acids in egg yolk. With further development in proteomics of egg yolk proteins, there will be great potential for identifying novel egg proteins and their biological functions.

2.4 Delipidated egg yolk protein and bioactive peptides

Delipidated egg yolk protein is a by-product remaining after extracting egg yolk lipids. Because solvents such as acetone and alcohol are used during extraction, egg yolk proteins are denatured, resulting in poor functionality. Applications of the delipidated egg yolk protein as a functional ingredient have been explored because egg yolk is predicted to contain a number of materials that are necessary to generate new cell growth, including growth of bone and feathers.

Leem et al. (2004) reported that delipidated egg yolk protein promotes longitudinal bone growth in adolescent rats. These bones grew significantly in rats fed with water-soluble yolk proteins compared to rats fed with milk proteins and casein. In addition, Kim et al. (2011) reported improved bone density in ovariectomized rats fed with chicken egg hydrolysates. Recently, Kobayashi et al. (2015) investigated the effect of egg yolk hydrolysates (bone pep [BP]) on bone metabolism in orchidectomized dogs. Serum bone alkaline phosphatase, which is the bone formation marker in dogs fed with BP, was significantly higher than in dogs without BP. This finding resulted in the commercial product Bonepep® (Pharma Foods International Co., Ltd., Kyoto, Japan), which is prepared using enzymatic treatment of delipidated water-soluble egg yolk proteins. Bonepep® is marketed to promote bone growth, bone density and bone strength, when taken with calcium.

Similarly, hen egg yolk-derived hair growth peptides (HGPs) were shown to promote hair growth activity *in vivo* (Harada et al., 2016). Hair loss is a serious problem not only in men but also in women in their 50s or 60s. The human hair growth cycle has three stages: anagen (growing phase; 2–6 years), catagen (transition stage; about 2 weeks) and telogen (resting stage; 3–4 months). Over time, hairs become weaker and thinner due to the decrease of the anagen stage. According to Harada et al. (2016) study, HGP mixed with feed was given to 7-week-old C57BL6 mice in the telogen stage. Hair follicles then entered the anagen stage. In another study described in the same paper, active fractions of HGP upregulated the expression of vascular endothelial growth factor (VEGF) in cultured human dermal papilla cells. Dermal papilla cells regulate hair-follicle formation and growth cycle. In anagen stage, a highly developed vascular network appears in the dermal papilla and the hair bulb. VEGF and its receptors are related to the formation of capillaries. The presence of VEGF is required to induce and maintain angiogenesis in the hair cycle. Finally, Harada et al. (2016) also reported the efficacy of a treatment to reduce hair loss in women, using a randomized and double-blind clinical trial. HGP was found to stimulate hair growth by thickening the hair shaft, suggesting that HGP is effective for preventing female-pattern hair loss.

In addition to promoting cell growth, hydrolysis products of delipidated egg yolk proteins have been studied as antioxidants. Zambrowicz et al. (2012) reported that enzymatic hydrolysates of delipidated egg yolk protein showed antioxidant and antimicrobial activities. The hydrolysates of egg yolk protein treated with trypsin and chymotrypsin demonstrated significantly stronger free radical 2,2-dephenyl-1-picrylhydrazyl (DPPH) scavenger activity and iron-ion-chelating capacity than peptic hydrolysates. Hydrolysates of egg yolk protein treated with trypsin possess weak antimicrobial activity after 4 h of hydrolysis.

Zambrowicz et al. (2015) also reported that hydrolysates of egg yolk protein by-product, obtained with the use of proteinase from *C. ficifolia*, showed antioxidant, alpha-glucosidase and dipeptidyl-peptidase-IV (DPP-IV) inhibitory activities. Alpha-glucosidase is a carbohydrate hydrolase present in the epithelium of the small intestine. It functions to facilitate glucose absorption in the small intestine by catalysing the hydrolytic cleavage of oligosaccharides and disaccharides into absorbable monosaccharides. DPP-IV degrades the glucagon-like peptide 1, which stimulates glucose-dependent insulin secretion in pancreatic beta cells. Sakanaka and Tachibana (2006) also reported on hydrolysates of delipidated egg yolk protein produced by adding food-grade protease from *Bacillus* sp. followed with Orientase (EC 3.4.21.62; HankyuBio-Industry, Osaka, Japan) and protease (EC 3.4.11.12; Amano Enzyme Inc., Nagoya, Japan). Egg yolk protein hydrolysates exhibited DPPH and hydroxyl radial scavenging activity on lipid oxidation in beef and tuna homogenates. In addition, Marwa and Nazlin (2015) reported antioxidant and angiotensin-converting enzyme (ACE) inhibitory activities of delipidated egg yolk peptides, in particular, the sequences of amino acid residues, SDNRNQGY and YPSPV.

Several other functions of delipidated egg yolk have also been reported, including improved solubility and inhibition of tumour colon cells. Wang and Wang (2008) reported that controlled enzymatic hydrolysis improved its solubility, resulting in improved foaming and emulsification properties. In addition, egg yolk proteins and their hydrolysates exhibited anti-tumour activity by inhibiting tumour cell proliferation in the colon (Ishikawa et al. 2009).

3 Egg yolk lipids

Lipids are found only in egg yolk and comprise nearly 30% of the yolk by weight (Li-Chan et al., 1995). Egg yolk lipids exist as lipoprotein, that is, combined with protein. Egg yolk lipids have been widely used in food, cosmetic and pharmaceutical industries as emulsifiers. Recently, in particular, medical and pharmaceutical fields have shown much interest in applications of PLs. Egg yolk lipid serves several critical functions for the developing chick embryo: it contributes to the composition of cell membranes, provides the main energy source for the embryo and is essential for brain development (Juneja, 1997). Therefore, it is expected that egg yolk lipids would play an important role in structuring human somatic cells and the composition of brain and nerve cell membranes. In this section, we will discuss components of egg yolk lipids and their features and health-promoting functions. Two topics will be discussed: first, the functions of eggs fortified with omega-3 polyunsaturated fatty acid (ω-3 PUFA) and secondly, recent applications of PLs in medical, pharmaceutical, cosmetic and food fields. Current information on cholesterol research will be discussed separately in Section 4.

3.1 Components of egg yolk lipids

Lipids comprise nearly 30% of egg yolk by weight (Li-Chan et al., 1995). Egg yolk lipid mainly consists of neutral lipids (65%), PLs (30%) and cholesterol (4%) (Juneja, 1997) (Fig. 3). Neutral lipids mainly consist of triacylglycerols. However, other constituents are also included in this category, such as diacylglycerols, monoacylglycerols, fatty acids, carotenoids and sterols. The major fatty acids of triacylglycerols are oleic acid (43.6%), palmitic acid (25.1%), linoleic acid (13.4%), stearic acid (8.6%), palmitoleic acid (3.6%), docosahexaenoic acid (1.8%) and arachidonic acid (1.7%) (Tesedo et al., 2006) (Table 2).

PLs have a glycerol–phosphate backbone and an amphiphilic structure, which consists of two hydrophobic fatty acids as tails and a hydrophilic portion as a head. Hydroxyl groups at sn-1 and sn-2 in glycerol are esterified with fatty acids. The fatty acid substituent at sn-1 is saturated while sn-2 is unsaturated. The hydroxyl group at sn-3 in glycerol is esterified with phosphoric acid, which is then further esterified with compounds

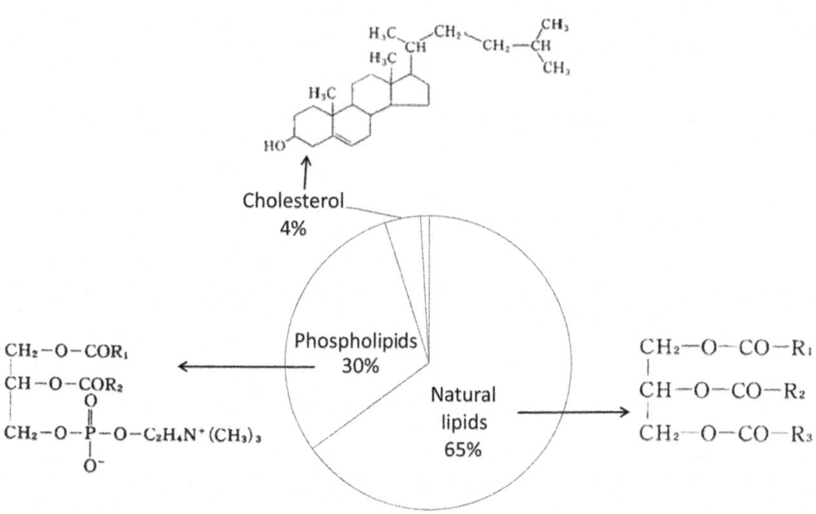

Figure 3 Composition of egg yolk lipids.

Table 2 Fatty acid composition of egg yolk lipids and their chemical structure [Reproduced from Hatta et al. 2008. Bioactive components in egg yolk. In Y. Mine (ed.), *Egg Bioscience and Biotechnology*. New Jersey: John Wiley & Sons, p. 194]

Fatty acid	C-chain	Classification	n-series	Constitution
Oleic acid	C18:1	Monounsaturated	n-9	$CH_3(CH_2)_7CH{=}CH(CH_2)_7COOH$
Palmitic acid	C16:0	Saturated	—	$CH_3(CH_2)_{14}COOH$
Linoleic acid	C18:2	Polyunsaturated	n-6	$CH_3(CH_2)_4CH{=}CHCH_2\text{-}CH{=}CH(CH_2)_7COOH$
Stearic acid	C18:0	Saturated	—	$CH_3(CH_2)_{16}COOH$
Palmitoleic acid	C16:1	Monounsaturated	n-9	$CH_3(CH_2)_7CH{=}CH(CH_2)_5COOH$
Docosahexaenoic acid	C22:6	Polyunsaturated	n-3	$CH_3CH_2CH{=}CH\text{-}CH{=}CHCH_2CH{=}CHCH_2CH{=}CHCH_2CH{=}CHCH_2CH{=}CH\text{-}(CH_2)_3COOH$
Arachidonic acid	C20:4	Polyunsaturated	n-6	$CH_3(CH_2)_4CH{=}CHCH_2CH{=}CHCH_2CH{=}CHCH_2CH{=}CH\text{-}(CH_2)_3COOH$
α-Linolenic acid	C18:3	Polyunsaturated	n-3	$CH_3CH_2CH{=}CHCH_2CH{=}CHCH_2CH{=}CH(CH_2)_7COOH$
Eicosapentaenoic acid	C20:5	Polyunsaturated	n-3	$CH_3CH_2CH{=}CHCH_2CH{=}CHCH_2CH{=}CHCH_2CH{=}CHCH_2CH{=}CH\text{-}(CH_2)_3COOH$

Table 3 Comparison of phospholipids

Name of phospholipids	Soya bean (%)	Egg yolk (%)
Phosphatidyl choline (PC)	33	84.3
Phosphatidyl ethanolamine (PE)	14.1	11.9
Phosphatidyl inositol (PI)	16.8	
Phosphatidic acid (PA)	6.4	
Sphingomyelin (SM)		1.9
Lysophosphatidyl choline (LPC)	0.9	1.9
Others	28.8	

such as choline, ethanolamine, serine or inositol to form phosphatidylcholines (PCs), phosphatidylethanolamines (PE), phosphatidylserines (PS) or phosphatidylinositol (PI), respectively (Juneja, 1997) (Table 3).

3.2 Nutritional characteristics of egg yolk lipids

Fatty acids are classified into three groups: saturated fatty acids (SFAs), monounsaturated fatty acids (MUFAs) and polyunsaturated fatty acids (PUFAs). Egg yolk lipids contain 34% SFA, 48% MUFA and 18% PUFA (Juneja, 1997). The ratio of PUFA to SFA (P/S) is considered important in influencing serum cholesterol. The ratio of P/S of egg yolk is 0.5, indicating that the content of SFA is more than double than that of PUFA. However, in one large ω-3 PUFA fortified egg, the ratio of PUFA to SFA becomes 1.

Eggs have two essential FAs: omega-3 fatty acids (ω-3 FAs) with a double bond in the *n*-3 position and omega-6 fatty acids (ω-6 FAs) with a double bond in the *n*-6 position. Major ω-3 and ω-6 FAs in egg are α-linolenic acid (LNA, ω-3), eicosapentaenoic acid (EPA, ω-3), docosahexaenoic acid (DHA, ω-3), linoleic acid (LA, ω-6) and arachidonic acid (AA, ω-6) (Juneja, 1997).

The human body cannot synthesize LNA (ω-3) and LA (ω-6). Therefore, these PUFAs need to be obtained from food. Figure 4 shows metabolic pathways of n-6 and n-3 fatty acids. In the body, LNA is converted into EPA and DHA, while LA is converted into AA by enzymes and hormones such as prostaglandin and leukotriene derived from these PUFAs. Hormone-like substances derived from AA play a role as inflammatory mediators and promote cancer, allergy and blood clots, while EPA and DHA have opposite effects. In the past, LA was believed to decrease blood cholesterol. However, this decrease of blood cholesterol is temporary and a high intake of LA increases AA, resulting in increases in inflammatory mediators and onset of lifestyle-related diseases. Therefore, the increase in the dietary ratio of ω-6 to ω-3 in modern diets has been proposed to contribute to the onset of inflammatory diseases. On the other hand, intake of ω-3 PUFA is reported to decrease the production of inflammatory mediators, resulting in prevention of cancer, hypercholesterolemia, thrombosis and immune hypersensitivity.

Increasing dietary intake of ω-3 FAs and bringing the ratio of ω-3 FA to ω-6 FA into a healthy balance would improve health. According to dietary guidelines in developed countries, the recommended ratio of ω-6 to ω-3 ratio is under 4. Currently, the ratio of ω-6 to ω-3 value in Japan is nearly 4. In Europe and the United States, the ratio of ω-6 to ω-3 has exceeded 10 due to consumption of more meat rather than seafood. In egg yolk, the ratio of ω-6 to ω-3 is approximately 7. Furthermore, in the ω-3 fortified egg, the ratio

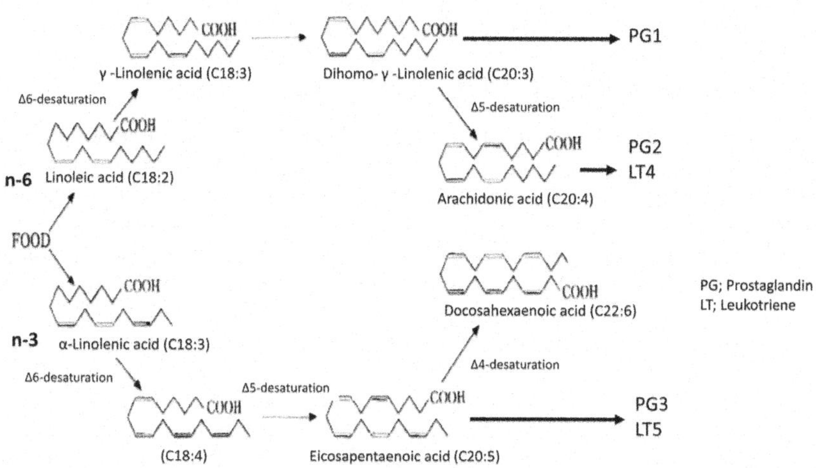

Figure 4 Metabolic pathways n-6 and n-3 fatty acids [Reproduced from Juneja, L. R. 1977. Egg yolk lipids. In T. Yamamoto, L. R. Juneja, H. Hatta and M. Kim (eds), *Hen Eggs, Their Basic and Applied Science*. New York: CRC Press, p. 89].

of ω-6 to ω-3 became lower than 1, suggesting that consuming ω-3-rich eggs will help to keep the preferable ratio of ω-6 to ω-3 in diet.

Egg yolk lipids have been used as a fat source for infant milk formula to closely mimic the fatty acid composition of human breast milk. Human breast milk contains balanced DHA and AA (AA = 0.28–0.60% and DHA = 0.22–1.0%) (Makrides et al., 2002) and both DHA and AA are essential for the development of infant brain and retina. DHA and AA from eggs can provide a rich source of lipids in infant formula, while AA and DHA levels in infant formula prepared with soya bean oil, which is often used as a fat source for infant formula, are insufficient.

3.3 Effects of feed

3.3.1 Overview

The composition of egg yolk lipids can be modified by selecting specific feed. Some vitamins, minerals and essential fatty acids have been fortified in eggs by feeding hens with these nutrients. However, some nutrients cannot be easily transferred into eggs. Water- and fat-soluble vitamins, minerals (iodine and manganese) and unsaturated fatty acids (LA, LNA, EPA and DHA) can be easily transferred into eggs, but calcium, magnesium, iron, vitamin C and amino acids have difficulty in being transferred. Generally, fat-soluble nutrients tend to be transferred mainly into egg yolk, while water-soluble nutrients are transferred into both egg yolk and egg white. Advanced unsaturated fatty acids can be incorporated into the composition of PLs in egg yolk.

Currently, about 1500 brands of value-added eggs have been produced by egg farmers and introduced into the Japanese market (see Egg Museum website). Among these value-added egg brands, ω-3 fortified eggs have been extensively studied both in animal and human clinical trials and their health benefits have been proven (Sim and Sunwoo, 2002). In this section, we will discuss ω-3 FA fortified eggs, as these have the broadest market recognition internationally and the strongest scientific support. Since folic acid is a water-soluble vitamin B, folic acid-fortified eggs will be discussed separately in Section 6.1.

3.3.2 Development of ω-3 fatty acid fortified eggs

Interest in functions of ω-3 FAs began with Bang et al. (1971). In their study, Inuit people from Greenland were found to have significantly lower than expected total cholesterol, LDL and triglyceride levels due to their high consumption of fish, which is a good source of ω-3 FAs. Since then, much research has been conducted on the relationship between ω-6/ω-3 FAs and cardiovascular diseases (Dyerberg and Bang, 1979; Kinsella et al., 1990; Simopoulos, 1991; Maehre et al., 2015). Among these studies, ω-3 FAs are reported to prevent cancer, heart attack and inflammation.

Recently, an excess of LA has been found to promote the accumulation of AA in the body and increase the level of prostaglandin E2, resulting in accelerated onset of lifestyle-related diseases. On the other hand, the consumption of LNA, EPA and DHA has the opposite effect. As a result, dietary guidelines on lipid consumption were re-evaluated in many countries. In order to decrease the ratio of ω-6 to ω-3, the consumption of ω-3 FAs from fish and fish oil is recommended in developed countries.

By mixing fish oils with hen's feed, it is possible to produce ω-3 FA fortified eggs (Nwokolo and Sim, 1989; Caston and Lesson, 1990; Hargis et al., 1991; Jiang et al., 1992a).

The composition of lipid in egg yolk is determined by the components both synthesized in the hen's body and those found in the hen's feed. In particular, LNA is the main ω-3 FA that is usually transferred into egg yolk through fortified feed.

Dr. Sim's research group at the University of Alberta, Canada, has pioneered the development of ω-3 FA fortified eggs. These ω-3 FA fortified eggs have received much attention as a good source of ω-3 FAs to replace fish and fish oil (Marshall et al., 1994). A series of studies has been done to develop a ω-3 FA fortified egg with an elevated ratio of PUFA/SFA (from 0.6 to 1.02) and a decreased ratio of ω-6/ω-3 (from 10:1 to 1.1) (Jiang et al., 1992b; Sim et al., 1992). This ω-3 FA fortified egg has about 600 mg of ω-3 FAs with balanced DHA and EPA and about 6 mg tocopherol. This amount is similar to the amount in 100 g of fish.

Sim and Guang-Ha (1995) investigated the effect of flaxseed (56% LNA) on the fat composition of egg yolk. The egg yolk was found to have increased levels of ω-3 FAs in the following order: LNA>DHA>EPA. The amount of AA (ω-6) decreased, resulting in a ω-6/ω-3 ratio of 1. The ω-6/ω-3 ratio of an average egg is 10. Furthermore, by adding tocopherol as an antioxidant, oxidation of PUFA in hen's feed was prevented. Tocopherol was transferred into egg yolk and showed antioxidant effects on the fat contained in egg yolk, resulting in decreased cholesterol oxidation and reduction in thiobarbituric acid values in eggs (Cherian et al., 1996a,b; Li et al., 1996).

Animal and human studies have demonstrated the clinical application of fortified eggs. Jiang and Sim (1992) investigated the effect of ω-3 fortified eggs on cholesterol and lipid in rat's blood. The ω-3 fortified egg was boiled and the egg yolk removed, dried and powdered. The dried egg yolk powder was mixed in feed (15% w/w) and fed to Sprague–Dawley rats for 4 weeks. The cholesterol and lipid in the rat's blood and liver decreased 20% and 37%, respectively. In the liver, EPA and DHA increased, while AA decreased, resulting in a decreased ratio of ω-6/ω-3.

Jiang and Sim (1993) also investigated the effect of ω-3 FA fortified eggs on blood levels of cholesterol and lipid in a clinical trial. Healthy male students (18–32 years) had two eggs at every breakfast for 18 days. The experimental group had no increase in total blood cholesterol, an increase of 5% in LDL levels and an increase of 12.5% HDL, compared to the control group that had increased total cholesterol (6%) and LDL (12.5%) and no change in HDL. Furthermore, serum-neutral lipid decreased 40% for the experimental group.

Cherian and Sim (1996) reported the effect of ω-3 FA fortified eggs on the lipid composition of breast milk and blood. Eight breastfeeding mothers ate two ω-3 FA fortified eggs for 6 weeks. The ω-3 FAs in breast milk increased from 1.9 to 3.6% and the ratio of ω-6/ω-3 decreased from 6.7 to 3.0. The content of AA was not changed. As a result, the ratio of DHA to AA increased from 0.75 to 1.2. There was no difference in total blood cholesterol or neutral lipids. A ω-3 FA fortified egg includes 165 mg of longer chain ω-3 FAs: EPA, DPA and DHA. A ω-3 FA fortified egg or its oil can have a positive effect on breast milk, resulting in a decrease of the ratio of ω-6/ω-3 in breast milk and healthy development of the baby.

3.4 Functions of egg yolk PLs

Egg yolk PLs, such as egg yolk lecithin, have been used in food cosmetics, pharmaceutical and medical areas (Baker, 1989; van Hoogevest and Wendel, 2014).

3.4.1 Application in pharmaceutical, medical and cosmetic areas

A primary area of recent research for pharmaceutical and medical applications of PLs is the treatment of Alzheimer's disease and addition of choline in infant formula. With rapid ageing of the global population, the number of Alzheimer patients is predicted to increase. As normal ageing progresses at the cellular level, vital organs lose some functions. Every day, in the average adult body, about 300–400 million cells are dying and new cells are generated. The maximum life expectancy of a human is around 120 years, but cells normally can divide only about 50–60 times before dying. During that time, cell membranes are renewed and keep the fluidity inside the cells. In Alzheimer patients, this fluidity is not maintained.

The main component of egg yolk PLs is PC. The choline is a precursor of acetylcholine, which is a neurotransmitter. Alzheimer patients show a low level of acetylcholine in their brains. The oral intake of egg yolk PC is expected to increase the concentration of choline and improve nerve cells. Masuda et al. (1998) reported improvement of brain function (65%) with a combination of egg yolk PLs and vitamin B-12, which increases the activity of acetylcholine synthesis enzyme. In addition, PC plays an important role in the development of infant brain function (Zeisel, 1981).

A second area of application of PLs is in drug administration. Since egg yolk lipid has excellent emulsification ability, it has been used as an ingredient for liposome and liposome-based drugs synthesis in pharmaceutical industries. Liposomes have been used as oral and intravenous carriers for enzymes, vaccines, drugs such as anti-tumour dugs, cell-modifying molecules and hormones (Tyrrell et al., 1976). Emulsification is discussed further in the next section.

PLs are very popular as an ingredient in cosmetic products: cleansing creams, moisturizing creams, hair products, aftershave lotions, bath lotions (rinse), body lotions, shampoo, conditioner, skin cream and sun protective cream (Baker, 1989). Furthermore, PLs can be used as an emulsifier, a liposome former, a solubilizer and a wetting agent (van Hoogevest and Wendel, 2014).

3.4.2 Applications in food industry

The chief function of egg yolk in the food industry is emulsification, which is used to make food more visually appealing and to improve taste and texture. Egg yolk is a key ingredient in a wide variety of food emulsion systems. Oil and water cannot mix together due to the interfacial tension of each fluid. In order to produce emulsion, an emulsifier is needed to decrease the surface tension of each liquid. Egg yolks are an effective emulsifier due to the properties of egg yolk LDL.

Egg yolk as an emulsifier

An emulsifier has both hydrophilic and lipophilic regions. The hydrophobic–lipophilic balance (HLB) determines each characteristic of the emulsifiers. For example, paraffin, which has no hydrophilic regions, has HLB 0. Polyethylene glycol, which has only hydrophilic regions, has HLB 20. Molecules with more hydrophilic regions have higher HLBs. HLB 8–18 indicates a water-soluble emulsifier which will produce oil-in-water (O/W) emulsion. The HLB of egg yolk is about 12. Thus, egg yolk has been used as an emulsifier in many common O/W emulsions such as mayonnaise and dressings.

Phospholipase has been used to catalyse PLs, producing free fatty acids and lysophospholipids, which show a higher solubility and emulsifying property than untreated PLs (Daimer and Kulozik, 2008; Daimer and Kulozik, 2009). Phospholipases are classified according to the cleavage of the ester bonds for which they are specific (De Maria et al., 2007). Phospholipase A1 (PLA1), phospholipase A2 (PLA2), phospholipase B (PLB) and phospholipase C (PLC) catalyse the hydrolysis of the ester bond at the sn-1, sn-2 and sn-3 carbon, respectively. Phospholipase D (PLD) hydrolyses other phosphodiester bonds. PLB cleaves both sn-1 and sn-2 ester bonds.

Porcine pancreatic PLA2 has been used extensively to modify egg yolk to enhance its emulsifying property or degumming of oil. In the past decades, much research has investigated the development of microbial-based phospholipase. Recently, Lecitase® Ultra, a protein-engineered PLA1 introduced by Novozymes, has received increased attention from researchers and food manufacturers as a biocatalyst for modifying lipids (Wang et al., 2010; Liu et al., 2011). Lecitase® Ultra was originally produced for degumming of vegetable oils. Lysophospholipids produced via PLA1 have better higher protein solubility, thermal stability and emulsifying properties (Jin et al., 2013). Furthermore, lysophospholipids influence a broad variety of processes: carcinogenesis, neurogenesis, immunity, vascular development or regulation of metabolic diseases, suggesting therapeutic potential (Grezelczyk and Gendaszewska-Darmach, 2013).

In the past, it was believed that high emulsifying capacity is caused by lecithin (PLs). Based on the study to investigate emulsifying capacity of each component in egg yolk, lipoprotein was identified to play an important role for its emulsifying capacity. The main component of egg yolk is lipoproteins – LDL in plasma and HDL in granules. The main component, LDL, has more than 90% of egg yolk lipids and has a neutral lipid as a core, surrounded on the surface by PLs and proteins. Generally, the emulsifying capacity of food proteins is influenced by the surface structure of proteins, especially surface hydrophobicity. Among food proteins, BSA has higher surface hydrophobicity and shows an excellent emulsifying capacity. Mizutani and Nakamura (1988) showed that the emulsifying capacity of LDL was much stronger than that of BSA and lecithin (PLs). LDL shows stronger absorbance on oil than BSA and forms a very tiny oil droplet. Oshida (1976) showed that LDL had a better emulsifying capacity than HDL due to a complex of proteins and PLs.

Egg yolk emulsifying ability and stability

An emulsifier is evaluated by emulsifying ability and emulsifying stability. Egg yolk sterilized at 62°C for 5 minutes increased emulsifying ability. However, egg yolk sterilized at 67°C, 73°C and 78°C decreased emulsifying ability slightly. Tsutsui (1988) investigated the effect of heating temperature and the emulsifying ability of purified LDL. Four per cent of LDL was heated for 5 min at various temperatures. At temperatures up to 75°C, both emulsifying ability and stability were stable and, between 75 and 80°C, emulsifying ability and stability decreased significantly.

The addition of salt stabilizes the emulsifying stability and the addition of vinegar decreases the emulsifying ability. With the addition of salt (6~10%), emulsifying ability increased 10%, and with the addition of vinegar (0~5%) with any levels of salt concentration, the egg yolk emulsifying ability decreased (Oshida, 1976).

When egg yolk is frozen, egg yolk apolipoproteins are denatured and the egg yolk becomes temporally viscous. During the thawing of egg yolk, it forms an irreversible gel. Once frozen, the emulsifying ability and emulsifying stability of egg yolk decrease.

Emulsifying ability becomes 67% after a week and it was maintained without further loss, 66% after 10 weeks (Hatta et al., 1997). The addition of sugar and salt can prevent denaturation of apolipoprotein in egg yolk. In mayonnaise and salad dressing, frozen egg yolk with 10% salt is used. In ice cream and pudding, frozen egg yolk with 10% sugar is used (Hatta et al., 1997).

In frozen foods with mayonnaise, heat-resistant and freeze-resistant mayonnaise is demanded. Generally, normal mayonnaise is not heat-resistant. After microwave heating, emulsifying ability is destroyed and oil separates. Egg yolk treated with PLA2 is usually used to make mayonnaise. The emulsifying ability increases and emulsion droplet is much smaller and the emulsion is heat-resistant.

Mayonnaise has more than 65% lipids, which are mainly neutral lipids. Along with consumer trends, low calorie mayonnaises have been on the market. With half the amount of oil, in order to have the same texture of normal mayonnaise, micro-emulsion techniques have been introduced. In addition, in the double emulsion technology, calories are reduced by 65% (see Ajinomoto website).

4 Egg yolk cholesterol

Egg yolk research has practical implications not only for the food industry but for government guidelines and ultimately for consumer practices. Egg yolk cholesterol is a prime example of this influence. For 30 years, it was believed that egg cholesterol (185 mg/large egg) (United Sates Department of Agriculture (USDA)) increased serum cholesterol, resulting in a high risk of cardiovascular diseases. As a result, Japan and the United States developed guidelines restricting egg consumption. Most other developed countries did not have a standard guideline on cholesterol intake. In 2015, at last, the United States and Japan removed the maximum daily cholesterol intake from their guidelines. In this section we will review basic roles of cholesterol, the history of cholesterol guidelines and current international guidelines on daily cholesterol intake.

4.1 Roles of cholesterol

The human body is made up of about 37 billion cells and contains about 150 g of cholesterol. A lipid bilayer is a component of all cell membranes. The principal components of the lipid bilayer are PLs, cholesterol and proteins. The function of cholesterol is to add firmness and integrity to the membrane and to maintain its fluidity inside membranes. Therefore, cholesterol facilitates the channelling or pumping of substances into and out of cells.

Cholesterol is particularly abundant in the tissues of the brain and nervous systems. Twenty-five per cent of the cholesterol in the body is found in the brain. Children with lower levels of cholesterol have higher rates of attention deficit and hyperactivity disorder. In the case of adults, rates of anxiety and depression increase, suggesting that lower brain cholesterol levels and a reduction in synapses may be features of mood disorders (Beasley et al., 2005).

The human body also needs cholesterol to make bile acids, hormones and vitamin D. Bile acids are synthesized from cholesterol in the liver, stored in the gall bladder and secreted into the duodenum. They work as detergents to emulsify fats and are therefore critical for digestion and absorption of fats and fat-soluble nutrients in the small intestine. Testosterone and estrogen are derived from cholesterol. Vitamin D, which plays an important role in calcium metabolism, is also synthesized from cholesterol when the skin is exposed to ultraviolet rays.

In the blood, cholesterol is a component of chylomicrons, remnants, LDL and HDL. LDL, sometimes referred to as the 'bad cholesterol', transports cholesterol from the liver to other tissues. HDL, sometimes referred to as the 'good cholesterol', transports cholesterol from other tissues back to the liver. In actuality, research has shown that both the 'good' and 'bad' cholesterols play an important role. A balance between LDL and HDL (i.e. the ratio of LDL to HDL) is important. Even people with a normal LDL level (104 mg/dl) can develop cardiovascular disease. When the LDL:HDL ratio is above 1.5, cholesterol is circulated in the blood and starts to attach to the interior walls of the blood vessels. Above a 2.0 ratio, it begins to form deposits. When the ratio is over 2.5, blood vessels show significant atherosclerosis (Fernandez and Webb, 2008).

About 70% of the cholesterol in the human body is synthesized, mainly in the liver, from carbohydrates and lipids. The remaining 30% comes from the diet. The level of cholesterol synthesis in the liver is regulated by the dietary intake of cholesterol in order to maintain a certain level of cholesterol in the body. Approximately 40–60% of cholesterol is absorbed in the small intestine. The human body needs 1.0–1.5 g cholesterol per day. Therefore, it is suggested that 500–600 mg of cholesterol is an adequate amount to gain from dietary intake. On average, Japanese people ingest 300–500 mg of cholesterol daily, mainly from egg, milk, meat and fish. Sixty per cent of cholesterol in the Japanese diet comes from eggs and egg dishes, resulting in about 180–300 mg cholesterol per day (see Ministry of Health, Labour and Welfare website).

Currently, it is recognized that the risk of cardiovascular disease has a positive relationship with intake of saturated and trans-fatty acids. Generally, food with a high cholesterol level tends to have a high level of SFAs. However, the egg is an exception and has much lower SFAs. According to Herron et al. (2003), when people (n = 40) who are very sensitive to cholesterol ate egg cholesterol (640 mg/day), both LDL and HDL increased, resulting in no change in the ratio of LDL to HDL.

In summary, recent research has contradicted past beliefs about the relationship between ingestion of eggs and elevated cholesterol in the body. In the next section, the history of international cholesterol guidelines is reviewed.

4.2 The United States and Japanese history of cholesterol issues

In 1968, the American Heart Association published a dietary reference guide, saying that cholesterol intake should be limited to less than 300 mg per day and that a maximum of three eggs per week was recommended (Lichtenstein et al. 2006). At that time, the number of deaths due to heart disease had increased and become a significant concern in the United States. An increase of serum cholesterol was believed to be the cause of cardiovascular disease. As a result, a 'low-cholesterol' and 'no-cholesterol' trend developed. From 1970 to 1995, the rate of individual annual egg consumption declined from 311 to 238. In 1984, the Egg Nutrition Center was launched to objectively analyse the relationship between egg cholesterol and cardiovascular disease. For twenty years, many animal studies, epidemiological studies and clinical studies had been conducted. Based on the results from these studies, it was concluded that there was no evidence of a correlation between intake of eggs and risk of cardiovascular disease. Therefore, in 2000, the American Heart Association announced that there was no more recommendation on how many eggs per day can be eaten.

In Japan, a 2010 dietary guideline set the daily cholesterol intake at 750 mg/day for men and 600 mg/day for women (Ministry of Health, Labour and Welfare). However, like the United States, Japan removed its limitation. Based on studies on the relationship between cholesterol intake and risk of cardiovascular diseases, the Japanese government recommended a low intake of cholesterol, but there is no evidence to set the desired amount in the guideline. In addition, the guidelines recommend that elderly people should be mindful of the amount of cholesterol in meat, but should not avoid meat as a source of nutrition. If people limit the amount of cholesterol by limiting meat intake too strictly, it may result in inadequate protein and energy intake.

In Japan, current production of egg is about 2.5 million tonnes annually, which is equivalent to about 42 billion eggs (International Egg Commission, 2014). Of this, 50% of eggs are for domestic consumption, 30% for industrial consumption and 20% for processed egg. Annual consumption per individual in Mexico is 347 eggs, in Malaysia, 331 eggs, and in Japan, 329 eggs (International Egg Commission, 2014). Japanese people eat 0.901 eggs per day (IEC, 2014).

4.3 International guidelines for daily cholesterol intake

Every five years since 1980, the USDA and Food and Drug Administration (FDA) have been publishing 'Dietary Guidelines for Americans'. As noted by Brownawell and Falk (2010), the 2010 guideline limited daily cholesterol intake for normal adults to less than 300 mg per day, and cholesterol intake for high LDL hypercholesterolemia patients was limited to less than 200 mg per day. On the other hand, guidelines in many developed countries such as Canada, Britain and Australia, do not set a daily maximum for cholesterol (Brownawell and Falk, 2010). Instead, they recommend decreasing intake of total lipids (e.g. saturated and trans-fatty acids) and increasing PUFA. In developing countries it is recommended to eat eggs daily due to the feed efficiency of egg production (see Food and Agricultural Organization of the United Nations, 2009).

Comparing international guidelines on daily cholesterol intake, a US dietary guideline advisory committee in 2008 started to re-evaluate the effect of current nutritional policies on dietary patterns and dietary guidelines (Brownawell and Falk, 2010). In February 2015, the committee concluded that there was no relationship between dietary cholesterol and serum cholesterol. This view was incorporated into the 2015 US national dietary guideline.

In Canada, in 1988, a conference on cholesterol intake was held and recommended a reduction in fat (<30% of energy as total fat and 10% as saturated fat) (McDonald, 2004). However, there was no recommended limitation on daily cholesterol intake. According to the Canadian dietary guideline published by Health Canada in 1990, a healthful diet includes a variety of foods and decreases the amount of fatty foods such as meat and lipids.

In summary, international dietary guidelines and committees recommend that cardiovascular diseases should be prevented by reducing the amount of lipids, especially SFA, instead of cholesterol. An egg is an important food and contains all major nutrients except vitamin C and dietary fibre. The egg is low in SFAs and calories. Egg production has good feed efficiency: 2 kg of feed provides 1 kg of eggs. This efficiency would be beneficial in any potential future food supply crisis. Without limitation of cholesterol intake from international food guidelines, it is predicted that one or two eggs per day can be consumed by a person without medical limitations.

5 Egg yolk antibody (IgY)

The human body produces immune proteins – specific antibodies – to defend itself from pathogenic antigens such as bacteria, viruses and non-self-components of hetero-proteins. When an antigen invades the body, an immune system is stimulated. Then an antibody, which has a specific binding ability to the particular antigen, produces specific quantities of antibodies to neutralize the toxicity and infectivity of the invading antigen.

An antibody is a glycoprotein of the group called immunoglobulin (Ig) and is found in living body fluids such as blood, saliva, nasal cavity liquid, lactation and so on. Antibodies in mammals are divided into five classes: IgG, IgM, IgA, IgD and IgE (Maddison and Reimer, 1976) (Fig. 5). While IgG, IgM and IgA exist in chicken blood serum, IgG exists only in egg yolk (Rose et al., 1974; Schade et al., 1991; Li et al., 1997). After the egg hatches, the IgG of the egg yolk is transferred into the blood serum of the chick. Both IgA and IgM go into the intestinal tract to prevent infections until the chick is able to make antibodies by itself (Fig. 6).

An antibody in the egg yolk is different from an IgG antibody in protein chemistry and immunochemistry (Leslie and Clem, 1969). Therefore, the egg yolk antibody is referred as IgY in the field of comparative immunology. Hen eggs are presently used as a source of IgY and are considered to be a large-scale antibody production system. Useful applications for IgY as an immunological tool have been studied in the fields of research,

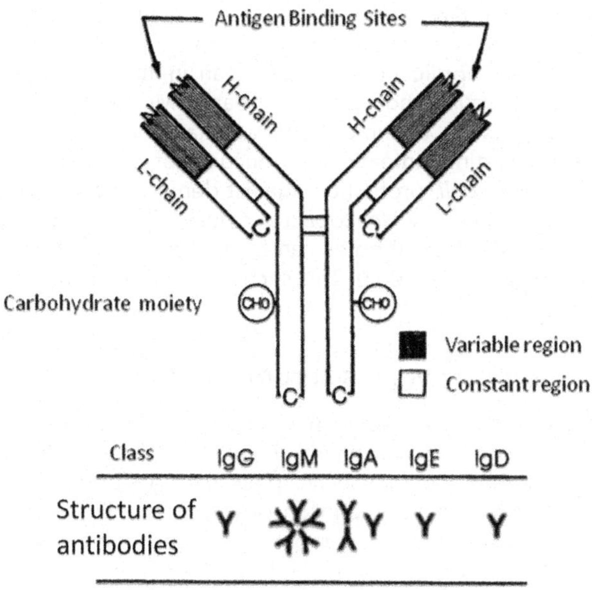

Figure 5 Basic structure of antibodies.

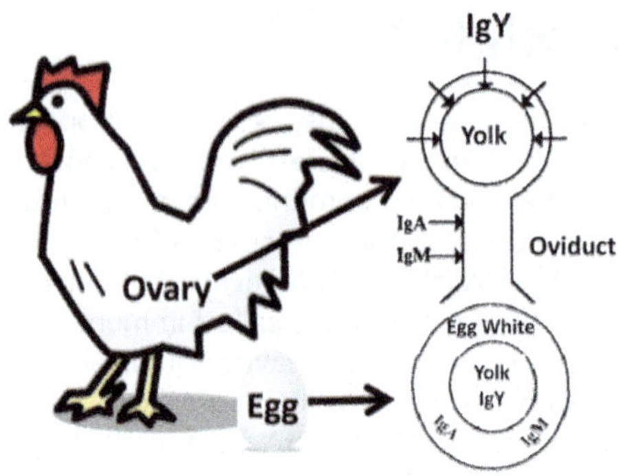

Figure 6 Distribution of immunoglobulin in hen.

diagnostics and therapy (Hatta et al., 2008; Xu et al., 2011). Recently, the applicability of IgY to prevent infectious diseases has received much attention due to the potential to produce specific antibodies in egg yolk. Antibodies have been produced against bursal disease (Eterradossi et al., 1997), porcine epidemic diarrhoea virus (Kweon et al., 2000), bovine coronavirus-induced diarrhoea (Ikemori et al., 1997), cryptosporidiosis by *Cryptosporidium* (Kobayashi et al., 2004) and swine respiratory disease caused by *Bordetella bronchiseptica*, *Pasteurella multocida* and *Actinobacillus pleuropneumoniae* (Shin et al., 2002).

In this section, recent applications of IgY as an immunological tool in the fields of diagnosis as well as pure and applied research will be discussed. In particular, applications for IgY against a number of bacteria and viruses will be discussed: inhibition of gastrointestinal infections, reduction of the incidence of diarrhoea caused by rotavirus or *E. coli*, protection against dental caries formed by *S. mutants*, prevention of stomach ulcer by *Helicobacter pylori* and prevention of fish diseases. Basic information about IgY antibodies can be found elsewhere (Hatta et al., 1997, 2008; Kovacs-Nolan and Mine, 2004). This section updates on the review provided by Hatta et al. (2008).

5.1 Egg yolk antibody versus serum antibody

Using chickens as a host for antigen immunization to generate IgY has shown several advantages over the conventional animal immunized method (Hatta et al., 1997). Figure 7 shows a preparation of specific polyclonal antibodies from chickens and small animals. Table 4 shows a comparison of yolk antibodies and serum antibodies of mammals. Small mammals such as rabbit, goat and guinea pig have been used to generate a specific antibody IgG from their blood serum. In this system, blood collection is required and, therefore, mass production is difficult. On the other hand, the immunized hen egg method is cost-effective for large-scale production of antibodies without drawing blood. Furthermore, the productivity of the specific antibodies derived via viral antigens and protein antigens using hen immunization is superior to mammalian immunization.

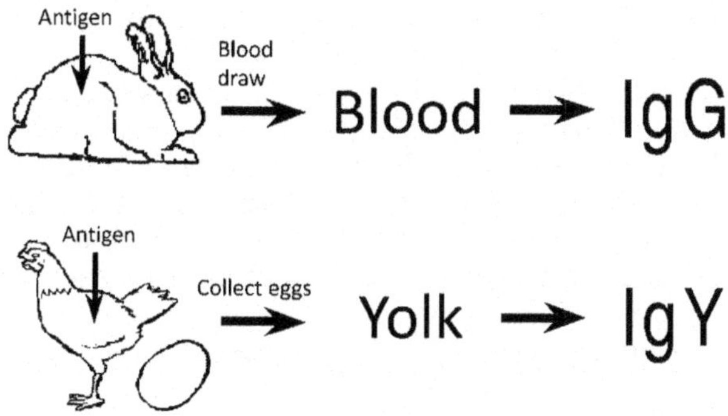

Figure 7 Preparation of specific polyclonal antibodies [Reproduced from Hatta et al. 1977. Egg yolk antibody IgY and its application. In T. Yamamoto, L. R. Juneja, H. Hatta, M. Kim (eds), *Hen Eggs, Their Basic and Applied Science*. New York: CRC Press, p. 152.

Table 4 Comparison of a specific antibody preparation

	Immunity egg way	Rabbit immunization
1. Extraction source of an antibody	Hen egg yolk	Blood of rabbit
2. A specific antibody preparation	Immunization to a chicken Egg collection and yolk separation Water-soluble protein separation Purification of the IgY	Immunization to rabbit Exsanguination Serum separation Purification of the IgG
3. Class of the antibody	Yolk includes only IgY and purification is easy	An IgG in addition to an IgA and an IgM are included in the serum.
4. Animal feeding	It is possible to increase in quantities (large-scale poultry farming)	Mass rearing is difficult
5. Immunization	An immunization is systematized for the purpose of chicken disease prevention (vaccination)	Rabbit is fixed and performed
6. Antibody manufacturing scale	Industrial-scale mass production is possible	The laboratory level

5.2 Purification of IgY

Purification of IgY from egg yolk involves labour-intensive steps to dissociate the yolk water-soluble protein from the yolk lipoprotein. Several separation methods have been developed for purification of IgY from egg yolks, including ultracentrifugation of lipoprotein and organic solvent delipidation, followed by IgY extraction (Kovacs-Nolan and Mine, 2004).

Large-scale preparation is rather difficult for conventional production of IgY due to problems related to food safety and cost constraints. Lipoprotein-coagulating agents such as polyethylene glycol, dextran sulphate sodium and polyacrylic resin have been used in the refining process for water-soluble proteins containing IgY. Earlier, alginate sodium, which is a food additive, was used to coagulate the yolk lipoprotein (Hatta et al., 1988). Later, more suitable and effective yolk-lipoprotein-coagulating agents were used due to their strong yolk lipoprotein coagulation activity. These agents include natural gums such as λ-carrageenan, κ-carrageenan, ι-carrageenan, xanthan gum and sodium alginate (Hatta et al., 1990).

5.3 Application of IgY as a diagnosis tool

IgY has been applied as a diagnostic tool for cancer, viruses, bacteria and so on. For example, IgY has been produced against the human thymidine kinase enzyme for the early diagnosis of cancer (Wu et al., 2003). IgY was also studied to screen human papilloma virus for the early detection of cervical cancer (Di Lonardo et al., 2001). IgY was used to detect the protein YRL-40, which is a marker for arthritis, cancer, atherosclerosis and liver fibrosis (De Ceuninck et al., 2001). Furthermore, various microbial pathogens have been detected using IgY, including rabies virus proteins produced in *E. coli* (Motoi et al., 2005), *E. coli* O157:H7 (Sunwoo et al., 2006) and *Loma salmonae* in Pacific salmon (Young et al., 2007).

Compared to the use of IgG from blood serum, there are several advantages to the use of egg yolk IgY for immunological assays in clinical tests; in particular, IgY can eliminate interference and false positives, which are normally experienced in the case of IgG (Larsson et al., 1991). IgG can activate the complement system in human serum, but IgY does not. In addition, IgY does not react with either rheumatoid factor (RF) or human anti-mouse IgG antibodies (HAMA), which often causes false reactions in immunological assays (Carlander et al., 1999). Finally, the Fc portion of the IgY antibody does not interact with the Fc receptor on many types of blood cells and bacteria.

Hens are excellent potential alternative animals due to their evolutional distance from mammals. IgY is more sensitive than rabbit IgG against highly conserved mammalian proteins. Evolutionary differences allow a minimum cross-reactivity (Carlander et al., 1999). For example, IgY against glutathione peroxidase (Yoshimura et al., 1991), human insulin (Lee et al., 1991) and RNA polymerase 2 (Carroll and Stollar, 1983) were successfully prepared with a high specific antibody affinity.

5.4 Application of IgY as a ligand of an immuno-adsorbent

In addition to its potential use in diagnostics, IgY is considered to be an excellent potential replacement for the polyclonal and monoclonal antibodies used in current immunoaffinity chromatography (Li-Chan, 2000). Immunoaffinity chromatography is used to purify proteins. Typically, rabbit IgG has been used as a ligand to attach an adsorbent such as cellulose or agarose. The main disadvantage of this system is that the protein is often denatured due to the acidic solution (less than pH 2) during the dissociation step. Another disadvantage is that the large-scale production of rabbit IgG is not economically feasible. Therefore, currently, immunoaffinity chromatography has been used to purify only specific proteins due to high cost.

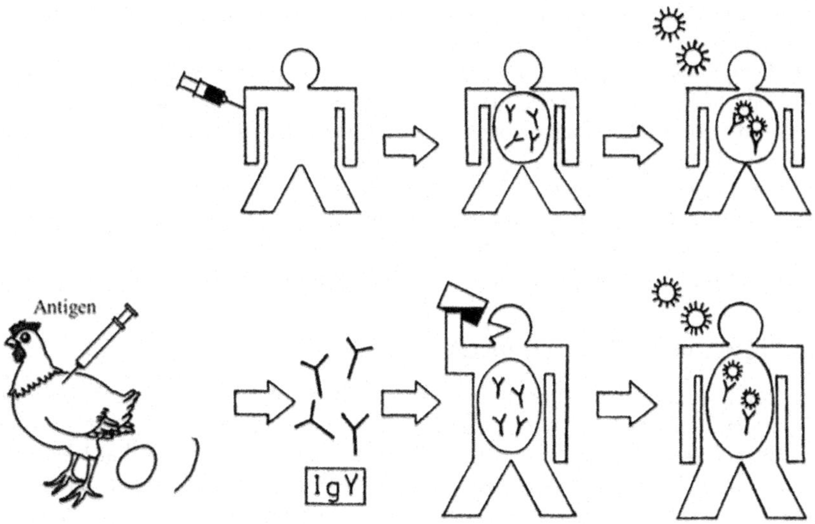

Figure 8 Differences between active immunization and passive immunization [Reproduced from Hatta et al. 2008. Bioactive components in egg yolk. In Y. Mine (ed.), *Egg Bioscience and Biotechnology*. New Jersey: John Wiley & Sons, p. 216].

Hen-derived IgY could be an effective alternative antibody as a ligand for the immunoadsorbent due to the less harsh elution condition required. Although IgY is more sensitive to low pH than IgG, it was found that an IgY immunoaffinity column was capable of retaining stability when subjected to standard affinity chromatography conditions and could be reused over 50 times without a significant decrease in binding capacity (Akita and Li-Chan, 1998).

5.5 Passive immunization

In passive immunization, pathogen-specific antibodies are obtained from animals immunized with the pathogen and are then transferred into other animals (Hatta et al., 1993a; Dias da Silva and Tambourgi, 2010; Kovacs-Nolan and Mine, 2012) (Fig. 8). The administrated specific antibody can prevent humans and animals from developing pathogen-related diseases. Anti-serum therapy has been used to treat humans with an injection of an anti-toxic serum containing anti-toxin antibodies and an antibacterial serum containing antibacterial antibodies.

5.5.1 Neutralizing toxins

Anti-serum therapies have been developed and made available for snake, scorpion, spider and jellyfish bites. A horse anti-serum is often used to neutralize toxins. However, impurities in the serum and IgG antibody of a horse anti-serum often cause inflammation as well as other side effects. The IgY antibody is regarded as a safe detoxification antibody to replace the conventional horse serum. A commercial preparation of highly purified IgY antibody from egg yolks has already been established; therefore, its use as an antibody for safer detoxification is expected. A specific IgY antibody has been directly injected into infected patients. Anti-venom IgY has a higher bioactivity than anti-venoms traditionally produced in horses (Talley and Carroll, 1990; Almeida et al., 1998). Furthermore, IgY does not bind with human alexin, avoiding the negative health effects of such binding.

Anti-enterotoxin IgY may also work as both prophylactic and therapeutic agent against lethal doses of *Staphylococcal aureus* enterotoxins (LeClaire et al., 2002). *Staphylococcus* enterotoxins are a family of bacterial super-antigens produced by *S. aureus*. They are associated with food poisoning, bacterial arthritis, toxic shock syndrome and other serious diseases (Fraser et al., 2000). Specific IgY was reported to inhibit the production of *S. aureus* enterotoxin-A and enterotoxin-B (Sugita-Konishi et al., 1996; LeClaire et al., 2002).

5.5.2 Newcastle disease virus

IgY has been studied in relation to the development of injectable vaccines to prevent various viral infections. Newcastle disease (ND) is one of the most dangerous infectious diseases in chickens. It is caused by the virulent strains of ND virus that belong to the *Avulavrus* genus of the Parmyxoiridae family (Alexander, 2000). The virus causes serious respiratory and neurological impairment and is always fatal, resulting in enormous economic losses. Therefore, preventive vaccination is currently mandated. Although ND was not previously thought to cause severe disease in humans, a recent report showed that it is possible for ND virus to cause respiratory disease in humans (Goebel et al., 2007). Box et al. (1969) described a preparation of ND-virus-specific IgY from egg yolks. The prevention of ND virus was confirmed by increased titre levels after injection and circulation of the IgY antibody.

5.5.3 Rabies virus

IgY could also serve as a possible alternative to currently available anti-rabies human or equine immunoglobulins (Fig. 9). Motoi et al. (2005) immunized layer chickens with part of

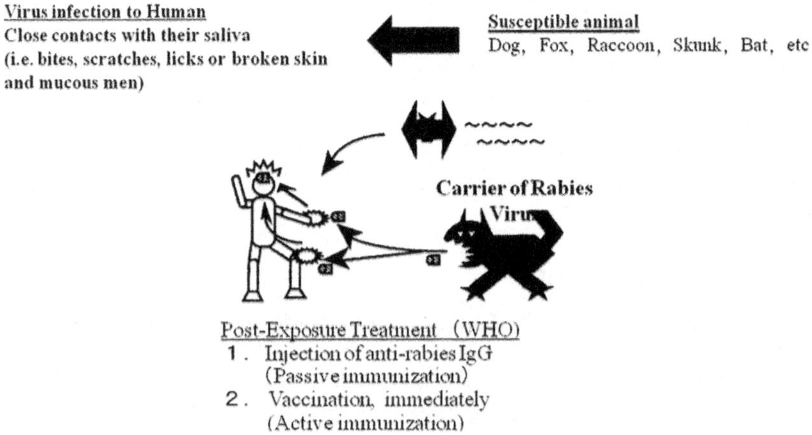

Figure 9 Rabies virus infection to human.

the G-protein of rabies virus expressed in *E. coli* to produce a more economical anti-rabies immunoglobulin for people living in developing countries. Injection of the antibody into mice infected with rabies virus reduced the mortality caused by the virus. IgY purified from the yolks of eggs laid by immunized chickens was found to bind specifically to virions as well as infected cells.

5.5.4 Xenotransplantation

Xenotransplantation, the transplantation of pigs' organs into humans, has received much attention due to the shortage of human organ donors. The barriers to xenotransplantation are considerable: hyper-acute rejection, complement fixation and rapid onset of intravascular coagulation are serious problems (Sandrin and McKenzie, 1994). IgY can be used to inhibit pig-to-human xenograft rejection.

Galactose α-1, 3-galactose (αGal) epitopes, which are found on the cell surfaces of all mammals, except humans, apes, cercopithecoidea and birds, react initially with human xenoreactive natural antibodies (XNAs), resulting in acute vascular xenograft rejection (Bouhours et al., 1988). Therefore, it is important to remove and/or block αGal antigens in xenotransplantation. Anti-αGal IgY can be used to inhibit the interactions that may cause xenograft rejection, because IgY does not bind to human complement or Fc receptors. Anti-αGal IgY has been shown to prevent human XNA binding to both porcine and rat tissues, and was also found to inhibit cell lyses of porcine cells by human serum (Fryer et al., 1999).

Another concern about xenotransplantation is the potential transmission of porcine endogenous retrovirus (PERV) into humans. There are three subtypes of PERV: PERV-A, PERV-B and PERV-C. Both PERV-A and -B can infect human cells. PERV-C does not infect human cells by itself, but can do so if combined with PERV-A. Leventhal et al. (2001) reported that anti-αGal IgY was able to significantly reduce infectivity of PERVs.

5.6 Oral administration of IgY

The oral administration of IgY can be used to prevent and treat infections caused by pathogenic bacteria and viruses in both humans and animals. This method has been well studied and has many advantages. Research on rotavirus diarrhoea, tooth decay, *H. pylori* and cultured fish infections will be addressed in the following sections.

5.6.1 Rotavirus diarrhoea

Human rotavirus (HRV) is a common infection that can be dangerous for infants because their immune systems are not fully developed. In developing countries, HRV infection causes millions of infant deaths every year due to diarrhoea (Clark et al., 1998). To prevent HRV infection, oral administration of an anti-HRV antibody is needed to restrain adherent infection of HRV in the intestinal tract (Bishop et al., 1973).

Thermal limitation and tolerance to the pH of digestive enzymes are very important factors in selecting the specific antibody to prevent HRV infection. IgY is denatured around 73.9°C and tends to lose its viral neutralizing activity, which was measured by neutralization titre of anti-HRV IgY, at pH less than 3.5 due to a structural change. However, IgY is relatively stable to trypsin and chymotrypsin digestion. Complete deactivation of pepsin occurs at pH 2.0. Antibody titre is activated and remains almost 50% active after being digested in the infant's stomach at pH 4.0 for 4 h. In a typical experiment, an anti-HRV IgY was orally introduced to an infant mouse (22.5 μg/mouse), which was then infected with HRV. HRV diarrhoea was prevented by the anti-HRV IgY (Hatta et al., 1993b).

The recombinant HRV coat protein VP8 is a cleavage product of the rotavirus spike protein VP4, which is involved in viral infectivity and neutralization of the virus. VP8 has been studied in relation to the induction of antibodies against HRV (Kovacs-Nolan et al., 2001). An anti-VP8 IgY exhibited significant neutralizing activity *in vitro* against the Wa strain of HRV. Conversely, Sarker et al. (2001) found that oral administration of anti-HRV egg yolk to children affected with HRV resulted in only a modest improvement in HRV-related symptoms. However, the authors note that further studies are required.

Rotavirus is also a problem for non-human mammals. For example, bovine rotavirus (BRV) causes neonatal calf diarrhoea, a significant cause of mortality in cattle (Lee et al., 1995). Orally administered anti-BRV IgY has been found to protect against two strains of BRV, using a mouse model (Kuroki et al., 1993). The passive protection of calves against BRV infection, using anti-BRV IgY, has also been demonstrated by Kuroki et al. (1994) and Vega et al. (2011, 2015).

5.6.2 Dental caries

Anti-cavity bacterial IgY antibodies can prevent adherent infection of dental caries bacteria on the surface of a tooth. Dental caries is caused by *Streptococcus mutans*,

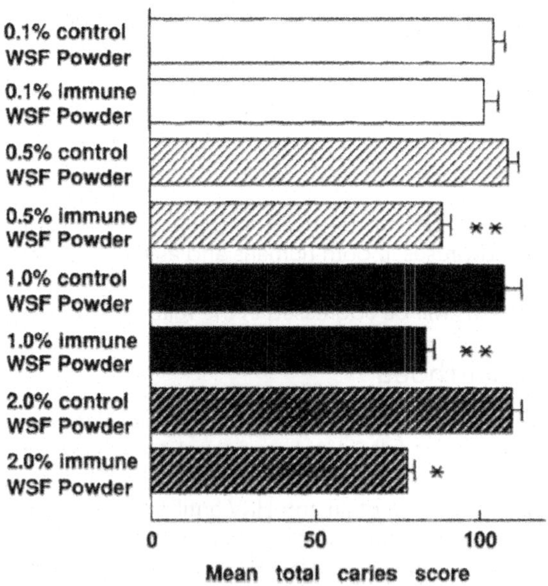

Figure 10 Mean caries scores in mandibular molars of specific pathogen-free rats infected with *S. mutans* M8148 (serotype c) and fed with diet 2000 containing WSF (water soluble fraction in egg yolk) for 58 days. Significance of difference: ** $P < 0.05$, * $P < 0.1$ [Otake et al. 1991. *J. Dent. Res.* 70(3), p. 164].
The *S. mutants* MT8148 (c) strain (c) is a serotype of the bacteria.
WSF is a water soluble fraction in egg yolk containing IgY.

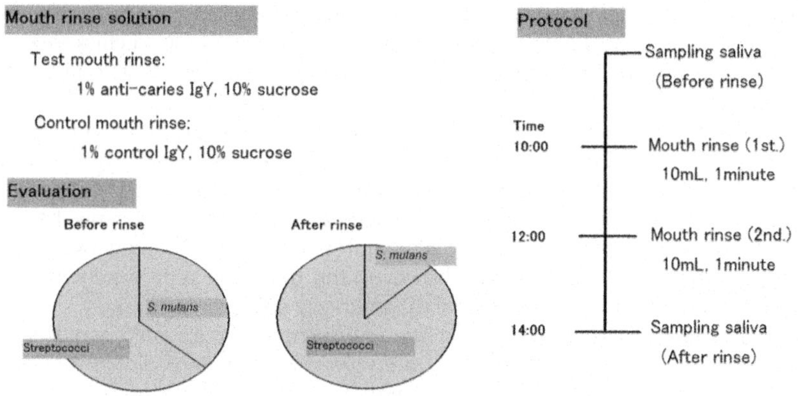

Figure 11 Differences between active immunization and passive immunization.

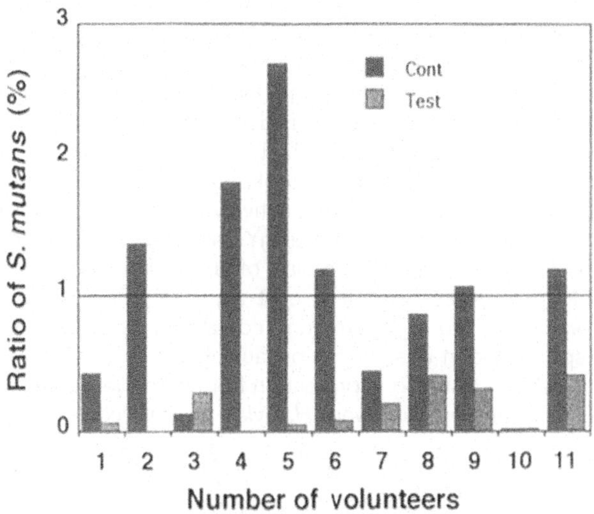

Figure 12 Changes ratio of *S. mutans* (%) in saliva of 11 volunteers after the mouth rinse test using sucrose solution containing immune IgY or control IgY [Hatta et al. 1997. *Caries Res.* 31(4), pp. 268–74].

which possesses glycosyltransferase (Hamada and Slade, 1980). This enzyme forms adhesive polysaccharides at the surface of the bacterial cell. Adhesion, also called plaque formation, helps cavity bacteria to attach firmly on the tooth surface. Generally, the lactic acid produced by lactobacillus deteriorates a tooth in the region of the plaque.

Several studies suggest that IgY may inhibit *S. mutans* accumulation to control plaque and associated oral health problems. For example, Otake et al. (1991) reported a preparation from egg yolks of immunized chicken of an IgY specific to the bacterial cell with an adhesive polysaccharide formation in human carious bacteria. They also prevented caries formation using a rat animal model as follows. Rats were infected with cavity bacteria in their teeth. In the control IgY group (derived from egg yolks from non-immunized chickens), a severe cavity was observed, while the degree of the cavity formation was reduced by the use of anti-carious bacterial IgY (Fig. 10).

Further, the effectiveness of IgY antibodies was also examined in humans by a mouth rinse test (Hatta et al., 1997). Approximately 1% of anti-carious bacterial IgY or control IgY was added to a 10% sucrose solution and a gargle test was performed (Fig. 11). Saliva was collected 4 h later and the ratio of all cavity bacteria to *Streptococcus* genus bacteria was estimated. The results revealed that, in all volunteers, the cavity bacteria were increased by gargling with the control solution; however, the cavity bacteria were diminished by gargling with the 1% anti-cavity bacterial IgY antibody solution (Fig. 12).

IgY may also interrupt the activity of *S. mutans* by decreasing the *S. mutans* glucan-binding protein B (GBP-B), which is believed to be involved in *S. mutans* biofilm development. Using a rat model, Smith et al. (2001) observed a decrease in both *S. mutans* and dental caries after treatment with anti-GBP-B IgY, compared to the control group.

5.6.3 *Helicobacter pylori*

Helicobacter pylori are gastric bacteria that selectively grow in the stomach and the intestine. Most people with *H. pylori* colonization do not show any symptoms (Dunn et al., 1997). However, infection with *H. pylori* can cause gastric ulcer disease when the bacteria penetrate the stomach's mucous lining and produce substances that neutralize stomach acids (Fig. 13). Furthermore, there is clear evidence that *H. pylori* infection significantly increases gastric cancer risk (Dunn et al., 1997; Uemura et al., 2001).

In order to prevent stomach cancer, daily sterilization by food or antibiotics is necessary. Urease, which is a bacterial surface enzyme, actively participates in adhesion of *H. pylori* to the gastric mucosa. Anti-*H. pylori* urease IgY antibody has been reported as a new sterilization method in order to prevent gluing of *H. pylori* to the gastric mucosa simply by drinking anti-*H. pylori* urease IgY (Horie et al., 2004) (Fig. 14). One per cent egg yolk IgY-urease was added to a yogurt drink containing *Lactobacillus acidophilus* and *Bifidobacterium* spp. A yogurt was chosen because live and active cultures may have an immunostimulatory effect, resulting in protection from pathogens (Matsuzaki et al., 1998). In a clinical trial, 42 volunteers who were *H. pylori* positive consumed 150 mL portions

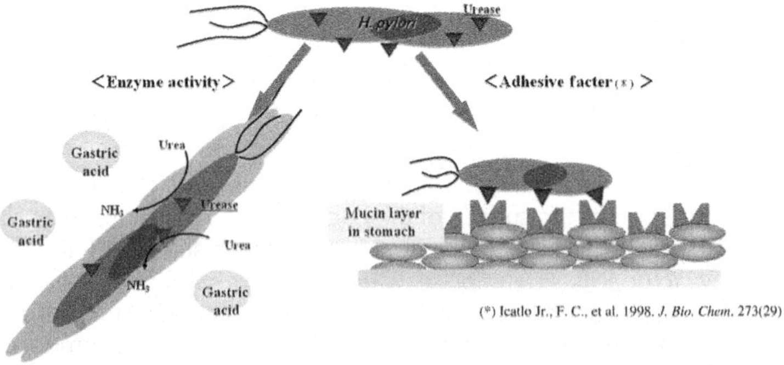

Figure 13 How *H. pylori* survives and infects in stomach.

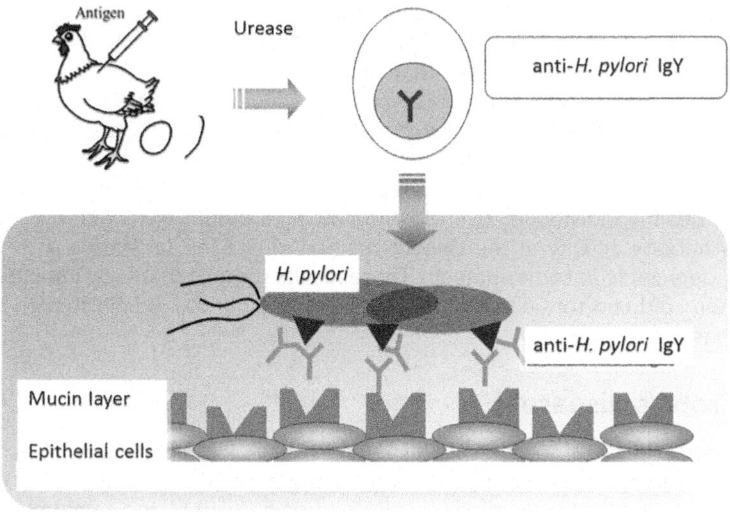

Figure 14 Competitive inhibition of *H. pylori* adherence by anti-*H-pylori* IgY [Reproduced from Hatta et al. 2008. Bioactive components in egg yolk. In Y. Mine (ed.), *Egg Bioscience and Biotechnology*. New Jersey: John Wiley & Sons, p. 222].

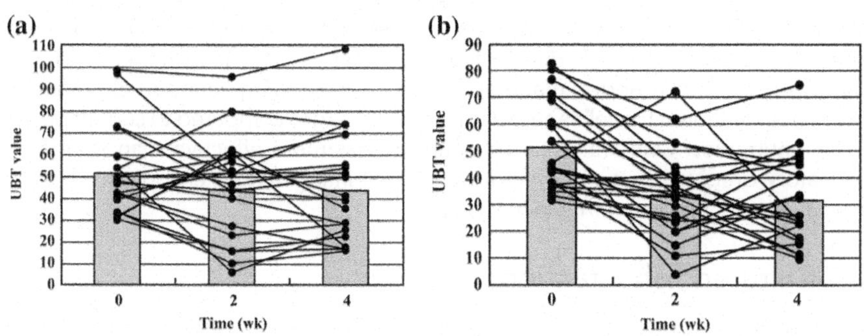

Figure 15 Effect of ingestion of (a) anti-*H. pylori* urease immunoglobulins (IgY-urease) – free drinking yogurt (control group) and (b) drinking yogurt containing egg yolk IgY-urease (test group) for 4 weeks in the presence of *H. pylori* represented by urea breath test (UBT) values. Circles indicate the values for each individual. Values of the same individual are connected by lines. The bars represent the means at 0, 2 and 4 weeks [Horie, K. et al. 2004. *J. Dairy Sci.* 87, pp. 4073–9].

three times daily for 4 weeks. The *H. pylori* antigen, as measured by a urea breath test, decreased substantially (Fig. 15).

5.6.4 Cystic fibrosis

Pseudomonas aeruginosa is the major pathogen in the cystic fibrosis (CF) lung (Stuart et al., 2010). Respiratory infection caused by *P. aeruginosa* leads to increased morbidity and mortality of CF patients. Once *P. aeruginosa* is colonized in patients, it is challenging to eliminate it due to its biofilm matrix of polysaccharide, protein and DNA. The bacteria biofilms show an increased tolerance to antibiotics.

IgY has been shown to inhibit the adhesion of *P. aeruginosa* to epithelial cells *in vitro*, but it did not inhibit its growth. Carlander et al. (2002) reported that the specific IgY against *P. aeruginosa* can interfere with the bacterial infection process and prevent colonization in patients. Carlander et al. (2002) also reported the stability of anti-*P. aeruginosa* IgY in saliva from healthy individuals, over time, following a mouth rinse with an aqueous IgY solution. Antibody activity in the saliva remained after 8 h, decreasing after 24 h, but remaining detectable in some subjects. These results suggest that local infections such as the common cold and tonsillitis could be treatable by the oral administration of specific IgY solutions.

5.6.5 Crohn's disease

Crohn's disease (CD) is a chronic inflammatory bowel disease (IBD), which involves chronic inflammation of all or part of the digestive system (Manuc et al., 2016). In general, symptoms include chronic diarrhoea, abdominal pain, fullness and rectal bleeding. The causes of IBD are not known. However, tumour necrosis factor (TNF) is implicated in the pathogenesis of IBD (Manuc et al., 2016). TNF can stimulate the release of secondary mediators including proinflammatory cytokines, hormones and small bioactive molecules, resulting in tissue inflammation and oedema.

Anti-inflammatory drugs such as sulfasalazine and dexamethasone have been applied as the standard medical care for CD. Passive immunization of anti-TNF antibody may be a better therapeutic approach due to fewer side effects and less immunosuppression. Worledge et al. (2000) reported that oral delivery of anti-TNF IgY antibodies is an effective experimental treatment for colitis and may have advantages compared to current parenteral anti-TNF antibodies. Oral anti-TNF IgY antibody was more effective than standard anti-inflammatory drugs and significantly decreased all inflammatory end points in both the acute and chronic phases of the animal model.

5.6.6 Cultured fish infections

The oral administration of specific IgY against fish pathogens would provide an alternative to antibiotic and chemotherapy treatment to prevent diseases in fish farms. IgY treatment is cost-effective and avoids slaughtering a stock of fish in order to stop the spread of diseases.

Infectious paracolo diseases caused by *Edwardsiella tarda* invade the Japanese eel, *Anguilla japonica*, from its intestinal tract. The infected eel has abscess formation either in the liver, kidney or both. Paracolo disease was well controlled until 1977 (Aoki et al., 1977). However, drug-resistant strains have appeared, causing mass mortalities in pond-reared fish. An antibiotic was used for prevention and treatment, but antibiotics are problematic due to bacterial resistance as well as residual tendency. In addition, a large volume of antibiotic must be used. Therefore, more effective control methods against paracolo disease are required.

Gutierrez et al. (1993) reported on the preparation of whole egg powder containing anti-*E. tarda* IgY antibody and its effectiveness in the prevention of the infectious paracolo diseases of infected eels. In this study, administration of *E. tarda* (10^{5-6} CFU/eel) caused severe mortalities with in 15 days of the experiment. All dead or surviving eels showed abscess formation in the liver or kidney indicating Edwardsiellosis. In contrast, the eels administered with anti-*E. tarda* IgY (whole egg powder 400 mg/eel) together with *E. tarda* (10^{5-6} CFU/eel) survived over the experimental period of 40 days without showing any symptoms of *Edwardsiellosis*

Yersinia ruckeri causes enteric red mouth disease, a systemic bacterial septicaemia of salmonid fish. *Y. ruckeri* can be persistent in carrier fish, creating a continuous source of infection as bacteria are shed in feces. Lee et al. (2000) studied the passive immunization of rainbow trout against *Y. ruckeri* using IgY and found that fish fed with anti-*Y. ruckeri* IgY showed lower mortality and infection rates. The infection rate appeared lower regardless of whether IgY was administered prior to, or after, challenge with *Y. ruckeri.*

Recently, Gan et al. (2015) reported the application of IgY against ulcer disease in koi carp (*Cyprinus carpio koi*), which is a popular and economically valuable ornamental fish. Ulcer disease is caused by an infection of atypical *Aeromonas salmonicida*, resulting in skin ulceration and disfigurement. The primary sites of the infection are the skin and gills. Therefore, immersion of fish into water containing specific IgY was suggested as a preventive treatment.

6 Other bioactive components derived from egg yolk

In this section, research on other important topics such as sialic acid and lutein will be discussed briefly. Folic acid fortified eggs will also be discussed here.

6.1 Folic acid fortified egg

Folic acid is one of the vitamin B groups. Folic acid participates in the synthesis of DNA and RNA and formation of red blood cells. A lack of folic acid causes megaloblastic anaemia and neural tube closure failure.

Neural tube defects (NTDs) are severe birth defects. The neural tube is the embryonic structure that develops into the brain and spinal cord. Neural tube closure occurs during early embryogenesis. The failure of neural tube closure is a relatively common congenital malformation. The most common and severe forms of NTDs are anencephaly and spina bifida. In case of spina bifida, the bottom of the neural tube fails to close properly. Therefore, the infant's spinal cord and nerves may be damaged. Even after a very complex operation, movement disorder and incontinence can remain. In the case of anencephaly, the upper part of the neural tube fails to form properly, resulting in increased rates of miscarriage and stillbirth.

Folic acid has been found to decrease rates of NTD. In 1998, the USFDA mandated that all enriched wheat flour was to be fortified with folic acid (140 μg/100 g). As a result, the number of NTDs decreased by 19% (Honein et al., 2001). In Canada, fortification of enriched flour with folic acid started in 1998. Among 1.9 million births, the prevalence of NTDs decreased from 1.58 per 1000 births before fortification to 0.86 per 1000 births during the full-fortification period (De Wals et al., 2007). Among developed countries, Japan is the only country with an increasing rate of NTDs. In 1993, 3.9 per 10000 births had NTDs. The number has been increasing gradually and, in 2003, 6.16 per 10000 births had NTDs (International clearing house for Birth Defects Surveillance and Research, 2006). In Japan, women of childbearing age do not get enough folic acid in their diets because of a concern over body weight and being thin (Kondo et al., 2009).

An average egg contains about 10 µg folic acid. Folic acid fortified eggs have been produced by feeding folic acid (pteroyl-L-glutamic acid) to hens. The folic acid fortified egg contains 60–80 µg folic acid. In general, folic acid is found in leafy green vegetables, beans and liver. However, due to its water solubility, folic acid in these foods is lost during cooking. For example, spinach loses 87% folic acid in water during boiling, but the folic acid fortified egg retains the acid after cooking, that is, it retains 90% of its folic acid after 20 min and 2 h of heating.

Dietary folic acid exists in monoglutamate and polyglutamate forms. Folic acid polyglutamate has to be deconjugated to the monoglutamate form by folate conjugase in order to be absorbed in the small intestine (Fig. 16). Vegetables are a source of polyglutamate folate, while the egg is a source of monoglutamate folate. In addition, about 90% of folic acid in an egg is 5-metyltetrahydrofolate, while folate in supplements is pteroylglutamic acid (Vahteristo et al., 1997) (Fig. 17). In blood, folic acid exists as 5-metyltetrahydrofolate. Egg yolk also contains high concentrations of 5-methyltetrahydrofolate. Therefore, the folic acid in the egg makes eggs a superior source of folic acid.

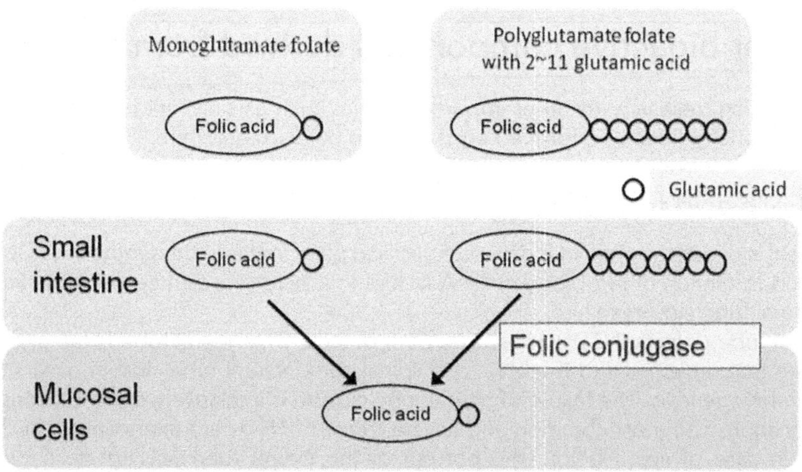

Figure 16 Absorption of monoglutamate folate and polyglutamate folate.

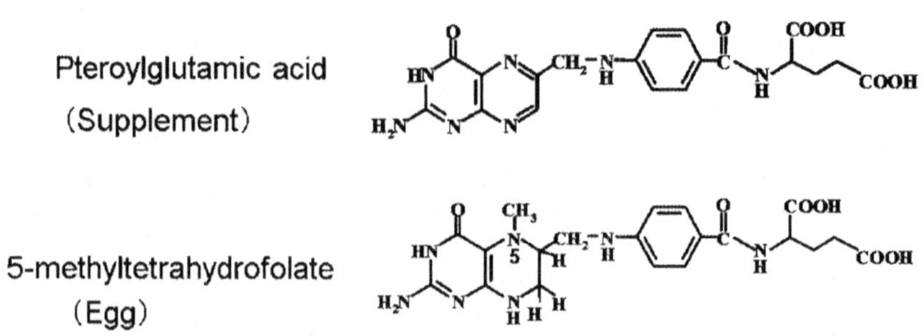

Figure 17 Folic acid and fortified egg.

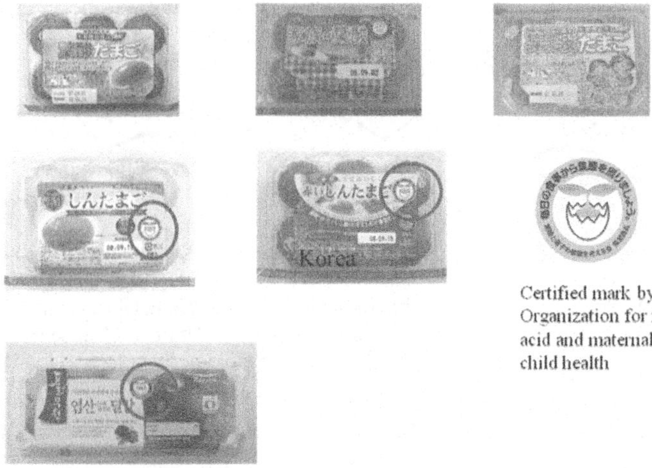

Figure 18 A variety of folic acid fortified eggs.

In 2005, Pharma Foods International Co., Ltd. (Kyoto, Japan) developed folic acid fortified eggs. Since then, such eggs have been popular in Japan. Recently, in Korea and China, folic acid fortified eggs have appeared on the market. Figure 18 shows several such eggs in Japan and Korea. In 2000, the Japanese Minister of Welfare recommended women planning pregnancy (from 1 month before pregnancy to 3 month's pregnancy) to take 400 μg of folic acid per day (Kondo et al., 2009). The upper limit is set at 1000 μg per day. However, more recently, a variety of folic acid fortified products have appeared on the market. As a result, excess intake of folic acid has become a problem.

6.2 Sialic acids

Sialic acids are a family of acidic nine-carbon monosaccharides, which are naturally located at the end of sugar chains in glycoconjugates attached to cell surfaces or soluble proteins. Sialylglycoconjugates such as gangliosides, sialyloligosaccharides and sialylglycoproteins have been reported to play important roles in biological functions: cell adhesion, viral infection and toxin neutralization (Schauer, 1982). Abnormalities of sialic acid metabolism lead to both physical and mental deterioration. Over the past few decades, sialic acid and sialyloligosaccharides have drawn the attention of the pharmaceutical and food industries due to these biological functions. In particular, the incorporation of hen egg sialylglycoconjugates into functional foods or pharmaceuticals has been investigated (Koketsu et al., 1992).

Sialic acids comprise a family of 43 naturally occurring derivatives of neuramic acid (5-amino-3,5-dideoxy-D-glycero-D-galacto-nonulsonic acid) (Schauer, 1982). Two major types of sialic acids are *N*-acetylneuraminic acid (Neu5Ac) and *N*-glycolylneuraminic acid (Neu5Gc) (Fig. 19). Neu5Ac is the biosynthetic precursor for most other types, while Neu5Gc is a modified form of Neu5Ac with the addition of single oxygen (Kelm and Schauer, 1997; Varki, 1997). The most common form of sialic acid in humans is Neu5Ac, which is found mainly in the brain and kidney. In many animals (but not chickens), Neu5Gc

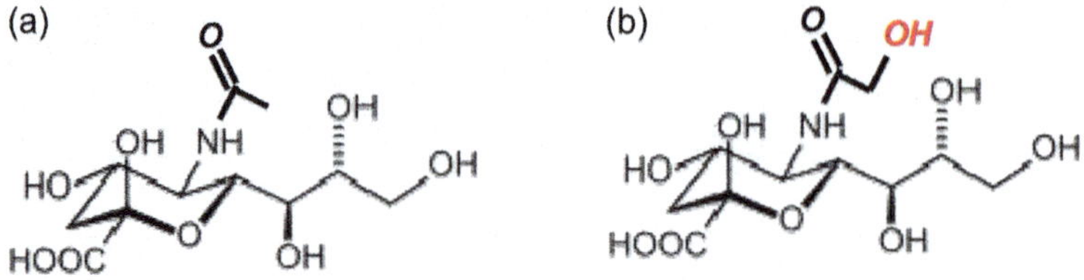

Figure 19 Illustration showing the structures of (a) *N*-acetylneuraminic acid (Neu5Ac) and (b) *N*-glycolylneuraminic acid (Neu5Gc) [Reproduced from Hatta et al. 2008. Bioactive components in egg yolk. In Y. Mine (ed.), *Egg Bioscience and Biotechnology*. New Jersey: John Wiley & Sons, p. 201].

coexists with Neu5Gc. High levels of sialic acid are found in the brain, the adrenal glands and the heart.

Hen egg is a potential source for pure sialic acid for industrial applications (Koketsu et al., 1992). Carbohydrate in egg yolk comprises 0.7–1.0%. Seventy per cent of these carbohydrates are oligosaccharides, which are mainly mannose and glucosamine, while 30% is glucose (Koketsu, 1997; Sugino et al., 1997). Sialic acids in egg yolk are bound in glycoproteins and glycolipids. The sialic acid found in hen eggs is Neu5Ac. Sialic acid is quantitatively present in several proteins of hen egg yolks: phosvitin, immunoglobulin and riboflavin-binding protein. Glycolipids in egg yolks also contain galactosyl ceramide, *N*-acetylneuraminosylgalactosylceramide, *N*-acetylneuraminosyl lactosylceramide and di-4*N*-acetylneuraminosyl lactosylceramide with reasonable yields (Li et al., 1978).

The content of sialic acid is estimated to be about 0.95 gram per kilogram of fresh hen egg yolk. Several sialyloligosaccharides in egg yolk have been isolated and their structures have been determined (Juneja, 1997). Several methods have been reported to isolate sialic acid with 97% purity and *N*-acetylneuraminic acid with 100% purity from hen egg yolks. The details can be found in Juneja (1997).

Oligosaccharides with sialic acids at the terminal end have shown unique functionality. They work as a receptor for viruses such as Sendai virus (Holmgren et al., 1980), influenza virus (von Itzstein et al., 1993) and corona virus (Schultz and Herrler, 1992). Sialic acid derivatives can be synthesized by chemical and enzymatic reactions. For example, the inhibitor of sialidase for influenza virus, 4-guanidino-NeuAc2en, has been sold commercially as Relenza®. Furthermore, sialylglycoconjugates on the cell surface serve as a receptor for microorganisms, toxins, inflammatory response and cancer metastasis (Fishman and Brady, 1976; Paulson et al., 1984; Smit et al., 1984; Kannagi et al., 1988; Morschhäuser et al., 1990; Phillips et al., 1990).

Sialyloligosaccharides have been applied to rotavirus, which is a well-known major pathogen of infectious gastroenteritis, diarrhoea and vomiting in infants (Tabassum et al., 1994). Clinical studies of vaccines for the prevention of rotavirus infections have been unsuccessful because of the difficulty in inducing a specific antibody in the intestinal tract of infants who have not yet developed general immunity (De Mol et al., 1986). The inhibitory effect of the sialyloligosaccharide fraction on rotavirus *in vivo* was investigated in mice (Koketsu, 1997). The group administered the sialyloligosaccharide fraction, which showed a significantly lower incidence of rotavirus diarrhoea. The incidence of diarrhoea was inspected 1, 3 and 5 days after oral administration of 2.5 mg/mouse sailyloligosaccharides and was found to be lowered by 24% and 43% on the 3rd day and the 5th day, respectively, compared to the control group.

6.3 Lutein

Lutein and zeaxanthin belong to the xanthophyll carotenoid group and are accumulated in the macular region of the human retina. In age-related macular degenerative disease, the small central portion of the retina, known as the macular region, deteriorates during the ageing process, resulting in progressive and irreversible loss of central region vision. It is reported that age-related macular degeneration may be prevented by eating supplements and/or food that are rich in lutein and zeaxanthin (Goodrow et al., 2006). Furthermore, in another study, lutein and other carotenoids were reported to have antioxidant activity, resulting in reduction of cancers (Slattery et al., 2000; Bertone et al., 2001; Holick et al., 2002; Sato et al., 2002).

Egg yolk is a richer source of lutein and zeaxanthin than other common dietary sources of carotenoids (Schaeffer et al, 1988; Rasmussen et al., 2012). Lutein and zeaxanthin are dispersed in the matrix along with other fat-soluble nutrients such as vitamins. Therefore, lutein and zeaxanthin of eggs are considered to be highly available than other common dietary sources of carotenoids. Furthermore, lutein and zeaxanthin are reported to be preferentially deposited in the retina, while β-carotene has limited accumulation in the retina (Leeson and Caston, 2004).

Egg yolk contains 0.3–0.5 mg total xanthophylls, with half present as lutein. Lutein is found mainly in plasma LDL due to its polarity. Even though a very small amount of lutein and zeaxanthin is found in eggs, they seem to be readily absorbed in the body. A recent study shows a three-to-eightfold increase in carotenoid absorption for salad containing three eggs, compared to the salads without eggs (Kim et al., 2015). Carotenoids in the salads fed to study subjects included β-carotene, α-carotene, lycopene, lutein and zeaxanthin. The lipid content of the eggs acted as the agent to enhance carotenoid absorption. According to van der Made et al. (2014), people who had early signs of macular degeneration disease had increased plasma lutein after drinking buttermilk containing lutein-enriched egg yolk for a year.

7 Conclusion

For humans, eggs have been consumed around the world since ancient times. For chicks, an egg is a capsule of life, wherein all ingredients are contained in sufficient amounts to develop an embryo and give rise to the birth of a chick. Weighing only about 60 g, an egg is a powerhouse that has various nutritional and physiological functions. Egg yolk provides not only essential energy for life but also necessary nutrients – essential amino acids, fatty acids, cholesterol, vitamins and minerals – to maintain and promote health. In fact, one reason that eggs are of interest is that they contain all nutrients necessary for human health, except dietary fibre and vitamin C.

To promote health and prevent diseases, much research has explored the nutritional and physiological functions of egg yolk. In this chapter, we focused on the bioactive components derived from egg yolk, issues related to egg cholesterol and recent applications of IgY.

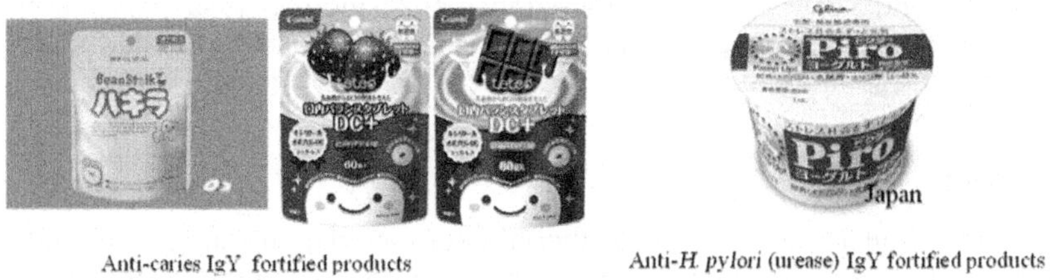

Figure 20 Commercial products fortified with egg yolk containing specific IgY.

Research on other areas such as sialic acids and lutein were also covered. Recently, some applications of egg yolk bioactive peptides have reached a market-ready stage. Useful applications for IgY as an immunological tool have been studied in the fields of diagnosis as well as in pure research. The applicability of IgY to prevent infectious diseases has received much attention due to the potential to produce specific antibodies in egg yolk. Several IgY-fortified products are already on the market in Japan, such as IgY-fortified chocolate for the prevention of dental caries and IgY-fortified yogurt for the prevention of *H. pylori* infection (Fig. 20).

Eggs are very familiar in daily life and many assumptions have been made about their properties. Egg research has investigated various topics and provided strong scientific knowledge about the nutritional, physical and physiological properties of eggs. As a result, we now know much about the nutritional and health benefits of eating eggs. In the future, we hope that a new research trend will emerge in which the egg is considered more holistically. The egg is a capsule of life. Although understanding its components is beneficial, research is needed into the functioning and benefits of the whole egg.

8 Where to look for further information

Readers can explore further information on nutritional and physiological functions of egg yolk components in the following sources:

1. General information on nutritional and physiological functions of eggs

   Egg Nutrition Center (http://www.eggnutritioncenter.org), the Science and Nutrition Education Division of the American Egg Board, provides following information in the following areas: cardiometabolic health, egg allergies, eggs across the lifespan, nutrients in eggs, nutritious dietary patterns, physical performance, and weight management and satiety.

2. Research and development on IgY

   Immunology Research Institute in Gifu (IRIG), Japan, has developed a variety of specific IgY against different pathogens and their current focus is practical application of IgY (http://www.igy-research.com/English/irig.html).

3. Preparation and application of monoclonal antibody IgY

The following three papers published by Dr. Matsuda's research group are suggested:

a) Miyamoto, K., Shimamoto, T., Aosasa, M., Kimura, S., Nakamura, N., Okubo, Y., Yokoyama, T., Horiuchi, H., Furusawa, S. and Matsuda, H. 2007. Development of recombinant chicken IgY from single chain fragment of variable region for diagnosis of BSE. *Biologicals* 35(1), pp. 31–4.
b) Miyamoto, K., Kimura, S., Nakamura, N., Yokoyama, T., Horiuchi, H., Furusawa, S. and Matsuda, H. 2007. Chicken antibody against a restrictive epitope of prion protein distinguishes normal and abnormal prion proteins. *Biologicals* 35(4), pp. 303–8.
c) Sato, Y., Nishimichi, N., Nakano, A., Takikawa, K., Inoue, N., Matsuda, H. and Sawamura, T. 2008. Determination of LOX-1-ligand activity in mouse plasma with a chicken monoclonal antibody for ApoB. *Atherosclerosis* 200(2), pp. 303–9.

4. Information on transgenic chickens and avian development, growth and differentiation

Japanese Society of Development Biologist published a special issue of 'Development, Growth & Differentiation', edited by R. Ladher, T. Suzuki and H. Nakamura in January 2013. Review articles on avian model system were included in this issue (http://onlinelibrary.wiley.com/doi/10.1111/dgd.2013.55.issue-1/issuetoc).

5. Proteomics

Mann, K. 2008. Proteomic analysis of the chicken egg vitelline membrane. *Proteomics* 8(11), pp. 2322–32.

6. Compositions of eggs and their physiological functions

Anton, M., Nau, F. and Nys, Y. 2006. Review articles: Bioactive egg component and their potential uses. *World's Poult. Sci. J.* 62(3), pp. 429–38.

9 Abbreviations

AA	Arachidonic acid
ACE	Angiotensin converting enzyme
BP	Bone pep
BRV	Bovine rotavirus
CD	Crohn's disease
CF	Cystic fibrosis
DHA	Docosahexaenoic acid
DPPH	2,2-Dephenyl-1-picrylhydrazyl
DPP-IV	Dipeptidyl-peptidase-IV
EPA	Eicosapentaenoic acid
FA	Fatty acid
HAMA	Human anti-mouse IgG antibody
HDL	High-density lipoprotein
HGP	Hair growth peptide
HLB	Lipophilic balance
HRV	Human rotavirus

IBD	Inflammatory bowel disease
IgY	Immunoglobulin Y
LA	Linoleic acid
LDL	Low-density lipoproteins
LNA	α-Linolenic acid
MUFA	Monounsaturated fatty acid
NTD	Neural tube defect
O/W	Oil in water
PC	Phosphatidylcholine
PE	Phosphatidylethanolamine
PERV	Porcine endogenous retrovirus
PI	Phosphatidylinositol
PLA1	Phospholipase A1
PLA2	Phospholipase A2
PLB	Phospholipase B
PLC	Phospholipase C
PLD	Phospholipase D
PS	Phosphatidylserines
PUFA	Polyunsaturated fatty acid
SFA	Saturated fatty acid
TNF	Tumour necrosis factor
VEGF	Vascular endothelial growth factor
VLDL	Very low-density lipoprotein
XNA	Xenoreactive natural antibody
αGal	α-1, 3-Galactose
ω-3 FA	Omega-3 fatty acid
ω-3 PUFA	Omega-3 polyunsaturated fatty acid
ω-6 FA	Omega-6 fatty acid

10 References

Ajinomoto. Available at: http://www.ajinomoto.co.jp/products/ [Accessed 26 April 2016].

Akita, E. M. and Li-Chan, E. C. 1998. Isolation of bovine immunoglobulin G subclasses from milk, colostrum, and whey using immobilized egg yolk antibodies. *J. Dairy Sci.* 81(1), pp. 54–63.

Alexander, D. J. 2000. Newcastle disease and other avian paramyxoviruses. *Rev. Sci. Tech.* 19(2), pp. 443–62.

Almeida, C. M., Kanashiro, M. M., Rangel Filho, F. B., Mata, M. F., Kipnis, T. L. and da Silva, W. D. 1998. Development of snake antivenom antibodies in chickens and the purification from yolk. *Vet. Rec.* 143(21), pp. 579–84.

Amoresano, A., Brancaccio, A., Andolfo, A., Perduca, M., Monaco, H. L. and Marino, G. 1999. The carbohydrates of the isoforms of three avian riboflavin-binding proteins. *Eur. J. Biochem.* 263(3), pp. 849–58.

Anton, M. 2013. Egg yolk: structures, functionalities and processes. *J. Sci. Food Agric.* 93(12), pp. 2871–80.

Aoki, T., Arai, T. and Egusa, S. 1977. Detection of R plasmids in naturally occurring fish pathogenic bacteria, *Edwardsiella tarda*. *Microbiol. Immunol.* 21, pp. 77–83.

Baker, C. 1989. Lecithin in cosmetics. In B. F. Szuhaj (ed.), *Lecithins: Sources, Manufacture and Uses.* Champaign, IL: American Oil Chemists Society, pp. 253–60.

Bang, H. O., Dyerberg, J. and Nielsen, A. E. 1971. Plasma lipid and lipoprotein pattern in Greenlandic west-coast eskimos. *Lancet* 297(7710), pp. 1143–46.

Beasley, C. L., Honer, W. G., Bergmann, K., Falkai, P., Lütjohann, D. and Bayer, T. A. 2005. Reductions in cholesterol and synaptic markers in association cortex in mood disorders. *Bipolar Disord.* 7(5), pp. 449–55.

Bertone, E. R., Hankinson, S. E., Newcomb, P. A., Rosner, B., Willet, W. C., Stampfer, M. J. and Egan, K. M. 2001. A population-based case–control study of carotenoid and vitamin A intake and ovarian cancer (United States). *Cancer Causes and Control* 12(1), pp. 83–90.

Bishop, R. F., Davidson, G. P., Holmes, I. H. and Ruck, B. J. 1973. Virus particles in epithelial cells of duodenal mucosa from children with acute non-bacterial gastroenteritis. *Lancet* 2(7841), pp. 1281–3.

Bouhours, J. F., Richard, C., Ruvoen, N., Barreau, N., Naulet, J. and Bouhours, D. 1998. Characterization of a polyclonal anti-Galalpha1–3Gal antibody from chicken. *Glycoconj. J.* 15(1), pp. 93–9.

Box, P. G., Stedman, R. A. and Singleton, L. 1969. Newcastle disease. I. The use of egg yolk dried for passive immunization of chickens. *J. Comp. Pathol.* 79(4), pp. 495–506.

Brownawell, A. M. and Falk, M. C. 2010. Cholesterol: where science and public health policy interest. *Nutr. Rev.* 68(6), pp. 355–64.

Carlander, D., Kollberg, H., Wejaker, P. E. and Larsson, A. 2002. Prevention of chronic *Pseudomonas aeruginosa* colonization by gargling with specific antibodies: a preliminary report. In J. S. Sim, S. Nakai and W. Guenter (eds), *Egg Nutrition and Biotechnology*. New York: CABI Publishing, pp. 371–4.

Carlander, D., Stålberg, J. and Larsson, A. 1999. Chicken antibodies: a clinical chemistry perspective. *Ups. J. Med. Sci.* 104(3), pp. 179–89.

Carroll, S. B. and Stollar, B. D. 1983. Antibodies to calf thymus RNA polymerase II form egg yolks of immunized hens. *J. Biol. Chem.* 258(1), pp. 24–6.

Caston, L. and Leeson, S. 1990. Research note: Dietary flaxseed and egg composition. *Poult. Sci.* 69(9), pp. 1617–20.

Cherian, G. and Sim, J. S. 1996. Changes in the breast milk fatty acids and plasma lipids of nursing mothers following consumption of n-3 polyunsaturated fatty acid enriched eggs. *Nutrition* 12(1), pp. 8–12.

Cherian, G., Wolfe, F. W. and Sim, J. S. 1996a. Dietary oils with added tocopherols: effects on egg or tissue tocopherols, fatty acids and oxidative stability. *Poult. Sci.* 75(3), pp. 423–31.

Cherian, G., Wolfe, W. and Sim, J. S. 1996b. Feeding dietary oils with tocopherols: Effects on internal qualities of eggs during storage. *J. Food Sci.* 61(1), pp. 15–8.

Clark, H. F., Glass, R. I. and Offit, P. A. 1998. Rotavirus vaccines. In S. A. Plotkin and W. A. Orenstein (eds), *Vaccines*, 3rd edn. Philadelphia: W.B. Saunders, pp. 987–1005.

Cook, W. H. and Martin, W. V. 1969. Egg lipoproteins. In E. Tria and A. M. Scanu (eds), *Structural and Functional Aspects of Lipoproteins in Living Systems*. London and New York: Academic Press, pp. 579–615.

Daimer, K. and Kulozik, U. 2008. Impact of a treatment with phospholipase A2 on the physiochemical properties of hen egg yolk. *J. Agri. Food Chem.* 56(11), pp. 4172–80.

Daimer, K. and Kulozik, U. 2009. Oil-in-water emulsion properties of egg yolk: Effect of enzymatic modification by phospholipase A_2. *Food Hydrocol.* 23(5), pp. 1366–73.

Damiano, M. G., Mutharasa, R. K., Tripathy, S., McMahon, K. M. and Thaxton, C. S. 2013. Templated high density lipoprotein nanoparticles as potential therapies and for molecular delivery. *Adv. Drug Deliv. Rev.* 65(5), pp. 649–62.

De Ceuninck, F., Pastoureau, P., Agnellet, S., Bonnet, J. and Vanhoutte, P. M. 2001. Development of an enzyme-linked immunoassay for the quantification of YKL-40 (cartilage gp-39) in guinea pig serum using hen egg yolk antibodies. *J. Immunol. Methods* 252(1–2), pp. 153–61.

De Maria, L., Vind, J., Oxenbøll, K. M., Svendsen, A. and Patkar, S. 2007. Phospholipases and their industrial applications. *Appl. Microbiol. Biotechnol.* 74(2), pp. 290–300.

De Mol, P., Zissis, G., Butzler, J. P., Mutwewingabo, A. and André, F. E. 1986. Failure of live, attenuated oral rotavirus vaccine. *Lancet* 2(8498), p. 108.

De Wals, P., Tairou, F., Van Allen, M. I., Uh, S. H., Lowry, R. B., Sibbals, B., Evans, J. A.,Van den Hof, M. C., Zimmer, P., Crowley, M., Fernandez, B., Lee, N. S. and Niyonsenga, T. 2007. Reduction in neural-tube defects after folic acid fortification in Canada. *N. Engl. J. Med.* 357(2), pp. 135–42.

Di Lonardo, A. D., Marcante, M. L., Poggiali, F., Hamsøíkovà, E. and Venuti, A. 2001. Egg yolk antibodies against the E7 oncogenic protein of human papillomavirus type 16. *Arch Virol.* 146(1), pp. 117–25.

Dias da Silva, W. and Tambourgi, D. V. 2010. IgY: a promising antibody for use in immunodiagnostic and in immunotherapy. *Vet. Immunol. Immunopathol.* 135(3–4), pp. 173–80.

Dunn, B. E., Cohen, H. and Blaser, M. J. 1997. *Helicobacter pylori. Clin. Microbiol. Rev.* 10(4), pp. 720–41.

Dyerberg, J. and Bang, H. O. 1979. Haemostatic function and platelet polyunsaturated fatty acid in Eskimos. *Lancet* 2(8140), pp. 433–5.

Egg museum. Available at: http://homepage3.nifty.com/takakis2/ [Accessed 20 April 2016] [In Japanese].

Eterradossi, N., Toquin, D., Abbassi, H., Rivallan, G., Cotte, J. P. and Guittet, M. 1997. Passive protection of specific pathogen free chicks against infectious bursal disease by in-ovo injection of semi-purified egg-yolk antiviral immunoglobulins. *Zentralbl Veterinarmed B.* 44(6), pp. 371–83.

Evans, A. J. and Burley, R. W. 1987. Proteolysis of apoprotein B during the transfer of very low density lipoprotein form hens' blood to egg yolk. *J. Biol. Chem.* 262(2), pp. 501–4.

Farinazzo, A., Restuccia, U., Bachi, A., Guerrier, L., Fortis, F., Boschetti, E., Fasoli, E., Citterio, A. and Righetti, P. G. 2009. Chicken egg yolk cytoplasmic proteome, mined via combinational peptide ligand libraries. *J. Chromatogr.* 1216(8), pp. 1241–52.

Fernandez, M. L. and Webb, D. 2008. The LDL to HDL cholesterol ratio as a valuable tool to evaluate coronary heart disease risk. *J. Am. Coll. Nutr.* 27, pp. 1–5.

Fishman, P. H. and Brady, R. O. 1976. Biosynthesis and function of gangliosides. *Science* 194(4268), pp. 906–15.

Food and Agricultural Organization of the United Nations. 2009. Food Based dietary guidelines by country. Available at: http://www.fao.org/ag/humannutrition/nutritioneducation/fbdg/en/

Fraser, J., Arcus, V., Kong, P., Baker, E. and Proft, T. 2000. Superantigens – powerful modifiers of the immune system. *Mol. Med. Today* 6(3), pp. 125–32.

Fryer, J. P., Firca, J., Leventhal, J. R., Blondin, B., Malcolm, A., Ivancic, D., Gandhi, R., Shah, A., Pao, W., Abecassis, M., Kaufman, D., Stuart, F. and Anderson, B. 1999. IgY antiporcine endothelial cell antibodies effectively block human antiprocine xenoantibody binding. *Xenotransplantation* 6(2), pp. 98–109.

Fuller, N. R., Sainsbury, A., Caterson, I. D. and Markovic, T. P. 2015. Egg consumption and human cardio-metabolic health in people with and without diabetes. *Nutrients* 7(9), pp. 7399–420.

Gan, H., He, H., Sato, A., Hatta, H., Nakao, M. and Somamoto, T. 2015. Ulcer disease prophylaxis in koi carp by bath immersion with chicken egg yolk containing anti-Aeromonas salmonicida IgY. *Res. Vet. Sci.* 99, pp. 82–6.

Goebel, S. J., Taylor, J., Barr, B. C., Kiehin, T. E., Castro-Malaspina, H. R., Hedvat, C. V., Rush-Wilson, K. A., Kelly, C. D., and Davis, S. W., Samsonoff, W. A., Hurst, K. R., Behr, M. J. and Maters, P. S. 2007. Isolation of avian parmyxovirus 1 from a patient with a lethal case of pneumonia. *J. Virol.* 81(22), pp. 12709–14.

Goodrow, E. F., Wilson, T. A., Houde, S. C., Vishwanathan, R., Scollin, P. A., Handelman, G. and Nicolosi, R. J. 2006. Consumption of one egg per day increases serum lutein and zeaxanthin concentrations in older adults without altering serum lipid and lipoprotein cholesterol concentrations. *J. Nutr.* 136(10), pp. 2519–24.

Grezelczyk, A. and Gendaszewska-Darmach, E. 2013. Novel bioactive glycerol-based lysophopholipids: new data – new insight into their function. *Bichimie.* 95(4), pp. 667–79.

Groche, D., Rashkovetsky, L. G., Falchuk, K. H. and Auld, D. S. 2000. Subunits composition of the zinc proteins alpha- and beta-lipovitellin from chicken. *J. Protein Chem.* 19(5), pp. 379–87.

Gutierrez, M. A., Miyazaki, T., Hatta, H. and Kim, M. 1993. Protective properties of egg yolk IgY containing anti-*Edwardshiella tarda* antibody against paracolo disease in the Japanese eel, *Anguilla japonica* Temmick & Schlegel. *J. Fish Dis.* 16(2), pp. 113–22.

Hamada, S. and Slade, H. D. 1980. Biology, immunology, and carcinogenicity of Streptococcus mutans. *Microbiol. Rev.* 44(2), pp. 331–84.

Harada, S., Nakamura, T. and Yamamura, H. 2016. The development of egg yolk-derived hair growth peptides and prospects for hair growth treatment. *Fragrance J.* 1, pp. 55–60.

Hargis, P. S., Van Elswyk, M. E. and Hargis, B. M. 1991. Dietary modification of yolk lipid with menhaden oil. *Poult. Sci.*, 70(4), pp. 874–83.

Hatta, H. 2011. Egg cholesterol and risk of cardiovascular diseases. *The NIwatori-no-kenkyu* 86(5), pp. 40–2 [In Japanese].

Hatta, H., Hagi, T. and Hiranok, K. 1997. Chemical and physicochemical properties of hen eggs and their application in foods. In T. Yamamoto, L. R. Juneja, H. Hatta and M. Kim (eds), *Hen Eggs: Their Basic and Applied Science*. New York: CRC press, pp. 117–33.

Hatta, H., Kapoor, M. P. and Juneja, L. R. 2008. Bioactive components in egg yolk. In Y. Mine (ed.), *Egg Bioscience and Biotechnology*. New Jersey: John Willey & Sons, Inc., pp. 185–237.

Hatta, H., Kim, M. and Yamamoto, T. 1990. A novel isolation method for hen egg yolk antibody, 'IgY'. *Agric. Biol. Chem.* 54(10), pp. 2531–5.

Hatta, H., Ozeki, M. and Tsuda, K. 1997. Egg yolk antibody IgG and its application. In T. Yamamoto, L. R. Juneja, H. Hatta and M. Kim (eds), *Hen Eggs: Their Basic and Applied Science*. New York: CRC Press, pp. 151–78.

Hatta, H., Sim, J. S. and Nakai, S. 1988. Separation of phospholipids from egg yolk and recovery of water-soluble proteins. *J. Food Sci.* 53(2), pp. 425–7.

Hatta, H., Tsuda, K., Akachi, S., Kim, M., Yamamoto, T. and Ebina, T. 1993b. Oral passive immunization effect of anti-human rotavirus IgY and its behavior against proteolytic enzymes. *Biosci. Biotechnol. Biochem.* 57(7), pp. 1077–81.

Hatta, H., Tsuda, K., Ozeki, M., Kim, M., Yamamoto, T., Otake, S., Hirosawa, M., Katz, J., Childers, N. K. and Michalek, S. M. 1997. Passive immunization against dental plaque formation in humans: Effect of a mouth rinse containing egg yolk antibodies (IgY) specific to *Streptococcus mutans. Caries Res.* 31(4), pp. 268–74.

Health Canada. Available at: http://www.hc-sc.gc.ca/fn-an/food-guide-aliment/context/fg_history-histoire_ga-eng.php [Accessed 17 April 2016].

Herron, K. L., Vega-Lopez, S., Conde, K., Ramjiganesh, T., Shachter, N. S. and Fernandez, M. L. 2003. Men classified as hypo- or hyperresponders to dietary cholesterol feeding exhibit differences in lipoprotein metabolism. *J. Nutr.* 133(4), pp. 1036–42.

Holick, C. N., Michaud, D. S., Stolzenberg-Solomon, R., Mayne, S. T., Pietine, P., Tayler, P. R., Virtamo, J. and Albanes, D. 2002. Dietary carotenoids, serum beta-carotene, and retinol and risk of lung cancer in the alpha-tocopherol, beta-carotene cohort study. *Am. J. Epidemiol.* 156(6), pp. 536–47.

Holmgren, J., Elwing, H., Fredman, P., Strannegård, O. and Svennerholm, L. 1980. Gangliosides as receptors for bacterial toxins and Sendai virus. *Adv. Exp. Med. Biol.* 125, pp. 453–70.

Honein, M. A., Paulozzi, L. J., Mathews, T. J., Erickson, D. and Wong, L. Y. 2001. Impact of folic acid fortification of the US food supply on the occurrence of neural tube defects. *JAMA* 285(23), pp. 2981–6.

Horie, K., Horie, N., Abdou, A. M., Yang, J. O., Yun, S. S., Chun, H. N., Park, C. K., Kim, M. and Hatta, H. 2004. Suppressive effect of functional drinking yogurt containing specific egg yolk immunoglobulin on Helicobacter pylori in humans. *J. Dairy Sci.* 87(12), pp. 4073–9.

International Clearinghouse for Birth Defects Surveillance and Research. 2006. Annual Report of 2006 with Data for 2004. Available at: http://http://www.icbdsr.org/filebank/documents/Report2006.pdf [Accessed 19 June 2016].

International Egg Commission. 2014. Annual Review 2014.

Ikemori, Y., Ohta, M., Umeda, K., Ichatlo, F. C. Jr., Kuroki, M., Yokoyama, H. and Kodama, Y. 1997. Passive protection of neonatal calves against bovine coronavirus-induced diarrhea by administration of egg yolk or colostrum antibody powder. *Vet. Microbiol.* 58(2–4), pp. 105–11.

Inglis, A. S., Strike, P. M. and Burley, R. W. 1982. Two low-molecular weight apoprotein (apovitellenins I and II) from a lipoprotein of goose's egg yolk: a comparison with related species. *Aust. J. Biol. Sci.* 35(3), pp. 263–9.

Ishikawa, S., Asano, T., Takenoshita, S., Nozawa, Y., Arihara, K. and Itoh, M. 2009. Egg yolk proteins suppress azoxymethane-induced aberrant crypt foci formation and cell proliferation in the colon of rats. *Ntur. Res.* 29(1), pp. 64–9.

Ishikawa, S., Yano, Y., Arihara, K. and Itoh, M. 2004. Egg yolk phosvitin inhibits hydroxyl radical formation from the fenton reaction. *Biosci. Biotechnol. Biochem.* 68(6), pp. 1324–31.

Ito, Y. and Fujii, T. 1962. Chemical compositions of the egg yolk lipoproteins. *J. Biochem.* 52(3), pp. 221–2.

Itoh T., Abe, Y. and Adachi, S. 1983. Comparative studies on the α and β-phosvitins from hen's egg yolk. *J. Food Sci.*, 48(6), pp. 1755–57.

Jiang, B. and Mine, Y. 2000. Preparation of novel functional oligophosphospeptides from hen egg yolk phosvitin. *J. Agric. Food Chem.* 48(4), pp. 990–4.

Jiang, B. and Mine, Y. 2001. Phosphopeptides derived from hen egg yolk phosvitin: Effect of molecular size on the calcium-binding properties. *Biosci. Biotechnol. Biochem.* 65, pp. 1187–90.

Jiang, Z., Ahn, D. U., Ladner, L. and Sim, J. S. 1992. Influence of feeding full-fat flax and sunflower seeds on internal and sensory qualities of eggs. *Poult. Sci.* 71, pp. 378–82.

Jiang, Z. and Sim, J. S. 1992. Effects of dietary n-3 fatty acid-enriched chicken eggs on plasma and tissue cholesterol and fatty acid compositions of rats. *Lipids* 27(4), pp. 279–84.

Jiang, Z. and Sim, J. S. 1993. Consumption of n-3 polyunsaturated fatty acid-enriched eggs and changes in plasma lipids of human subjects. *Nutrition* 9(6), pp. 513–8.

Jin, Y.-G., Huang, D., Ding, T., Ma, M.-H., Oh, D.-H. 2013. Effect of phospholipase A1 on the physiochemical and functional properties of hen's egg yolk, plasma and granules. *J. Food Biochem.* 37(1), pp. 70–9.

Juneja, L. R. 1997. Egg yolk lipids. In T. Yamamoto, L. R. Juneja, H. Hatta and M. Kim (eds), *Hen Eggs: Their Basic and Applied Sciences*. New York: CRC Press, pp. 73–98.

Kannagi, R., Kitahara, A., Itai, S., Zenita, K., Shigeta, K., Tachikawa, T., Noda, A., Hirano, H., Abe, M. and Shin, S., et al. 1988. Quantitative and qualitative characterization of human cancer-associated serum glycoprotein antigens expressing epitopes consisting of sialyl or sialyl-fucosyl type 1 chain. *Cancer Res.* 48(13), pp. 3856–63.

Kelm, S. and Schauer, R. 1997. Sialic acids in molecular and cellular interactions. *Int. Rev. Cytol.*, 175, pp. 137–240.

Kim, H. K., Lee, S. and Leem, K. H. 2011. Protective effects of egg yolk peptide on bone metabolism. *Menopause* 18(3), pp. 307–13.

Kim, J. E., Gordon, S. L., Ferruzzi, M. G. and Campbell, W. W. 2015. Effects of egg consumption on carotenoid absorption from co-consumed, raw vegetables. *Am. J. Clin. Nutr.* doi: 10.3945, pp. 1–9.

Kinsella, J. E., Lokesh, B. and Stone, R. A. 1990. Dietary n-3 polyunsaturated fatty acids and amelioration of cardiovascular disease: possible mechanism. *Am. J. Clin. Nutr.* 52 (1), pp. 1–28.

Kobayashi, C., Yokoyama, H., Nguyen, S. V., Kodama, Y., Kimata, T. and Izeki, M. 2004. Effect of egg yolk antibody on experimental Cryptosporidium parvum infection in scid mice. *Vaccine* 23(2), pp. 232–5.

Kobayashi, T., Koie, H., Watanabe, A., Ino, A., Watabe, K, Kim, M., Kanayama, K. and Otsuji, K. 2015. Effects of food enriched with egg yolk hydrolysate (bone peptide) on bone metabolism in orchidectomized dogs. *J. Vet. Med. Sci.* 77(4), pp. 503–6.

Koketsu, M. 1997. Glycochemistry of hen eggs. In T. Yamamoto and L. R. Juneja, H. Hatta and M. Kim (eds), *Hen Eggs: Their Basics and Applied Science*. New York: CRC Press, pp. 99–115.

Koketsu, M., Juneja, L. R., Kawanami, H., Kim, M. and Yamamoto, T. 1992. Preparation of N-acetylneuraminic acid from delipidated egg yolk. *Glycoconj.* J. 9(2), pp. 70–4.

Kondo, A., Kamihira, O. and Ozawa, H. 2009. Neutral tube defects: Prevalence, etiology and prevention. *Int. J. Urol.* 16, pp. 49–57.

Kovacs-Nolan, J. and Mine, Y. 2004. Avian egg antibodies: basic and potential applications. *Avian Poultry Biol. Rev.* 15(1), pp. 25–46.

Kovacs-Nolan, J. and Mine, Y. 2012. Egg yolk antibodies for passive immunity. *Annu. Rev. Food Sci. Technol.* 3, pp. 163–82.

Kovacs-Nolan, J., Sasaki, E., Yoo, D. and Mine, Y. 2001. Cloning and expression of human rotavirus spike protein, VP8*, in *Escherichia coli*. *Biochem. Biophys Res. Commun.* 282(5), pp. 1183–88.

Kritchevsky, S. B. 2004. A review of scientific research and recommendations regarding eggs. *J. Am. Coll. Nutr.* 23(6 Suppl), pp. 596S–600S.

Kurisaki, J., Yamauchi, K., Isshiki, H. and Ogiwara, S. 1981. Difference between α- and β-lipovitellin form hen egg yolk. *Agric. Biol. Chem.* 45(3), pp. 699–704.

Kuroki, M., Ikemori, Y., Yokoyama, H., Peralta, R. C., Icatlo Jr, F. C. and Kodama, Y. 1993. Passive protection against bovine rotavirus-induced diarrhea in murine model by specific immunoglobulins from chicken egg yolk. *Vet. Microbiol.* 37(1–2), pp. 135–46.

Kuroki, M., Ohta, Y, Ikemori, Y., Peralta, R. C., Yokoyama, H. and Kodama, Y. 1994. Passive protection against bovine rotavirus in calves by specific immunoglobulins from chicken egg yolk. *Arch Virol.* 138(1–2), pp. 143–8.

Kweon, C. H., Kwon, B. J., Woo, S. R., Kim, J. M., Woo, G. H., Son, D. H., Hur, W. and Lee, Y. S. 2000. Immunoprophylactic effect of chicken egg yolk immunoglobulin (IgY) against porcine epidemic diarrhea virus (PEDV) in piglets. *J. Vet. Med. Sci.* 62(9), pp. 961–4.

Larsson, A., Karlsson-Parra, A. and Sjöquist, J. 1991. Use of chicken antibodies in enzyme immunoassays to avoid interference by rheumatoid factors. *Clin. Chem.* 37(3), pp. 411–4.

LeClaire, R. D., Hunt, R. E. and Bavari, S. 2002. Protection against bacterial superantigen Staphylococcal enterotoxin B by passive vaccination. *Infect Immun.* 70(5), pp. 2278–81.

Lee, J., Babiuk, L. A., Harland, R., Gibbons, E., Elazhary, Y. and Yoo, D. 1995. Immunological response to recombinant VP8* subunit protein of bovine rotavirus in pregnant cattle. *J. Gen. Virol.* 76 (Pt 10), pp. 2477–83.

Lee, K., Ametani, A., Shimizu, M., Hatta, H., Yamamoto, T. and Kaminogawa, S. 1991. Production and characterization of anti-human insulin antibodies in the hen's egg. *Agric. Biol. Chem.* 55(8), pp. 2141–3.

Lee, S. B., Mine, Y. and Stevenson, R. M. 2000. Effects of hen egg yolk immunoglobulin in passive protection of rainbow trout against Yersinia ruckeri. *J. Agric. Food Chem.* 48(1), pp. 110–5.

Leem, K. H., Kim, M. G., Kim, H. M., Kim, M., Lee, Y. J. and Kim, H. K. 2004. Effect of egg yolk proteins on the longitudinal bone growth of adolescent male rat. *Biosci. Biotechnol. Biochem.* 68(11), pp. 2388–90.

Leeson, S. and Caston, L. 2004. Enrichment of eggs with lutein. *Poult. Sci.* 83(10), pp. 1709–12.

Leslie, G. A. and Clem, L. W. 1969. Phylogeny of immunoglobulin structure and function. III. Immunoglobulins of the chicken. *J. Exp. Med.* 130(6), pp. 1337–52.

Leventhal, J. R., Su, A., Kaufman, D. B., Abecassis, M. I., Stuart, F. P., Anderson, B. and Fryer, J. P. 2001. Altered infectivity of porcine endogenous retrovirus by 'protective' avian antibodies: implications for pig-to-human xenotransplantation. *Transplant Proc.* 33(1–2), pp. 690.

Li, S. C., Chien, J. L., Wan, C. C. and Li, Y. T. 1978. Occurrence of glycosphingolipids in chicken egg yolk. *Biochem. J.* 173(2), pp. 697–9.

Li, S. X., Cherian, G. and Sim, J. S. 1996. Cholesterol oxidation in egg yolk powder during storage and heating as affected by dietary oils and tocopherol. *J. Food Sci.* 61(4), pp. 721–5.

Li, X., Nakano, T., Sunwoo, H. H., Paek, B. H., Chae, H. S. and Sim, J. S. 1997. Effects of egg and yolk weights on yolk antibody (IgY) production in laying chickens. *Poult. Sci.* 77(2), pp. 266–70.

Li-Chan, E. C. Y. 2000. Applications of egg immunoglobulins in immunoaffinity chromatography. In J. S. Sim and S. Nakai and W. Guenter (eds), *Egg Nutrition and Biotechnology*. New York: CAB International, pp. 323–39.

Li-Chan, E. C. Y., Powrie, W. D. and Nakai, S. 1995. The chemistry of eggs and egg products. In W. J. Stadelman and O. J. Cotterilll (eds), *Egg Science and Technology*, 4th ed. New York: Haworth Press, pp. 105–75.

Lichtenstein, A. H., Appel, L. J., Brands, M., Camethon, B., Daniels, S., Franch, H. A., et al. 2006. Summary of American heart accusation diet and life style recommendations revision. *Atherioscler Throbo Vasc. Biol.* 26, pp. 2186–91.

Liu, N., Wang, Y., Zhao, Q., Zhang, Q. and Zhao, M. 2011. Fast synthesis of 1,3-DAG by Lecitase® Ultra-catalyzed esterification in solvent-free system. *Eur. J. Lipid Sic. Technol.* 113(8), pp. 973–9.

Liu, X., Suo, R., Xiong, S. L., Zhang, Q. H. and Yi, G. H. 2013. HDL drug carries for targeted therapy. *Clin. Chim. Acta* 415, pp. 94–100.

Luthi, A. J., Patel, P. C., Ko, C. H., Mutharasan, R. K., Mirkin, C. A. and Thaxton, C. S. 2010. Nanotechnology for synthetic high-density lipoproteins. *Trends Mol. Med.* 16(12), pp. 553–60.

Maddison, S. E. and Reimer, C. B. 1976. Normative values of serum immunoglobulins by single radial immunodiffusion: a review. *Clin. Chem.* 22(5), pp. 594–601.

Maehre, H. K., Jensen, I. J., Elvevoll, E. O. and Eilertsen, K. E. 2015. ω-3 fatty acids and cardiovascular disease: effects, mechanisms and dietary relevance. *Int. J. Mol. Sci.* 16, pp. 22636–61.

Makrides, M., Hawkes, J. S., Neumann, M. A. and Gibson, R. A. 2002. Nutritional effect of including egg yolk in the weaning diet of breast-fed and formula-fed infants: a randomized controlled trial. Am. *J. Clin. Nutr.* 75(6), pp. 1084–92.

Mann, K. 2007. The chicken egg white proteome. *Proteomics* 7(19), pp. 3558–68.

Mann, K. and Mann, M. 2008. The chicken egg yolk plasma and granule proteomes. *Proteomics* 8(1), pp. 178–91.

Mann, K., Olsen, J. V., Macek, B., Gnad, F. and Mann, M. 2008. Identification of new chicken egg proteins by mass spectrometry-based proteomic analysis. *Worlds Poult. Sci. J.* 64(2), pp. 209–18.

Manuc, T. E. M., Manuc, M. M. and Diculescu, M. M. 2016. Recent insights into the molecular pathogenesis of Crohn's disease: a review of emerging therapeutic targets. *Clin. Exp. Gastoenterol.* 9, pp. 59–70.

Marshall, A. C., Kubena, K. S., Hinton, K. R., Hargis, P. S. and Van Elswyk, M. E. 1994. n-3 fatty acid enriched table eggs: A survey of consumer acceptability. *Poult. Sci.* 73(8), pp. 1334–40.

Martin, W. G., Augstyniak, J. and Cook, W. H. 1964. Fractionation and characterization of the low density lipoproteins of hen's egg yolk. *Biochem. Biophs. Acta* 84(6), pp. 714–20.

Marwa, Y. and Nazlin, H. 2015. Antioxidant and ACE inhibitory bioactive peptides purified from egg yolk proteins. *Int. J. Mol. Sci.* 16(12), pp. 29161–78.

Masuda, Y., Kokubu, T., Yamashita, M., Ikeda, H. and Inoue, S. 1998. Egg phosphatidylcholine combined with vitamin B12 improved memory impairment following lesioning of nucleus basalis in rats. *Life Sci.* 62(9), pp. 813–22.

Matsuzaki, T., Yamazaki, R., Hashimoto, S. and Yokokura, T. 1998. The effect of oral feeding of Lactobacillus casei strain Shirota on immunoglobulin E production in mice. *J. Dairy Sci.* 81(1), pp. 48–53.

McDonald, B. E. 2004. The Canadian experience: why Canada decided against an upper lipid for cholesterol. *J. Am. Coll. Nutr.* 23(6 Suppl), pp. 616S–620S.

Ministry of Health, Labor and Welfare. Available at: http://www.mhlw.go.jp/english/ [Accessed 16 April 2016].

Ministry of Health, Labour and Welfare. Available at: http://www.mhlw.go.jp/bunya/kenkou/sessyu-kijun.html [Accessed 28 April 2016].

Ministry of Health, Labour and Welfare. Available at: http://www.mhlw.go.jp/bunya/kenkou/dl/kenkou_eiyou_chousa_tokubetsushuukei_h22.pdf.

Mizutani, R. and Nakamura, R. 1988. Analysis of emulsion property of LDL in egg yolk. *New Food Industry*, pp. 49–58; 69–74 [In Japanese].

Morschhäuser, J., Hoschützky, H., Jann, K. and Hacker, J. 1990. Functional analysis of the sialic acid-binding adhesion SfaS of pathogenic *Escherichia coli* by site-specific mutagenesis. *Infect. Immun.* 58(7), pp. 2133–8.

Motoi, Y., Inoue, S., Hatta, H., Sato, K., Morimoto, K. and Yamada, A. 2005. Detection of rabies-specific antigens by egg yolk antibody (IgY) to the recombinant rabies virus proteins produced in Escherichia coli. *Jpn. J. Infect. Dis.* 58(2), pp. 115–8.

Muniyappa, K. and Adiga, P. R. 1981. Nature of the thiamin-binding protein from chicken egg yolk. *Biochem J.* 193(3), pp. 679–85.

Murphy, A. J. and Woollard, K. J. 2010. High-density lipoprotein: a potent inhibitor of inflammation. *Clin. Exp. Pharmacol. Physiol.* 37, pp. 710–18.

National Chicken Council. 2010. National chicken council animal welfare guidelines and audit checklist for broiler [online]. Washington, DC: National Chicken Council. Available at: http://www.nationalchickencouncil.org/wp-content/uploads/2012/01/NCC-Animal-Welfare-Guidelines-2010-Revision-BROILERS [Accessed 16 April 2016].

Navidghasemizad, S., Temelli, F. and Wu, J. 2015. Phase separation behavior of egg yolk suspensions after anionic polysaccharides addition. *Carbohydr. Polym.* 117, pp. 297–303.

Ng, K. K., Lovell, J. F. and Zheng, G. 2011. Lipoprotein-inspired nanoparticles for cancer. *Acc. Chem. Res.* 44(10), pp. 1105–113.

Niskanen, E. A., Hytönen, V. P., Grapputo, A., Nordlund, H. R., Kulomaa, M. S. and Laitinen, O. H. 2005. Chicken genome analysis reveals novel genes encoding biotin-binding proteins related to avidin family. *BMC Genomics* 6(1), p. 41.

Nwokolo, E. and Sim, J. S. 1989. Barley and full-fat canola seeds in layer diets. *Poult. Sci.* 68(11), pp. 1485–9.

Oshida, K. 1976. Basic study on manufacturing of mayonnaise Part 4: Effects of salt and vinegar on emulsion stability of low density fraction and high density fraction of egg yolk. *Nippon Shokuhin Kogyo Gakkaishi* 23(6), pp. 250–6 [In Japanese].

Otake, S., Nishihara, Y., Makimura, M., Hatta, H., Kim, M., Yamamoto, T. and Hirasawa, M. 1991. Protection of rats against dental caries by passive immunization with hen-egg-yolk antibody (IgY). *J. Dent. Res.* 70(3), pp. 162–6.

Parkinson, T. L. 1966. The chemical composition of eggs. *J. Sci. Food Agri.* 17(3), pp. 101–11.

Paulson, J. C, Rogers, G. N., Carroll, S. M., Higa, H. H., Pritchett, T., Milks, G. and Sabesan, S. 1984. Selection of influenza virus variants based on sialyoligosaccharide receptor specificity. *Pure Appl. Chem.* 56(7), pp. 797–805.

Phillips, M. L., Nudelman, E., Gaeta, F. C. A., Perez, M., Singhal, A. K., Hakomori, S. and Paulson, J. C. 1990. ELMA-1 mediates cell adhesion by recognition of a carbohydrate ligand, sialyl-Lex. *Science* 250(4984), pp. 1130–2.

Rasmussen, H. M., Muzhingi, T., Eggert, E. M. R. and Johnson, E. J. 2012. Lutein, zeazanthin, mezzo-zeaxanthin content in egg yolk and their absence in fish and seafood. *J. Food Comp. Anal.* 27(2), pp. 139–44.

Robbins, J. M., Petrone, A. B., Ellison, R. C., Hunt, S. C., Carr, J. J., Heiss, G., Arnett, D. K, Gaziano, J. M. and Djoussé, L. 2014. Association of egg consumption and calcified atherosclerotic plaque in the coronary arteries: the NHLBI Family Heart Study. *ESPEN J.* 9, pp. e131–e135.

Rong, Y., Chen, L., Zhu, T., Song, Y., Yu, M., Shan, Z., Sands, A., Hu, F. B. and Liu, L. 2013. Egg consumption and risk of coronary heart disease and stroke: dose-response meta-analysis of prospective cohort studies. *BMJ*, 346, p. e8539.

Rose, M. E., Orlans, E. and Buttress, N. 1974. Immunoglobulin classes in the hen's eggs: their segregation in yolk and white. *Eur. J. Immunol.* 49(7), pp. 521–3.

Sakanaka, S. and Tachibana, Y. 2006. Active oxygen scavenging activity of egg-yolk protein hydrolysates and their effects on lipid oxidation in beef and tuna homogenates. *Food Chem.* 95(2), pp. 243–9.

Sandrin, M. S. and McKenzie, I. F. 1994. Gal alpha (1,3)Gal, the major xenoantigens(s) recognised in pigs by human natural antibodies. *Immunol. Rev.* 141, pp. 169–90.

Sarker, S. A., Casswall, T. H. Juneja, L. R., Hoq, E., Hossain, I., Fuchs, G. J. and Hammarström, L. 2001. Randomized, placebo-controlled, clinical trial of hyperimmunized chicken egg yolk immunoglobulin in children with rotavirus diarrhea. *J. Pediatr. Gastroenterol. Nutr.* 32(1), pp. 19–25.

Sato, R., Helzlsouer, K. J. Alberg, A. J., Hoffman, S. C., Norkus, E. P. and Comstock, G. W. 2002. Prospective study of carotenoids, tocopherols, and retinoid concentrations and the risk of breast cancer. *Cancer Epidemiol. Biomarkers Prev.* 11(5), pp. 451–7.

Schade, R., Pfister, C., Halatsch, R. and Henklein, P. 1991. Polyclonal IgY antibodies from chicken egg yolk - an alternative to the production of mammalian IgG type antibodies in rabbits. *ATLA* 19(4), pp. 403–19.

Schaeffer, J. L., Tyczkowski, J. K., Parkhurst, C. R. and Hamilton, P. B. 1988. Carotenoid composition of serum and egg yolks of hens fed diets varying in carotenoid composition. *Poult. Sci.* 67(4), pp. 608–14.

Schauer, R. 1982. Chemistry, metabolism, and biological function of sialic acids. *Adv. Carbohydr. Chem. Biochem.* 40, pp. 131–234.

Schultz, B. and Herrler, G. 1992. Bovine coronavirus uses N-acetyl-9-O-acetylenuraminic acid as a receptor determinant to initiate the infection of cultured cells. *J. Gen. Virol.* 73(Pt 4), pp. 901–6.

Shin, N. R., Choi, I. S., Kim, J. M., Hur, W. and Yoo, H. S. 2002. Effective methods for the production of immunoglobulin Y using immunogens of Bordetella bronchiseptica, Pasteurella multocida and Achinobacillus pleuropneumoniae. *J. Vet. Sci.* 3(1), pp. 45–57.

Sim, J. S., Cherian, G. and Jiang, Z. 1992. Alpha-linolenic acid metabolism: the chicken and the egg. *Nutrition* 8(3), pp. 221–2.

Sim, J. S. and Qi, G-H. 1995. Designing poultry products using flaxseed. In L. U. Thompson and S. Cunnane (eds), *Flaxseed in Human Nutrition*. American Oil Chemist's Society Press (AOCS), pp. 315–33.

Sim, J. S. and Sunwoo, H. H. 2002. Designer eggs: Nutritional and functional significant. In R. R. Watson (ed.), *Eggs and Health Promotion*. New Jersey: Joh Wiley & Sons, pp. 19–38.

Simopoulos, A. P. 1991. Omega-3 fatty acids in health and disease and in growth and development. *Am. J. Clin. Nutr.* 54(3), pp. 438–63.

Slattery, M. L., Benson, J., Curtin, K., Ma, K. N., Schaeffer, D. and Potter, J. D. 2000. Carotenoids and colon cancer. *Am. J. Clin. Nutr.* 71(2), pp. 575–82.

Smit, H., Gaastra, W., Kamerling, J. P., Vliegenthart, J. F. and de Graaf, F. K. 1984. Isolation and structural characterization of the equine erythrocyte receptor for enterotoxigenic *Escherichia coli* K99 fimbrial adhesion. *Infect. Immun.* 46(2), pp. 578–84.

Smith, D. J. King, W. F. and Godiska, R. 2001. Passive transfer of immunoglobulin Y to Streptococcus mutans glucan binding protein B can confer protection against experimental dental caries. *Infect. Immun.* 69(5), pp. 3135–42.

Stuart, B., Lin, J. H. and Morgayzel Jr., P. J. 2010. Early eradication of Pseudomonas aeruginosa in patients with cystic fibrosis. *Paediatr. Respir. Rev.* 11(3), pp. 177–84.

Sugino, H., Nitoda, T. and Juneja, L. R. 1997. General chemical composition of hen eggs. In T. Yamamoto, L. R. Juneja, H. Hatta and M. Kim (eds), *Hens Egg: The Basic and Applied Science*. New York: CRC Press, pp. 13–24.

Sugita-Konishi, Y., Shibata, K., Yun, S. S., Hara-Kudo, Y., Yamaguchi, K. and Kumagai, S. 1996. Immune functions of immunoglobulin Y isolated from egg yolk of hens immunized with various infectious bacteria. *Biosci. Biotechnol. Biochem.* 60(5), pp. 886–8.

Sunwoo, H. H., Wang, W. W. and Sim, J. S. 2006. Detection of Escherichia coli O157:H7 using chicken immunoglobulin Y. *Immunol. Lett.* 106(2), pp. 191–3.

Tabassum, S., Shears, P. and Hart, C. A. 1994. Genomic characterization of rotavirus strains obtained from hospitalized children with diarrhea in Bangladesh. *J. Med. Virol.* 43(1), pp. 50–6.

Talley, B. S. and Carroll, S. B. 1990. Rattlesnakes and scorpion antivenoms from the egg yolk of immunized hens. *Biotechnol.* 8(10), pp. 934–8.

Tesedo, J., Barrado, E., Sanz, M. A., Tesedo, A. and de la Rosa, F. 2006. Fatty acid profiles of processed chicken egg yolk. *J. Agric. Food Chem.* 54 (17), pp. 6255–60.

Tsutsui, T. 1988. Functional properties of heat-treated egg yolk low density lipoproteins. *J. Food Sci.* 53(4) pp. 1103–6.

Tyrrell, D. A., Health, T. D., Colley, C. M. and Ryman, B. E. 1976. New aspects of liposomes. *Biochim. Biphys. Acta.* 457(3–4), pp. 259–302.

Uemura, N., Okamoto, S., Yamamoto, S., Mathumura, N., Yamaguchi, S., Yamakido, M., Taniyama, K., Sasaki, N. and Schlemper, R. J. 2001. Helicobacter pylori infection and the development of gastric cancer. *N. Eng. J. Med.* 345(11), pp. 784–9.

USDA. Available at: http://ndb.nal.usda.gov/ndb/foods/show/112 [Accessed 25 April 2016].
Vahteristo, L. T., Ollilainen, V. and Varo, P. J. 1997. Liquid chromatographic determination of folate monoglutamates in fish meat, egg, and dairy products consumed in Finland. *J. AOAC Int.* 8092, pp. 373–8.
Van der Made, S. M., Kelly, E. R., Berendschot, T. T., Kijlstra, A., Lütjohann, D. and Plat, J. 2014. Consuming a buttermilk drink containing lutein-enriched egg yolk daily for 1 year increased plasma lutein but did not affect serum lipid or lipoprotein concentrations in adults with early signs of age-related macular degeneration. *J. Nutr.* 144, pp. 1370–7.
van Hoogevest, P. and Wendel, A. 2014. The use of natural and synthetic phospholipids as pharmaceutical excipients. *Eur. J. Lipid Sci. Technolo.* 116(9), pp. 1088–107.
Varki, A. 1997. Sialic acid as ligands in recognition phenomena. *FASEB J.* 11(4), pp. 248–55.
Vasudevan, N., Bahadur, U. and Kondaiah, P. 2001. Characterization of chicken riboflavin carrier protein gene structure and promoter regulation by oestrogen. *J. Biosci.* 26(1), pp. 39–46.
Vega, C., Bok, M., Chacana, P., Saif, L., Fernandez, F. and Parreno, V. 2011. Egg yolk IgY: Protection against rotavirus induced diarrhea and modulatory effect on the systematic and mucosal antibody response in newborn calves. *Vet. Immunol. Immunolopathol.* 142(3–4), pp. 156–69.
Vega, C., Bok, M., Saif, L., Fernandez, F. and Parreno, V. 2015. Egg yolk IgY antibodies: A therapeutic intervention against group A rotavirus in calves. *Res.Vet. Sci.* 103, pp. 1–10.
Virtanen, J. K., Mursu, J., Virtanen, H. E. K., Fogelholm, M., Salonen, J. T., Koskinen, T., Voutilainen, S. and Tuomainen, T.-P. 2016. Associations of egg and cholesterol intakes with carotid intima-media thickness and risk of incident coronary artery disease according to apolipoprotein E phenotype in men: the Kuopio Ischamemic Heart Disease Risk Factor Study. *Am. J. Clin. Nutr.* doi: 103945/ajcn.115.122317.
von Itzstein, M., Wu, W. Y., Kok, G. B., Pegg, M. S., Dyason, J. C., Jin, B., Van Phan, T., Smythe, M. L., White, H. F., Oliver, S. W., et al. 1993. Rational design of potent sialidase-based inhibitors of influenza virus replication. *Nature* 363(6428), pp. 418–23.
Wallance, R. A. and Morgan, J. P. 1986. Chromatographic resolution of chicken phosvitin: Multiple macromolecular species in a classic vitellogenin-derived phosphoprotein. *Biochem. J.* 240(3), pp. 871–8.
Wang, G. and Wang, T. 2008. Egg yolk protein modification by controlled enzymatic hydrolysis for improved functionalities. *Int. J. Food Sci. Technol.* 44(4), pp. 763–9.
Wang, Y., Zhao, M., Song, K., Wang, L., Tang, S. and Riley, W. W. 2010. Partial hydrolysis of soybean oil by phospholipase A1 (Lecitase Ultra). *Food Chem.* 121(4), pp. 1066–72.
Williams, J. 1962. Serum proteins and the livetins of hen's egg yolk. *Biochem. J.* 83(2), pp. 346–55.
Worledge, K. L., Godiska, R., Barrett, T. A. and Kink, J. A. 2000. Oral administration of avian tumor necrosis factor antibodies effectively treats experimental colitis in rats. *Dig. Dis. Sci.* 45(12), pp. 2298–305.
Wu, C., Yang, R., Zhou, J., Bao, S., Zoum, L., Zhang, P., Mao, Y., Wu, J. and He, Q. 2003. Production and characterization of a novel chicken IgY antibody raised against C-terminal peptide from human thymidine kinase 1. *J. Immunol. Methods* 277(1–2), pp. 157–69.
Xu, Y., Li, X., Jin, L., Zhen, Y., Lu, Y., Li, S., You, J. and Wang, L. 2011. Application of chicken egg yolk immunoglobulins in the control of terrestrial and aquatic animal disease: a review. *Biotechnol. Adv.* 29(6), pp. 860–8.
Yoshimura, S., Watanabe, K., Suemizu, H., Onozawa, T., Mizoguchi, J., Tsuda, K., Hatta, H. and Moriuchi, T. 1991. Tissue specific expression of the plasma glutathione peroxidase gene in rat kidney. *J. Biochem.* 109(6), pp. 918–23.
Young, C. A., Silversides, F. G. and Jones, S. R. M. 2007. Chicken-derived IgY recognizes developing and mature stages of Loma salmonae (Microsporidia) in Pacific salmon, Oncorhynchus spp. *Aquaculture* 273(4), pp. 398–404.
Zambrowicz, A., Eckert, E., Pokora, M., Bobak, L., Dabrowska, A., Szottysik, M., Trziszka, T. and Chrzanowska, J. 2015. Antioxidant and antidiabetic activities of peptides isolated from a hydrolysate of an egg-yolk protein by-product prepared with a proteinase from Asian pumpkin (Cucurbita ficifolia). *RSC Adv.* 5, pp. 10460–7.

Zambrowicz, A., Pokora, M., Eckert, E., Szoltysik, M., Dabrowska, A., Chrzanowska, J. and Trziszka, T. 2012. Antioxidant and antimicrobial activity of lecithin free egg yolk protein preparation hydrolysates obtained with digestive enzymes. *FFHD*. 2(12), pp. 487–500.

Zeisel, S. H. 1981. Dietary choline: Biochemistry, physiology and pharmacology. *Ann. Rev. Nutr.* 1, pp. 95–121.

Part 2

Safety

Chapter 4

Pathogens affecting table eggs

Kapil Chousalkar, University of Adelaide, Australia and Kylie Hewson, Australian Chicken Meat Federation, Australia

1 Introduction

The unique structure of the avian eggshell enables the egg to maintain its structural integrity and also withstand biological and mechanical trauma. Previous studies on the microbiology of eggs focused predominantly on fertile eggs in order to study hatchability (De Reu et al., 2008). With the subsequent rise in egg consumption and a simultaneous increase in egg-related foodborne outbreaks, there was a need to understand the microflora on the eggshell and within the egg contents. Eggs can be contaminated by pathogenic microorganisms in a number of ways, and there is a range of microorganisms that have the ability to penetrate the eggshell and survive in egg contents until consumption. However, the single largest food safety threat that has been related to the consumption of eggs is the presence of different serovars of *Salmonella enterica*. There are several factors which can influence the level of *Salmonella* contamination of eggs such as production system, farm management, handling of eggs, storage conditions, packaging, transport conditions, food preparation and general hygiene conditions from farm to fork. This chapter focuses particularly on pathogenic microorganisms present on table eggs.

2 Pathogen contamination of eggs: an overview

The egg yolk provides an ideal growth medium for human pathogenic microorganisms. All laying hens have a single common opening for the intestine, urinary and reproductive

http://dx.doi.org/10.19103/AS.2016.0012.04

tracts, which can contribute to external eggshell contamination as the egg passes through this region. The contact of the faecal material with the eggshell is often unavoidable and can potentiate the entry of microorganisms into the egg. It could be argued that there is a biological reason for a hen to pass her gut flora onto the outside of the eggshell, as the gut of her offspring would become populated with maternal flora after hatching. However, with the rise in consumption of infertile hen eggs, humans are now becoming exposed to these microorganisms, which are not pathogenic to the hen, but maybe to humans.

Globally, *Salmonella* spp., *Bacillus cereus, Listeria monocytogenes* and *Staphylococcus aureus* are the major foodborne pathogens in relation to commercial egg production. Bacteria from different genera found on the eggshell surface include *Micrococcus, Streptococcus, Sarcina, Proteus, Serratia, Aeromonas, Staphylococcus, Aerobacter, Arthrobacter, Escherichia, Bacillus, Cytophaga, Pseudomonas, Flavobacterium, Acinetobacter* and *Alcaligens* (Board and Tranter, 1995). Besides these bacteria, Moats (1980) reported the presence of *Kurthia, Moraxella, Propionibacterium* and yeast on eggs.

Eggs in contact with faeces can be exposed to other enteropathogens like *Escherichia coli*, which can also contaminate the eggshell (Adesiyun et al., 2007). Cooking can destroy or reduce the level of bacteria in eggs; however, it is common practice in most parts of the world, including Australia, to eat raw or lightly cooked egg. Even after cooking an egg, the persistence of heat-stable toxin-producing bacteria like *E. coli* and *S. aureus* could pose a threat to human health. The internal properties of eggs favour survival and growth of contaminating microorganisms which are Gram-negative, such as *E. coli* and *Salmonella*. These types of bacteria have relatively simple nutritional requirements and have the ability to continue multiplying at low temperatures (Board and Tranter, 1995), which makes limiting the risk to human health more complex than simple, through chain refrigeration.

There are two ways to consider the quality of an egg – the microbiological quality (i.e., low level such that the egg presents a small risk of foodborne illness) and the phenotypic quality characteristics such as shell and yolk colour, albumen height, etc. It is not often that these two measures of quality are considered together. Eggshell quality can be influenced by many other factors such as age, strain of hen, temperature, disease (Roberts, 2004), and these impact on the measures of egg quality. Gram-negative bacteria, and occasionally a few Gram-positive microorganisms, are usually present in spoiled table eggs and commonly these are *Serratia, Pseudomonas flurorescens, Pseudomonas putida, Alcaligens, Proteus, Escherichia, Xanthomonas, Maltophia, Aeromonas, Hafnia, Citrobacter, Pseudomonas aeroginosa, Acinetobacter, Cytophaga, Bacillus, Micrococcus, Streptococcus* and *Arthrobacter*. Gram-negative bacteria are also considered the primary culprits for egg spoilage (Musgrove, 2011). Thus, the environment of the internal contents of egg favours the growth of Gram-negative bacteria (Board and Tranter, 1995).

The Enterobacteriaceae present on the eggshell surface of unwashed and washed eggs have been characterized using biochemical tests (Musgrove et al., 2004; Gole et al., 2013) (Table 1). These authors reported the presence of *E. coli, Enterobacter cloacae, Enterobacter sakazakii, Enterobacter* spp., *Citrobacter youngae, Klebsiella pneumoniae, Klebsiella* spp., *Serratia odorifera, Serratia* spp., *Kluyvera* spp., *Providencia rettgeri, Providencia* spp., *Pantoea* spp., *Rahnella aquatilis, Salmonella* spp., *Yersinia* spp., *Flavimonas oryzihabitans and Xanthomonas maltophilia* on unwashed eggs. Gole et al. (2013) reported that the average Enterobacteriaceae count on the eggshell surface was 1.46 log CFU/eggshell. However, these low levels of bacterial contamination have the potential to contaminate egg processing and packing plants and can multiply easily. Similarly, they can also act as a source of contamination of egg washing equipment.

Table 1 Enterobacteriaceae isolated from the eggshell surface of unwashed eggs

Flock number	Age (weeks)	Enterobacteriaceae reported (number isolated from 60 eggs collected from each flock)
1	22	*Yersinia enterocolitica* (2), *Klebsiella pneumoniae* (1), *Escherichiacoli* 1 (1)
2	24	*Serratia plymuthica* (1), *E.coli* 1 (1)
3	24	*E. coli* 1 (8), *K. pneumoniae* (1)
4	26	*S. plymuthica* (1)
5	27	*E. coli* 1 (1), *K. pneumoniae* (1)
6	27	*E. coli* 1 (10)
7	30	*E. coli* 1 (8), *E. coli* 2 (2), *Salmonella* Infantis (1)
8	30	*E. coli* 1 (1), *E. coli* 2 (1)
9	31	*Enterobacter* spp. (2), *Kluyvera* spp. (1), *Serratia* spp. (1), *E. coli* 1 (2), *Leclercia adecarboxylata* (1)
10	34	*E. coli* (3), *Enterobacter carcinogenous* (1), *Serratia odorifera* (1)
11	41	*E. coli 1* (5), *E. coli* 2 (2), *Escherichia fergusonii* (1), *K. pneumoniae* (1), *S.* Infantis (1)
12	45	*E. coli 1* (2), *Pantoea* spp. 3 (1), *Pantoea* spp. 2 (1)
13	45	*E. coli* 1 (1)
14	46	*E. coli* 1 (8), *E. fergusonii* (1), *S.* Infantis (1)
15	50	–
16	52	*Kluyvera* (1), *K. pneumoniae* (1), *E. coli* 1 (2), *S.* Infantis (2)
17	53	*S. plymuthica* (1), *Pantoea* spp. (1)
18	55	*E. coli* 1 (1)
19	56	*E. coli* 1 (5), *Pantoea* spp. (1), *E. coli 1* (2)
20	58	*E. coli* 1 (5), *Escherichia* genus (1), *S. odorifera* (1), *Enterobacter cloacae* (2), *S.* Infantis (4)
21	59	*Y. enterocolitica* (1), *S. plymuthica* (1), *Kluyvera* spp. (1)
22	62	*Cedecea* spp. (1), *E. coli 1* (1), *Citrobacter freundii* (1), *Enterobacter* (2)
23	63	*E. coli* 1 (1)
24	64	*L. adecarboxylata* (1), *E. coli* 2 (1), *E. cloacae* (1)
25	66	*Enterobacter* (1), *E. fergusonii* (1), *E. coli* 1 (3), *E. coli* 2 (1), *S.* Infantis (1)
26	66	*Y. enterocolitica* (2), *E. coli* 1 (1)
27	66	*E. coli* 1 (1)
28	68	*E. cloacae* (1), *E. coli* 1 (5), *E. coli* 1 (1)
29	72	*Pantoea* spp. 3 (1), *Pantoea* spp. 1 (1), *Y. enterocolitica* (1), *E. cloacae* (1)
30	74	*L. adecarboxylata* (1), *S.* Infantis (2), *Salmonella enterica* subsp. *enterica* serovar 4,12:d: (2)
31	80	*E. coli* 1 (3), *E. cloacae* (1), *Enterobacter aerogenes* (1), *S. odorifera* (1)

The bacterial load on an eggshell can fluctuate over time, and the average bacterial load on an unwashed egg is considered to be 10^4 CFU (Board, 1966). However, this is also significantly dependent on the husbandry and egg handling practices from farm to fork. Visually contaminated eggs (termed 'soiled eggs') are expected to have a high bacterial load but some studies found relatively high loads of *Salmonella* spp. on the eggshell surface of visually clean unwashed eggs (Gole et al., 2013, 2014c). Samiullah et al. (2014) observed bacterial levels in eggshell pores of 1.27 log CFU in free-range eggs and 1.21 log CFU in caged eggs. The egg has natural defence barriers such as the cuticle on the eggshell, shell membranes and egg albumen. Differences in bacterial load on eggs could be due to differences in eggshell quality and variable cuticle deposition on eggs. Egg white is a poor medium for the growth of bacteria because of its antibacterial contents such as lysozyme, ovotransferrin, vitamin chelating proteins, protease inhibitors and alkaline pH (Alabdeh et al., 2011; Baron et al., 2016). Moreover, the viscosity of egg white (Lucisano et al., 1996) and mineral composition (such as the absence of iron) are also responsible for inhibiting the growth of bacteria in egg white. There are three physical barriers which protect the egg from the bacterial penetration: cuticle, crystalline eggshell and the membranes. The mechanism of bacterial penetration will be discussed in a separate chapter of this book but, briefly, bacterial strain, load of bacteria, temperature, pH of the inoculum, eggshell quality, moisture, relative humidity and eggshell porosity can influence the eggshell penetration by microorganisms present on the eggshell. Interestingly, different pore structures, the total number of pores and shell thickness do not appear to play a role in the horizontal infection of eggs by the *Salmonella* Typhimurium strain (Ray et al., 2015).

Although there is the potential for vertical transmission of some *Salmonella* types from bird to egg (as has been shown to occur for *Salmonella* Enteritidis), it is generally accepted that horizontal transmission is the most likely source of contamination of eggshells by most bacteria. Small defects in the eggshell may provide the means for the predominant bacterial species on the eggshell to penetrate and move into the egg contents, and the internal properties of eggs favour survival and growth of contaminating microorganisms which are Gram-negative, such as *E. coli* and *Salmonella*.

3 *Salmonella* spp.

The presence of human pathogenic *Salmonella* spp. is considered to be the most significant risk present with the consumption of table eggs. *Salmonella* present on and/or in eggs pose a varied risk to human health, which is primarily dependent on how the eggs are used by the consumer. Thorough cooking of eggs will destroy most, if not all, bacteria present, including *Salmonella,* and will pose an almost non-existent risk to human health. However, if egg preparations are raw or lightly cooked, this will not destroy all *Salmonella,* if they are present. A rough estimate of the presence of *Salmonella* (all species, not just pathogenic) on eggs is greater than 1 in 20 000 (Ebel and Schlosser, 2000). Considering the estimated production of eggs in Australia in 2015 was 421.5 million dozen, the risk of foodborne illness in general is quite low for humans consuming eggs. However, it has been demonstrated in a number of regions and studies that consumption of raw or lightly cooked eggs/egg preparations significantly increases the risk of human foodborne illness. Therefore, there is a need for awareness of the risks that *Salmonella* and raw/lightly

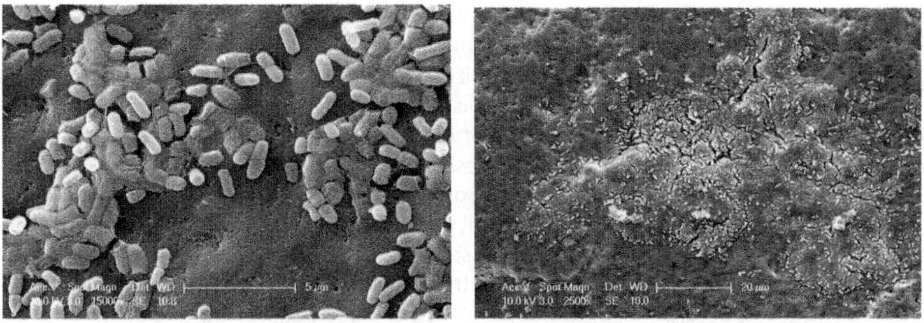

Figure 1 Scanning electron micrograph of *S.* Typhimurium (phage type 9) biofilm formation on the eggshell surface.

cooked eggs pose to every person, from the producer to the consumer. Even though the frequency of egg contamination by *Salmonella* spp. is low and depends upon the shedding levels of the flock, due to the large number of eggs consumed, even a very low rate of infection is important for population health. The egg yolk is an ideal growth medium for microorganisms and contamination of the yolk quickly results in rapid growth of bacteria, due to the nutritious environment (Chousalkar et al., 2015). A recent investigation also demonstrated that *Salmonella* serovars are capable of forming biofilm on the eggshell surface; however, this behaviour is influenced by temperature and serovar (Pande et al., 2016 unpublished) (Fig. 1a and b).

In Europe and the USA, *S.* Enteritidis PT4 (SE) is of greatest concern because of its transovarian transmission route into the egg. SE has not been detected in the Australian egg industry; however, other *Salmonella* serotypes are still of interest. External or internal contamination of an egg can vary considerably between *Salmonella*-positive flocks (Humphrey et al., 1989; Davies and Breslin, 2004; De Buck et al., 2004) and the shedding dynamics on-farm. Many different *Salmonella* serovars (such as Enteritidis, Typhimurium, Agona, Derby, Bredney, Virchow, Infantis, Livingstone, Mbandaka, Newport) have been isolated from the eggshells (Martelli and Davies, 2012). However, *S.* Enteritidis and *S.* Typhimurium are considered the most important serovars in terms of public health significance.

S. Enteritidis has been the major cause of the foodborne salmonellosis in humans in various countries (not including Australia). Eggs can be contaminated with *S.* Enteritidis on the eggshell surface and internally (Gantois et al., 2009). Internal contamination can be horizontal (through eggshell) or vertical (by direct contamination of egg contents from infection of the reproductive organs) (Gantois et al., 2009). Although the egg albumen has many antibacterial properties, *S.* Enteritidis has the ability to multiply in the albumen and several relevant genes associated with survival of SE in egg white have been identified (Clavijo et al., 2006). It has been demonstrated that the virulence of SE is unaffected by the environment within the egg albumen (Baron et al., 2004).

Previous studies have also suggested that *S.* Typhimurium phage types have a similar ability to survive in the egg albumen, up to 21 days after infection, as well as survival on the eggshell surface (Gole et al., 2014a). Genes associated with the survival of *Salmonella* in egg albumen have been detected in *S.* Typhimurium definitive types (McWhorter and Chousalkar, 2015; McWhorter et al., 2015). It is not clear whether the virulence of

S. Typhimurium is affected by the environment within the egg albumen. Based on a recent study by Pande et al. (2016), *S.* Typhimurium was detected in different oviduct segments but not in egg white. It could be hypothesized that some *S.* Typhimurium strains do not have the capacity to leave the oviduct cells and engage in being incorporated in egg white. Also, *S.* Typhimurium strains may not have the cluster of genes for survival in egg white; however, further investigation is necessary.

The presence of *Salmonella* on the eggshell underlines the importance of proper handling of eggs in the food industry as well as in the kitchen environment to avoid cross-contamination of other food items. Other serovars such as *S.* Livingstone, *S.* Singapore and *S.* Adelaide have the capacity to penetrate the eggshell (Gole et al., 2014b). These findings suggest that eggshell penetration is not a characteristic isolated to *S.* Typhimurium or *S.* Enteritidis. Studies regarding egg penetration indicated that *Salmonella* penetration across the eggshell was influenced by eggshell quality and cuticle deposition (Gole et al., 2014a,b; De Reu et al., 2006). Eggshell quality can be affected by any disease that can potentially lead to damage of the reproductive tract of the hen.

Besides *S.* Enteritidis and *S.* Typhimurium, a number of other *Salmonella* serovars including *S.* Infantis have been isolated from eggs in the worldwide surveys conducted on the prevalence of *Salmonella* in poultry industry (Martelli and Davies, 2012). These serovars were primarily isolated from eggshells and also from egg contents.

4 *Campylobacter* spp. and other bacteria

4.1 *Campylobacter* spp.

Campylobacter is the leading cause of bacterial foodborne illness in many high-income countries (Sahin et al., 2003; Messelhausser et al., 2011). The sources of infection and the modes of transmission of the bacteria on chicken farms is not well understood (Messelhausser et al., 2011). It has been commonly assumed that horizontal transmission from the environment is the major source of *Campylobacter* spp., and vertical transmission from reproductive organs is unlikely (Shanker et al., 1986). However, as *Campylobacter* colonizes the gut of laying hens, it is reasonable to suggest that eggshells and egg contents can become contaminated and, moreover, the presence of thermotolerant *Campylobacter* spp. on the eggshell could pose the risk of cross-contamination in the kitchen environment (Messelhausser et al., 2011). *Campylobacter* spp. have been detected on the eggshell surface of table eggs in various studies (Parisi et al., 2015; Doyle, 1984). In contrast, field studies conducted in different housing systems suggested that, although a large proportion of hens were colonized with *Campylobacter* spp., eggs produced from such hens were not *Campylobacter* positive (Jones et al., 2016; Jones and Musgrove, 2007a). A few studies that have been conducted on eggs have focused on free-range production systems; usually, they are focused on chicken meat breeder hen eggs and eggs from caged layer flocks. Chicken meat breeder studies demonstrated a very low level of bacterial isolation from the outer and inner surface of the eggs, and no contamination of internal contents, even when *Campylobacter* was actively shed in the faeces of the hens (Sahin et al., 2003; Doyle, 1984). Studies by Sahin et al. (2003) found that, if *Campylobacter* was present on newly laid eggs, after proper storage of the eggs for 7 days, *Campylobacter* was no longer present. It was also shown that, even if penetration into the internal contents had occurred, it is unlikely that the bacteria survived in the egg during the storage period (Neill et al., 1985). This was also the case with the study conducted by Doyle (1984). Although *Campylobacter* can be found on the eggshell, unlike *Salmonella*, it does not survive well on the eggshell for a prolonged period due to the dry environment on the eggshell (Sexton, 2016).

4.2 Other bacteria

Foodborne pathogens such as *S. aureus*, *B. cereus*, *L. monocytogenes*, *E. coli*, *Yersinia enterocolitica* have been isolated and/or detected from the eggshell surface.

A high degree of egg contamination by *Staphylococcus* spp. has been reported (Stepien-Pysniak et al., 2009), with *S. aureus* as a predominant species; however, studies also suggested that cuticle and outer eggshell protein extracts could inhibit the growth of *S. aureus*. The contamination of the eggshell by *Staphylococcus* spp. could be attributed to environmental or handling contamination, as well as from the hen.

B. cereus are endospore-forming bacteria that are able to survive in harsh environmental conditions, and frequently contaminate the raw material used in food production because of their ubiquitous distribution in nature (Jan et al., 2011). These spore-forming bacteria are adhesive and can attach to the surfaces of equipment used in the food industry, where bacteria might sporulate and multiply under favourable conditions (Tauveron et al., 2006). Ovotransferrin in the egg white has been shown to have inhibitory effects on the survival of *B. cereus* (Baron et al., 2014).

Hens can also carry *L. monocytogenes*, although whether they contribute to foodborne illness via eggs is not clear (Pohjola et al., 2016). There are some reports on the detection of *L. monocytogenes* from the eggshell surface (Guzman-Gomez et al., 2013; Jones and Musgrove, 2007b); however, ready-to-eat seafood, raw meat, vegetables, cheese, soft served ice-creams are listed as high-risk food for *Listeria* spp. Although the bacteria can be found in the poultry environment (Schwaiger et al., 2010), its association and/or attribution to the consumption of eggs is not as high as for other foodborne pathogens such as *Salmonella* spp. Moreover, *L. monocytogenes* does not seem to thrive well on the eggshell surface at 5 and 20°C (Brackett and Beuchat, 1992).

Studies performed on eggs collected from caged layer flocks or commercial egg grading facilities reported that *Escherichia* was the most frequently detected genus (60.8%) (Gole et al., 2013; Musgrove et al., 2004). While *E. coli* is the most frequently isolated Enterobacteriaceae and is an important human pathogen, the human form is either absent in chickens or present at very low levels and is not considered a significant contributor to human foodborne illness with *E. coli*.

Imai (1980) reported low levels of enterococci on the eggshell surface but suggested that these bacteria had greater resistance to egg washing disinfectants. However, enterococci-related food poisoning has not been widely reported.

Studies have suggested that the presence of C-type lysozyme, ovotransferrin and ovocalyxin-32 in eggshell cuticle extract can reduce the contamination of bacteria such as *Pseudomonas*, *Staphylococcus* and *Bacillus* spp. (Wellman-Labadie et al., 2008).

5 Yeasts and moulds

Yeats and moulds can grow on eggs if eggs are packed wet after washing or stored in a humid environment and these microorganisms are able to survive and grow on eggshells at refrigeration temperature (Musgrove et al., 2008). The growth of yeasts and moulds appears to be dependent upon the pseudomonads for the release of nutrients from the cuticle. Although egg washing can reduce the level of yeast and mould population on the eggshell (Jones et al., 2004), washing is not uniformly practised all over the world. Eggs laid on the floor in a free-range production system or barn production system are highly likely to be contaminated with yeasts and moulds. Yeast and mould levels were reported to be significantly higher in free-range floor eggs than in free-range nest box eggs (Jones et al., 2011).

There is a dearth of detailed literature regarding the virulence and pathogenicity of moulds or fungi; however, it can be deduced that cases of yeast infections attributed to eggs are rare and hence such pathogens are of low risk to human health through egg consumption. However, these pathogens, because of their survivability in the relatively harsh environment of the eggshell, can spoil eggs and reduce the shelf life if eggs are not dried post-washing or stored properly.

6 Conclusions and future trends

Although there has been a substantial effort across the international egg industry to implement a range of on-farm and grading floor risk mitigation strategies to reduce the bacterial load on eggs, continuous efforts for improvement need to be maintained; the risks associated with the food service sector or through the egg supply chain cannot be ignored. *In vitro* work on *Salmonella* virulence characterization has demonstrated that the invasive capacity of *S.* Typhimurium isolated from egg wash, to a human gut cell-line, was low, although post-enrichment of this same *S.* Typhimurium resulted in a ten-fold increase in its invasive capacity (McWhorter et al., 2015). Therefore, if there is any stage through the supply chain and food preparation or storage that encourages the growth or enrichment of *Salmonella* within the food item, the risk of potential food poisoning increases. It has been reported in the literature that some specific virulence genes from specific pathogenicity Island 1 (SPI 1) of *Salmonella* were highly expressed after enrichment (Patterson et al., 2012). A recent comparative analysis of the whole genome sequence of virulent and non-virulent non-typhoidal *Salmonella* serovars indicated that there may be defects in genes of SPI 1, 2, 3, 4 and 5 responsible for variation in intestinal epithelial invasion and replication within the cell and macrophages (McWhorter and Chousalkar, 2015).

It is important to note that there is no single *Salmonella* intervention strategy that will provide an absolute control of the risk that *Salmonella* associated with eggs may pose to human health. There needs to be ongoing efforts from farm to fork to further reduce the risk of human illness from the consumption of eggs and/or egg products.

7 References

Adesiyun, A., Offiah, N., Seepersadsingh, N., Rodrigo, S., Lashley, V. and Musai, L. 2007. Antimicrobial resistance of Salmonella spp. and Escherichia coli isolated from table eggs. *Food Control*, 18, 306–11.

Alabdeh, M., Lechevalier, V., Nau, F., Gautier, M., Cochet, M. F., Gonnet, F., Jan, S. and Baron, F. 2011. Role of incubation conditions and protein fraction on the antimicrobial activity of egg white against Salmonella Enteritidis and Escherichia coli. *J. Food Prot.*, 74, 24–31.

Baron, F., Briandet, R., Lesne, J., Humbert, F., Ablain, W. and Gautier, M. 2004. Influence of a nonfavorable environment, egg white, on resistance to heat and disinfectant, adhesion, and virulence of Salmonella enteritidis. *J. Food Prot.*, 67, 2269–73.

Baron, F., Jan, S., Gonnet, F., Pasco, M., Jardin, J., Giudici, B., Gautier, M., Guerin-Dubiard, C. and Nau, F. 2014. Ovotransferrin plays a major role in the strong bactericidal effect of egg white against the Bacillus cereus group. *J. Food Prot.*, 77, 955–62.

Baron, F., Nau, F., Guerin-Dubiard, C., Bonnassie, S., Gautier, M., Andrews, S. C. and Jan, S. 2016. Egg white versus Salmonella Enteritidis! A harsh medium meets a resilient pathogen. *Food Microbiol.*, 53, 82–93.

Board, R. G. 1966. Review article: the course of microbial infection of the hen's egg. *J. Appl. Bacteriol.*, 29, 319–41.

Board, R. G. and Tranter, H. S. 1995. *The Microbiology of Eggs,* New York, Food Products Press, NY.

Brackett, R. E. and Beuchat, L. R. 1992. Survival of listeria-monocytogenes on the surface of egg-shells and during frying of whole and scrambled eggs. *J. Food Prot.*, 55, 862–5.

Chousalkar, K. K., Sexton, M., McWhorter, A., Hewson, K., Martin, G., Shadbolt, C. and Goldsmith, P. 2015. Salmonella typhimurium in the australian egg industry: multidisciplinary approach to addressing the public health challenge and future directions. *Crit. Rev. Food Sci. Nutr.*

Clavijo, R. I., Loui, C., Andersen, G. L., Riley, L. W. and Lu, S. 2006. Identification of genes associated with survival of Salmonella enterica serovar Enteritidis in chicken egg albumen. *Appl. Environ. Microbiol.*, 72, 1055–64.

Davies, R. and Breslin, M. 2004. Observations on Salmonella contamination of eggs from infected commercial laying flocks where vaccination for Salmonella enterica serovar Enteritidis had been used. *Avian Pathol.*, 33, 135–46.

De Buck, J., Van Immerseel, F., Haesebrouck, F. and Ducatelle, R. 2004. Colonization of the chicken reproductive tract and egg contamination by Salmonella. *J. Appl. Microbiol.*, 97, 233–45.

De Reu, K., Grijspeerdt, K., Messens, W., Heyndrickx, M., Uyttendaele, M., Debevere, J. and Herman, L. 2006. Eggshell factors influencing eggshell penetration and whole egg contamination by different bacteria, including Salmonella enteritidis. *Int. J. Food Microbiol.*, 112, 253–60.

De Reu, K., Messens, W., Heyndrickx, M., Rodenburg, T. B., Uyttendaele, M. and Herman, L. 2008. Bacterial contamination of table eggs and the influence of housing systems. *Worlds Poult. Sci. J.*, 64, 5–19.

Doyle, M. P. 1984. Association of Campylobacter jejuni with laying hens and eggs. *Appl. Environ. Microbiol.*, 47, 533–6.

Ebel, E. and Schlosser, W. 2000. Estimating the annual fraction of eggs contaminated with Salmonella enteritidis in the United States. *Int. J. Food Microbiol.*, 61, 51–62.

Gantois, I., Ducatelle, R., Pasmans, F., Haesebrouck, F., Gast, R., Humphrey, T. J. and Van Immerseel, F. 2009. Mechanisms of egg contamination by Salmonella Enteritidis. *FEMS Microbiol. Rev.*, 33, 718–38.

Gole, V. C., Chousalkar, K. K. and Roberts, J. R. 2013. Survey of Enterobacteriaceae contamination of table eggs collected from layer flocks in Australia. *Int. J. Food Microbiol.*, 164, 161–5.

Gole, V. C., Chousalkar, K. K., Roberts, J. R., Sexton, M., May, D., Tan, J. and Kiermeier, A. 2014a. Effect of egg washing and correlation between eggshell characteristics and egg penetration by various Salmonella Typhimurium strains. *PLoS ONE*, 9, e90987.

Gole, V. C., Roberts, J. R., Sexton, M., May, D., Kiermeier, A. and Chousalkar, K. K. 2014b. Effect of egg washing and correlation between cuticle and egg penetration by various Salmonella strains. *Int. J. Food Microbiol.*, 18–25; 182–3.

Gole, V. C., Torok, V., Sexton, M., Caraguel, C. G. and Chousalkar, K. K. 2014c. Association between indoor environmental contamination by Salmonella enterica and contamination of eggs on layer farms. *J. Clin. Microbiol.*, 52, 3250–8.

Guzman-Gomez, G., Valdovinos, M. A. A., Cabrera-Diaz, E., Perez-Montano, J. A., Munoz-Valle, J. F., Torres-Vitela, M. R. and Ruiz-Quezada, S. L. 2013. Frequency of salmonella and listeria monocytogenes in five commercial brands of chicken eggs using a combined method of enrichment and nested-PCR. *J. Food Prot.*, 76, 429–34.

Humphrey, T. J., Baskerville, A., Chart, H. and Rowe, B. 1989. Infection of egg-laying hens with Salmonella enteritidis PT4 by oral inoculation. *Vet. Rec.*, 125, 531–2.

Imai, C. 1980. Behavior of enterococci in egg processing operations. *Poult. Sci.*, 59, 1767–72.

Jan, S., Brunet, N., Techer, C., Le Marechal, C., Kone, A. Z., Grosset, N., Cochet, M. F., Gillard, A., Gautier, M., Puterflam, J. and Baron, F. 2011. Biodiversity of psychrotrophic bacteria of the Bacillus cereus group collected on farm and in egg product industry. *Food Microbiol.*, 28, 261–5.

Jones, D. R., Anderson, K. E. and Musgrove, M. T. 2011. Comparison of environmental and egg microbiology associated with conventional and free-range laying hen management. *Poult. Sci.*, 90, 2063–8.

Jones, D. R., Guard, J., Gast, R. K., Buhr, R. J., Fedorka-Cray, P. J., Abdo, Z., Plumblee, J. R., Bourassa, D. V., Cox, N. A., Rigsby, L. L., Robison, C. I., Regmi, P. and Karcher, D. M. 2016. Influence of commercial laying hen housing systems on the incidence and identification of Salmonella and Campylobacter. *Poult. Sci.*, 95, 1116–24.

Jones, D. R. and Musgrove, M. T. 2007a. Pathogen prevalence and microbial levels associated with restricted shell eggs. *J. Food Prot.*, 70, 2004–7.

Jones, D. R. and Musgrove, M. T. 2007b. Pathogen prevalence and microbial levels associated with restricted shell eggs. *J. Food Prot.*, 70, 2004–7.

Jones, D. R., Musgrove, M. T. and Northcutt, J. K. 2004. Variations in external and internal microbial populations in shell eggs during extended storage. *J. Food Prot.*, 67, 2657–60.

Lucisano, M., Hidalgo, A., Comelli, E. M. and Rossi, M. 1996. Evolution of chemical and physical albumen characteristics during the storage of shell eggs. *J. Agric. Food Chem.*, 44, 1235–40.

Martelli, F. and Davies, R. H. 2012. Salmonella serovars isolated from table eggs: An overview. *Food Res. Int.*, 45, 745–54.

McWhorter, A. R. and Chousalkar, K. K. 2015. Comparative phenotypic and genotypic virulence of Salmonella strains isolated from Australian layer farms. *Front Microbiol.*, 6, 12.

McWhorter, A. R., Davos, D. and Chousalkar, K. K. 2015. Pathogenicity of Salmonella strains isolated from egg shells and the layer farm environment in australia. *Appl. Environ. Microbiol.*, 81, 405–14.

Messelhausser, U., Tharigen, D., Elmer-Englhard, D., Bauer, H., Schreiner, H. and Holler, C. 2011. Occurrence of thermotolerant Campylobacter spp. on eggshells: a missing link for food-borne infections? *Appl. Environ. Microbiol.*, 77, 3896–7.

Moats, W. A. 1980. Classification of bacteria from commercial egg washers and washed and unwashed eggs. *Appl. Environ. Microbiol.*, 40, 710–4.

Musgrove, M. T. 2011. Microbiology and safety of table eggs. In Immerseel, F. V., Nys, Y. and Bain, M. (eds), *Improving the Safety and Quality of Eggs and Egg Products*. Cambridge: Woodhead Publishing Limited.

Musgrove, M. T., Jones, D. R., Hinton Jr., A., Ingram, K. D. and Northcutt, J. K. 2008. Identification of yeasts isolated from commercial shell eggs stored at refrigerated temperatures. *J. Food Prot.*, 71, 1258–61.

Musgrove, M. T., Jones, D. R., Northcutt, J. K., Cox, N. A. and Harrison, M. A. 2004. Identification of Enterobacteriaceae from washed and unwashed commercial shell eggs. *J. Food Prot.*, 67, 2613–6.

Neill, S. D., Campbell, J. N. and Obrien, J. J. 1985. Egg penetration by Campylobacter-Jejuni. *Avian Pathol.*, 14, 313–20.

Pande, V. V., Devon, R., Sharma, P. S., Mcwhorter, A. and Chousalkar, K. 2016. Study of Salmonella Typhimurium infection in laying hens. *Front. Microbiol.*, 7.

Parisi, M. A., Northcutt, J. K., Smith, D. P., Steinberg, E. L. and Dawson, P. L. 2015. Microbiological contamination of shell eggs produced in conventional and free-range housing systems. *Food Control*, 47, 161–5.

Patterson, S. K., Borewicz, K., Johnson, T., Xu, W. and Isaacson, R. E. 2012. Characterization and differential gene expression between two phenotypic phase variants in Salmonella enterica serovar Typhimurium. *PLoS ONE*, 7, e43592.

Pohjola, L., Nykasenoja, S., Kivisto, R., Soveri, T., Huovilainen, A., Hanninen, M. L. and Fredriksson-Ahomaa, M. 2016. Zoonotic public health hazards in backyard chickens. *Zoonoses Public Health*, 63, 420–30.

Ray, A., Roberts, J. R., Flavel, R. and Chousalkar, K. K. 2015. Eggshell penetration by Salmonella Typhimurium in table eggs: Examination of underlying eggshell structures by micro-computed tomography and scanning electron microscopy. *Food Res. Int.*, 78, 34–40.

Roberts, J. R. 2004. Factors affecting egg internal quality and egg shell quality in laying hens. *J. Poult. Sci.*, 41, 161–77.

Sahin, O., Kobalka, P. and Zhang, Q. 2003. Detection and survival of Campylobacter in chicken eggs. *J. Appl. Microbiol.*, 95, 1070–9.

Samiullah, Roberts, J. R. and Chousalkar, K. K. 2014. Effect of production system and flock age on egg quality and total bacterial load in commercial laying hens. *J.Appl. Poult. Res.*, 23, 59–70.

Schwaiger, K., Schmied, E. M. and Bauer, J. 2010. Comparative analysis on antibiotic resistance characteristics of Listeria spp. and Enterococcus spp. isolated from laying hens and eggs in conventional and organic keeping systems in Bavaria, Germany. *Zoonoses Public Health*, 57, 171–80.

Shanker, S., Lee, A. and Sorrell, T. C. 1986. Campylobacter jejuni in broilers: the role of vertical transmission. *J. Hyg (Lond)*, 96, 153–9.

Stepien-Pysniak, D., Marek, A. and Rzedzicki, J. 2009. Occurrence of bacteria of the genus Staphylococcus in table eggs descended from different sources. *Pol. J. Vet. Sci.*, 12, 481–4.

Tauveron, G., Slomianny, C., Henry, C. and Faille, C. 2006. Variability among Bacillus cereus strains in spore surface properties and influence on their ability to contaminate food surface equipment. *Int. J. Food Microbiol.*, 110, 254–62.

Wellman-Labadie, O., Picman, J. and Hincke, M. T. 2008. Antimicrobial activity of cuticle and outer eggshell protein extracts from three species of domestic birds. *British Poultry Science*, 49, 133–43.

Chapter 5

Mechanisms for the transmission of pathogens into eggs

Sophie Jan and Florence Baron, Agrocampus Ouest-INRA, France

1 Introduction

The egg ensures the nutritional needs of the chicken embryo and the preservation of a sterile environment during its development. It thus possesses a remarkable arsenal of antimicrobial defences. However, the conditions of egg production and, in particular, the health status of the livestock and the conditions of egg collection, packaging and storage may lead to microbial invasion. Therefore, the egg can convey and transmit certain microorganisms that may cause human food poisoning. *Salmonella* Enteritidis is considered the pathogen currently representing the major risk of eggborne diseases in the European Union (EU) (EFSA BIOHAZ Panel, 2014). The number of human salmonellosis cases in the EU in 2010 was estimated to be 5.4 million with 57%, 17%, 11% and 2.6% of cases attributed to pig, laying hen, broiler and turkey reservoirs, respectively.

In France, between 1996 and 2005, the most common microorganisms isolated from outbreaks caused by egg and egg products were *Salmonella* (86%), *Staphylococcus aureus* (7.4%), *Clostridium perfringens* (2%) and other agents (4%) (Delmas et al., 2006).

http://dx.doi.org/10.19103/AS.2016.0012.05

Salmonella spp. also accounted for more than 85% of strong-evidence outbreaks caused by eggs and egg products in the EU in 2012. The serovar *S.* Enteritidis was reported in 67% of these outbreaks. Two outbreaks were caused by *Bacillus cereus* (0.6%) and staphylococcal (0.6%) toxins (EFSA and ECDC, 2014).

This chapter particularly focuses on *S.* Enteritidis because of its particular ability to invade the hen's genital tract and to resist the egg's physical and mechanical defence mechanisms. A focus will be given on the two main mechanisms for pathogen transmission into the egg's internal compartments. The first route of transmission is named vertical contamination or primary contamination. It corresponds to the contamination of the egg during its formation in the hen's genital tract. The second one, named secondary contamination, is due to penetration of the microorganisms present on the eggshell surface once the egg is laid (horizontal contamination).

2 Microbial hazards associated with egg consumption

2.1 *S.* Enteritidis

Several serovars and subspecies of *Salmonella*, such as *S.* Enteritidis, *S.* Typhimurium and *S.* Heidelberg can cause paratyphoid disease in chickens (Shivaprasad et al., 2013). These serovars can also cause foodborne infections in humans (Rabsch et al., 2001; Foley et al., 2011). In the paratyphoidal diseases, infection is mainly located in the intestinal epithelium, leading to faecal excretion (Shivaprasad et al., 2013). *S.* Enteritidis is particularly effective at spreading from the gastrointestinal tract to various organs, notably the ovary (Gast and Beard, 1990; Gast, 1994; Peralta et al., 1994; Thiagarajan et al., 1996; Rajashekara et al., 2000; Rank et al., 2009; Berghaus et al., 2011), causing a systemic infection (Humphrey et al., 1989; Keller et al., 1995; Okamura et al., 2001, 2010; Shivaprasad et al., 2013). The frequent association of *S.* Enteritidis with egg infection is ascribed to its spread and persistence in the parental breeder flock population (Thorns, 2000), its long-term unique ability to colonize the ovary and oviduct of laying hens (EFSA BIOHAZ Panel, 2010) and its exceptional survival capability within the harsh conditions provided by egg white (see Baron et al., 2016 for a review).

In humans, the essential characteristic of *S.* Enteritidis pathogenicity is its ability to enter host cells and to remain as a facultative intracellular parasite (Finlay and Falkow, 1989).

The process of invasion begins in the gastrointestinal tract. *Salmonella* is ingested and, after passing through the digestive tract, penetrates the intestinal cells in the ileum. Bacteria adhere through adhesin receptors that specifically recognize certain eukaryotic epithelial cells. These adhesins may be associated with fimbriae, flagella or the outer membrane (Finlay and Falkow, 1989). The internalization of *Salmonella* involves the induction of various signals. Bacteria interact with the epithelial cells, form appendages called invasomes (Ginocchio et al., 1994) and cause effacement of the microvilli and of the brush border. The latter is then regenerated and bacteria are observed in vacuoles, allowing them to multiply, sheltered from the immune system. There is then a systemic stage where bacteria are phagocytosed by host macrophages. Bacteria are adapted to grow inside the macrophages after a latency period. If this is the case, infected macrophages can spread to various organs via the lymphatic and blood circulatory systems. Bacteria are found in abundance in the lymph nodes and in the macrophages of liver and spleen (Nnalue et al., 1992; Uzzau et al., 2000), organs for which they have a particular tropism. The secretion of an enterotoxin is partly responsible for gastroenteritis-like symptoms, if the infection remains localized in the intestine. This toxin causes destruction of the villi of the intestinal mucosa. The intestinal absorption functions are disturbed and, a significant inflammation increases secretions.

Dynamics of *Salmonella* infection in the laying hen flock depend, among other contributing factors, on the housing system, the flock management (flock size, flock age, type of housing system, age and size of the farm, hygiene practices), the breed of hen, the *Salmonella* strain, the time of exposure and the presence of vectors (rodents, flies, litter beetles, red mites) (Howard et al., 2012; Wales and Davies, 2013; Van Hoorebeke et al., 2011). Transmission of *Salmonella* inside the flock occurs via hen-to-hen contact and ingestion of materials contaminated with infected faeces (Holt et al., 1998), feed and/or water (Nakamura et al., 1994, 1997; Holt, 1995) and/or air (dust and aerosols) (Baskerville et al., 1992; Nakamura et al., 1997; Gast et al., 1998; Holt et al., 1998). In the EU, initiatives to control *Salmonella* infection in commercial flocks of laying hens started in some Member States between the end of the 1980s and the beginning of the 1990s (AFSSA, 2009 and Defra, 2007). These initiatives have led to a marked decrease in the incidence of *S.* Enteritidis infection in laying hen flocks, and of human infections due to this serotype (EFSA BIOHAZ Panel, 2014). Detection of *Salmonella* in eggs remains difficult due to the low prevalence of positive eggs, even when produced by the most infected flocks, and due to the inhibitory effect of egg albumen (EFSA BIOHAZ Panel, 2014). Environmental sampling, including dust sampling (Riemann et al., 1998; Arnold et al., 2014; Davies and Wray, 1996), has been recognized as an effective way to detect zoonotic serotypes of *Salmonella* in a poultry flock (Aho, 1992; Johansson et al., 1996; Musgrove and Jones, 2005). Samples taken from egg conveyors and grading equipment are recognized as providing more sensitive detection of contamination of batches of eggs that have been handled (EFSA BIOHAZ Panel, 2014). Testing of *Salmonella* is mandatory in France. It involves the analysis of fresh caecal droppings and surface swabs applied on poultry house surfaces. If there is a suspicion of infection, the flock is eradicated (Viénot and Le Boucher, 2007) and egg sale is placed under the official supervision of the competent veterinary authorities. The only authorized purchaser of *Salmonella*-contaminated eggs is the egg-processing industry. Upon receipt, the eggs coming from infected flocks must undergo thermal treatment.

According to Van Immerseel et al. (2005), vaccination of laying hens against *S.* Enteritidis, although not fully protective, is likely to reduce faecal excretion and ovarian transmission, thereby directly contributing to public health by reducing the level of on-farm contamination, including the level of contamination of table eggs (Davies and Breslin, 2004; Gantois et al., 2006). Among other practices, vaccination is thereby considered the most effective measure for reduction of eggborne infection in humans, particularly in the European Member States exhibiting high farm prevalence (more than 10%) of *Salmonella* infection (EFSA, 2007; EFSA BIOHAZ Panel, 2014).

The *Salmonella* control programmes now in place in the EU may have contributed to the 42% reduction in the number of *Salmonella* spp. foodborne outbreaks caused by eggs reported from 2007 to 2011. This decrease correlates with the reduction of *S.* Enteritidis in laying hen flocks and of *Salmonella* spp. in table eggs (from 0.8 to 0.1%) (EFSA BIOHAZ Panel, 2014).

Since January 2012, the new EU welfare legislation has led to the ban of conventional 'battery' cages in favour of decommissioned or refurbished cages. These alternative systems, involving smaller flocks, are also thought to be less conducive to *Salmonella* infection (Van Hoorebeke et al., 2011, 2012; Galiş et al., 2013; EFSA BIOHAZ Panel, 2014).

2.2 Pathogens other than *Salmonella*

Among other microbial hazards associated with egg consumption, a small number of outbreaks due to *B. cereus* group (0.6%; two outbreaks) and *S. aureus* (0.6%) toxins have been reported in the EU in 2012 (EFSA and ECDC, 2013).

The role of the *B. cereus* group in human infection due to egg consumption remains unclear. These spore formers are recognized as being poorly represented on the eggshell surface, provided that biosecurity practices are correctly implemented on the farms (Koné et al., 2013). They are, rather, present in the egg-processing environment due to their selection by the industrial processes of egg product stabilization, that is, pasteurization and chilling (Baron et al., 2007; Jan et al., 2011) and by their ability to develop in the form of biofilms on the stainless steel processing equipment (Andersson et al., 1995). In pasteurized liquid egg products, psychrotrophic *B. cereus* group bacteria are often isolated, and they mainly lead to spoilage events even at refrigeration temperatures (Baron et al., 2007; Jan et al., 2011; Techer et al., 2014). The involvement of this bacterial group in the rare cases of foodborne diseases associated with egg consumption may be put in perspective by the fact that mixed foods are the most commonly implicated *B. cereus* food vehicles (EFSA and ECDC, 2014). Other ingredients could have therefore conveyed these bacteria in the reported cases in the EU in 2012 (EFSA and ECDC, 2013) and data are insufficient to identify the specific contribution of the egg component in these outbreaks (EFSA BIOHAZ Panel, 2014). Moreover, the possible confusion between the symptoms caused by the *B. cereus* group toxins and those caused by *C. perfringens* and *S. aureus* toxins may be treated with caution (Cadel Six et al., 2012). Otherwise, foodborne intoxications due to this bacterial group are recognized as generally mild and rapidly resolved, and clinical diseases are mainly associated with high levels of exposure.

The role of eggs as a potential source of infections by antibiotic-resistant *Staphylococcus* strains in humans has been considered in several studies (Manie et al., 1998; Abulreesh and Organji, 2011). According to Bahrouz and Al-Jaff (2005), staphylococci present on the eggshell surface could invade the egg content by crossing the eggshell during passage through the cloaca or after it has been laid. In the UK, between 1969 and 1990, 3.5% of cases of staphylococcal food poisoning were attributed to the consumption of eggs contaminated with *Staphylococcus* (Wieneke et al., 1993), and 11% in France between 1999 and 2000 (Haeghebaert et al., 2002). In localized outbreaks of food poisoning and foodborne infection in Poland, 6.9% cases in 2001 (Przybylska, 2002, 2003) and 25% cases in 2009 (Baumann-Popczyk and Sadkowska-Todys, 2011) were linked to eggs contaminated with *S. aureus*.

Other pathogens, such as *Listeria monocytogenes*, can be found on or in eggs, but eggs are not reported as being a cause of outbreaks (EFSA BIOHAZ Panel, 2014). Regarding *Campylobacter*, although its ability to colonize the hen's genital tract has been reported, there is no evidence for transmission of this pathogen to the egg (Sahin et al., 2003; Callicott et al., 2006; Fonseca et al., 2006), and eggshell contamination with *Campylobacter* has not been widely investigated. Finally, although colonization of the chicken intestinal tract with *Escherichia coli* O157:H7 has been described (Berry et al., 1985; Schoeni and Doyle,

1994; Kabir, 2010), no outbreaks have been reported due to the contamination of the eggshell or the egg content with this pathogenic strain.

3 Vertical transmission

At the time of ovodeposition, the egg content is generally sterile, provided that it is produced in a healthy environment. In the case of an infection of the hen's genital tract, the egg content may be contaminated at the time of ovodeposition. This is what is called vertical or primary contamination. It mainly concerns *Salmonella*, but several studies also suggest vertical contamination with bacteria of the species *Campylobacter jejuni* and the avian influenza virus.

The contamination of the egg content following artificial inoculation with *S.* Enteritidis has been tested in different studies. However, this transmission route is sporadic and occurs when high bacterial concentrations are inoculated (about 8–9 log cells per hen). Depending on the studies, the percentage of infected eggs varies from 0 to 8.1% (Shivaprasad et al., 1990; Humphrey et al., 1991; Gast and Beard, 1992; Keller et al., 1995; Gast and Holt, 2000a; Gast et al., 2002). These results are in accordance with those obtained for eggs collected in poultry flocks naturally infected with *Salmonella*: 0.06% according to Poppe et al. (1991), 0.6% according to Lock and Board (1992) and 0.3% according to Elson et al. (2005). Whatever the route of inoculation, the eggs are contaminated at low levels, ranging from 1 to 400 cells per egg and rarely above 40 cells per egg (Gast and Beard, 1990, 1992; Gast and Holt, 2000a; Gast et al., 2002). For eggs produced by hens naturally infected with *S.* Enteritidis, the content of positive eggs was found to contain fewer than 10 (Humphrey et al., 1989) and fewer than 20 cells of *S.* Enteritidis per egg (Humphrey et al., 1991).

The deposition site of *Salmonella* in the egg is recognized as depending on the site of colonization in the reproductive tract. Contamination of the magnum, isthmus and uterus may result in contamination of the albumen, the eggshell membranes and the eggshell itself, respectively (Gantois et al., 2009).

When eggs contaminated with *S.* Enteritidis are analysed at the time of ovodeposition, bacteria are most frequently isolated from the content of the egg rather than on the eggshell. This observation suggests that the ovary or the upper part of the oviduct (magnum) may be the source of contamination of the egg and that egg contamination has taken place prior to the deposition of the eggshell (Gast and Beard, 1990; Suzuki, 1994). Several studies show that egg white is far more frequently contaminated than egg yolk (Humphrey et al., 1989; Gast and Beard, 1990, 1992; Shivaprasad et al., 1990; Thorns, 2000). Moreover, bacteria are mainly detected in the magnum, the site of egg white deposition during egg's formation (Shivaprasad et al., 1990; Gast and Beard, 1990; Keller et al., 1995). The yolk membrane is also recognized as a favoured site for infection by *Salmonella* (Humphrey et al., 1989; Timoney et al., 1989; Shivaprasad et al., 1990; Gast and Holt, 2000a, 2001; Gast et al., 2002, 2013). Direct contamination inside the yolk is a rare event (Gast and Holt, 2000a, 2001; Gast et al., 2002, 2003). Eggshell and shell membranes can also become contaminated during egg development (De Buck et al., 2004), if isthmus, shell gland and/or vagina are infected by *Salmonella* either via a descending route of infection from the bloodstream and the ovaries or via an ascending route of infection from the cloaca (EFSA BIOHAZ Panel, 2014).

According to Hopper and Mawer (1988), the low prevalence of contaminated eggs laid by infected hens may be ascribed to the fact that both infected and healthy eggs can coexist within a single ovary. Other authors believe that the high frequency of shell membrane contamination could be related to the contamination of the distal portion of the oviduct (Okamura et al., 2001; De Buck et al., 2004). Although the eggshell is weakly contaminated at the time of ovodeposition compared with the internal egg compartments, the hypothesis of an ascending vaginal colonization by *Salmonella* has been proposed by Mizumoto et al. (2005), representing a putative route of eggshell infection. This event could be favoured if artificial insemination is carried out (Reiber et al., 1995). Under such conditions, the contamination is confined to the eggshell and remains sporadic. Keller et al. (1995) observed, after oral inoculation, that the number of contaminated eggs was higher when they were collected in the oviduct (28.25%) than at the time of ovodeposition (0–0.6%). These authors suggest that inhibiting factors present in the egg contents, such as antibodies, antibacterial proteins, protease inhibitors and proteins chelating essential elements, would be responsible for the decrease of the contamination during the egg formation in the genital tract. This hypothesis is supported by Barrow and Lovell (1991), Okamura et al. (2001) and Keller et al. (1997), who observed that the presence of *Salmonella* in the ovaries of infected hens did not necessarily lead to the production of contaminated freshly laid eggs. Another explanation is based on the assumption that the contaminated follicles degenerate before the completion of egg formation, due to extensive multiplication of *Salmonella* in this nutrient-rich environment at the chicken body temperature of 42°C (Gantois et al., 2009).

The review of the literature reveals differences between the mortality rate of infected chickens, the clinical symptoms, the time of faecal carriage, the prevalence of infected eggs and the most involved organs or tissues. These differences could be explained by the variety of the strains studied, their phage type, the inoculum size, the method of hen inoculation (oral, intravenous or intraperitoneal), the age and lineage of the hen and the mode of egg collection and analysis.

Shivaprasad et al. (1990) have shown differences between three strains of *S.* Enteritidis isolated from humans, egg yolk or hen's ovary and orally inoculated to hens. The first strain of human origin was shown to cause diarrhoea, anorexia, reduced egg production and, in some cases, death. Bacteria were found on some eggshells but never in the egg contents. The strain isolated from egg yolk caused diarrhoea, anorexia and a slight decrease of egg production. The contamination concerned mainly the eggshell, and sometimes the egg white. The strain isolated from hen's ovary caused no abnormal symptoms in inoculated hens and did not affect egg production, despite colonizing the liver, pancreas and oviduct. The bacteria were localized in egg yolk and egg white, but this contamination involved a small number of eggs.

Barrow (2007) also reported differences in the hen mortality depending on the phage type of the inoculated strain. Similarly, Mizumoto et al. (2005) showed that *S.* Enteritidis has various abilities to colonize the vagina, depending on the profile of the surface lipopolysaccharides of the strains tested.

Regarding the effect of hen lineage, Keller et al. (1995) have observed that the tissues of ISA brown hens were more contaminated than those of Leghorn hens by the bacterium tested in their study. The age of the hen is also important: the resistance of the hens against *Salmonella* generally increases with age, probably in connection with the development of a mature intestinal flora and an effective immune system (Suzuki, 1994). However, between 27 and 60 weeks, there would be no effect of hen's age on the level of eggshell contamination (Protais et al., 2003).

The colonization of reproductive tissues has also been described for other *Salmonella enterica* serotypes, such as *S.* Heidelberg (Gast et al., 2004, 2007a), *S.* Typhimurium (Keller et al., 1997; Reiber et al., 1995), *S.* Pullorum (Wigley et al., 2005), *S.* Hadar and *S.* Infantis (Martelli and Davies, 2012). These serotypes can occasionally contaminate eggs, but internal contamination is not as great as for *S.* Enteritidis (EFSA BIOHAZ Panel, 2014; Chemaly et al., 2009; Martelli and Davies, 2012).

Considering non-*Salmonella* pathogens, whereas the ability of *Campylobacter* to colonize reproductive tracts has been reported, there is a lack of evidence for vertical transmission of this pathogen into eggs (Sahin et al., 2003; Callicott et al., 2006; Fonseca et al., 2006).

A vertical contamination is possible with the avian influenza virus (type A), which causes viral disease in several species of birds, including chickens. The type A influenza viruses are classified into subtypes depending on the antigenic characteristics of two surface glycoproteins: the neuraminidase (9 N-subtypes) and the haemagglutinin (16 H-subtypes). Each subtype can be divided into many more or less pathogenic variants. Highly pathogenic strains are distinguished from low pathogenic strains. While all subtypes appear to infect all bird species, five of them (H1Nx, H2Nx, H3Nx, HxN1 and HxN2) have been shown to have features more suitable for human infection. Most strains have a tropism for the respiratory and the digestive systems, but some highly pathogenic strains have the ability to spread rapidly in all tissues of infected birds.

Although infected birds stop laying, eggs produced in the early stage of the disease can potentially be contaminated. The virus has been recovered in up to 7% of egg white and 57% of egg yolk in eggs collected in contaminated farms during the 1983–4 American episode. However, it was not found in the 9930 eggs analysed during the 1996–8 episode, although it was detected in the oviduct of the hens (Swayne and Beck, 2004). For the 2001–3 episode, the virus was not detected in the eggs analysed by Lu et al. (2004), although it was detected in the cloacal swabs. However, the virus was isolated from eggs produced by experimentally infected hens. Swayne and Beck (2004) have detected 85–100% positive eggs with viral titres of around 4.9 log EID_{50}/ml (EID_{50} is the median infective dose necessary to infect half of the embryos). However, this virus is not considered as a food safety risk and, in the event of an epidemic, specific measures would be implemented by the competent authorities.

To conclude on vertical contamination, *Salmonella*-infected hens can produce *Salmonella*-contaminated eggs. However, the number of internally contaminated eggs produced by a *Salmonella*-infected flock remains sporadic. In such an eventuality, the storage of infected eggs at refrigeration temperatures and their consumption after cooking would reduce the risk.

4 Horizontal transmission

The contamination of the surface of the eggshell after oviposition occurs through contact of the egg with manure or with microorganisms of the housing environment or, downstream, in the packing centres. The level and types of contamination of the eggshell surface depend on the health status of the livestock and the housing environment.

Depending on the study, the level of the aerobic mesophilic flora reported on the eggshell is between 3.8 and 6.3 log colony-forming units (CFU) per egg, with an average level of around 4.5 log CFU per egg (Moats, 1980; Lucore et al., 1997; Favier et al., 2000; Knape et al., 2002; Jones et al., 2004; De Reu et al., 2005, 2006a; Musgrove et al., 2005). Regarding the type of contamination, Moats (1980) isolated Gram-positive bacteria of the genera *Streptococcus*, *Aerococcus*, *Micrococcus* and *Staphylococcus*, as well as the species *E. coli*. De Reu (2006) showed that the major contaminants present on the eggshell surface were the species *E. coli* (4.7 log CFU per egg) and the genus *Staphylococcus* (4.6 log CFU per egg). Depending on the study, the concentration of enteric bacteria varies from 1.2 to 4.6 log CFU per egg (De Reu et al., 2005, 2006a; Musgrove et al., 2005). Enterobacteria were identified as belonging to the species *E. coli* and the genera *Yersinia*, *Enterobacter*, *Klebsiella*, *Serratia*, *Citrobacter* and *Salmonella* (Musgrove et al., 2004).

Although *Salmonella* are not always detected, the majority of the published studies focus on this genus because of its involvement in egg pathogenicity. The hens represent a primary source of dissemination of *Salmonella*. In studies where hens were orally infected, they were shown to shed the bacteria in their faeces for several weeks after infection (Humphrey et al., 1991; Shivaprasad et al., 1990; Gast and Beard, 1990). Oral contamination of the hens, even at low dose (<2 log), has also been shown to cause intermittent shedding for at least 18 weeks (Van Immerseel et al., 2004).

Another distinctive feature of *Salmonella* is its ability to survive in the housing environment, leading to the contamination of the entire flock or between two different flocks. An illustration of this particular resilience is given by the demonstration of the persistence of *S.* Enteritidis in the housing environment 26 months after the removal of the infected hens (Davies and Beslin, 2003). *Salmonella* can better resist desiccation than other coliforms (Hiramatsu et al., 2005) and is very sensitive to light (Nyeleti et al., 2004). While the poultry houses were empty, Davies and Beslin (2003) detected *Salmonella* for more than 8 months in the dry litter and in the light-protected areas.

Biosecurity of the breeding practices is essential, considering the contribution of many species of wildlife as reservoirs and vectors of *S.* Enteritidis in farms (Liebana et al., 2003). *S.* Enteritidis can be conveyed by arthropods, contributing to rapid spread and survival in the housing environment between two flocks of laying hens. For instance, 50% of the flies can be contaminated after 2 days in a farm containing infected hens (Holt et al., 2007). Poultry lice (*Dermanyssus gallinae*) have direct access to the blood of laying hens and are widespread in the European farms. They can contract *Salmonella* in litter or on their host and disseminate rapidly (Valente Moro et al., 2007). The biting of the hens by infected lice may represent a safety issue since intravenous inoculation of 5–7 log of *Salmonella* to hens was shown to lead to the contamination of the egg content (Gast et al., 2002). Rodents represent another safety issue since they are recognized as potential reservoirs and vectors of microorganisms. They can directly feed in the hen feeders and contaminate the food and the environment. Lapuz et al. (2007) showed very similar typing profiles between *Salmonella* strains isolated from rats and from eggs collected in the same farm.

The feed for laying hens is another potential source of *Salmonella* dissemination, knowing that the same bacterial cellular structures are recognized as being involved in the virulence of *Salmonella* in animals and in the adherence of the bacterium to plant tissues (Barak et al., 2005). The feed may also represent a source of dust. This latter is recognized as participating in the overall level of egg horizontal contamination. It varies according to the activity and the behaviour of the hens (including the spawning behaviour), the use of cages and their type (conventional or enriched), the housing equipment and the bedding material (De Reu et al., 2005). A seasonal effect was observed with poorer contamination during the winter period (De Reu et al., 2005; Wales et al., 2007).

Finally, egg transport and packing represent other risks of horizontal contamination. The eggshell surface may be contaminated by infected environments or through transmission from one egg to another, especially in the case of spreading of egg contents from cracked eggs in the grading machine or upstream. A study carried out with sterilized eggs shows that the packing step can infect at least 0.3% of table eggs (Davies and Breslin, 2003).

5 Secondary contamination: trans-shell penetration

Even if the eggshell surface is contaminated, the egg has different barriers that prevent penetration of microorganisms into the internal egg compartments and access to the nutrient-rich egg yolk compartment. These barriers include, from the outside in, the cuticle, the eggshell and the shell membranes, the albumen and the vitelline membrane.

The cuticle is a film of mucoproteins, which represents a physical barrier impermeable to water and microorganisms (Vadehra et al., 1970; Tung et al., 1979; Haigh and Betts, 1991; Miyamoto et al., 1997; Allen and Griffiths, 2001; De Reu et al., 2006b,c). Some studies show a positive correlation between the thickness of the cuticle and its ability to prevent penetration of microorganisms (De Reu et al., 2006a,b). Others refute this hypothesis (De Reu et al., 2006c; Allen and Griffiths, 2001). The efficiency of this protective layer is limited in time, because of its cracking under the effect of desiccation or treatments (Vadehra et al., 1970; Tung et al., 1979; De Reu et al., 2006b). Haigh and Betts (1991) have observed a strong resistance capacity of the cuticle of fresh eggs (<100 h after laying), whose shell had been subjected to a treatment as strong as application of concentrated nitric acid. The decrease in the protective action of the cuticle may begin from the fourth day of storage at ambient temperature (EFSA BIOHAZ Panel, 2014).

The eggshell is a calcified protein matrix that does not represent an effective physical barrier due to the presence of pores that can represent a route of bacterial penetration. However, there is no evidence of a correlation between the number of pores and the level of trans-shell penetration (Messens et al., 2005; De Reu et al., 2006b). Some pores would not traverse the whole section of eggshell and others would be obstructed by the cuticle (De Reu et al., 2006b). However, Haigh and Betts (1991) have shown that brown eggs better prevent penetration of microorganisms than white eggs, due to lower occurrence of pores. Apart from tangible defects such as cracks, micro-cracks or calcification problems (Fajardo et al., 1995; Ernst and Fuka, 1998; Allen and Griffiths, 2001; Bain et al., 2006), trans-shell penetration would be independent of the quality of the shell (Vadehra et al., 1970; Jones and Musgrove, 2005; de Reu et al., 2006a). Furthermore, the eggshell contains the well-known lysozyme and ovotransferrin antimicrobial proteins, which can play an important physiological role in trans-shell penetration (Hincke et al., 2000; Gautron et al., 2001).

The outer (which adheres to the shell) and internal (which adheres to the albumen) shell membranes are much more effective than the eggshell itself. They consist of glycoprotein fibres organized in networks and functioning as antibacterial filters (Tung et al., 1979). They contain lysozyme, β-N-acetylglucosaminidase and ovotransferrin, whose antimicrobial activities can hinder the progression of bacteria already trapped in the glycoprotein network (Hincke et al., 2000; Gautron et al., 2001; Ahlborn and Sheldon, 2005). The inner shell membrane, although thinner than the outer shell membrane, is more effective than the outer membrane, probably due to its higher lysozyme content (Vadehra et al., 1970;

Hincke et al., 2000). However, isolated inner shell membranes were shown to be permeable to *E. coli* (Haigh and Betts, 1991). As for the eggshell and cuticle, the effectiveness of the protection provided by the shell membranes is questioned for eggs with cracks or showing defects in calcification (Allen and Griffiths, 2001; Ernst and Fuqua, 1998).

The different methods used to study trans-shell penetration sometimes lead to conflicting conclusions. The concentrations of bacterial solutions put in contact with the eggs also vary from one study to another, ranging from 3 to 9 log bacteria/ml. One of the most used methods is to put the egg in contact with a bacterial suspension and to quantify the internal bacterial concentration after egg drying and incubation under various conditions of temperature and humidity. Another method consists of replacing the internal contents of the egg by a nutrient-rich agar medium containing a substrate whose consumption acts as an indicator of microbial growth, causing colouration visible to the naked eye (Haigh and Betts, 1991). Other studies suggest the addition of tetrazolium in the agar medium culture (De Reu et al., 2006b; Messens et al., 2006, 2007). Egg candling allows tracking the growth of bacteria in the form of red colonies, attesting the reduction of tetrazolium into formamide due to bacterial growth. Under these conditions, bacterial penetration is observed 2 days after inoculation. Less invasive methods propose monitoring the penetration of eggs by luminescent strains (Allen and Griffiths, 2001). They offer the advantage of not affecting the integrity of the eggs.

The studies that aim at characterizing the ability of different bacteria to cross the eggshell focus mainly on *S.* Enteritidis. However, *in vitro* studies of the penetration of *S.* Heidelberg (Gast et al., 2005a, 2007a), *S.* Infantis, *S.* Montevideo (Murase et al., 2006), *S.* Typhimurium (Miyamoto et al., 1998; Guan et al., 2006), *Escherichia* (Haigh and Betts, 1991), *Alcaligenes*, *Carnobacterium*, *Staphylococcus*, *Acinetobacter*, *Serratia* (De Reu et al., 2006b), *Campylobacter* (Allen and Griffiths, 2001) and *Pseudomonas* (Vadehra et al., 1970; De Reu et al., 2006b) also proved their ability to penetrate the eggshell. However, mobile Gram-negative bacteria are most likely to penetrate the eggshell and contaminate the egg contents (De Reu et al., 2006b).

Although the models are different, all the results of *in vitro* studies should be considered with caution since the level of artificial eggshell surface contamination is generally greater than the ones observed *in vivo*. Many factors have also been re-examined because previous data were contradictory (Messens et al., 2007). Thus, there is no clear evidence of a correlation between the egg weight, the thickness of the eggshell (Messens et al., 2005, 2007), the number of pores, the quality of the cuticle (Messens et al., 2005), the presence of condensed water on the eggshell surface (Allen and Griffiths, 2001; Ernst and Fuqua, 1998) and the level of bacterial trans-shell penetration. It was proposed that surface moisture, associated with variations in temperature, would promote bacterial penetration by causing retraction of the egg contents and uptake of putative infected water (Miyamoto et al., 1998; Berrang, 1999). However, the conditions of storage of eggs, within the limits of temperature and humidity covering the atmospheric conditions of the farms, the packaging and distribution centres in Western Europe, do not influence their infection rates (Messens et al., 2006). Whereas the age of the hen influences the characteristics of the eggshell, there is no evidence of an effect of this parameter on bacterial trans-shell penetration (Messens et al., 2005; De Reu et al., 2006b). The capacity of penetration seems to be independent of the hen lineage and of the housing system. The type of feed could affect bacterial penetration, but this theory has yet to be confirmed.

To conclude, the key factors influencing trans-shell penetration appear to be the level of contamination of the egg surface (Chen et al., 1996; Miyamoto et al., 1998; Messens et al., 2006, 2007) and the presence of cracks or defects of eggshell calcification (Allen and Griffiths, 2001; Ernst and Fuqua, 1998).

6 The effect of antimicrobial molecules on the behaviour of microorganisms in egg white

The egg white is another line of defence against egg microbial invasion. Unlike the animal's immune system that produces bactericidal compounds depending on the level of the risk exposure to bacteria, egg white can effectively resist microbial invasion for a long time in the absence of innate and systemic immune systems. Many studies have shown that egg white does not offer favourable conditions for microbial growth. It may allow a weak growth but can also have bacteriostatic or bactericidal effects, depending on the microorganism concerned and on the environmental conditions. In addition to the activity of antimicrobial molecules, physicochemical and/or environmental factors are nowadays clearly recognized as being a crucial factor in the antimicrobial efficiency of egg white. In this section, we will discuss the effect of antimicrobial molecules in resisting microbial establishment in the egg. Section 7 will review the influence of the physicochemical and environmental conditions in the egg white, and Section 8 will discuss the migration and behaviour of microorganisms in egg yolk.

The major molecular factors involved in the antibacterial activities of egg white may differ among the bacterial species considered. The role of protection of egg white against microbial invasion underlies a broad spectrum of antibacterial activities. From a microbiological point of view, egg white could be compared to an intracellular fluid protecting the developing embryo in fertilized eggs. However, some pathogens, and more specifically *S.* Enteritidis, develop more or less effective protective responses. The following paragraphs aim at presenting the major types of molecules recognized as involved in the antimicrobial effect of egg white and also the resistance mechanisms of *Salmonella* incubated in this stressful physiological fluid.

Several proteins, such as lysozyme, ovotransferrin and proteinase inhibitors, have all been commonly suggested to explain the ability of egg white to prevent bacterial growth. These molecules are naturally abundant in egg white, but it is noteworthy that some minor proteins and peptides recently revealed by high-throughput approaches might also play a role in the defence against bacterial contamination (Baron et al., 2016).

6.1 Lysozyme

Lysozyme is a ubiquitous protein that can be found in many biological fluids and tissues of a large number of living organisms. This protein is particularly abundant in hen egg white (around 3.5 g/l), where its ability to control the growth of susceptible bacteria represents an important component of the non-specific defence mechanisms of the egg. Several mechanisms are reported to explain its bactericidal activity. The best known activity is the degradation of the β-1,4-glycosidic linkage between the N-acetylglucosamine and the N-acetylmuramic acid residues of the bacterial peptidoglycan of Gram-positive bacteria, leading to the formation of spheroplasts. These are later lysed under low osmotic strength conditions, or when the rate of the synthesis and polymerization process for new peptidoglycan formation is slower than the degradation by lysozyme.

Gram-negative bacteria are generally resistant to the lytic action of lysozyme, because they are protected by the outer membrane (Nikaido, 2003). However, some Enterobacteriaceae, including *S.* Enteritidis, express lysozyme inhibitors in the periplasm (Callewaert et al., 2008; Deckers et al., 2004). Considering the widespread occurrence and effectiveness of lysozyme as a component of innate immunity in animals, it is not surprising that pathogenic bacteria have developed such mechanisms of resistance to hydrolysis by lysozyme. However, the enzymatic activity of lysozyme to the defence against infections is debatable under the alkaline egg white conditions that prevail in the egg a few days after laying (pH around 9). Ibrahim et al. (1996) showed an absence of lytic activity for lysozyme at alkaline pH.

At high pH, lysozyme is thus enzymatically inactive but remains substantially bactericidal presumably due to non-specific and non-hydrolytic mechanisms, such as membrane disruption. Several studies have shown that lysozyme devoid of enzymatic activity retains antibacterial activity against Gram-negative and Gram-positive bacteria (Wang and Shelef, 1991; Düring et al., 1999; Ibrahim et al., 1996). A catalytically inactive site-directed mutant of lysozyme was shown to exhibit bactericidal properties against *Bacillus subtilis* and *S. aureus* (Ibrahim et al., 2001a). Other experiments, showing the antimicrobial activity of peptides issued from lysozyme hydrolysis, suggest the presence of non-enzymatic bacteriostatic domains in the lysozyme (Düring et al., 1999; Ibrahim et al., 1996; Mine et al., 2004; Masschalck and Michiels, 2003). In addition, induction of pore formation in the outer membrane of *E. coli* has recently been ascribed to native lysozyme (Derde et al., 2013), which was shown to insert into the outer bacterial lipopolysaccharide monolayer, leading to the reorganization of the latter. Pore formation was observed for the outer membrane, but not for the inner membrane. However, the membrane potential of the inner membrane was modified by the presence of native lysozyme (Derde et al., 2013). These observations support the hypothesis that disruption of membranes is also a major antibacterial mechanism for lysozyme. Another antimicrobial mechanism of lysozyme was suggested by Pellegrini et al. (2000), who provided evidence that lysozyme inhibits DNA and RNA synthesis in *E. coli*. Finally, Ibrahim et al. (2001b) have hypothesized the expression of a lysozyme-mediated autolysin induction mechanism in *B. subtilis*. However, this mechanism has never been described in Gram-negative bacteria.

6.2 Ovotransferrin

Ovotransferrin is a glycoprotein of 77.7 kDa, consisting of a 686 residue single polypeptide chain, and is widely distributed in physiological fluids. Its natural concentration is of 13 g/l in egg white. This protein is a member of the transferrin family, an extended family of metal-binding transport proteins with an in vivo preference for iron (Wu and Acero-Lopez, 2012). It is composed of two homologous lobes named the amino-terminal lobe (N-lobe) and the carboxyl-terminal lobe (C-lobe), respectively. Each lobe has the ability to chelate multivalent ions, such as iron, zinc, copper and manganese (Abdallah and Chahine, 1999; Ibrahim et al., 2007). The protein is predominantly iron-free under the natural conditions of egg white and all the iron present in egg white is bound to ovotransferrin, considering the low concentration of iron in egg white (around 25 mmol/l) and the high concentration of ovotransferrin (1.7 mmol/l) (Thapon et al., 1994). This leads to an iron-deficient environment, which is deleterious for bacterial growth (Baron et al., 1997; Garibaldi, 1970). This effect can be reversed by the addition of iron in amounts sufficient for ovotransferrin saturation (Lock and Board, 1992). In order to counter the iron restriction conditions encountered in biological fluids such as egg white, many bacteria are able to use intracellular iron reserves, or to synthesize extracellular ferric chelators, called siderophores (Neilands, 1995), which can efficiently compete with iron-binding proteins (Andrews et al., 2003).

In addition to its well-described iron restriction activity in egg white, the metal-binding property of ovotransferrin is also assumed to have an effect on the outer and/or inner membrane integrity of Gram-negative bacteria (Ellison et al., 1988; Valenti et al., 1986; Aguilera et al., 2003), leading to bacteriostasis. Strikingly, Baron et al. (2014) showed the presence of a strong bactericidal effect of native ovotransferrin on Gram-positive bacteria belonging to the *B. cereus* group, under the natural alkaline pH conditions of egg white. Ovotransferrin was shown to provoke a perturbation of the membrane electrochemical potential and membrane dysfunction was strongly suggested to be involved in the bactericidal effect (Baron et al., 2014). Such lethal activity of native ovotransferrin has never been reported for Gram-negative bacteria.

6.3 Vitamin-chelating proteins

Egg white contains three proteins, riboflavin-binding protein (binding riboflavin), avidin (binding biotin) and thiamine-binding protein (binding thiamine) that would be part of the defence system of egg white against microorganisms. However, only avidin has been shown so far to express antibacterial activity. This minor protein of egg white (0.05 g/l) is a cationic homotetrameric glycoprotein that has a very high affinity for biotin (dissociation constant of 6.10^{-16} mol/l) (Green, 1990). By depriving bacteria of biotin, avidin may exert a bacteriostatic effect (Banks et al., 1986). Biotin is used as a coenzyme for biotin-dependent carboxylases that catalyse the transfer of carboxylate groups. In *E. coli*, the acetyl-CoA carboxylase EC 6.4.1.2 (ACC) is the only biotin-dependent carboxylase described so far (Beckett, 2007). This enzyme is involved in the first step of the synthesis of fatty acids (Streit and Entcheva, 2003). According to Beckett (2007), avidin would also affect cellular transcription. However, the biotin-deprivation activity of avidin may be limited for some bacteria such as *E. coli*, *S.* Typhimurium and *B. subtilis*, recognized as being able to synthesize biotin when needed (Beckett, 2007).

6.4 Protease inhibitors

Egg white contains various protease inhibitors that could limit the proteolytic potential of bacteria. However, there is no sufficient evidence to conclude on their actual role in the natural defence system of egg white compared to the strong efficiency of ovotransferrin towards egg white invasion by Gram-negative bacteria (Baron et al., 1997; Garibaldi, 1970; Kang et al., 2006).

The protease inhibitors identified to date in egg white are ovostatin, cystatin, ovoinhibitor and ovomucoid.

Ovostatin is a minor egg white protein (around 0.5 g/l) that belongs to the α2-macroglobulin family. It consists of a large homotetramer with four disulfide linked subunits (4 × 95 kDa). It is known to bind and inhibit various proteases with a better efficiency towards metalloproteases (Réhault et al., 2007). Its antimicrobial effect has already been demonstrated towards pathogenic bacteria (Molla et al., 1987).

Cystatin, present in egg white at very low concentration (0.05 g/l), is a non-glycosylated small protein (13 kDa) containing two disulfide bonds. Mostly active on cysteine proteases, it has been shown to be bactericidal towards *S.* Typhimurium (Nakai, 2000) and *E. coli* (Wesierska et al., 2005).

Ovomucoid is a member of the Kazal family of protease inhibitors. It is the second most abundant protein in egg white (13 g/l) after ovalbumin. This 28-kDa protein is highly glycosylated (25% of its weight is represented by sialyl-oligosaccharides and sulphated oligosaccharides), and contains nine disulfide bonds. It expresses a wide spectrum of antiprotease properties, especially antitrypsin activity (Kato et al., 1987). However, its involvement in the whole egg antimicrobial activity has never been demonstrated so far. Ovoinhibitor is a 48-kDa glycoprotein present at low concentration in egg white (around 1.5 g/l). It possesses seven Kazal-like domains, each containing an inhibitory site, and 21 disulfide bonds. Although shown to be active against a large spectrum of serine proteases such as trypsin, chymotrypsin and elastase (Guérin-Dubiard et al., 2010), ovoinhibitor has not yet been demonstrated as participating in the protection of egg white against bacterial contamination.

In 2005, a novel protein belonging to the serpin family was detected in egg white, the ovalbumin-related protein X (OVAX) (Guérin-Dubiard et al., 2005). It is a 45-kDa glycosylated heparin-binding protein that is estimated to be present at low concentration in egg white (around 0.5 g/l). Serpins are recognized as expressing serine antiprotease activities. In egg white, only OVAX has been recently demonstrated as expressing an antimicrobial activity towards *S.* Enteritidis (Réhault-Godbert et al., 2013). A cluster of cationic amino acid residues present at the surface of the protein is supposed, through interaction with heparin, to take part in this antibacterial mechanism (Réhault-Godbert et al., 2013).

6.5 Other proteins and peptides

Thanks to the development of sensitive analytical tools based on mass spectrometry, the number of proteins identified in egg white has increased considerably over the past few years (Mann and Mann, 2011). Among the 158 proteins identified nowadays, several show functional homologies to recognized antimicrobial proteins. However, their antimicrobial activity remains often putative and their efficiency under the natural conditions of egg white remains to be demonstrated. The most noticeable examples are (i) the bactericidal permeability increasing protein/lipopolysaccharide-binding protein (BPI/LBP) family-like proteins, such as TenP (Guérin-Dubiard et al., 2006), which could lead to the inhibition of bacterial growth (Elsbach and Weiss, 1998), and BPIL2 and similar-to-BPI proteins (D'Ambrosio et al., 2008; Mann, 2007); (ii) egg white defensin-related molecules, the avian β-defensin 11 (AvBD11) (Mann, 2007), and gallin. The defensins are cysteine-rich cationic peptides involved in host innate defence in many living species through membrane disturbing activities (White et al., 1995) against a broad spectrum of cells, including bacteria, fungi and viruses (Epand and Vogel, 1999). The AvBD11 collected from the vitelline membrane (but not from egg white) has been shown to reduce the growth of *S.* Enteritidis, *S.* Typhimurium, *E. coli*, *L. monocytogenes* and *S. aureus* in rich medium (Hervé-Grépinet et al., 2010). Gallin is a member of the ovodefensins, a subfamily of β-defensins found in the egg white of different avian species. A recombinant form of this peptide has been shown to inhibit *E. coli* growth in phosphate buffered saline (PBS) (Gong et al., 2010). Gallin Ex-FABP (also called ch21), detected in egg white by Guérin-Dubiard et al. (2006), has been shown to inhibit the growth of *E. coli* and *B. subtilis* (Correnti et al., 2011).

7 The influence of the physicochemical and environmental conditions on the egg white

7.1 Effect of inoculum state

There is a general agreement on the need to avoid inoculating eggs with bacterial inoculum that could provide iron or other nutrients to egg white. Bacterial inoculum are therefore prepared in physiological solution (Ringer), PBS or egg white filtrate (solvent phase obtained by ultrafiltration of egg white) in order to minimize the change in the ionic composition of egg white (Baron et al., 1997).

The inoculation level varies widely from one study to another, ranging from 0.5 to 7 log cells/ml of egg white. In most studies, low levels (<3 log cells/ml) are inoculated in order to approximate the natural conditions. However, several studies were designed to study the effect of the inoculum size on bacterial behaviour in egg white. Cogan et al. (2001) have shown that, at 20°C, the growth of *Salmonella* occurs only for inoculum levels >0.5 cells/ml. For incubations in egg white at 37°C for 50 h, Kang et al. (2006) observed a bactericidal effect for an inoculum of 3 log cells/ml. The whole population was undetectable after 50 h of incubation. A bacteriostatic effect was observed for an inoculum of 5 log cells/ml and weak growth (+1 log cell/ml in 50 h) was observed for an inoculum of 7 cells/ml. The authors conclude that high bacterial concentrations allow bacteria to overcome the antibacterial properties of egg white, either through saturation of the antimicrobial factors or release of nutrients from the lysed cells.

The location of inoculation represents another factor to be considered. In several studies, egg white is inoculated inside intact eggs. In other studies, the incubation takes place after egg breaking and egg white collection separate from egg yolk. Both methods give the same results provided that the vitelline membrane prevents egg yolk migration towards egg white. Gast and Holt (2000b) observed higher growth of *Salmonella* inoculated in egg white near the vitelline membrane rather than away from it.

7.2 Effect of egg freshness

Egg freshness is another parameter that influences the level of antibacterial activity of egg white. Duboccage et al. (2001) have shown that the growth of *Salmonella* in egg white is promoted at 20°C for eggs of <1 week, compared to eggs stored at this temperature for 2–3 weeks. After 3 weeks of storage, growth would again be favoured in egg white, probably due to the alteration of the vitelline membrane and to the decrease in viscosity that would favour the migration of egg yolk nutrients towards egg white, or the migration of *Salmonella* towards egg yolk (Humphrey and Whitehead, 1993; Chen et al., 2005). These results can be explained by the ageing process of the egg white during storage. The loss of carbon dioxide through the pores of the eggshell leads to an increase in pH, from 7.6 to about 9.3, several days after oviposition. This alkalization is associated with a liquefaction phenomenon that could also modulate the antibacterial activity of egg white.

7.3 Effect of environmental temperature

The environmental conditions of temperature are known to strongly impact egg white antimicrobial activities against *Salmonella* and other microorganisms, whether in shell eggs or in the egg white medium used for in vitro analyses.

Below 8°C, no growth occurs in egg white (Ruzckova, 1994; Schoeni et al., 1995). A bactericidal effect can even be observed, which can cause the disappearance of the population at 4°C in a few days (Chen et al., 2005; Schoeni et al., 1995; Lock and Board, 1992). At 10°C, contrasting results were obtained by Clay and Board (1991) and Chen et al. (2005), who observed the absence of growth, and by Schoeni et al. (1995), who observed a slight growth.

Between 20°C and 30°C, growth is permitted (Cogan et al., 2001; Schoeni et al., 1995; Humphrey and Whitehead, 1993; Chen et al., 2005; Clay and Board, 1991; Gast and Holt, 2001; Duboccage et al., 2001; Murase et al., 2005), from 1 to 4 log/ml depending on temperature and incubation time. Ruzickova (1994) observed an increase of 1.5 log CFU/ml at 21°C for 48-h incubation. Baron et al. (1997) observed one log CFU/ml increase at 30°C for 24-h incubation, and another log CFU/ml increase for 4–6 days incubation. Lock and Board (1992) showed an increase of 4 log CFU/ml at 30°C for 15-day incubation.

Above 37°C, a more or less significant bactericidal effect is observed depending on temperature and time of incubation. At 37°C, Ruzickova (1994) and Bradshaw et al. (1990) showed one log CFU/ml decrease in 48 and 72 h, respectively. Clavijo et al. (2006) observed 0.6 log CFU/ml decrease in 24 h, using an inoculum of 3 log bacteria/ml. Kang et al. (2006) observed a decrease of 1–3 log CFU/ml in 50 h for the same inoculum size (3 log CFU/ml). These authors observed a bactericidal effect of incubation at 42°C for 24 h. These results are in accordance with the findings of Tranter and Board (1984), who showed that the antimicrobial activity of egg white is more drastic at 39.5°C than at 30°C for Gram-negative bacteria. These authors suggest that certain egg white enzymatic activities would be greater at temperatures above 37°C and could cause the death of bacteria. The temperature also seems to play a crucial role on the survival of other species. Tranter and Board (1984) showed the death of Gram-positive bacteria, even if they are resistant to lysozyme lytic effect, and yeasts incubated in egg white at a temperature not exceeding 30°C. Dolman and Board (1992) have studied the behaviour of a mixture of *Pseudomonas*, *Staphylococcus*, *Enterococcus*, *E. coli* and S. Enteritidis injected in the egg air chamber of shell eggs incubated at different temperatures. They showed the predominance of *Salmonella* and *E. coli* at 37°C, whereas *Pseudomonas* predominated at 4°C, 15°C or 20°C. At 37°C, *Salmonella* and *E. coli* would be more competitive than the other species inoculated. This observation is in agreement with the study of Clavijo et al. (2006), which shows higher ability of S. Enteritidis to survive in egg white at 37°C, even compared to *E. coli*. The differences observed between species or between *Salmonella* serotypes may, however, depend on the temperature. Hanninen et al. (1984) observed, for *Campylobacter*, a bactericidal effect which was temperature- and time-dependent. Bacteria were killed within 48 h at 4°C, whereas 24 h was sufficient at 37°C and 20°C. Finally, Alabdeh et al. (2011), investigating the effect of temperature from 37°C to 48°C on the survival of S. Enteritidis, demonstrated that egg white became bactericidal at temperatures >42°C. At 45°C, a drastic destruction was observed. This phenomenon was not thought to be a simple temperature effect because bacteria were able to grow in optimum medium tryptic soy broth (TSB) at 45°C. This suggests that certain antibacterial activities, leading to bacterial death, are enhanced at temperatures >37°C.

7.4 Effect of pH

Immediately after laying, egg white pH value increases within a few days, from 7.6 up to 9.3, due to carbon dioxide being lost through the pores of the eggshell. The increase in pH value has potential consequences on bacteria. Baron (1998) has shown that S. Enteritidis lacks flagella at alkaline pH, which may limit its access to nutrients. The role of alkaline pH on *Salmonella* has been highlighted by several studies. Above pH 8.8, a bactericidal effect can be observed, in contrast to pH 7.5 or 8 (Kang et al., 2006; Messens et al., 2004; Tranter and Board, 1984).

The increase in pH value would also enhance the antimicrobial activity of egg white proteins. As far as ovotransferrin is concerned, alkaline pH may further limit the ability of bacteria to obtain sufficient iron (Tranter and Board, 1984; Halbrooks et al., 2005; Okamoto et al., 2004).

Regarding the effect of alkaline pH on lysozyme activity, results are inconsistent. Levashov et al. (2010) have shown an increase in the activity of the protein at alkaline pH, whereas Ibrahim et al. (2001a) found that it inhibited the catalytic activity of lysozyme, without changing its other bactericidal activities. Moreover, high pH values have been found to promote the bactericidal effects of some antibacterial peptides. Similar effects may be expected within egg white, as suggested by Abriouel et al. (1998).

7.5 Synergetic effects leading to the death of *S.* Enteritidis in egg white

Many studies published in literature on the antimicrobial activities of egg white components fail to utilize conditions reflective of those found in the egg white. Baron et al. (1997) and Alabdeh et al. (2011) have focused on the efficiency of egg white antimicrobial systems, considering the intrinsic properties of egg white, environmental conditions and potential interactions and synergy between egg white molecules. According to these authors, *S.* Enteritidis destruction is observed in egg white at elevated temperatures. This killing effect is mainly assumed to be due to stress accumulation: alkaline pH, iron deprivation and activity of antimicrobial compounds. Some of the conditions leading to the death of *S.* Enteritidis in egg white are assumed to approach those of the natural conditions encountered during egg formation in the hen genital tract (42°C), and during hatching (temperature around 42°C and alkaline pH). In addition to biochemical studies (research of egg white antimicrobial activities), genetic studies (research of genes essential for the survival of *S.* Enteritidis in egg white) have helped understand the mechanisms of cell death in egg white at temperature >37°C (Clavijo et al., 2006; Kang et al., 2006). This dual approach has helped converging towards the assumption that egg white inhibits bacterial growth through iron chelation by ovotransferrin, disruption of bacterial membranes and penetration of some components into the cell that kill bacteria through yet unidentified mechanisms (Baron et al., 2016).

To conclude, it appears unquestionable that the antibacterial effect of egg white is multicomponent and multifactorial. Egg white is a harsh and complex biological fluid containing an arsenal of natural antimicrobial molecules that may interact with bacteria in a synergistic fashion. Several physicochemical parameters, such as the natural alkaline pH, the incubation temperature and the ionic environment, are also thought to modulate the activity of egg white antimicrobial components and increase the overall stress levels impacting upon bacterial physiology.

8 Migration and behaviour of microorganisms in egg yolk

Egg white is an effective barrier to prevent migration of microorganisms into the yolk, if it is held in a centre position by chalazae. However, chalazae and egg white liquefy over time and the gradual migration of the egg yolk to the upward point of the egg reduces the distance between the egg yolk and the eggshell, promoting access of microorganisms to egg yolk.

The vitelline membrane represents the last barrier to the penetration of microorganisms in the egg yolk. It is composed of two layers separated by a thin continuous layer. The outer membrane, in contact with egg white, contains lysozyme and ovomucin, and other specific proteins (Chen et al., 2005). During storage, increase in pH leads to the weakening of this membrane, through destructuring of the ovomucin–lysozyme complex in a way that approaches the phenomenon of egg white liquefaction (Baron et al., 2010). In addition to the degradation of the structural integrity of the vitelline membrane along storage at ambient temperature, the increase in the egg yolk volume participates in the weakening of this barrier (Fromm, 1964; Back et al., 1982; Kirunda and McKee, 2000; Chen et al., 2005; Jones and Musgrove, 2005; Berardinelli et al., 2008; Leleu et al., 2011). The consequences are an increasing risk of bacterial invasion inside the egg due to the diffusion of components of egg yolk into egg white (mainly iron and amino acids) (Grijspeerdt et al., 2005; Gast and Holt, 2000; Murase et al., 2005), resulting in a dramatic decrease of the antimicrobial egg white defences.

The penetration of microorganisms through the vitelline membrane varies depending on the study due to the use of various strains (avian or reference strains, strains re-isolated from eggs laid by artificially inoculated hens, strains isolated from foodborne disease, etc.), various sizes of inoculum, various culture media and various incubation temperatures. The event of bacterial penetration is estimated at 10–25% for temperatures between 25°C and 30°C for 3–6 days and for an inoculum size of approximately 100 cells per egg (Murase et al., 2005, 2006; Guan et al., 2006). At the same temperatures, Gast et al. (2005b, 2006 and 2007b) and Gast and Holt (2001) showed higher penetration levels (up to 75% depending on the study and the strain used). However, it may be ascribed to nutrient supply at the time of bacteria inoculation since it was carried out in the bacterial culture medium. On the other hand, Fleischman et al. (2003) showed no penetration of the membrane 2 days after inoculation at these temperatures. It is worth noting that bacterial penetration through the vitelline membrane does not occur at 42°C, the temperature of egg formation in the hen's genital tract (Guan et al., 2006).

At storage temperatures of 4°C and perhaps 10°C, the ageing process of eggs delays, preserving the egg white antimicrobial activities and maintaining the egg structures (Gast and Holt, 2001; Gast et al., 2006; Chen et al., 2005). Low-temperature storage, therefore, has a significant impact on the safety and overall egg quality (EFSA BIOHAZ Panel, 2014).

If the bacteria reach egg yolk, growth is very quick. Above 25°C, the bacterial population can reach 8–9 log CFU/ml within 10–48 h for an inoculum of around 2 log CFU/ml (Gast and Holt, 2000b, 2001; Schoeni et al., 1995; Ruzckova, 1994; Gumadavelli et al., 2007). Growth of *Salmonella* is reduced at 10°C (Gast and Holt, 2000b; Gumadavelli et al., 2007) and would not be possible for temperatures <7°C (Gumadavelli et al., 2007).

9 Conclusion

Sporadic cases of vertical contamination of the egg may occur if the hen's reproductive tract is infected prior to egg development. *Salmonella* is the main microorganism involved in this route of contamination, and specifically the serotype *S.* Enteritidis.

Once the egg is laid, horizontal contamination is systematic. It concerns the surface of the egg and can occasionally lead to secondary contamination if the level of the flora present on the eggshell surface is high or when eggs are newly-laid and/or stored under humid conditions. However, provided that hygiene practices are correctly implemented on the farms and in the egg conditioning centres, the physico-chemical barriers of the egg together with the multicomponent and multifactorial antibacterial activities of egg white represent harsh conditions for bacterial penetration and development, reducing the overall risk of eggborne disease. Therefore, it may be considered that the risk in relation to *Salmonella* in eggs is significantly affected by egg sources and poor hygiene and cooking practices.

10 Where to look for further information

https://www.food.gov.uk/news-updates/consultations/2016/microbiological-risk-from-shell-eggs-and-their-products

11 References

Abdallah, F. B. and Chahine, J. M. (1999), 'Transferrins, the mechanism of iron release by ovotransferrin', *Eur. J. Biochem. FEBS*, 263, 912–20.

Abriouel, H., Valdivia, E., Galvez, A. and Maqueda, M. (1998), 'Response of *Salmonella choleraesuis* LT2 spheroplasts and permeabilized cells to the bacteriocin AS-48', *Appl. Environ. Microbiol.*, 64, 4623–6.

Abulreesh, H. H. and Organji, S. R. (2011), 'The prevalence of multidrug-resistant staphylococci in food and the environment of Makkah, Saudi Arabia', *Res. J. Microbiol.*, 6, 510–23.

AFSSA (2009), 'Opinion of the French food safety agency concerning two draft amendments to orders for controlling salmonellae in the *Gallus gallus* species', Request no. 2009-SA-0182.

Aguilera, O., Quiros, L. M. and Fierro, J. F. (2003), 'Transferrins selectively cause ion efflux through bacterial and artificial membranes', *FEBS Lett.*, 548, 5–10.

Ahlborn, G. and Sheldon, B. W. (2005), 'Enzymatic and microbiological inhibitory activity in eggshell membranes as influenced by layer strains and age and storage variables', *Poult. Sci.*, 84(12), 1935–41.

Aho, M. (1992), 'Problems of *Salmonella* sampling', *Int. J. Food Microbiol.*, 15, 225–35.

Alabdeh, M., Lechevalier, V., Nau, F., Gautier, M., Cochet, M.-F., Gonnet, F., Jan, S. and Baron, F. (2011), 'Role of incubation conditions and protein fraction on the antimicrobial activity of egg white against *Salmonella* Enteritidis and *Escherichia coli*', *J. Food Prot.*, 74, 24–31.

Allen, K. J. and Griffiths, M. W. (2001), 'Use of luminescent *Campylobacter jejuni* ATCC 33291 to assess eggshell colonization and penetration in fresh and retail eggs', *J. Food Prot.*, 64(12), 2058–62.

Andersson, A., Ronner, U. and Granum, P. E. (1995), 'What problems does the food industry have with the spore-forming pathogens *Bacillus cereus* and *Clostridium perfringens*?' *Int. J. Food Microbiol.*, 28, 145–55.

Andrews, S. C., Robinson, A. K. and Rodríguez-Quinones, F. (2003), 'Bacterial iron homeostasis', *FEMS Microbiol. Rev.*, 27, 215–37.

Arnold, M. E., Martelli, F., McLaren, I. and Davies, R. H. (2014), 'Estimation of the sensitivity of environmental sampling for detection of *Salmonella* in commercial layer flocks post-introduction of national control programmes', *Epidemiology and Infection*, 142, 1061–9.

Back, J. F., Bain, J. M., Vadehra, D. V. and Burley, R. W. (1982), 'Proteins of the outer layer of the vitelline membrane of hens eggs', *Biochim. Biophys. Acta*, 705, 12–19.

Bahrouz, M. and Al-Jaff, A. (2005), 'The risk of bacterial contamination in hen eggs of Sulaimani poultries', *J. Zankoy Sulaimani*, 8, 63–71.

Bain, M. M., MacLeod, N., Thomson, R. and Hancock, J. W. (2006), 'Microcracks in eggs', *Poult. Sci.*, 85(11), 2001–8.

Banks, J. G., Board, R. G. and Sparks, N. H. (1986), 'Natural antimicrobial systems and their potential in food preservation of the future', *Biotechnol. Appl. Biochem.*, 8, 103–47.

Barak, J. D., Gorski, L., Naraghi-Arani, P. and Charkowski, A. O. (2005), '*Salmonella enterica* virulence genes are required for bacterial attachment to plant tissue', *Appl. Environ. Microbiol.*, 71(10), 5685–91.

Baron, F., Gautier, M. and Brule, G. (1997), 'Factors involved in the inhibition of growth of *Salmonella* Enteritidis in liquid egg white', *J. Food Prot.*, 60, 1318–23.

Baron, F. (1998), 'Etude du comportement de *Salmonella* Enteridis dans le blanc d'œuf', PhD Thesis, Ecole Nationale Supérieure d'Agronomie, Rennes, France.

Baron, F., Cochet, M. F., Grosset, N., Madec, M. N., Briandet, R., Dessaigne, S., Chevalier, S., Gautier, M. and Jan, S. (2007), 'Isolation and characterization of a psychrotolerant toxin producer, *Bacillus weihenstephanensis*, in liquid egg products', *J. Food Prot.*, 70, 2782–91.

Baron, F., Jan, S. and Jeantet, R. (2010) 'Qualité microbiologique des ovoproduits', in Nau, F., Guérin-Dubiard, C., Baron, F. and Thapon, J. L. (Eds), *Science et technologie de l'œuf, vol 2: de l'œuf aux ovoproduits*, 321–49. Lavoisier, Editions TEC & DOC.

Baron, F., Jan, S., Gonnet, F., Pasco, M., Jardin, J., Giudici, B., Gautier, M., Guérin-Dubiard, C. and Nau, F. (2014), 'Ovotransferrin plays a major role in the strong bactericidal effect of egg white against the *Bacillus cereus* group', *J. Food Prot.*, 77, 955–62.

Baron, F., Nau, F., Guérin, C., Bonnassie-Rouxin, S., Gautier, M., Andrews, S. C. and Jan, S. (2016), 'Egg white versus *Salmonella* Enteritidis! A harsh medium meets a resilient pathogen', *Food Chem.*, 53, 82–93.

Barrow, P. A. and Lovell, M. A., (1991), 'Experimental infection of egg-laying hens with *Salmonella*-Enteritidis phage type-4', *Avian Pathol.*, 20, 335–48.

Barrow, P. A. (2007), '*Salmonella* infections: immune and non-immune protection with vaccines', *Avian Pathol.*, 36(1), 1–13.

Baskerville, A., Humphrey, T. J., Fitzgeorge, R. B., Cook, R. W., Chart, H., Rowe, B. and Whitehead, A. (1992), 'Airborne infection of laying hens with *Salmonella enteritidis* phage type 4', *Vet. Rec.*, 130, 395–8.

Baumann-Popczyk, A. and Sadkowska-Todys, M. (2011), 'Foodborne infections and intoxications in Poland in 2009', *Przeglad Epidemiol.*, 65, 227–34.

Beckett, D. (2007), 'Biotin sensing: universal influence of biotin status on transcription', *Annu. Rev. Genet.*, 41, 443–64.

Berardinelli, A., Ragni, L., Giunchi, A., Gradari, P. and Guarnieri, A. (2008), 'Physical-mechanical modifications of eggs for food-processing during storage', *Poult. Sci.*, 87, 2117–25.

Berghaus, R. D., Thayer, S. G., Maurer, J. J. and Hofacre, C. L. (2011), 'Effect of vaccinating breeder chickens with a killed *Salmonella* vaccine on *Salmonella* prevalences and loads in breeder and broiler chicken flocks', *J. Food Prot.*, 74, 727–34.

Berrang, M. E. (1999), 'Bacterial penetration of the eggshell and shell membranes of the chicken hatching egg: a review', *J. Appl. Poult. Res.*, 8, 499–504.

Berry, J. T., Doyle, M. P. and Schoeni, J. L. (1985), 'Colonization of chicken cecae by *Escherichia coli* associated with hemorrhagic colitis', *Appl. Environ. Microbiol.*, 49, 310–15.

Bradshaw, J., Shah, D. B., Forney, E. and Madden, J. M. (1990), 'Growth of *Salmonella enteritidis* in yolk of shell eggs from normal and seropositive hens', *J. Food Prot.*, 53, 1033–6.

Cadel Six, S., De Buyser, M. L., Vignaud, M. L., Dao, T. T., Messio, S., Pairaud, S., Hennekinne, J. A., Pihier, N. and Brisabois, A. (2012), 'Toxi-infections alimentaires collectives à *Bacillus cereus*: bilan de la caractérisation des souches de 2006 à 2010', *Bulletin Epidémiologique Hebdomadaire*, Hors-série/9 mai 2012, 45–9.

Callewaert, L., Aertsen, A., Deckers, D., Vanoirbeek, K. G. A., Vanderkelen, L., Van Herreweghe, J. M., Masschalck, B., Nakimbugwe, D., Robben, J. and Michiels, C. W. (2008), 'A new family of lysozyme inhibitors contributing to lysozyme tolerance in gram-negative bacteria', *PLoS Pathog.*, 4, e1000019.

Callicott, K. A., Friethriksdottir, V., Reiersen, J., Lowman, R., Bisaillon, J. R., Gunnarsson, E., Berndtson, E., Hiett, K. L., Needleman, D. S. and Stern, N. J. (2006), 'Lack of evidence for vertical transmission of *Campylobacter* spp. in chickens', *Appl. Environ. Microbiol.*, 72, 5794–8.

Chemaly, M., Huneau-Salaun, A., Labbe, A., Houdayer, C., Petetin, I. and Fravalo, P. (2009), 'Isolation of *Salmonella enterica* in laying-hen flocks and assessment of eggshell contamination in France', *J. Food Prot.*, 72, 2071–7.

Chen, J., Clarke, R. C. and Griffiths, M. W. (1996), 'Use of luminescent strains of *Salmonella enteritidis* to monitor contamination and survival in eggs', *J. Food Prot.*, 59(9), 915–21.

Chen, J., Shallo Thesmar, H. and Kerr, W. L. (2005), 'Outgrowth of Salmonellae and the physical property of albumen and vitelline membrane as influenced by egg storage conditions', *J. Food Prot.*, 68(12), 2553–8.

Clavijo, R. I., Loui, C., Andersen, G. L., Riley, L. W. and Lu, S. (2006), 'Identification of genes associated with survival of *Salmonella enterica* Serovar Enteritidis in chicken egg albumen', *Appl. Environ. Microbiol.*, 72(2), 1055–64.

Clay, C. E. and Board, R. G. (1991), 'Growth of *Salmonella* Enteritidis in artificially contaminated hens' shell eggs', *Epidemiol. Infect.*, 106(2), 271–81.

Cogan, T. A., Domingue, G., Lappin-Scott, H. M., Benson, C. E., Woodward, M. J. and Humphrey, T. J. (2001), 'Growth of *Salmonella* Enteritidis in artificially contaminated eggs: the effects of inoculum size and suspending media', *Int. J. Food Microbiol.*, 70(1–2), 131–41.

Correnti, C., Clifton, M. C., Abergel, R. J., Allred, B. T., Hoette, T. M., Ruiz, M., Cancedda, R., Raymond, K. N., Descalzi, F. and Strong, R. K. (2011), 'Galline Ex-FABP is an antibacterial siderocalin and a lysophosphatidic acid sensor functioning through dual ligand specificities', *Structure*, 19, 1796–806.

D'Ambrosio, C., Arena, S., Scaloni, A., Guerrier, L., Boschetti, E., Mendieta, M. E., Citterio, A. and Righetti, P. G. (2008), 'Exploring the chicken egg white proteome with combinatorial peptide ligand libraries', *J. Proteome Res.*, 7, 3461–74.

Davies, R. H. and Wray, C. (1996), 'Determination of an effective sampling regime to detect *Salmonella enteritidis* in the environment of poultry units', *Vet. Microbiol.*, 50, 117–27.

Davies, R. H. and Breslin, M. (2003), 'Investigations into possible alternative decontamination methods for *Salmonella enteritidis* on the surface of table eggs', *J. Vet. Med. B. Infect. Dis. Vet. Public Health*, 50(1), 38–41.

Davies, R. and Breslin, M. (2004), 'Observations on *Salmonella* contamination of eggs from infected commercial laying flocks where vaccination for *Salmonella enterica* serovar Enteritidis had been used', *Avian Pathol.*, 33, 133–44.

De Buck, J., Van Immerseel, F., Haesebrouck, F. and Ducatelle, R. (2004), 'Colonization of the chicken reproductive tract and egg contamination by *Salmonella*', *J. Appl. Microbiol.*, 97, 233–45.

De Reu, K., Grijspeerdt, K., Heyndrickx, M., Zoons, J., De Baere, K., Uyttendaele, M., Debevere, J. and Herman L. (2005), 'Bacterial eggshell contamination in conventional cages, furnished cages and aviary housing systems for laying hens', *Br. Poult. Sci.*, 46(2), 149–55.

De Reu, K. (2006), 'Bacteriological contamination and infection of shell eggs in the production chain', PhD Thesis, Faculty of Bioscience Engineering, University of Ghent, Belgium.

De Reu, K., Grijspeerdt, K., Heyndrickx, M., Uyttendaele, M., Debevere, J. and Herman, L. (2006a), 'Bacterial shell contamination in the egg collection chains of different housing systems for laying hens', *Br. Poult. Sci.*, 47(2), 163–72.

De Reu, K., Grijspeerdt, K., Messens, W., Heyndrickx, M., Uyttendaele, M., Debevere, J. and Herman, L. (2006b), 'Eggshell factors influencing eggshell penetration and whole egg contamination by different bacteria, including *Salmonella enteritidis*', *Int. J. Food Microbiol.*, 112(3), 253–60.

De Reu, K., Grijspeerdt, K., Heyndrickx, M., Messens, W., Uyttendaele, M., Debevere, J. and Herman, L. (2006c), 'Influence of eggshell condensation on eggshell penetration and whole egg contamination with *Salmonella enterica* serovar *enteritidis*', *J. Food Prot.*, 69, 1539–45.

Deckers, D., Masschalck, B., Aertsen, A., Callewaert, L., Van Tiggelen, C. G. M., Atanassova, M. and Michiels, C. W. (2004), 'Periplasmic lysozyme inhibitor contributes to lysozyme resistance in *Escherichia coli*', *Cell. Mol. Life Sci.*, 61, 1229–37.

Defra (2007), UK national control programme for *Salmonella* in layers (*Gallus gallus*). https://www.gov.uk/government/uploads/system/uploads/attachment_data/file/183077/salmonella-layers.pdf.

Delmas, G., Gallay, A., Espié, E., Haeghebaert, S., Pihier, N., Weill, F. X., De Valk, H., Vaillant, V. and Désenclos, J. C. (2006), 'Les toxi-infections alimentaires collectives en France entre 1996 et 2005', *Bulletin Epidémiologique Hebdomadaire* (51–2), 418–21.

Derde, M., Lechevalier, V., Guerin-Dubiard, C., Cochet, M.-F., Jan, S., Baron, F., Gautier, M., Vie, V. and Nau, F. (2013), 'Hen egg white lysozyme permeabilizes *Escherichia coli* outer and inner membranes', *J. Agric. Food Chem.*, 61, 9922–9.

Dolman, J. and Board, R. G. (1992), 'The influence of temperature on the behaviour of mixed bacterial contamination of the shell membrane of the hen's egg', *Epidemiol. Infect.*, 108(1), 115–21.

Duboccage, L., Heyndrickx, M., Grijspeerdt, K. and Herman, L. (2001), 'Growth of *Salmonella* in egg white', *Meded Rijksuniv Gent Fak Landbouwkd Toegep Biol Wet*, 66(3b), 531–4.

Düring, K., Porsch, P., Mahn, A., Brinkmann, O. and Gieffers, W. (1999), 'The nonenzymatic microbicidal activity of lysozymes', *FEBS Lett.*, 449, 93–100.

EFSA (2007), Report of the Task Force on Zoonoses Data Collection on harmonising the reporting of foodborne outbreaks through Community reporting system in accordance with Directive 2003/99/EC, *EFSA J.*, 123r, 16 pp.

EFSA BIOHAZ Panel (EFSA Panel on Biological Hazards) (2010), Scientific Opinion on monitoring and assessment of the public health risk of 'Salmonella Typhimurium-like' strains, *EFSA J.*, 8(10), 1826, 48 pp.

EFSA and ECDC (European Food Safety Authority and European Centre for Disease Prevention and Control) (2013), The European Union Summary Report on Trends and Sources of Zoonoses, Zoonotic Agents and Food-borne Outbreaks in 2011, *EFSA J.*, 11(4), 3129, 250 pp.

EFSA and ECDC (2014), The European Union Summary Report on Trends and Sources of Zoonoses, Zoonotic Agents and Food-borne Outbreaks in 2012. *EFSA J.*, 12(2), 3547, 312 pp.

EFSA BIOHAZ Panel (EFSA Panel on Biological Hazards) (2014), Scientific Opinion on the public health risks of table eggs due to deterioration and development of pathogens, *EFSA J.*, 12, 3782, 147 pp.

Ellison, R. T., Giehl, T. J. and LaForce F. M. (1988), 'Damage of the outer membrane of enteric gram-negative bacteria by lactoferrin and transferrin', *Infect. Immun.*, 56, 2774–81.

Elsbach, P. and Weiss, J. (1998), 'Role of the bactericidal/permeability-increasing protein in host defence', *Curr. Opin. Immunol.*, 10, 45–9.

Elson, R., Little, C. L. and Mitchell, R. T. (2005), 'Salmonella and raw shell eggs: results of a cross-sectional study of contamination rates and egg safety practices in the United Kingdom catering sector in 2003', *J. Food Prot.*, 68(2), 256–64.

Epand, R. M. and Vogel, H. J. (1999), 'Diversity of antimicrobial peptides and their mechanisms of action', *Biochim. Biophys. Acta*, 1462, 11–28.

Ernst, R. A. and Fuqua, L. (1998), 'Effect of sweating on shell penetration of *Salmonella enteritidis*', *J. Appl. Poult. Res.*, 7, 81–4.

Fajardo, T. A., Anantheswaran, R. C., Puri, V. M. and Knabel, S. J. (1995), 'Penetration of *Salmonella enteritidis* into eggs subjected to rapid cooling', *J. Food Prot.*, 58(5), 473–7.

Favier, G. I., Escudero, M. E., Mattar, M. A. and de Guzman, A. M. (2000), 'Survival of *Yersinia enterocolitica* and mesophilic aerobic bacteria on eggshell after washing with hypochlorite and organic acid solutions', *J. Food Prot.*, 63(8), 1053–7.

Finlay, B. B. and Falkow, S. (1989), 'Common themes in microbial pathogenicity', *Microbiol. Rev.*, 53(2), 210–30.

Fleischman, G. J., Napier, C. L., Stewart, D. and Palumbo, S. A. (2003), 'Effect of temperature on the growth response of *Salmonella* Enteritidis inoculated onto the vitelline membranes of fresh eggs', *J. Food Prot.*, 66(8), 1368–73.

Foley, S. L., Nayak, R., Hanning, I. B., Johnson, T. J., Han, J. and Ricke, S. C. (2011), 'Population dynamics of *Salmonella enterica* serotypes in commercial egg and poultry production', *Appl. Environ. Microbiol.*, 77, 4273–9.

Fonseca, B. B., Soncini, R. A., Gimaraes, A. R. and Rossi, D. A. (2006), '*Campylobacter* sp. in eggs from cloacal swab positive breeder hens', *Braz. J. Microbiol.*, 37, 573–5.

Fromm, D. (1964), 'Strength distribution, weight and some histological aspects of the vitelline membrane of the hen's egg yolk', *Poult. Sci.*, 43, 1240–7.

Galiş, A. M., Marcq, C., Marlier, D., Portetelle, D., Van, I., Beckers, Y. and Théwis, A. (2013), 'Control of *Salmonella* contamination of shell eggs – Preharvest and postharvest methods: A review', *Compr. Rev. Food Sci. Food Safety*, 12, 155–82.

Gantois, I., Ducatelle, R., Timbermont, L., Boyen, F., Bohez, L., Haesebrouck, F., Pasmans, F. and van Immerseel, F. (2006), 'Oral immunisation of laying hens with the live vaccine strains of TAD *Salmonella* vac E and TAD *Salmonella* vac T reduces internal egg contamination with *Salmonella* Enteritidis', *Vaccine*, 24, 6250–5.

Gantois, I., Ducatelle, R., Pasmans, F., Haesebrouck, F., Gast, R., Humphrey, T. J. and Van Immerseel, F. (2009), 'Mechanisms of egg contamination by *Salmonella* Enteritidis', *FEMS Microbiol. Rev.*, 33, 718–38.

Garibaldi, J. A. (1970), 'Role of microbial iron transport compounds in bacterial spoilage of eggs', *Appl. Microbiol.*, 20, 558–60.

Gast, R. K. and Beard, C. W. (1990), 'Isolation of *Salmonella enteritidis* from internal organs of experimentally infected hens', *Avian Dis.*, 34, 991–3.

Gast, R. K. and Beard, C. W. (1992), 'Detection and enumeration of *Salmonella*-Enteritidis in fresh and stored eggs laid by experimentally infected hens', *J. Food Prot.*, 55, 152–6.

Gast, R. K. (1994), 'Understanding *Salmonella enteritidis* in laying chickens: the contributions of experimental infections', *Int. J. Food Microbiol.*, 21, 107–16.

Gast, R. K., Mitchell, B. W. and Holt, P. S. (1998), 'Airborne transmission of *Salmonella enteritidis* infection between groups of chicks in controlled-environment isolation cabinets', *Avian Dis.*, 42, 315–20.

Gast, R. K. and Holt, P. S. (2000a), 'Influence of the level and location of contamination on the multiplication of *Salmonella enteritidis* at different storage temperatures in experimentally inoculated eggs', *Poult. Sci.*, 79(4), 559–63.

Gast, R. K. and Holt, P. S. (2000b), 'Deposition of phage type 4 and 13a *Salmonella* Enteritidis strains in the yolk and albumen of eggs laid by experimentally infected hens', *Avian Dis.*, 44(3), 706–10.

Gast, R. K. and Holt, P. S. (2001), 'Assessing the frequency and consequences of *Salmonella enteritidis* deposition on the egg yolk membrane', *Poult. Sci.*, 80(7), 997–1002.

Gast, R. K., Guard-Petter, J. and Holt, P. S. (2002), 'Characteristics of *Salmonella enteritidis* contamination in eggs after oral, aerosol, and intravenous inoculation of laying hens', *Avian Dis.*, 46(3), 629–35.

Gast, R. K., Shivaprasad, H. L. and Barrow, P. A. (2003), '*Salmonella* infections', *Dis. Poult.*, 11, 567–613.

Gast, R. K., Guard-Bouldin, J. and Holt P. S. (2004), 'Colonization of reproductive organs and internal contamination of eggs after experimental infection of laying hens with *Salmonella* Heidelberg and *Salmonella* Enteritidis', *Avian Dis.*, 48(4), 863–9.

Gast, R. K., Guard-Bouldin, J. and Holt, P. S. (2005a), 'The relationship between the duration of fecal shedding and the production of contaminated eggs by laying hens infected with strains of *Salmonella* Enteritidis and *Salmonella* Heidelberg', *Avian Dis.*, 49(3), 382–6.

Gast, R. K., Holt, P. S. and Murase, T. (2005b), 'Penetration of *Salmonella* Enteritidis and *Salmonella* Heidelberg into egg yolks in an in vitro contamination model', *Poult. Sci.*, 84(4), 621–5.

Gast, R. K., Holt, P. S. and Guraya, R. (2006), 'Effect of refrigeration on in vitro penetration of *Salmonella* Enteritidis through the egg yolk membrane', *J. Food Prot.*, 69(6), 1426–9.

Gast, R. K., Guraya, R., Guard-Bouldin, J. and Holt P. S. (2007a), 'In vitro penetration of egg yolks by *Salmonella* Enteritidis and *Salmonella* Heidelberg strains during thirty-six-hour ambient temperature storage', *Poult. Sci.*, 86(7), 1431–5.

Gast, R. K., Guraya, R., Guard-Bouldin, J., Holt, P. S. and Moore, R. W. (2007b), 'Colonization of specific regions of the reproductive tract and deposition at different locations inside eggs laid by hens infected with *Salmonella* Enteritidis or *Salmonella* Heidelberg', *Avian Dis.*, 51(1), 40–4.

Gast, R. K., Guraya, R., Jones, D. R. and Anderson, K. E. (2013), 'Colonization of internal organs by *Salmonella* Enteritidis in experimentally infected laying hens housed in conventional or enriched cages', *Poult. Sci.*, 92, 468–73.

Gautron, J., Hincke, M. T., Panheleux, M., Garcia-Ruiz, J. M., Boldicke, T. and Nys, Y. (2001), 'Ovotransferrin is a matrix protein of the hen eggshell membranes and basal calcified layer', *Connect. Tissue Res.*, 42(4), 255–67.

Ginocchio, C. C., Olmsted, S. B., Wells, C. L. and Galan, J. E. (1994), 'Contact with epithelial cells induces the formation of surface appendages on *Salmonella typhimurium*', *Cell*, 76(4), 717–24.

Gong, D., Wilson, P. W., Bain, M. M., McDade, K., Kalina, J., Hervé-Grepinet, V., Nys, Y. and Dunn, I. C. (2010), 'Gallin an antimicrobial peptide member of a new avian defensin family, the ovodefensins, has been subject to recent gene duplication', *BMC Immunol.*, 11(1), 12.

Green, N. M. (1990), 'Avidin and streptavidin', *Methods Enzymol.*, 184, 51–67.

Grijspeerdt, K., Kreft, J. U. and Messens, W. (2005), 'Individual-based modelling of growth and migration of *Salmonella* Enteritidis in hens' eggs', *Int. J. Food Microbiol.*, 100(1–3), 323–33.

Guan J., Grenier C., Brooks B. W. (2006), 'In vitro study of *Salmonella* Enteritidis and *Salmonella* Typhimurium definitive type 104: survival in egg albumen and penetration through the vitelline membrane', *Poult. Sci.*, 85(9), 1678–81.

Guérin-Dubiard, C., Pasco, M., Hietanen, A., Quiros del Bosque, A., Nau, F. and Croguennec, T. (2005), 'Hen egg white fractionation by ion-exchange chromatography', *J. Chromatogr.*, A1090, 58–67.

Guérin-Dubiard, C., Pasco, M., Mollé, D., Désert, C., Croguennec, T. and Nau, F. (2006), 'Proteomic analysis of hen egg white', *J. Agric. Food Chem.*, 54, 3901–10.

Guérin-Dubiard, C., Anton, M., Gautron, J., Nys, Y. and Nau, F. (2010), 'Composition de l'œuf', in Nau, F., Guérin-Dubiard, C., Baron, F. and Thapon, J. L. (Eds), *Science et Technologie de L'œuf*, 1–89. Lavoisier, Paris.

Gumudavelli, V., Subbiah, J., Thippareddi, H., Velugoti, P. R. and Froning, G. (2007), 'Dynamic predictive model for growth of *Salmonella* Enteritidis in egg yolk', *J. Food Sci.*, 72(7), M254–62.

Haeghebaert, S., Le Querrec, F., Gallay, A., Bouvet, P., Gomez, M. and Vaillant, V. (2002), 'Les toxi-infections alimentaires collectives en France, en 1999 et 2000', *Zoonoses Alimentaires*, 23, 161–9.

Haigh, T. and Betts, W. B. (1991), 'Microbial barrier properties of hen egg shells', *Microbios*, 68(276–7), 137–46.

Halbrooks, P. J., Giannetti, A. M., Klein, J. S., Björkman, P. J., Larouche, J. R., Smith, V. C., MacGillivray, R. T. A., Everse, S. J. and Mason, A. B. (2005), 'Composition of pH-sensitive triad in C-lobe of human serum transferrin. Comparison to sequences of ovotransferrin and lactoferrin provides insight into functional differences in iron release', *Biochemistry (Mosc.)*, 44, 15451–60.

Hanninen, M. L., Korkeala, H. and Pakkala, P. (1984), 'Growth and survival characteristics of *Campylobacter jejuni* in liquid egg', *J. Hyg.*, 92(1), 53–8.

Hervé-Grépinet, V., Réhault-Godbert, S., Labas, V., Magallon, T., Derache, C., Lavergne, M., Gautron, J., Lalmanach, A.-C. and Nys, Y. (2010), 'Purification and characterization of avian beta-defensin 11, an antimicrobial peptide of the hen egg', *Antimicrob. Agents Chemother.*, 54, 4401–9.

Hincke, M. T., Gautron, J., Panheleux, M., Garcia-Ruiz, J., McKee, M. D. and Nys, Y. (2000), 'Identification and localization of lysozyme as a component of eggshell membranes and eggshell matrix', *Matrix Biol.*, 19(5), 443–53.

Hiramatsu, R., Matsumoto, M., Sakae, K., Miyazaki, Y. (2005), 'Ability of Shiga toxin-producing *Escherichia coli* and *Salmonella* spp. to survive in a desiccation model system and in dry foods', *Appl. Environ. Microbiol.*, 71(11), 6657–63.

Holt, P. S. (1995), 'Horizontal transmission of *Salmonella-Enteritidis* in molted and unmolted laying chickens', *Avian Dis.*, 39, 239–49

Holt, P. S., Mitchell, B. W. and Gast, R. K. (1998), 'Airborne horizontal transmission of *Salmonella* Enteritidis in molted laying chickens', *Avian Dis.*, 42, 45–52.

Holt, P. S., Geden, C. J., Moore, R. W. and Gast R. K. (2007), 'Isolation of *Salmonella enterica* serovar Enteritidis from houseflies (*Musca domestica*) found in rooms containing *Salmonella* serovar Enteritidis-challenged hens', *Appl. Environ. Microbiol.*, 73(19), 6030–5.

Hopper, S. A. and Mawer, S. (1988), '*Salmonella* Enteritidis in a commercial layer flock', *Vet. Res.*, 123(13), 351.

Howard, Z. R., O'Bryan, C. A., Crandall, P. G. and Ricke S. C. (2012), '*Salmonella* Enteritidis in shell eggs: Current issues and prospects for control', *Food Res. Int.*, 45, 755–64.

Humphrey, T. J., Baskerville, A., Mawer, S., Rowe, B. and Hopper, S. (1989), '*Salmonella enteritidis* phage type 4 from the contents of intact eggs: a study involving naturally infected hens', *Epidemiol. Infect.*, 103, 415–23.

Humphrey, T. J., Whitehead, A., Gawler, A. H., Henley, A. and Rowe, B. (1991), 'Numbers of *Salmonella enteritidis* in the contents of naturally contaminated hens' eggs', *Epidemiol. Infect.*, 106, 489–96.

Humphrey, T. J. and Whitehead, A. (1993), 'Egg age and the growth of *Salmonella enteritidis* PT4 in egg contents', *Epidemiol. Infect.*, 111(2), 209–19.

Ibrahim, H. R., Higashiguchi, S., Juneja, L. R., Kim, M. and Yamamoto, T. (1996), 'A structural phase of heat-denatured lysozyme with novel antimicrobial action', *J. Agric. Food Chem.*, 44, 1416–23.

Ibrahim, H. R., Matsuzaki, T. and Aoki, T. (2001a), 'Genetic evidence that antibacterial activity of lysozyme is independent of its catalytic function', *FEBS Lett.*, 506, 27–32.

Ibrahim, H. R., Thomas, U. and Pellegrini, A. (2001b), 'A helix-loop-helix peptide at the upper lip of the active site cleft of lysozyme confers potent antimicrobial activity with membrane permeabilization action', *J. Biol. Chem.* 276, 43767–74.

Ibrahim, H. R., Hoq, M. I. and Aoki, T. (2007), 'Ovotransferrin possesses SOD-like superoxide anion scavenging activity that is promoted by copper and manganese binding', *Int. J. Biol. Macromol.*, 41, 631–40.

Jan, S., Brunet, N., Techer, C., Le Maréchal, C., Koné, A. Z., Grosset, N., Cochet, M. F., Gillard, A., Gautier, M,. Puterflam, J. and Baron, F. (2011), 'Biodiversity of psychrotrophic bacteria of the *Bacillus cereus* group collected on farm and in egg product industry', *Food Microbiol.*, 28, 261–5.

Johansson, T. M. L., Schildt, R., AliYrkko, S., Siitonen, A. and Maijala, R. L. (1996), 'The first *Salmonella enteritidis* phage type 1 infection of a commercial layer flock in Finland', *Acta Vet. Scand.*, 37, 471–9.

Jones, D. R., Musgrove, M. T. and Northcutt, J. K. (2004), 'Variations in external and internal microbial populations in shell eggs during extended storage', *J. Food Prot.*, 67(12), 2657–60.

Jones, D. R. and Musgrove, M. T. (2005), 'Correlation of eggshell strength and *Salmonella enteritidis* contamination of commercial shell eggs', *J. Food Prot.*, 68(10), 2035–8.

Kabir, S. M. L. (2010), 'Avian colibacillosis and salmonellosis: a closer look at epidemiology, pathogenesis, diagnosis, control and public health concerns', *Int. J. Environ. Res. Public Health*, 7, 89–114.

Kang, H., Loui, C., Clavijo, R. I., Riley, L. W. and Lu, S. (2006), 'Survival characteristics of *Salmonella enterica* serovar Enteritidis in chicken egg albumen', *Epidemiol. Infect.*, 134(5), 967–76.

Kato, I., Schrode, J., Kohr, W. J. and Laskowski, M. (1987), 'Chicken ovomucoid: determination of its amino acid sequence, determination of the trypsin reactive site, and preparation of all three of its domains', *Biochemistry (Mosc.)*, 26, 193–201.

Keller, L. H., Benson, C. E., Krotec, K. and Eckroade, R. J. (1995), '*Salmonella enteritidis* colonization of the reproductive tract and forming and freshly laid eggs of chickens', *Infect. Immun.*, 63, 2443–9.

Keller, L. H., Schifferli, D. M., Benson, C. E., Aslam, S. and Eckroade, R. J. (1997), 'Invasion of chicken reproductive tissues and forming eggs is not unique to *Salmonella enteritidis*', *Avian Dis.*, 41, 535–9.

Kirunda, D. F. and McKee, S. R. (2000), 'Relating quality characteristics of aged eggs and fresh eggs to vitelline membrane strength as determined by a texture analyzer', *Poult. Sci.*, 79, 1189–93.

Knape, K. D., Chavez, C., Burgess, R. P., Coufal, C. D. and Carey, J. B. (2002), 'Comparison of eggshell surface microbial populations for in-line and off-line commercial egg processing facilities', *Poult. Sci.*, 81(5): 695–8.

Koné, A. Z., Jan, S., Le Maréchal, C., Grosset, N., Gautier, M., Puterflam, J. and Baron, F. (2013), 'Identifying risk factors for eggshell contamination by *Bacillus cereus* group bacteria in French laying farms', Brit. *Poult. Sci.*, 54, 298–305.

Lapuz, R., Tani, H., Sasai, K., Shirota, K. and Katoh, H. (2007), 'An epidemiological analysis of *Salmonella enteritidis* contamination in a rat-infested chicken layer farm, an egg processing facility, and liquid egg samples by pulsed-field gel electrophoresis', *J. Vet. Med. Sci.*, 69(6), 649–52.

Leleu, S., Messens, W., De Reu, K., De Preter, S., Herman, L., Heyndrickx, M., De Baerdemaeker, J., Michiels, C. W. and Bain, M. (2011), 'Effect of egg washing on the cuticle quality of brown and white table eggs', *J. Food Prot.*, 74, 1649–54.

Levashov, P. A., Sedov, S. A. and Shipovskov, S. (2010), 'Quantitative turbidimetric assay of enzymatic gram-negative bacteria lysis', *Anal. Chem.*, 82, 2161–3.

Liebana, E., Garcia-Migura, L., Clouting, C., Clifton-Hadley, F. A., Breslin, M., et al. (2003), 'Molecular fingerprinting evidence of the contribution of wildlife vectors in the maintenance of *Salmonella* Enteritidis infection in layer farms', *J. Appl. Microbiol.*, 94(6), 1024–9.

Lock, J. L. and Board, R. G. (1992), 'Persistence of contamination of hens' egg albumen in vitro with *Salmonella* serotypes', *Epidemiol. Infect.*, 108(3), 389–96.

Lu, H., Dunn, P. A., Wallner-Pendleton, E. A., Henzler, D. J., Kradel, D. C., et al. (2004), 'Investigation of H7N2 avian influenza outbreaks in two broiler breeder flocks in Pennsylvania, 2001–02', *Avian Dis.*, 48(1), 26–33.

Lucore, L., Jones, F., Anderson, K. E. and Curtis, P. A. (1997), 'Internal and external bacterial counts from shells of eggs washed in a commercial-type processor at various wash-water temperatures', *J. Food Prot.*, 60(11), 1324–8.

Manie, T., Khan, S., Brozel, V. S., Veith, W. J. and Gouws, P. A. (1998), 'Antimicrobial resistance of bacteria isolated from slaughtered and retail chickens in South Africa', *Let. Appl. Microbiol.*, 26, 253–8.

Mann, K. (2007), 'The chicken egg white proteome', *Proteomics*, 7, 3558–68.

Mann, K. and Mann, M. (2011), 'In-depth analysis of the chicken egg white proteome using an LTQ Orbitrap Velos', *Proteome Sci.*, 9 (7).

Martelli, F. and Davies, R. H. (2012), '*Salmonella* serovars isolated from table eggs: An overview', *Food Res. Int.*, 45, 745–54.

Masschalck, B. and Michiels, C. W. (2003), 'Antimicrobial properties of lysozyme in relation to foodborne vegetative bacteria', *Crit. Rev. Microbiol.*, 29, 191–214.

Messens, W., Duboccage, L., Grijspeerdt, K., Heyndrickx, M. and Herman, L. (2004), 'Growth of *Salmonella* serovars in hens' egg albumen as affected by storage prior to inoculation', *Food Microbiol.*, 21, 25–32.

Messens, W., Grijspeerdt, K. and Herman, L. (2005), 'Eggshell characteristics and penetration by *Salmonella enterica* serovar Enteritidis through the production period of a layer flock', *Br. Poult. Sci.*, 46(6), 694–700.

Messens, W., Grijspeerdt, K. and Herman, L. (2006), 'Eggshell penetration of hen's eggs by *Salmonella enterica* serovar Enteritidis upon various storage conditions', *Br. Poult. Sci.*, 47(5), 554–60.

Messens, W., Grijspeerdt, K., De Reu, K., De Ketelaere, B. and Mertens, K. (2007), 'Eggshell penetration of various types of hens' eggs by *Salmonella enterica* serovar Enteritidis', *J. Food Prot.*, 70(3), 623–8.

Mine, Y., Ma, F. and Lauriau, S. (2004), 'Antimicrobial peptides released by enzymatic hydrolysis of hen egg white lysozyme', *J. Agric. Food Chem.*, 52, 1088–94.

Miyamoto, T., Baba, E., Tanaka, T., Sasai, K. and Fukata, T. (1997), '*Salmonella* Enteritidis contamination of eggs from hens inoculated by vaginal, cloacal, and intravenous routes', *Avian Dis.*, 41(2), 296–303.

Miyamoto, T., Horie, T., Baba, E., Sasai, K. and Fukata, T. (1998), '*Salmonella* penetration through eggshell associated with freshness of laid eggs and refrigeration', *J. Food Prot.*, 61(3), 350–3.

Mizumoto, N., Sasai, K., Tani, H. and Baba, E. (2005), 'Specific adhesion and invasion of *Salmonella* Enteritidis in the vagina of laying hens', *Vet. Microbiol.*, 111(1–2), 99–105.

Moats, W. A. (1980), 'Classification of bacteria from commercial egg washers and washed and unwashed eggs', *Appl. Environ. Microbiol.*, 40(4), 710–14.

Molla, A., Matsumura, Y., Yamamoto, T., Okamura, R. and Maeda, H. (1987), 'Pathogenic capacity of proteases from *Serratia marcescens* and *Pseudomonas aeruginosa* and their suppression by chicken egg white ovomacroglobulin', *Infect. Immun.*, 55, 2509–17.

Murase, T., Holt, P. S. and Gast, R. K. (2005), 'Growth of *Salmonella enterica* serovar Enteritidis in albumen and yolk contents of eggs inoculated with this organism onto the vitelline membrane', *J. Food Prot.*, 68(4), 718–21.

Murase, T., Fujimoto, K., Nakayama, R. and Otsuki, K. (2006), 'Multiplication and motility of *Salmonella enterica* serovars *enteritidis*, *infantis*, and *montevideo* in in vitro contamination models of eggs', *J. Food Prot.*, 69(5), 1012–16.

Musgrove, M. T., Jones, D. R., Northcutt, J. K., Cox, N. A. and Harrison, M. A. (2004), 'Identification of Enterobacteriaceae from washed and unwashed commercial shell eggs', *J. Food Prot.*, 67(11), 2613–16.

Musgrove, M. T. and Jones, D. R. (2005), 'Recovery of *Salmonella* from nest run egg cart shelves', *Poult. Sci.*, 84, 77.

Musgrove, M. T., Jones, D. R., Northcutt, J. K., Cox, N. A. and Harrison, M. A. (2005), 'Shell rinse and shell crush methods for the recovery of aerobic microorganisms and Enterobacteriaceae from shell eggs', *J. Food Prot.*, 68(10), 2144–8.

Nakai, S. (2000), 'Molecular modifications of egg proteins for functional improvement', in Sim, J. S. and Guenter, W. (Eds), *Egg Nutrition and Biotechnology*, 205–17. CAB International, Oxon.

Nakamura, M., Nagamine, N., Takahashi, T., Suzuki, S., Kijima, M., Tamura, Y. and Sato, S. (1994), 'Horizontal transmission of *Salmonella-Enteritidis* and effect of stress on shedding in laying hens', *Avian Dis.*, 38, 282–8.

Nakamura, M., Takagi, M., Takahashi, T., Suzuki, S., Sato, S. and Takehara, K. (1997), 'The effect of the flow of air on horizontal transmission of *Salmonella enteritidis* in chickens', *Avian Dis.*, 41, 354–60.

Neilands, J. B. (1995), 'Siderophores: structure and function of microbial iron transport compounds', *J. Biol. Chem.*, 270, 26723–6.

Nikaido, H. (2003), 'Molecular basis of bacterial outer membrane permeability revisited', *Microbiol. Mol. Biol. Rev.*, 67, 593–656.

Nnalue, N. A., Shnyra, A., Hultenby, K. and Lindberg, A. A. (1992), '*Salmonella choleraesuis* and *Salmonella typhimurium* associated with liver cells after intravenous inoculation of rats are localized mainly in Kupffer cells and multiply intracellularly', *Infect. Immun.*, 60(7), 2758–68.

Nyeleti, C., Cogan, T. A. and Humphrey, T. J. (2004), 'Effect of sunlight on the survival of *Salmonella* on surfaces', *J. Appl. Microbiol.*, 97(3), 617–20.

Okamoto, I., Mizutani, K. and Hirose, M. (2004), 'Iron-binding process in the amino- and carboxyl-terminal lobes of ovotransferrin: quantitative studies utilizing single $Fe3^{+}$-binding mutants', *Biochemistry (Mosc.)*, 43, 11118–25.

Okamura, M., Miyamoto, T., Kamijima, Y., Tani, H., Sasai, K. and Baba, E. (2001), 'Differences in abilities to colonize reproductive organs and to contaminate eggs in intravaginally inoculated hens and in vitro adherences to vaginal explants between *Salmonella enteritidis* and other *Salmonella* serovars', *Avian Dis.*, 45, 962–71.

Okamura, M., Sonobe, M., Obara, S., Kubo, T., Nagai, T., Noguchi, M., Takehara, K. and Nakamura, M. (2010), 'Potential egg contamination by *Salmonella enterica* serovar Typhimurium definitive type 104 following experimental infection of pullets at the onset of lay', *Poult. Sci.*, 89, 1629–34.

Pellegrini, A., Thomas, U., Wild, P., Schraner, E., and Von Fellenberg, R. (2000), 'Effect of lysozyme or modified lysozyme fragments on DNA and RNA synthesis and membrane permeability of *Escherichia coli*', *Microbiol. Res.* 155, 69–77.

Peralta, R. C., Yokoyama, H., Ikemori, Y., Kuroki, M. and Kodama, Y. (1994), 'Passive immunisation against experimental salmonellosis in mice by orally administered hen egg-yolk antibodies specific for 14-kDa fimbriae of *Salmonella enteritidis*', *J. Med. Microbiol.*, 41, 29–35.

Poppe, C., Irwin, R. J., Messier, S., Finley, G. G. and Oggel, J. (1991), 'The prevalence of *Salmonella enteritidis* and other *Salmonella* sp. among Canadian registered commercial chicken broiler flocks', *Epidemiol. Infect.*, 107(1), 201–11.

Protais, J., Queguiner, S., Boscher, E., Piquet, J. C. and Nagard, B. (2003), 'Effect of housing systems on the bacterial flora of the air', *Br. Poult. Sci.*, 44(5), 778–9.

Przybylska, A. (2002), 'Foodborne infections and intoxications in Poland in 2000' (article in Polish), *Przeglad Epidemiologiczny*, 56, 293–304.

Przybylska, A. (2003), 'Foodborne infections and intoxications in Poland in 2001' (article in Polish), *Przeglad Epidemiologiczny*, 57, 85–98.

Rabsch, W., Tschape, H. and Baumler, A. J. (2001), 'Non-typhoidal salmonellosis: emerging problems', 'Microbes and Infection/Institut Pasteur', 3, 237–47.

Rajashekara, G., Munir, S., Alexeyev, M. F., Halvorson, D. A., Wells, C. L. and Nagaraja, K. V. (2000), 'Pathogenic role of SEF14, SEF17, and SEF21 fimbriae in *Salmonella enterica* serovar *enteritidis* infection of chickens', *Appl. Environ. Microbiol.*, 66, 1759–63.

Rank, D. L., Saeed, M. A. and Muriana, P. M. (2009), 'Cloning of *Salmonella enterica* serovar Enteritidis fimbrial protein SefA as a surface protein in *Escherichia coli* confers the ability to attach to eukaryotic cell lines', *Appl. Environ. Microbiol.*, 75, 6622–5.

Réhault, S., Anton, M., Nau, F., Gautron, J. and Nys, Y. (2007), 'Les activités biologiques de l'œuf', *INRA Prod. Anim.*, 20, 337–48.

Réhault-Godbert, S., Labas, V., Helloin, E., Hervé-Grépinet, V., Slugocki, C., Berges, M., Bourin, M.-C., Brionne, A., Poirier, J. C., Gautron, J., Coste, F. and Nys, Y. (2013), 'Ovalbumin-related protein X is a heparin-binding ov-serpin exhibiting antimicrobial activities', *J. Biol. Chem.*, 288, 17285–95.

Reiber, M. A., Conner, D. E. and Bilgili, S. F. (1995), '*Salmonella* colonization and shedding patterns of hens inoculated via semen', *Avian Dis.*, 39(2), 317–22.

Riemann, H., Himathongkham, S., Willoughby, D., Tarbell, R. and Breitmeyer, R. (1998), 'A survey for *Salmonella* by drag swabbing manure riles in California egg ranches', *Avian Dis.*, 42, 67–71.

Ruzickova, V. (1994), 'Growth and survival of *Salmonella* Enteritidis in selected egg foods', *Vet. Med.*, 39(4), 187–95.

Sahin, O., Kobalka, P. and Zhang, Q. (2003), 'Detection and survival of *Campylobacter* in chicken eggs', *J. Appl. Microbiol.*, 95, 1070–9.

Schoeni, J. L. and Doyle, M. P. (1994), 'Variable colonization of chickens perorally inoculated with *Escherichia coli* O157:H7 and subsequent contamination of eggs', *Appl. Environ. Microbiol.*, 60, 2958–62.

Schoeni, J. L., Glass, K. A., McDermott, J. L. and Wong, A. C. (1995), 'Growth and penetration of *Salmonella enteritidis*, *Salmonella heidelberg* and *Salmonella typhimurium* in eggs', *Int. J. Food Microbiol.*, 24, 385–96.

Shivaprasad, H. L. L, Timoney, J. F., Morales, S., Lucio, B. and Baker, R. C. (1990), 'Pathogenesis of *Salmonella enteritidis* infection in laying chickens. I. Studies on egg transmission, clinical signs, fecal shedding, and serologic responses', *Avian Dis.*, 34, 548–57.

Shivaprasad, H. L., Methner, U. and Barrow, P. (2013), '*Salmonella* Infections in the Domestic Fowl', in Barrow, P. and Methner, U. (Eds), *Salmonella in Domestic Animals*, 162–92. CAB International, University of Nottingham, UK.

Streit, W. R. and Entcheva, P. (2003), 'Biotin in microbes, the genes involved in its biosynthesis, its biochemical role and perspectives for biotechnological production', *Appl. Microbiol. Biotechnol.*, 61, 21–31.

Suzuki, S. (1994), 'Pathogenicity of *Salmonella* Enteritidis in poultry', *Int. J. Food Microbiol.*, 21(1–2), 89–105.

Swayne, D. E. and Beck, J. R. (2004), 'Heat inactivation of avian influenza and Newcastle disease viruses in egg products', *Avian Pathol.*, 33(5), 512–18.

Techer, C., Baron, F., Delbrassinne, L., Belaid, R., Brunet, N., Gillard, A., Gonnet, F., Cochet, M. F., Grosset, N., Gautier, M., Andjelkovic, M., Lechevalier, V. and Jan, S. (2014), 'Global overview of the risk linked to the *Bacillus cereus* group in the egg product industry: identification of food safety and food spoilage markers', *J. Appl. Microbiol.*, 116, 1344–58.

Thapon, J. L., Audiot, V., Nys, Y., Protais, J. and Sauveur, B. (1994), 'Présentation générale de l'œuf', in Thapon, J. L. and Bourgeois, C. M. (Eds), *L'œuf et les ovoproduits* (Chapter 1), Collection Sciences et Techniques Agroalimentaires, 1–108. Lavoisier Tec et Doc. Lavoisier, Paris.

Thiagarajan, D., Thacker, H. L. and Saeed, A. M. (1996), 'Experimental infection of laying hens with *Salmonella enteritidis* strains that express different types of fimbriae', *Poult. Sci.*, 75, 1365–72.

Thorns, C. J. (2000), 'Bacterial food-borne zoonoses', *Revue Scientifique et Technique*, 19, 226–39.

Timoney, J. F., Shivaprasad, H. L., Baker, R. C. and Rowe, B. (1989), 'Egg transmission after infection of hens with *Salmonella enteritidis* phage type 4', *Vet. Rec.*, 125, 600–1.

Tranter, H. S. and Board, R. G. (1984), 'The influence of incubation temperature and pH on the antimicrobial properties of hen egg albumen', *J. Appl. Bacteriol.*, 56(1), 53–61.

Tung, M. A., Garland, M. R. and Gill, P. K. (1979), 'A scanning electron microscope study of bacterial invasion in hen's egg shell', *Inst. Can. Sci. Technol. Aliment.*, 12(1), 16–22.

Uzzau, S., Brown, D. J., Wallis, T., Rubino, S., Leori, G., Bernard, S., Casadesus, J., Platt, D. J. and Olsen, J. E. (2000), 'Host adapted serotypes of *Salmonella enterica*', *Epidemiol. Infect.*, 125, 229–55.

Vadehra, D. V., Baker, R. C. and Naylor, H. B. (1970), 'Role of cuticule in spoilage of chicken eggs', *J. Food Sci.*, 35, 5–6.

Valente, Moro, C., Fravalo, P., Amelot, M., Chauve, C., Zenner, L. and Salvat, G. (2007), 'Colonization and organ invasion in chicks experimentally infected with *Dermanyssus gallinae* contaminated by *Salmonella Enteritidis*', *Avian Path*, 36, 307–11.

Valenti, P., Visca, P., Antonini, G. and Orsi, N. (1986), 'Interaction between lactoferrin and ovotransferrin and *Candida* cells', *FEMS Microbiol. Lett.* 33, 271–5.

Van Hoorebeke, S., Van Immerseel, F., Haesebrouck, F., Ducatelle, R. and Dewulf, J. (2011), 'The Influence of the Housing System on *Salmonella* Infections in Laying Hens: A Review', *Zoonoses and Public Health*, 58, 304–11.

Van Hoorebeke, S., Dewulf, J., Immerseel, F. V., Jorgensen, F., Sandilands, V. and Hocking, P. (2012), 'Production systems for laying hens and broilers and risk of human pathogens', in Sandilands, V. and Hocking, P. M. (Eds), *Alternative Systems for Poultry: Health, Welfare and Productivity* (Chapter 5), 77–96. CABI.

Van Immerseel, F., De Buck, J., Pasmans, F., Bohez, L., Boyen, F., Haesebrouck F. and Ducatelle, R. (2004), 'Intermittent long-term shedding and induction of carrier birds after infection of chickens early posthatch with a low or high dose of *Salmonella enteritidis*', *Poult. Sci.*, 83, 1911–16.

Van Immerseel, F., Methner, U., Rychlik, I., Nagy, B., Velge, P., Martin, G., Foster, N. and Ducatelle, R. and Barrow, P. A. (2005), 'Vaccination and early protection against non-host-specific *Salmonella* serotypes in poultry: exploitation of innate immunity and microbial activity', *Epidemiol. Infect.*, 133, 959–78.

Viénot, E. and Le Boucher G (2007), 'Le guide de l'éleveur de pondeuses', Filières avicoles.

Wales, A., Breslin, M., Carter, B., Sayers, R. and Davies, R. (2007), 'A longitudinal study of environmental *Salmonella* contamination in caged and free-range layer flocks', *Avian Pathol.*, 36(3),187–97.

Wales, A. and Davies, R. H. (2013), 'Environmental aspects of *Salmonella*', in Barrow, P. A. and Nethner, U. (Eds), *Salmonella in Domestic Animals* (Chapter 20), 399–425. CABI.

Wang, C. and Shelef, L. A. (1991), 'Factors contributing to antilisterial effects of raw egg albumen', *J. Food Sci.*, 56, 1251–4.

Wesierska, E., Saleh, Y., Trzizka, T., Kopec, W., Siewinski, M. and Korzekwa, L. (2005), 'Antimicrobial activity of chicken egg white cystatin', *World J. Microbiol. Biotechnol.*, 21, 59–64.

White, S. H., Wimley, W. C. and Selsted, M. E. (1995), 'Structure, function, and membrane integration of defensins', *Curr. Opin. Struct. Biol.*, 5, 521–7.

Wieneke, A. A., Roberts, D. and Gilbert, R. J. (1993), 'Staphylococcal food poisoning in the United Kingdom, 1969–90', *Epidemiol. Infect.*, 110, 519–31.

Wigley, P., Hulme, S. D., Powers, C., Beal, R. K., Berchieri Jr., A., Smith, A. and Barrow, P. (2005), 'Infection of the reproductive tract and eggs with *Salmonella enterica* serovar *pullorum* in the chicken is associated with suppression of cellular immunity at sexual maturity', *Infect. Immun.*, 73, 2986–90.

Wu, J. and Acero-Lopez, A. (2012), 'Ovotransferrin: structure, bioactivities, and preparation', *Food Res. Int.*, 46, 480–7.

Chapter 6

Sampling and detection of *Salmonella* in eggs

Richard K. Gast, United States Department of Agriculture, USA

1 Introduction: bacterial pathogens in eggs

Diverse microorganisms, including pathogens such as *Campylobacter jejuni*, *Listeria monocytogenes* and *Yersinia enterocolitica*, are sometimes found on eggshells and can survive or grow in the edible liquid contents of eggs (Jones et al., 2015; Ricke et al., 2015). However, the vast majority of egg-associated human illnesses are attributed to bacteria of the genus *Salmonella*. Before the passage of the 1970 Egg Products Inspection Act in the United States, infections with a wide variety of *Salmonella* serovars had a long history of association with contaminated eggs. This legislation, which prohibited the commercial marketing of cracked or dirty table eggs and instituted dependably effective pasteurization standards for liquid egg products, resulted in very substantial reductions in the frequency of egg-transmitted human salmonellosis in the years which followed.

1.1 *Salmonella* Enteritidis

In the mid-1980s, an emerging international public health issue led to a revival of concerns about the role of eggs in the foodborne transmission of *Salmonella* infections (St. Louis et al., 1988). However, unlike the older historical precedents, this newly identified problem was attributed almost entirely to clean, intact, grade A table eggs, and was caused almost exclusively by a single serovar, *Salmonella* Enteritidis. During the subsequent decades, a continuing high prevalence of *S.* Enteritidis infections around the world has been

http://dx.doi.org/10.19103/AS.2016.0012.14

principally linked with the consumption of contaminated eggs (Braden, 2006; Jackson et al., 2013; Pires et al., 2014). The frequency of human illness due to this pathogen has been epidemiologically correlated with its prevalence in commercial egg-laying chickens (Havelaar et al., 2013; Arnold et al., 2014). For example, egg-laying hens were implicated in 42% of all cases of human salmonellosis in a recent multinational European study (De Knegt et al., 2015). A sustained commitment to comprehensive risk reduction and testing programmes for egg-laying poultry flocks has led to several reports of documented progress in reducing both egg contamination and human illness (Esaki et al., 2013; O'Brien, 2013; Wright et al., 2016).

1.2 Other Salmonellae

Although eggborne human illness has most often been attributed to *S.* Enteritidis, other serovars have also been implicated. *S.* Heidelberg is among the most common salmonellae found in North American laying flocks and their housing environments, and has been identified as the source of a low incidence of egg-transmitted human disease outbreaks (Chittick et al., 2006). Likewise, the presence of *S.* Typhimurium in Australian laying flocks has been linked to sporadic disease transmission via eggs (Slinko et al., 2009). In a report from the European Union (EU), serovars other than *S.* Enteritidis (particularly *S.* Typhimurium) caused 22% of egg-associated disease outbreaks (Threlfall et al., 2014). Other salmonellae (such as serovar Kentucky in the United States and serovar Sofia in Australia) are commonly detected in commercial poultry flocks, but have not been linked to the production of contaminated eggs. Focusing *Salmonella* testing in laying flocks on the epidemiologically pre-eminent serovar Enteritidis is a cost-effective public health strategy, but continuing to monitor for the presence of other serovars protects against the emergence of previously infrequent or inconsequential strains.

2 Contamination of eggs with *Salmonella*

2.1 Mechanisms of egg contamination

Eggshells can become contaminated by bacteria as a result of exposure to faeces or other environmental sources in laying houses (Board and Fuller, 1994), or in egg processing facilities when sanitation standards are inadequate (Davies and Breslin, 2003). Extensive microbial contamination on freshly laid eggs is uncommon, but pathogens which are not removed during processing can persist and later be transferred to the edible liquid contents of eggs when their shells are broken for use or consumption (Messens et al., 2006). Bacteria can also reach the interior contents by penetration through eggshells. The porous eggshell is not an effective barrier to bacterial penetration from areas of moist faecal contamination, especially in the large end where the shell membranes separate to form an air cell (Berrang et al., 1999), although the two underlying shell membranes provide additional protection (Burley and Vadehra, 1989; Ricke et al., 2015). Eggs are routinely washed in some nations (including the United States, Canada, Australia and Japan) to remove pathogens and spoilage organisms from the exterior surface. However, washing may remove the proteinaceous cuticle, which plugs pores in the shell, and improper temperature control during egg processing can produce a pressure gradient that facilitates microbial penetration through the shell membranes (Stadelman and Cotterill, 1995; Berrang et al., 1999). Diverse salmonellae can be found on eggshells, but only a few serovars are commonly isolated from egg contents. *S.* Enteritidis has been observed to cause a higher frequency of egg contents contamination, but a lower frequency of eggshell contamination than other serovars (Arnold et al., 2014).

Chickens typically become infected with *Salmonella* via oral ingestion from their environment. Colonization of the gastrointestinal tract by the pathogen (notably the crop and ceca) is sometimes followed by invasion of mucosal epithelial cells and systemic dissemination. In laying hens, *S.* Enteritidis infection can involve colonization of the ovary (site of egg yolk maturation) and oviduct (site of albumen secretion around yolks as they descend towards oviposition) (Gast and Holt, 2000a; De Buck et al., 2004). Invasion of reproductive organs by *S.* Enteritidis is evidently the principal direct cause of contamination of the edible contents of eggs (Gantois et al., 2009; Gast et al., 2011a). *S.* Enteritidis has been found inside both preovulatory ovarian follicles and forming eggs removed from hens' oviducts (Keller et al., 1995). Nevertheless, even though systemic infection is evidently a necessary precursor for the production of internally contaminated eggs, experimental infection studies have reported an inconsistent relationship between the frequency or magnitude of organ invasion and the incidence of egg contamination (Gast et al., 2007b).

2.2 Quantitative aspects of egg contamination: frequency and magnitude

Experimental infection of laying hens with *S.* Enteritidis has been shown to cause reproductive organ colonization and egg contamination for a few weeks after oral inoculation, at frequencies that depend on the pathogen dose administered (Gast et al., 2011a, 2013). However, even very large oral doses of *S.* Enteritidis have generally led to relatively infrequent egg contamination (Gast and Holt, 2000a; Gast et al., 2013). Commercial laying flocks are likely to be only sporadically exposed to small infecting doses of *Salmonella* from environmental sources, and accordingly have been reported to produce contaminated eggs at a very low prevalence (Ebel and Schlosser, 2000; DeWinter et al., 2011; Esaki et al., 2013). The occurrence of egg contamination over time in commercial flocks can be intermittent as infection spreads gradually among co-housed laying hens. Naturally contaminated eggs (laid by infected laying hens) have typically been reported to harbour very small initial concentrations of *Salmonella* cells when tested immediately after oviposition. Most contaminated eggs contain less than ten cells each and pathogen levels exceeding a few hundred cells per egg are unusual, even when very large oral doses of *S.* Enteritidis are experimentally administered (Humphrey et al., 1989, 1991; Gast and Beard, 1992; Gast and Holt, 2000a; Gast et al., 2013). Nevertheless, bacterial growth to high levels has been observed in a small percentage of contaminated eggs (Humphrey et al., 1991; Cogan et al., 2004).

2.3 Location, multiplication and migration of *Salmonella* in eggs

The potential for small *Salmonella* populations in eggs to multiply to more dangerous levels is influenced by the initial magnitude and location of contamination, opportunities for the movement of bacteria or nutrients within eggs during storage and the rate at which growth-restricting temperatures are achieved. Colonization of different regions of the reproductive tract (ovary or oviduct) in systemically infected laying hens can result in deposition of *S.* Enteritidis in either the yolk or albumen of developing eggs (Bichler et al., 1996; Gast and Holt, 2000a; Gast et al., 2004, 2007b). The relatively small cell numbers of salmonellae that can be found inside fresh eggs have been interpreted as suggesting that the initial site of contamination may be more often in the growth-limiting albumen than in the nutrient-rich yolk (Humphrey, 1994; Gantois et al., 2009). Likewise, in eggs laid by experimentally inoculated hens, *S.* Enteritidis has been deposited far more frequently in association with either the albumen or vitelline (yolk) membrane than inside the interior yolk contents (Gast and Beard, 1990; Gast and Holt, 2001a; Gast et al., 2003a).

Avian eggs have several physical and biochemical barriers to protect developing embryos against exposure to pathogens, including albumen proteins which restrict microbial multiplication by limiting iron availability or disrupting bacterial membranes (Baron et al., 2016). Despite the presence of these antimicrobial constituents in albumen, *S.* Enteritidis is able to survive and sometimes even grow slowly (Gast and Holt, 2000b, 2001; Kang et al., 2006; Chen and Thesmar, 2008; Okamura et al., 2008). Bacterial lipopolysaccharide synthesis appears to be crucial for survival in egg albumen (Raspoet et al., 2014). *S.* Enteritidis multiplication proceeds faster in fresh than in stored albumen, perhaps as a result of increasing pH during storage (Messens et al., 2004). Egg yolk contains abundant nutrients that can support rapid and prolific bacterial growth and lacks any antimicrobial constituents (Humphrey and Whitehead, 1993; Braun and Fehlhaber, 1995; Gurtler and Conner, 2009). Accordingly, very small initial concentrations of *S.* Enteritidis cells can multiply to dangerously high levels within a single day after inoculation into egg yolk (Gast and Holt, 2000b; Gast and Guraya, 2013).

As eggs age during storage, their ability to restrict or support microbial growth changes. The gradual degradation of vitelline membrane integrity, accelerated by high storage temperatures, allows the release of yolk nutrients into the albumen, facilitating the multiplication of *S.* Enteritidis (Humphrey, 1994; Hara-Kudo et al., 2001b; Latimer et al., 2002). In experimentally inoculated eggs, *S.* Enteritidis growth in regions of the albumen surrounding the yolk increased with the age of the eggs, although rapid multiplication in albumen only occurred after 3 weeks of storage at 20°C (Humphrey and Whitehead, 1993). In addition, the antibacterial activity of egg white proteins was shown to decrease after storage at 37°C for 14 days (Rehault-Godbert et al., 2010).

Temperature has a principal influence on the likelihood of *Salmonella* growth in eggs. A risk assessment model concluded that holding eggs at 18.3°C for 5.5 days after laying would increase the risk of human salmonellosis by 25-fold (Pouillot et al., 2014). The temperature required to support the growth of *S.* Enteritidis has been estimated to be in the range of 7.4–45.2°C for liquid egg white and 9.9–46.8°C for liquid egg yolk (Huang, 2015). *S.* Enteritidis has been shown to multiply to high cell densities at 15°C or higher, but growth is considerably slower at 10°C and ceases at around 4°C (Kim et al., 1989; Saeed and Koons, 1993; Schoeni et al., 1995; Gast and Holt, 2000b; Gurtler and Conner, 2009). When eggs are transferred into refrigerated storage, their interior contents begin to cool gradually to equilibrate with the ambient temperature of the cooler, but *Salmonella* multiplication may continue in egg yolks during this transitional cooling period (Chen et al., 2002; Gast and Guraya, 2013). *S.* Enteritidis survived for at least 8 weeks when egg yolk was stored at 6°C (Lublin et al., 2008).

Salmonellae can persist for weeks on eggshells stored under favourable conditions of temperature and relative humidity, sometimes penetrating into the interior contents

(Messens et al., 2005a, 2006; Lublin et al., 2015). The potential for *Salmonella* penetration through shells differs between eggs from different lines of hens (Jones et al., 2004; Rathgeber et al., 2013). Bacterial penetration is retarded by the proteinaceous cuticle in freshly laid eggs (De Reu et al., 2006; Ray et al., 2015), but is apparently unaffected by eggshell thickness or pore characteristics (Jones and Musgrove, 2005; Muñoz et al., 2015).

Even after initial deposition on the exterior surface of the vitelline (yolk) membrane or in nearby areas of the albumen, *S.* Enteritidis can migrate into the interior yolk contents and begin to multiply (Murase et al., 2005; Gast et al., 2007a, 2008, 2010a). Chemotaxis may draw albumen contaminants towards nutrients inside the yolk (Grijspeerdt et al., 2005). Moreover, extensive bacterial multiplication outside the vitelline membrane may both precede and exceed subsequent multiplication inside the yolk contents (Gast et al., 2008). The migration of *S.* Enteritidis across the yolk membrane increases with the level of contamination, storage temperature and egg's age (Braun and Fehlhaber, 1995; Gast and Holt, 2000b). *In vitro* egg contamination models have shown substantial *Salmonella* migration across yolk membranes during storage at warm temperatures (Braun and Fehlhaber, 1995; Gast and Holt, 2000b; Murase et al., 2005), but this migration occurs much less frequently at 20°C and is entirely absent at 7.2°C (Gast et al., 2005b, 2006, 2007a, 2010a). Although evidence of bacterial penetration into yolks has been reported as early as two hours into storage (Gast et al., 2006), the relatively modest numbers of *S.* Enteritidis cells found inside nutritionally abundant yolks after inoculated eggs were incubated for 24 hours at 30°C suggests that migration into yolks proceeds relatively slowly during this first day of storage (Gast et al., 2005b). The likelihood of growth inside yolks following vitelline membrane penetration increases significantly when inoculated eggs are held for several days at warm temperatures (Guan et al., 2006; Murase et al., 2006), perhaps due to declining albumen viscosity and vitelline membrane integrity (Messens et al., 2004; Chen et al., 2005).

2.4 Serovar and strain differences

The public health consequences of *Salmonella* infections in egg-laying chickens can vary significantly between different bacterial serovars. There is a uniquely strong epidemiological relationship between human *S.* Enteritidis infections and contaminated eggs, although strains of some other serovars can cause some of the same outcomes in infected poultry. Egg-contaminating *S.* Enteritidis strains have been observed to adhere to reproductive tract mucosa and survive in forming eggs better than other environmental salmonellae, perhaps related to the expression of very long lipopolysaccharide O-antigen (Wales and Davies, 2011; Coward et al., 2013). Experimental oral infection of hens with some strains of serovars Heidelberg and Typhimurium has also led to colonization of reproductive tissues at high frequencies (Gantois et al., 2008), although the associated frequencies of egg contamination were much lower than for *S.* Enteritidis (Gast et al., 2005a, 2007b, 2011b).

Significant differences in both their growth properties in eggs (Gast and Holt, 2001b; Cogan et al., 2004; Gast and Guraya, 2013) and their propensity for migration across yolk membranes (Gast et al., 2007a; Gantois et al., 2008) have sometimes distinguished individual *Salmonella* strains. Isolates of *S.* Enteritidis were better able to survive in egg albumen than strains of other serovars (De Vylder et al., 2012), but *S.* Enteritidis strains that were sensitive to both acidic and oxidative stress had impaired albumen survival and growth properties (Shah et al., 2012). Moreover, no consistent differences in growth properties in eggs have been attributed to the various phage types within serovar Enteritidis (Gast and

Holt, 2001b; Gast and Guraya, 2013). The multiplication of bacterial contaminants is also potentially influenced by differences between eggs from different sources, as illustrated by the experimental observation that *S.* Enteritidis migration into and growth inside egg yolks varied significantly between eggs from several genetically distinct commercial lines of laying hens (Gast et al., 2010b).

3 Detection of *Salmonella* on or inside eggshells: sampling, pooling and incubation

3.1 Sampling eggshells

Although the presence of salmonellae in the edible liquid contents of eggs is of more direct epidemiological significance, eggshells are also sometimes tested for external surface contamination (García et al., 2011). The external surface of intact eggs can be sampled by immersing them in 30–50 ml of enrichment broth in a plastic bag, and soaking or manual rubbing for at least 30 s before the egg is removed. Eggshells can also be manually crushed in enrichment broth to allow a complete access of the media to internal surfaces (Waltman and Gast, 2016). Sampling by shell crush methods has reportedly yielded higher frequencies of *Salmonella* recovery than rinse methods (Musgrove et al., 2005; Kawasaki et al., 2008; Webb et al., 2014).

3.2 Pooling and incubation of egg content samples

The detection of *Salmonella* in the edible internal contents of eggshells provides the most incontrovertible and epidemiologically relevant evidence that laying flocks pose a threat to public health, so egg testing is an essential step in most risk reduction programmes for *S.* Enteritidis. In many instances, including the national regulatory plan in the United States, laying flocks are first screened for the presence of the pathogen in the poultry house environment, and eggs from presumptively infected flocks are then cultured to confirm their status (U.S. Food and Drug Administration, 2009). However, the diagnostic sensitivity of egg testing can sometimes be limited by the infrequent, sporadic and transient occurrence of egg contamination (Gantois et al., 2009). For example, testing eggs from an Austrian laying farm that was strongly implicated as the source of an egg-associated *S.* Enteritidis human disease outbreak yielded no positive results, but later environmental testing was positive and the farm's eggs would subsequently be implicated in another outbreak (Schmid et al., 2007). Liquid, frozen and dried egg products are also routinely tested for *Salmonella* contamination by methods which are described elsewhere (Ricke et al., 2015).

The principal concern in handling eggshells before culturing their contents for *Salmonella* is to avoid inadvertent contamination of the intended sample by organisms present on the shell (Ricke et al., 2015). Refrigerated eggs should be tempered at room temperature before sampling so that any moisture forming on the shell surface as the egg warms (and which might increase opportunities for the contents to become contaminated as the shell is broken) will evaporate. Shells are disinfected, most often by immersing them for at least 10 seconds in either 70% alcohol or a mixture of three parts 70% alcohol and one part iodine/potassium iodide solution (Ricke et al., 2015), and then air dried. Eggs are handled with sterile gloves, the shells are cracked against a sterile, sharp-edged surface, and the contents are transferred into sterile containers such as disposable beakers or Stomacher bags. The entire liquid contents (yolk plus albumen) can be tested together or a sterile egg separator or spoon can be used to sample the components individually. Samples are mixed (most often in a Stomacher bag or using a sterile utensil) until they are homogeneous.

The evolution of *Salmonella* detection methods has been profoundly influenced by some of the salient characteristics of egg contamination with this pathogen. *S.* Enteritidis deposition in eggs occurs at a very low prevalence in commercial laying flocks and most contaminated eggs typically contain very low bacterial cell concentrations, so large numbers of eggs must be sampled to achieve a high probability of detection (Gast, 1993a, 2007). In order to test the necessary high volume of eggs without completely overwhelming laboratory facilities and resources, the contents of 10–20 eggs are often pooled together for sampling. Regulations for eggshell production in the United States specify that batches of 1000 eggs must be collected from laying houses which are environmentally positive for *S.* Enteritidis and tested for contamination in 50 pooled samples of 20 eggs each (U.S. Food and Drug Administration, 2009). However, this pooling process further dilutes the already low *Salmonella* concentrations present in contaminated individual eggs. Accordingly, egg pools are sometimes incubated after mixing (overnight at 37°C or for several days at room temperature) to encourage the multiplication of salmonellae to more consistently detectable levels before continuing with subsequent bacteriological enrichment culture steps (Gast, 1993a; Gast and Holt, 2003). The necessary duration of an egg incubation step depends on both the temperature used and the sensitivity of tests, which are subsequently applied to the pools (Gast, 1993c). Supplementing incubating egg pools with concentrated sources of iron and other nutrients (often in the form of concentrated broth enrichment media) can improve the growth rate of salmonellae (Gast and Holt, 1995, 1998; Chen et al., 2001). Trypticase soy broth (with or without added iron) has been identified as an especially useful additive to promote *Salmonella* growth in incubating egg pools (Gast and Holt, 1998; Zhang et al., 2013a).

4 Detection of *Salmonella* on or inside eggshells: conventional culture methods

Although traditional methods generally prescribe the use of both non-selective broth pre-enrichment and selective broth enrichment, directly plating incubated egg pools onto plates of selective agar can prospectively reduce the time, media and labour requirements of the overall egg culturing process. Direct plating relies on the multiplication of salmonellae during pool incubation to generate sufficient cell numbers to make consistent detection possible (Gast, 1993b,c; Gast and Holt, 1995, 1998). However, although direct plating of egg pools reduces culturing time by several days, it yields significantly poorer *Salmonella* detection sensitivity (Gast, 1993b; Gast and Holt, 1995). The multiplication of *S.* Enteritidis during the incubation of egg pools must produce a concentration of approximately 10^5 cfu/ml to allow consist detection by direct plating (Gast and Holt, 1995). Incubation of egg pools for 24 hours at 37°C (or 72 hours at 25°C) has been shown to allow the detection of small initial *Salmonella* concentrations by direct plating (Gast, 1993a,c; Gast and Holt, 1995). Supplementing incubating egg pools with a source of iron or concentrated trypticase soy broth can significantly increase the growth of *S.* Enteritidis, thereby increasing the likelihood of detection by direct plating (Cudjoe et al., 1994; Chen et al., 2001; Gast and Holt, 1995, 1998, 2003).

The high detection sensitivity available from standard enrichment-based methods is usually preferred for testing eggs when time is not the most critical consideration (Gast, 1993a,c; Gast and Holt, 1995). Traditional multiple-enrichment culture methods can detect <10 cfu/ml of *S.* Enteritidis in egg pools (Gast and Holt, 1995; Hara-Kudo et al., 2001a). Following the incubation of egg pools, the next culturing step is non-selective pre-enrichment (Gast and Holt, 1995; Ricke et al., 2015). For example, federal egg safety regulations in the United States specify that a 25 ml portion of each incubated 20-egg pool is transferred into 225 ml of trypticase soy broth for pre-enrichment (U.S. Food and Drug Administration, 2009). Pre-enrichment can substantially increase the sensitivity of detecting salmonellae in comparison with direct selective enrichment, perhaps by supporting the multiplication of very small *Salmonella* populations before exposure to harsher selective enrichment conditions (Gast, 1993b). Pre-enrichment of a large volume of pooled liquid egg in trypticase soy broth was shown to provide sensitive *Salmonella* detection without a prior egg pool incubation step (Zhang et al., 2013b). Buffered peptone water is another popular pre-enrichment medium. Pre-enrichment cultures are usually incubated for 20–24 hours at 35–37°C.

Following pre-enrichment, selective enrichment supports further expansion of the *Salmonella* population in each sample while simultaneously suppressing the growth of other competing organisms. The most commonly used selective broth media for *Salmonella* isolation are tetrathionate (TT) broth and Rappaport-Vassiliadis (RV) broth. The regulatory testing programme in the United States mandates inoculation of both of these media from each pre-enriched egg sample (U.S. Food and Drug Administration, 2009). TT broth, commercially available in several variant formulations, generally includes brilliant green dye as a selective ingredient and is further supplemented with an iodine solution. TT broth is inoculated at a 1:10 ratio of sample-to-enrichment broth and then incubated for 20–24 hours at 35–37°C. RV broth is typically inoculated at a 1:100 ratio of sample-to-enrichment broth and incubated for 20–24 hours at 41–42°C. An alternative approach to selective enrichment involves the use of plates of semi-solid modified semisolid RV (MSRV) medium. MSRV is inoculated either on or below the surface with small volumes of pre-enrichment cultures and incubated for 24–48 hours at 41–42°C. Motile salmonellae are detected in this medium by their migration from the point of inoculation.

After selective enrichment, samples are transferred to plates of selective agar media to produce isolated bacterial colonies, each derived from a single cell. A variety of available agar media allow the differentiation of *Salmonella* colonies from other competing organisms. The most commonly used plating media for egg testing include brilliant green (BG), xylose-lysine-deoxycholate (XLD), xylose-lactose-tergitol 4 (XLT4), bismuth sulphite (BS) and Hektoen enteric (HE) agars (Waltman and Gast, 2016). These standard agar media can generally inhibit most coliform bacteria, although species of *Proteus, Pseudomonas, Citrobacter, Providencia and Morganella* sometimes produce colonies that resemble *Salmonella* and can only be differentiated via further biochemical testing of isolated colonies.

BG agar has been widely and successfully used to isolate salmonellae from diverse poultry sources. This medium is often supplemented with antimicrobial compounds (such as sulphapyridine, sulphadiazine or novobiocin) to increase its selectivity, particularly against *Proteus* species. Salmonella colonies on BG agar are typically transparent pink to deep fuchsia. Another group of agar plating media with proven efficacy for detecting salmonellae are XLD and XLT4 (the latter distinguished by the inclusion of the surfactant, tergitol 4). These media inhibit *Proteus* and *Pseudomonas*. Novobiocin is sometimes added to XLD medium to augment its anti-*Proteus* selectivity. *Salmonella* colonies on XLD or XLT4 plates are typically red with black centres, although they sometimes appear nearly completely black. Other plating media that are sometimes advocated for detecting *Salmonella* include BS and HE agars, both of which are widely used for human clinical applications (U.S. Food and Drug Administration, 2014). However, these media lack both the sensitivity and the specificity of BG-novobiocin and XLT4. Modified lysine iron agar incorporates novobiocin and is similar in performance to BG-novobiocin or XLD-novobiocin. Rambach agar, a chromogenic medium, is effective but considerably more expensive than the other available options.

To maximize the bacterial recovery, samples should always be plated on at least two different agar media, preferably with biochemically dissimilar differential indicator systems. The combination of BG-novobiocin and XLT4 is often recommended. The complementary differential properties of these media (BG identifies salmonellae as negative for lactose fermentation and XLT4 identifies them as positive for hydrogen sulphide production) should improve the probability of detecting biochemically variant *Salmonella* strains. Alternatively, in the testing requirement for eggshell production in the United States (U.S. Food and Drug Administration, 2009), egg samples are plated onto BS, HE and XLD agars. Agar plating media are normally incubated for 20–24 hours at 35–37°C.

Visually characteristic *Salmonella* colonies on selective agar plates must be tested further to confirm their genus identity and determine their serovar. The combination of triple sugar iron (TSI) agar and lysine iron agar (LIA) provides efficient and effective presumptive identification of salmonellae. These media are especially useful for differentiating salmonellae from common poultry-associated genera such as *Proteus* and *Citrobacter*. Three or more well-separated colonies are selected from each plate and transferred to TSI and LIA tubes by stabbing into the butt of the agar and streaking across the slant. These tubes are incubated for 24 hours at 37°C. Salmonellae typically produce an alkaline (red) slant and acid (yellow) butt in TSI, with gas bubbles and sometimes a blackening in the butt due to H_2S production, which can obscure the acid reaction. *Salmonella* causes lysine decarboxylation in LIA, evident as a deep purple (alkaline) slant and alkaline or neutral butt with slight blackening due to H_2S production. Additional biochemical tests are sometimes applied to further confirm an isolate as *Salmonella*. The observed pattern of fermentation with a battery of six carbohydrates (dextrose, lactose, sucrose, mannitol, maltose and dulcitol) can provide further differentiation of PT isolates from other organisms (Cox and Williams, 1976). Salmonellae typically produce acid and gas in all of these except lactose and sucrose. Other biochemical properties that are useful for identifying *Salmonella* include the production of indole, urease and β-galactosidase (salmonellae are negative for all of these).

After biochemical confirmation of the identity of a presumptive *Salmonella* isolate, its serogroup can be determined by slide agglutination tests with commercially available polyvalent antisera directed to somatic O antigens. Next, its specific serovar can be determined by slide agglutination testing with monovalent antisera to specific O antigens and tube agglutination tests with antisera to flagellar H antigens. In many instances, isolates are assigned to a serogroup at the diagnostic laboratory and then submitted to a reference laboratory for definitive serovar determination. In recent years, the availability of primers for serovar-specific *Salmonella* genes has facilitated the development and application of molecular typing methods. Molecular serotyping can be accomplished by methods including a multiplex, bead-based, suspension array assay (Fitzgerald et al., 2007) or 5S intergenic sequence ribotyping (Guard et al., 2012).

5 Detection of *Salmonella* on or inside eggshells: rapid detection methods and egg yolk antibody detection

5.1 Rapid detection methods

Conventional culture methods to detect *Salmonella* require several days to obtain negative results and up to 1 week to confirm positive results. More rapid technologies have been developed in recent years, although these assays have not yet supplanted traditional culturing in most testing protocols. Rapid methods usually generate results at least 1–2 days faster than conventional culturing and are often adaptable to some degree of automation. Nevertheless, many rapid methods are also characterized by a high cost per test and an inability to consistently detect *Salmonella* contamination without at least one prior enrichment culturing step (Gast and Holt, 2003). Rapid tests have seldom been demonstrated to achieve detection sensitivities matching those of multiple-enrichment conventional culture methods. Additionally, although negative results via rapid methods can be considered definitive, positive results are sometimes regarded as only presumptive and thus may require confirmation by conventional culture methods. A wide variety of rapid testing formats have become commercially available, but the most common approaches are based on the detection of salmonellae by specific antibodies in an enzyme immunoassay format (Holt et al., 1995) or by DNA probes following polymerase chain reaction (PCR) amplification (McElroy et al., 1996; Cook et al., 2002). Several rapid screening methods for *Salmonella* detection have been recognized as acceptable substitutes for conventional culturing by the U.S. Food and Drug Administration (2014).

A frequently cited justification for the development of more rapid alternatives to conventional culture methods is that they may provide an opportunity to make decisions about the microbiological safety of eggs within the constraints of marketing timetables. Ideally, this objective would best be achieved with the use of as few (and as brief) enrichment steps as possible before initiation of a rapid test. In general, this makes rapid assays highly dependent on the multiplication of *Salmonella* during a single egg pool incubation or pre-enrichment step. Nutrient supplementation of incubating egg pools to increase bacterial growth may be essential for rapidly achieving detection thresholds. For example, one study determined that two different rapid tests were unable to consistently detect *S.* Enteritidis concentrations below 10^7 cfu/ml in egg pools (Gast et al., 2003b). The previously reported sensitivities of enzyme immunoassay (Price et al., 1995) and PCR-based DNA hybridization (Burkhalter et al., 1995; McElroy et al., 1996) techniques are similar. Experimental inoculation of egg pools has shown that this value is unlikely to be attained within a single working day and only within 24 hours when the pools are supplemented with trypticase soy broth or an iron source (Chen et al., 2001; Gast and Holt, 1995, 1998, 2003).

Several non-enrichment options have been also investigated to address the detection sensitivity problems inherent in applying relatively insensitive rapid diagnostic technologies to the very low numbers of *Salmonella* contaminants found in eggs. These methods are faster than enrichment culturing, but introduce significant additional cost to each test. Prominent among these is immunomagnetic separation (IMS), which employs small magnetic beads coated with antibodies specific to *Salmonella* target antigens. The antibody-carrying beads are introduced to samples and mixed to allow an opportunity for them to encounter and bind target antigens. After a magnetic field is applied to recover the beads from the sample matrix, a secondary test (such as an enzyme immunoassay for the target antigen) is applied to detect salmonellae bound by antibodies on the beads. IMS concentration can facilitate the detection of small numbers of *S*. Enteritidis cells in pooled egg contents by both culturing and rapid technologies (Cudjoe et al., 1994; Holt et al., 1995; Tu et al., 2007). Another approach to concentrating low numbers of salmonellae in pooled egg contents utilized centrifugation after enzymatic digestion and chemical reduction of egg albumen (McElroy et al., 1995, 1996).

Antibodies specific to *Salmonella* target antigens have been harnessed in a variety of enzyme immunoassays and applied to egg samples. Antibodies directed to lipopolysaccharide, flagella or outer membrane protein antigens have successfully found salmonellae in a wide range of poultry-related samples, including eggs (Hassan et al., 1991; Keller et al., 1993; Blais and Martinez-Perez, 2008). Antibody-based methods are characteristically less sensitive than standard culture methods (Tate et al., 1992; Tan et al., 1997), and generally require an initial enrichment of samples to attain *Salmonella* detection threshold values of 10^5–10^7 cells/ml (Hassan et al., 1991; Keller et al., 1993). These tests are sometimes prone to yield false-positive results caused by competing flora able to grow in enrichment media (Beckers et al., 1988). Other antibody-dependent assay formats, including fluorescence polarization and lateral flow immunodiffusion, have detected *S*. Enteritidis in incubated egg pools in <15 minutes (Gast et al., 2003b; Seo et al., 2003).

In recent years, the most prominent rapid *Salmonella* testing technology has been developed to identify genus- or serovar-specific genetic sequences via the hybridization of specific probes with target DNA extracted from samples. The detection sensitivity threshold of DNA hybridization is generally similar to enzyme immunoassays, thus also requiring one or more initial enrichment culturing steps (Lopes et al., 2004). However, the development of PCR systems to amplify specific target DNA sequences has significantly increased the sensitivity of DNA hybridization for detecting salmonellae. After enrichment culturing, PCR-based methods have detected initial contaminant levels of <10 *Salmonella* cells in eggs, corresponding to the sensitivity of conventional culturing (Seo et al., 2004). A method combining initial enrichment with PCR-based DNA hybridization found salmonellae in more pooled eggshell samples than did traditional culturing (Guzmán-Gómez et al., 2013). A quantitative PCR method was reported to provide rapid and reliable enumeration of *Salmonella* contaminants in a variety of egg products (Jakočiūnė et al., 2014). Multiplex PCR assays offer the potential for simultaneous detection of more than one serovar in a sample (Hong et al., 2009). Other DNA amplification strategies have likewise been shown to provide a high sensitivity of *Salmonella* detection in egg samples (Ohtsuka et al., 2005; Yang et al., 2013).

5.2 Egg yolk antibody detection

In addition to their crucial role as the subjects of bacteriological testing to detect *Salmonella* contamination, eggs can provide important additional information about the *Salmonella* infection status of laying flocks. Infected laying hens produce antibodies to several prominent surface antigens of salmonellae, including lipopolysaccharide, flagella and outer membrane proteins. These antibodies are present in the blood and deposited in egg yolks, sometimes at high concentrations, for many weeks after exposure to the pathogen (Gast and Beard, 1991). Testing for antibodies in egg yolks offers several significant advantages over testing for serum antibodies, including the ease of sample collection for workers, the absence of stress for hens and the opportunity to conduct both bacteriological and serological assays on the same sets of samples. Antibodies have been found in egg yolks from hens infected experimentally with as few as 10^3 cells of *S.* Enteritidis, and the detection of egg yolk antibodies has strongly correlated with the production of internally contaminated eggs (Gast et al., 1997a,b). Egg yolk antibody assays have been reported to detect infected commercial flocks with a sensitivity similar to culturing environmental samples (Klinkenberg et al., 2011). Although tests for antibody production are highly sensitive for identifying infected laying flocks, their use has been limited by concerns about the possibility of antigenic cross-reactivity between different serovars and the continuing presence of detectable antibodies long after active infection and egg contamination have ceased.

6 Egg testing and *Salmonella* control: conclusions and challenges for future research

Testing to detect *S.* Enteritidis in eggs, laying hens and the poultry house environment has played a critical but sometimes controversial role in programmes to control the foodborne transmission of this pathogen. Testing programmes that are intended to be the basis for decisions about the fate of laying flocks or their eggs involve considerable uncertainty because of limitations regarding assay sensitivity, fluctuations over time in the prevalence of *S.* Enteritidis within flocks and inside their eggs and the nearly continuous opportunities for re-introduction of the pathogen into poultry houses from diverse environmental sources. However, sensitive and dependable tests for *Salmonella* in eggs remain essential for achieving long-term public health objectives. Although flock and environmental testing have useful predictive values for identifying a potential for egg contamination to occur (Gast, 2007), only actual egg testing can definitively determine whether or not consumers are at risk. Considerable research effort has made tangible progress in explaining how eggs are contaminated by *Salmonella*, where the pathogen is deposited and how it potentially multiplies inside eggs during storage and what types of assays are best able to find bacterial contaminants. Although an international consensus on the most efficacious approaches for detecting salmonellae in eggs is gradually emerging, two principal hurdles still constrain effective utilization of egg testing results. First, the nature of *Salmonella* contamination in eggs poses an especially severe challenge to the sensitivity of detection assays. Egg contamination by this pathogen occurs at an extremely low prevalence and at low cell concentrations, and infected flocks may lay contaminated eggs sporadically over time. Pooling eggs together for sampling is logistically necessary but also introduces a substantial further dilution of the already scarce salmonellae. Accordingly, a useful assay for providing assurance that public health is being protected must be unusually sensitive in comparison to many other applications. Second, eggs are perishable and they are quickly processed to reach market channels. Although testing eggs to identify the flocks which caused human outbreaks has value for avoiding further outbreaks from the same sources, directly preventing the consumption of contaminated eggs would be far more powerful. Attaining this objective will require rapid and efficient methods that offer true real-time results without requiring a commitment of 1 or 2 days to preliminary enrichment culturing as is currently necessary to reach the detection thresholds of all available testing strategies.

7 Where to look for further information

Several book chapters can be consulted for overviews of relatively broad topics. A detailed consideration of *Salmonella* as a poultry health issue is available in an overview of the subject by Gast (2013). Approved standard protocols for detecting pathogens in eggs in the United States are discussed by Ricke et al. (2015). Methods used for identifying salmonellae in poultry flocks and eggs, including media options, are covered in detail by Waltman and Gast (2016). Some review papers also broadly consider a range of public health and animal production issues related to egg-transmitted *Salmonella*, including Howard et al. (2012) and Whiley and Ross (2015). The most thorough review of the mechanisms by which *Salmonella* is deposited inside eggs laid by infected hens was provided by Gantois et al. (2009). Other topics that are addressed in useful review papers include the epidemiology of different *Salmonella* serovars as sources of egg-associated illness (Threlfall et al., 2014), eggshell penetration by salmonellae (Messens et al., 2005b), natural defences of eggs against pathogens (Baron et al., 2016), conventional bacteriological methods for detecting *Salmonella* in eggs (Carrique-Mas and Davies, 2008) and rapid *Salmonella* detection methods in poultry and poultry products (Park et al., 2013).

8 References

Arnold, M. E., Martelli, F., McLaren, I. and Davies, R. H. (2014), 'Estimation of the rate of egg contamination from *Salmonella*-infected chickens', *Zoonoses Pub. Health*, 61, 18–27.

Baron, F., Nau, F., Guérin-Dubiard, C., Bonnassie, S., Gautier, M., Andrews, S. C. and Jan, S. (2016), 'Egg white versus *Salmonella* Enteritidis! A harsh medium meets a resilient pathogen', *Food Microbiol.*, 53, 82–93.

Beckers, H. J., Tips, P. D., Soentoro, P. S. S., Delfgou-Van Asch, E. H. M. and Peters, R. (1988), 'The efficacy of enzyme immunoassays for the detection of salmonellas', *Food Microbiol.*, 5, 147–56.

Berrang, M. E., Cox, N. A., Frank, J. F. and Buhr, R. J. (1999), 'Bacterial penetration of the eggshell and shell membranes of the chicken hatching egg: a review', *J. Appl. Poult. Res.*, 8, 499–504.

Bichler, L. A., Nagaraja, K. V. and Halvorson, D. A. (1996), '*Salmonella enteritidis* in eggs, cloacal swab specimens, and internal organs of experimentally infected White Leghorn chickens', *Am. J. Vet. Res.* 57, 489–95.

Blais, B. W. and Martinez-Perez, A. (2008), 'Detection of group D salmonellae including *Salmonella* Enteritidis in eggs by polymyxin-based enzyme-linked immunosorbent assay', *J. Food Prot.*, 71, 392–6.

Board, R. G. and Fuller, R. (1994), *Microbiology of the Avian Egg*, London, Chapman and Hall.

Braden, C. R. (2006), '*Salmonella enterica* serotype Enteritidis and eggs: a national epidemic in the United States', *Clin. Infect. Dis.*, 43, 512–17.

Braun, P. and Fehlhaber, K. (1995), 'Migration of *Salmonella enteritidis* from the albumen into the egg yolk', *Int. J. Food Microbiol.*, 25, 95–9.

Burkhalter, P. W., Müller, C., Lüthy, J. and Candrian, U. (1995), 'Detection of *Salmonella* spp. in eggs: DNA analyses, culture techniques, and serology', *J. Assoc. Off. Anal. Chem. Int.*, 78, 1531–7.

Burley, R. W. and Vadehra, D. V. (1989), *The Avian Egg*, New York, John Wiley and Sons.
Carrique-Mas, J. J. and Davies, R. H. (2008), 'Bacteriological detection of *Salmonella* Enteritidis in eggs: a review', *Rev. Sci. Tech. Off. Int. Epiz.*, 27, 657–64.
Chen, H., Anantheswaran, R. C. and Knabel, S. J. (2001), 'Optimization of iron supplementation for enhanced detection of *Salmonella* Enteritidis in eggs', *J. Food Prot.*, 64, 1279–85.
Chen, H., Anantheswaran, R. C. and Knabel, S. J. (2002), 'Effect of rapid cooling on the growth and penetration of *Salmonella enteritidis* into egg contents', *J. Food Safety*, 22, 255–71.
Chen, J. and Thesmar, H. S. (2008), 'Populations of *Salmonella* Enteritidis in artificially inoculated chicken eggs as influenced by the temperatures under which eggs might be held from the day of lay until the day of processing', *J. Food Prot.*, 71, 2073–7.
Chen, J., Thesmar, H. S. and Kerr, W. L. (2005), 'Outgrowth of Salmonellae and the physical property of albumen and vitelline membrane as influenced by egg storage conditions', *J. Food Prot.*, 68, 2553–8.
Chittick, P., Sulka, A., Tauxe, R. V. and Fry, A. M. (2006), 'A summary of national reports of foodborne outbreaks of *Salmonella* Heidelberg infections in the United States: clues for disease prevention', *J. Food Prot.*, 69, 1150–3.
Cogan, T. A., Jørgensen, F., Lappin-Scott, H. M., Benson, C. E., Woodward, M. J. and Humphrey, T. J. (2004). 'Flagella and curli fimbriae are important for the growth of *Salmonella enterica* serovars in hen eggs', *Microbiol.*, 150, 1063–71.
Cook, N., Ellison, J., Kurdziel, A. S., Simpkins, S. and Hays, J. P. (2002), 'A nucleic acid sequence-based amplification method to detect *Salmonella enterica* serotype Enteritidis strain PT4 in liquid whole egg', *J. Food Prot.*, 65, 1177–8.
Coward, C., Sait, L., Cogan, T., Humphrey, T. J. and Maskell, D. J. (2013), 'O-antigen repeat number in *Salmonella enterica* serovar Enteritidis is important for egg contamination, colonisation of the chicken reproductive tract and survival in egg albumen', *FEMS Microbiol. Lett.*, 343, 169–76.
Cox, N. A. and Williams, J. E. (1976), 'A simplified biochemical system to screen salmonella isolates from poultry for serotyping', *Poult. Sci.*, 55, 1968–71.
Cudjoe, K. S., Krona, R., Grøn, B. and Olsen, E. (1994), 'Use of ferrous sulphate and immunomagnetic separation to recover *Salmonella enteritidis* from raw eggs', *Int. J. Food Microbiol.*, 23, 149–58.
Davies, R. H. and Breslin, M. (2003), 'Investigation of *Salmonella* contamination and disinfection in farm egg-packing plants', *J. Appl. Microbiol.*, 94, 191–6.
De Buck, J., Pasmans, F., Van Immerseel, F., Haesebrouck, F. and Ducatelle, R. (2004), 'Tubular'glands of the isthmus are the predominant colonization site of *Salmonella* Enteritidis in the upper oviduct of laying hens', *Poult. Sci.*, 83, 352–8.
De Knegt, L. V., Pires, S. M. and Hald, T. (2015), 'Attributing foodborne salmonellosis in humans to animal reservoirs in the European Union using a multi-country stochastic model', *Epidemiol. Infect.*, 143, 1175–86.
De Reu, K., Grijspeerdt, K., Messens, W., Heyndrickx, M., Uyttendaele, M., Debevere, J. and Herman, L. (2006), 'Eggshell factors influencing eggshell penetration and whole egg contamination by different bacteria, including *Salmonella enteritidis*', *Int. J. Food Microbiol.*, 112, 253–60.
De Vylder, Raspoet, R., Dewulf, J., Haesebrouck, F., Ducatelle, R. and Van Immerseel, F. (2012), '*Salmonella* Enteritidis is superior in egg white survival compared with other *Salmonella* serotypes', *Poult. Sci.*, 92, 842–5.
DeWinter, L. M., Ross, W. H., Couture, H. and Farber, J. F. (2011), 'Risk assessment of shell eggs internally contaminated with *Salmonella* Enteritidis', *Int. Food Risk Anal. J.*, 1, 40–81.
Ebel, E. and Schlosser, W. (2000), 'Estimating the annual fraction of eggs contaminated with *Salmonella enteritidis* in the United States', *Int. J. Food Microbiol.*, 61, 51–62.
Esaki, H., Shimura, K., Yamazaki, Y., Eguchi, M. and Nakamura, M. (2013), 'National surveillance of *Salmonella* Enteritidis in commercial eggs in Japan', *Epidemiol. Infect.*, 141, 941–3.
Fitzgerald, C., Collines, M., van Duyne, S., Mikoleti, M., Brown, T. and Fields, P. (2007), 'Multiplex, bead-based suspension array for molecular determination of common *Salmonella* serogroups', *J. Clin. Microbiol.*, 45, 3323–34.

Gantois, I., Ducatelle, R., Pasmans, F., Haesebrouck, F., Gast, R., Humphrey, T. J. and Van Immerseel, F. (2009), 'Mechanisms of egg contamination by *Salmonella* Enteritidis', *FEMS Microbiol. Rev.*, 33, 718–38.

Gantois, I., Eeckhaut, V., Pasmans, F., Haesebrouck, F., Ducatelle, R. and Van Immerseel, F. (2008), 'A comparative study on the pathogenesis of egg contamination by different serotypes of *Salmonella*', *Avian Pathol.*, 37, 399–406.

García, J., Soriano, M., Benítez, V. and Catalá-Gregori, P. (2011), 'Assessment of *Salmonella* spp. in feces, cloacal swabs, and eggs (eggshell and content separately) from a laying hen farm', *Poult. Sci.*, 90, 1581–5.

Gast, R. K. (1993a), 'Detection of *Salmonella enteritidis* in experimentally infected laying hens by culturing pools of egg contents', *Poult. Sci.*, 72, 267–74.

Gast, R. K. (1993b), 'Evaluation of direct plating for detecting Salmonella enteritidis in pools of egg contents', *Poult. Sci.*, 72, 1611–14.

Gast, R. K. (1993c), 'Recovery of *Salmonella enteritidis* from inoculated pools of egg contents', *J. Food Prot.*, 56, 21–4.

Gast, R. K. (2007), 'Serotype-specific and serotype-independent strategies for preharvest control of food-borne *Salmonella* in poultry', *Avian Dis.*, 51, 817–28.

Gast, R. K. (2013), 'Paratyphoid infections', in Swayne, D. E. (ed.), *Diseases of Poultry*, 13th edition, pp. 693–713, 718–33. Wiley-Blackwell Publishing, Ames, Iowa.

Gast, R. K. and Beard, C. W. (1990), 'Production of Salmonella enteritidis-contaminated eggs by experimentally infected hens', *Avian Dis.*, 34, 438–46.

Gast, R. K. and Beard, C. W. (1991), 'Detection of *Salmonella* serogroup D-specific antibodies in the yolks of eggs laid by hens infected with *Salmonella enteritidis*', *Poult. Sci.*, 70, 1273–6.

Gast, R. K. and Beard, C. W. (1992), 'Detection and enumeration of *Salmonella enteritidis* in fresh and stored eggs laid by experimentally infected hens', *J. Food Prot.*, 55, 152–6.

Gast, R. K., Guard-Bouldin, J. and Holt, P. S. (2004), 'Colonization of reproductive organs and internal contamination of eggs after experimental infection of laying hens with Salmonella Heidelberg and Salmonella enteritidis', *Avian Dis.*, 48, 863–9.

Gast, R. K., Guard-Bouldin, J. and Holt, P. S. (2005a), 'The relationship between the duration of fecal shedding and the production of contaminated eggs by laying hens infected with strains of Salmonella Enteritidis and Salmonella Heidelberg', *Avian Dis.*, 49, 382–6.

Gast, R. K., Guard-Petter, J. and Holt, P. S. (2003a), 'Effects of prior serial in vivo passage on the frequency of Salmonella enteritidis contamination in eggs from experimentally infected laying hens', *Avian Dis.*, 47, 633–9.

Gast, R. K. and Guraya, R. (2013), 'Continuing multiplication of *Salmonella enteritidis* in egg yolk during refrigeration at 7.2°C', *Int. J. Poult. Sci.*, 12, 622–7.

Gast, R. K., Guraya, R. and Guard, J. (2013), '*Salmonella* Enteritidis deposition in eggs after experimental infection of laying hens with different oral doses', *J. Food Prot.*, 76, 108–13.

Gast, R. K., Guraya, R., Guard, J. and Holt, P. S. (2010a), 'Multiplication of *Salmonella* Enteritidis in egg yolks after inoculation outside, on, and inside vitelline membranes and storage at different temperature', *J. Food Prot.*, 73, 1902–6.

Gast, R. K., Guraya, R., Guard, J. and Holt, P. S. (2011a), 'Frequency and magnitude of internal organ colonization following exposure of laying hens to different oral doses of Salmonella enteritidis', *Int. J. Poult. Sci.*, 10, 325–31.

Gast, R. K., Guraya, R., Guard-Bouldin, J. and Holt, P. S. (2007a), 'In vitro penetration of egg yolks by *Salmonella* Enteritidis and *Salmonella* Heidelberg strains during thirty-six-hour ambient temperature storage', *Poult. Sci.*, 86, 1431–5.

Gast, R. K., Guraya, R., Guard-Bouldin, J. and Holt, P. S. (2008), 'Multiplication of *Salmonella* Enteritidis on the yolk membrane and penetration to the yolk contents at 30°C in an in vitro egg contamination model', *J. Food Prot.*, 71, 1905–9.

Gast, R. K., Guraya, R., Guard-Bouldin, J., Holt, P. S. and Moore, R. W. (2007b), 'Colonization of specific regions of the reproductive tract and deposition at different locations inside eggs laid by hens infected with Salmonella enteritidis or Salmonella Heidelberg', *Avian Dis.*, 51, 40–4.

Gast, R. K., Guraya, R. and Holt, P. S. (2011b), 'Frequency and persistence of fecal shedding following exposure of laying hens to different oral doses of *Salmonella enteritidis*', *Int. J. Poult. Sci.*, 10, 750–6.

Gast, R. K. and Holt, P. S. (1995), 'Iron supplementation to enhance the recovery of *Salmonella enteritidis* from pools of egg contents', *J. Food Prot.*, 58, 268–72.

Gast, R. K. and Holt, P. S. (1998), 'Supplementing pools of egg contents with concentrated enrichment media to improve rapid detection of *Salmonella enteritidis*', *J. Food Prot.*, 61, 107–9.

Gast, R. K. and Holt, P. S. (2000a), 'Deposition of phage type 4 and 13a Salmonella enteritidis strains in the yolk and albumen of eggs laid by experimentally infected hens', *Avian Dis.*, 44, 706–10.

Gast, R. K. and Holt, P. S. (2000b), 'Influence of the level and location of contamination on the multiplication of *Salmonella enteritidis* at different storage temperatures in experimentally inoculated eggs', *Poult. Sci.*, 79, 559–63.

Gast, R. K. and Holt, P. S. (2001a), 'Assessing the frequency and consequences of *Salmonella enteritidis* deposition on the egg yolk membrane', *Poult. Sci.*, 80, 997–1002.

Gast, R. K. and Holt, P. S. (2001b), 'Multiplication in egg yolk and survival in egg albumen of *Salmonella enterica* serotype Enteritidis strains of phage types 4, 8, 13a, and 14b', *J. Food Prot.*, 64, 865–8.

Gast, R. K. and Holt, P. S. (2003), 'Incubation of supplemented egg contents pools to support rapid detection of *Salmonella enterica* serovar Enteritidis', *J. Food Prot.*, 66, 656–9.

Gast, R. K., Holt, P. S. and Guraya, R. (2006), 'Effect of refrigeration on in vitro penetration of *Salmonella* Enteritidis through the egg yolk membrane', *J. Food Prot.*, 69, 1426–9.

Gast, R. K., Holt, P. S. and Murase, T. (2005b), 'Penetration of *Salmonella enteritidis* and *Salmonella Heidelberg* into egg yolks in an in vitro contamination model', *Poult. Sci.*, 84, 621–5.

Gast, R. K., Holt, P. S., Nasir, M. S., Jolley, M. E. and Stone, H. D. (2003b), 'Detection of *Salmonella enteritidis* in incubated pools of egg contents by fluorescence polarization and lateral flow immunodiffusion', *Poult. Sci.*, 82, 687–90.

Gast, R. K., Jones, D. R., Anderson, K. E., Guraya, R., Guard, J. and Holt, P. S. (2010b), 'In vitro penetration of *Salmonella* Enteritidis through yolk membranes of eggs from 6 genetically distinct commercial lines of laying hens', *Poult. Sci.*, 89, 1732–6.

Gast, R. K., Porter Jr., R. E. and Holt, P. S. (1997a), 'Applying tests for specific yolk antibodies to predict contamination by Salmonella enteritidis in eggs from experimentally infected laying hens', *Avian Dis.*, 41, 195–202.

Gast, R. K., Porter Jr., R. E. and Holt, P. S. (1997b), 'Assessing the sensitivity of egg yolk antibody testing for detecting Salmonella enteritidis infections in laying hens', *Poult. Sci.*, 76, 798–801.

Grijspeerdt, K., Kreft, J.-U. and Messens, W. (2005), 'Individual-based modelling of growth and migration of *Salmonella enteritidis* in hens' eggs', *Int. J. Food Microbiol.*, 100, 323–33.

Guan, J., Grenier, C. and Brooks, B. W. (2006), 'In vitro study of *Salmonella enteritidis* and *Salmonella typhimurium* definitive type 104: survival in egg albumen and penetration through the vitelline membrane', *Poult. Sci.*, 85, 1678–81.

Guard, J., Sanchez-Ingunza, R., Morales, C., Stewart, T., Liljebjelke, K., Van Kessel, J., Ingram, K., Jones, D., Jackson, C., Fedorka-Cray, P., Frye, J., Gast, R. and Hinton Jr., A. (2012), 'Comparison of *dkgB*-linked intergenic sequence ribotyping to DNA microarray hybridization for assigning serotype to *Salmonella enterica*', *FEMS Microbiol. Lett.*, 337, 61–72.

Gurtler, J. B. and Conner, D. E. (2009), 'Survival and growth of *Salmonella* Enteritidis in liquid egg products varying by temperature, product composition, and carbon dioxide concentration', *Foodborne Pathogens Dis.*, 6, 561–7.

Guzmán-Gómez, G., Valdovinos, M. A. A., Cabrera-Díaz, E., Pérez-Montaño, J. A., Muñoz-Valle, J. F., Torres-Vitela, M. R. and Ruiz-Quesada, S. L. (2013), 'Frequency of *Salmonella* and *Listeria monocytogenes* in five commercial brands of chicken eggs using a combined method of enrichment and nested-PCR', *J. Food Prot.*, 76, 429–34.

Hara-Kudo, Y., Kumagai, S., Masuda, T., Goto, K., Ohtsuka, K., Masaki, H., Tanaka, H., Tanno, K., Miyahara, M. and Konuma, H. (2001a), 'Detection of *Salmonella enteritidis* in shell and liquid eggs using enrichment and plating', *Int. J. Food Microbiol.*, 64, 395–9.

Hara-Kudo, Y., Sakakibara, Y., Konuma, H., Sawada, T. and Kumagai, S. (2001b), 'Laying season and egg shell cracks on the growth of *Salmonella* Enteritidis in the egg albumen during storage', *J. Food Prot.*, 64, 1134–7.

Hassan, J. O., Mockett, A. P. A., McLeod, S. and Barrow, P. A. (1991), 'Indirect antigen-trap ELISAs using polyclonal antisera for detection of group B and D Salmonellas in chickens', *Avian Pathol.*, 20, 271–82.

Havelaar, A. H., Ivarsson, S., Löfdahl, M. and Nauta, M. J. (2013), 'Estimating the true incidence of campylobacteriosis and salmonellosis in the European Union, 2009', *Epidemiol. Infect.*, 141, 293–302.

Holt, P. S., Gast, R. K. and Greene, C. R. (1995), 'Rapid detection of *Salmonella enteritidis* in pooled liquid egg samples using a magnetic bead-ELISA system', *J. Food Prot.*, 58, 967–72.

Hong, Y., Liu, T., Lee, M. D., Hofacre, C. L., Maier, M., White, D. G., Ayers, S., Wang, L., Berghaus, R. and Maurer, J. (2009), 'A rapid screen of broth enrichments for Salmonella enterica serovars Enteritidis, Hadar, Heidelberg, and Typhimurium by using an allelotyping multiplex PCR that targets O- and H-antigen alleles', *J. Food Prot.*, 72, 2198–201.

Howard, Z. R., O'Bryan, C. A., Crandall, P. G. and Ricke, S. C. (2012), '*Salmonella* Enteritidis in shell eggs; current issues and prospects for control', *Food Res. Int.*, 45, 755–64.

Huang, L. (2015), 'Direct construction of predictive models for describing growth of *Salmonella* Enteritidis in liquid eggs – a one-step approach', *Food Control*, 57, 76–81.

Humphrey, T. J. (1994), 'Contamination of egg shell and contents with *Salmonella enteritidis*: a review', *Int. J. Food Microbiol.*, 21, 31–40.

Humphrey, T. J., Baskerville, A., Mawer, S., Rowe, B. and Hopper, S. (1989), '*Salmonella enteritidis* phage type 4 from the contents of intact eggs: a study involving naturally infected hens', *Epidemiol. Infect.*, 103, 415–23.

Humphrey, T. J. and Whitehead, A. (1993), 'Egg age and the growth of *Salmonella enteritidis* in egg contents', *Epidemiol. Infect.*, 111, 209–91.

Humphrey, T. J., Whitehead, A., Gawler, A. H. L., Henley, A. and Rowe, B. (1991), 'Numbers of *Salmonella enteritidis* in the contents of naturally contaminated hens' eggs', *Epidemiol. Infect.*, 106, 489–96.

Jackson, B. R., Griffin, P. M., Cole, D., Walsh, K. A. and Chai, S. J. (2013), 'Outbreak-associated *Salmonella enterica* serotypes and food commodities, United States, 1998-2008', *Emerg. Infect. Dis.*, 19, 1239–44.

Jakočiūné, D., Pasquali, F., da Silva, C. S., Löfström, C., Hoorfar, J., Klein, G., Manfreda, G. and Olsen, J. E. (2014), 'Enumeration of salmonellae in table eggs, pasteurized egg products, and egg-containing dishes by using quantitative real-time PCR', *Appl. Environ. Microbiol.*, 80, 1616–22.

Jones, D. R., Cox, N. A., Guard, J., Fedorka-Cray, P. J., Buhr, R. J., Gast, R. K., Abdo, Z., Rigsby, L. L., Plumblee, J. R., Karcher, D. M., Robison, C. I., Blatchford, R. A. and Makagon, M. M. (2015), 'Microbiological impact of three commercial laying hen housing systems', *Poult. Sci.*, 94, 544–51.

Jones, D. R., Curtis, P. A., Anderson, K. E. and Jone, S F. T. (2004), 'Microbial contamination in inoculated shell eggs: II. Effects of layer strain and egg storage', *Poult. Sci.*, 83, 95–100.

Jones, D. R. and Musgrove, M. T. (2005), 'Correlation of eggshell strength and *Salmonella* Enteritidis contamination of commercial shell eggs', *J. Food Prot.*, 68, 2035–8.

Kang, H., Loui, C., Clavijo, R. I., Riley, L. W. and Lu, S. (2006), 'Survival characteristics of *Salmonella enterica* serovar Enteritidis in chicken egg albumen', *Epidemiol. Infect.* 134, 967–76.

Kawasaki, T., Musgrove, M. T., Murata, M., Tominaga, N. and Kawamoto, S. (2008), 'Comparative study of shell swab and shell crush methods for the recovery of *Salmonella* from shell eggs', *J. Food Safety*, 28, 482–98.

Keller, L. H., Benson, C. E., Garcia, V., Nocks, E., Battenfelder, P. and Eckroade, R. J. (1993), 'Monoclonal antibody-based detection system for Salmonella enteritidis', *Avian Dis.* 37, 501–7.

Keller, L. H., Benson, C. E., Krotec, K. and Eckroade, R. J. (1995), '*Salmonella enteritidis* colonization of the reproductive tract and forming and freshly laid eggs of chickens', *Infect. Immun.*, 63, 2443–9.

Kim, C. J., Emery, D. A., Rinke, H., Nagaraja, K. V. and Halvorson, D. A. (1989), 'Effect of time and temperature on growth of Salmonella enteritidis in experimentally inoculated eggs', *Avian Dis.*, 33, 735–42.

Klinkenberg, D., Thomas, E., Artavia, F. F. C. and Bouma, A. (2011), '*Salmonella* Enteritidis surveillance by egg immunology: impact of the sampling scheme on the release of contaminated table eggs', *Risk. Anal.*, 31, 1260–70.

Latimer, H. K., Jaykus, L. A., Morales, R. A., Cowen, P. and Crawford-Brown, D. (2002), 'Sensitivity analysis of *Salmonella enteritidis* levels in contaminated shell eggs using a biphasic growth model', *Int. J. Food Microbiol.*, 75, 71–87.

Lopes, V. C., Velayudhan, B. T., Halvorson, D. A., Lauer, D. C., Gast, R. K. and Nagaraja, K. V. (2004), 'Comparison of methods for differentiation of Salmonella enterica serovar Enteritidis phage type 4 isolates', *Am. J. Vet. Res.*, 65, 538–43.

Lublin, A., Maler, I., Mechani, S., Pinto, R. and Sela-Saldinger, S. (2015), 'Survival of *Salmonella enterica* serovar Infantis on and within stored table eggs', *J. Food Prot.*, 78, 287–92.

Lublin, A. and Sela, S. (2008), 'The impact of temperature during the storage of table eggs on the viability of *Salmonella enterica* serovars Enteritidis and Virchow in the eggs', *Poult. Sci.*, 87, 2208–14.

McElroy, A. P., Cohen, N. D. and Hargis, B. M. (1995), 'Evaluation of a centrifugation method for the detection of *Salmonella enteritidis* in experimentally contaminated chicken eggs', *J. Food Prot.*, 58, 931–3.

McElroy, A. P., Cohen, N. D. and Hargis, B. M. (1996), 'Evaluation of the polymerase chain reaction for the detection of *Salmonella enteritidis* in experimentally inoculated eggs and eggs from experimentally challenged hens', *J. Food Prot.*, 59, 1273–8.

Messens, W., Duboccage, L., Grijspeerdt, K., Heyndrickx, M. and Herman, L. (2004), 'Growth of *Salmonella* serovars in hens' egg albumen as affected by storage prior to inoculation', *Food Microbiol.*, 21, 25–32.

Messens, W., Grijspeerdt, K. and Herman, L. (2005a), 'Eggshell characteristics and penetration by *Salmonella enterica* serovar Enteritidis through the production period of a layer flock', *Brit. Poult. Sci.*, 46, 694–700.

Messens, W., Grijspeerdt, K. and Herman, L. (2005b), 'Eggshell penetration by *Salmonella*: a review', *World's Poult. Sci. J.*, 61, 71–85.

Messens, W., Grijspeerdt, K. and Herman, L. (2006), 'Eggshell penetration of hen's eggs by *Salmonella enterica* serovar Enteritidis upon various storage conditions', *Brit. Poult. Sci.*, 47, 554–60.

Muñoz, A., Dominguez-Gasca, N., Jimenez-Lopez, C. and Rodriguez-Navarro, A. B. (2015), 'Importance of eggshell cuticle composition and maturity for avoiding trans-shell *Salmonella* contamination in chicken eggs', *Food Control*, 55, 31–8.

Murase, T., Fujimoto, K., Nakayama, R. and Otsuki, K. (2006), 'Multiplication and motility of *Salmonella enterica* serovars Enteritidis, Infantis, and Montevideo in in vitro contamination models of eggs', *J. Food Prot.*, 69, 1012–16.

Murase, T., Holt, P. S. and Gast, R. K. (2005), 'Growth of *Salmonella enterica* serovar Enteritidis in albumen and yolk contents of eggs inoculated with this organism onto the vitelline membrane', *J. Food Prot.*, 68, 718–21.

Musgrove, M. T., Jones, D. R., Northcutt, J. K., Harrison, M. A., Cox, N. A., Ingram, K. D. and Hinton Jr., A. J. (2005), 'Recovery of *Salmonella* from commercial shell eggs by shell rinse and shell crush methodologies', *Poult. Sci.*, 84, 1955–8.

O'Brien, S. J. (2013), 'The "decline and fall" of nontyphoidal *Salmonella* in the United Kingdom', *Clin. Infect. Dis.*, 56, 705–10.

Ohtsuka, K., Yanagawa, K., Takatori, K. and Hara-Kudo, Y. (2005), 'Detection of *Salmonella enterica* in naturally contaminated liquid eggs by loop-mediated isothermal amplification, and characterization of *Salmonella* isolates', *Appl. Environ. Microbiol.*, 71, 673–6735.

Okamura, M., Kikuchi, S., Suzuki, A., Tachizaki, H., Takehara, K. and Nakamura, M. (2008), 'Effect of fixed or changing temperatures during prolonged storage on the growth of *Salmonella enterica* serovar Enteritidis inoculated artificially into shell eggs', *Epidemiol. Infect.*, 136, 1210–16.

Park, S. H., Aydin, M., Khatiwara, A., Dolan, M. C., Gilmore, D. F., Bouldin, J. L., Ahn, S. and Ricke, S. C. (2013), 'Current and emerging technologies for rapid detection and characterization of *Salmonella* in poultry and poultry products', *Food Microbiol.*, 38, 250–62.

Pires, S. M., Vieira, A. R., Hald, T. and Col, D. (2014), 'Source attribution of human salmonellosis: an overview of methods and estimates', *Foodborne Pathogens Dis.*, 11, 667–76.

Pouillot, R., Hoelzer, K., Ramirez, G. A., de Graft-Hanson, J. and Denni, S. B. (2014), 'Assessment of the risk of salmonellosis from internally contaminated shell eggs following initial storage at 18°C (65°F), compared with 7°C (45°F)', *Food Microbiol.*, 43, 16–19.

Price, K. A., Keller, L. H., Davison, S. and Eckroade, R. J. (1995), 'Selection of egg component and optimal parameters of incubation for detection of *Salmonella enteritidis* contamination in grade A table eggs by monoclonal antibody-based ELISA', *J. Vet. Diagn. Invest.*, 7, 265–8.

Raspoet, R., Shearer, N., Appia-Ayme, C., Haesebrouck, F., Ducatelle, R., Thompson, A. and Van Immerseel, F. (2014), 'A genome-wide screen identifies *Salmonella* Enteritidis lipopolysaccharide biosynthesis and the HtrA heat shock protein as crucial factors involved in egg white persistence at chicken body temperatures', *Poult. Sci.*, 93, 1263–9.

Rathgeber, B. M., McCarron, P. and Budgell, K. L. (2013), '*Salmonella* penetration through eggshells of chickens of different genetic backgrounds', *Poult. Sci.*, 92, 2457–62.

Ray, A., Roberts, J. R., Flavel, R. and Chousalka, K. K. (2015), 'Eggshell penetration by *Salmonella* Typhimurium in table eggs: examination of underlying eggshell structures by micro-computed tomography and scanning electron microscopy', *Food Res. Int.*, 78, 34–40.

Rehault-Godbert, S., Baron, F., Mignon-Grasteau, S., Labas, V., Gautier, M., Hincke, M. T. and Nys, Y. (2010), 'Effect of temperature and time of storage on protein stability and anti-*Salmonella* activity of egg white', *J. Food Prot.*, 73, 1604–12.

Ricke, S. C., Jones, D. R. and Gast, R. K. (2015), 'Eggs and egg products', in Salfinger, Y. and Tortorello, M. L. (eds), *Compendium of Methods for the Microbiological Examination of Foods*, 5th edition, pp. 633–43. American Public Health Association, Washington, DC.

Saeed, A. M. and Koons, C. W., (1993), 'Growth and heat resistance of *Salmonella enteritidis* in refrigerated and abused eggs', *J. Food Prot.*, 56, 927–31.

Schmid, D., Luckner-Hornishcer, A., Holzhammer, G., Rokita, D., Federspiel, M., Lassnig, H., Pichler, A.-M., Lederer, I., Beranek, A., Kornschober, C., Berghold, C. and Allerberger, F. (2007), 'Lessons learned from a Salmonella Enteritidis phage type 4 outbreak in Austria, 2005', *J. Food Prot.*, 70, 35–9.

Schoeni, J. L., Glass, K. A., Mcdermott, J. L. and Wong, A. C. L. (1995), 'Growth and penetration of *Salmonella enteritidis, Salmonella Heidelberg* and *Salmonella typhimurium* in eggs', *Int. J. Food Microbiol.*, 24, 385–96.

Seo, K. H, Holt, P. S., Stone, H. D. and Gast, R. K. (2003), 'Simple and rapid methods for detecting *Salmonella enteritidis* in raw eggs', *Int. J. Food Microbiol.*, 887, 139–44.

Seo, K. H., Valentin-Bon, I. E., Brackett, R. E. and Holt. P. S. (2004), 'Rapid, specific detection of *Salmonella* Enteritidis in pooled eggs by real-time PCR', *J. Food Prot.*, 67, 864–9.

Shah, D. H., Casavant, C., Hawley, Q., Addwebi, T., Call, D. R. and Guard, J. (2012), 'Salmonella Enteritidis strains from poultry exhibit differential responses to acid stress, oxidative stress, and survival in the egg albumen', *Foodborne Pathogens Dis.*, 9, 258–65.

Slinko, V. G., McCall, M. J., Stafford, R. J., Bell, R. J., Hiley, L. A., Sandberg, S. M., White, S. A. and Bell, K. M. (2009), 'Outbreaks of *Salmonella* Typhimurium phage type 197 of multiple genotypes linked to an egg producer', *Commun. Dis. Intell.*, 33, 419–25.

Stadelman, W. J. and Cotterill, O. J. (eds) (1995), *Egg Science and Technology, 4th edition*, New York, Haworth Press.

St. Louis, M. E., Morse, D. L., Potter, M. E., DeMelfi, T. M., Guzewich, J. J., Tauxe, R. V and Blake, P. A. (1988), 'The emergence of Grade A eggs as a major source of *Salmonella enteritidis* infections. New implications for the control of salmonellosis', *J. Am. Med. Assoc.*, 259, 2103–7.

Tan, S., Gyles, C. L. and Wilkie, B. N. (1997), 'Comparison of an LPS-specific competitive ELISA with a motility enrichment culture method (MSRV) for detection of Salmonella typhimurium and S. enteritidis in chickens', *Vet. Microbiol.*, 56, 79–86.

Tate, C. R., Miller, R. G. and Mallinson, E. T. (1992), 'Evaluation of two isolation and two nonisolation methods for detecting naturally occurring Salmonellae from broiler flock environmental drag-swab samples', *J. Food Prot.*, 55, 964–7.

Threlfall, E. J., Wain, J., Peters, T., Lane, C., De pinna, E., Little, C. L., Wales, A. D. and Davies, R. H. (2014), 'Egg-borne infections of humans with salmonella: not only an S. enteritidis problem', *World's Poultry Sci. J.*, 70, 15–26.

Tu, S.-I., Gehring, A. and Paoli, G. C. (2007), 'Detection of *Salmonella* Enteritidis in shell egg contents by immunomagnetic capture and time-resolved fluorescence', *J. Rap. Meth. Automat. Microbiol.*, 15, 107–19.

U.S. Food and Drug Administration. (2009), 'Prevention of Salmonella Enteritidis in shell eggs during production, storage, and transportation; final rule', *Fed. Reg.*, 74, 3039–101.

U.S. Food and Drug Administration. (2014), 'Bacteriological Analytical Manual, Chapter 5, Salmonella', http://www.fda.gov/Food/FoodScienceResearch/LaboratoryMethods/ucm070149.htm, Accessed 16 June 2016.

Wales, A. D. and Davies, R. H. (2011), 'A critical review of *Salmonella* Typhimurium infection in laying hens', *Avian Pathol.*, 40, 429–36.

Waltman, W. D. and Gast, R. K. (2016), 'Salmonella enterica', in Williams, S. M., Dufour-Zavala, L., Jackwood, M. W., Lee, M. D., Lupiani, B., Reed, W. M., Spackman, E. and Woolcock, P. R. (eds), *A Laboratory Manual for the Isolation and Identification of Avian Pathogens*, 6th edition, pp. 103–12. American Association of Avian Pathologists, Athens, Georgia.

Webb, M. L., Spickler, J. L., Bourassa, D. V., Cox, N. A., Wilson, J. L. and Buhr, R. J. (2014), 'Recovery of *Salmonella* serovar Enteritidis from inoculated broiler hatching eggs using shell rinse and shell crush sampling method', *Poult. Sci.*, 93, 2117–22.

Whiley, H. and Ross, K. (2015), '*Salmonella* and eggs: from production to plate', *Int. J. Environ. Res. Public Health*, 12, 2543–56.

Wright, A. P., Richardson, L., Mahon, B. E., Rothenberg, R. and Cole, D. J. (2016), 'The rise and decline in *Salmonella enterica* serovar Enteritidis outbreaks attributed to egg-containing foods in the United States, 1973–2009', *Epidemiol. Infect.*, 144, 810–19.

Yang, Q., Chen, S. and Ge, B. (2013), 'Detecting *Salmonella* serovars in shell eggs by loop-mediated isothermal amplification', *J. Food Prot.*, 76, 1790–6.

Zhang, G., Brown, E. W. and Hammack, T. S. (2013a), 'Comparison of different preenrichment broths, egg: preenrichment broth ratios, and surface disinfection for the detection of *Salmonella enterica* ssp. *enterica* serovar Enteritidis in shell eggs', *Poult. Sci.*, 92, 3010–16.

Zhang, G., Thau, E., Brown, E. W. and Hammack, T. S. (2013b), 'Comparison of a novel strategy for the detection and isolation of *Salmonella* in shell eggs with the Food and Drug Administration Bacteriological Analytical Manual method', *Poult. Sci.*, 92, 3266–74.

Chapter 7

Understanding the natural antibacterial defences of egg white and their regulation

Nicolas Guyot, Sophie Réhault-Godbert, Yves Nys, INRA, France; and Florence Baron, INRA – Agrocampus Ouest, France

1 Introduction

The avian egg white (albumen) is an aqueous solution of proteins. It is composed of 84–89% of water, the dry matter being constituted mostly of proteins (90%) and small amounts of carbohydrates, vitamins and minerals. The major function of these constituents is to provide nutrients and to protect the developing embryo and the highly nutritive egg yolk against invading bacteria. Egg white is a biological fluid secreted into the oviduct lumen by tubular glands and epithelial cells of the magnum, a specific segment of the oviduct. The egg white surrounds the egg yolk before being covered by eggshell membranes and the eggshell. Egg white is poor in nutrients available for the microorganisms (e.g. iron) and contains a great number of molecules that contribute to its remarkable antimicrobial properties. It is well established that avian egg white possesses antibacterial properties. The first demonstration of bactericidal effect was probably done by Wurtz (1890) on *Bacillus anthracis* and other bacteria; these findings were confirmed later by Laschtschenko (1909). A thermolabile enzyme was suspected to be involved in this activity. This bactericidal molecule was first identified by Fleming (1922) and named lysozyme. Thereafter, additional egg white proteins with antibacterial properties were reported.

http://dx.doi.org/10.19103/AS.2016.0012.07

In addition to the shell and eggshell membranes, the egg white participates in the natural egg defences by protecting the developing embryo in hatching eggs from bacterial contamination and by ensuring the hygienic quality of table eggs. In spite of such an efficient system, eggs and egg contents can be contaminated by a variety of bacterial species, including foodborne bacteria. The most well known is *Salmonella enterica* Enteritidis, which is the predominant serotype, and is responsible for 90% of foodborne diseases (salmonellosis) due to consumption of contaminated egg or egg products (EFSA 2009). The consumption of raw or undercooked eggs constitutes the main risk factor. The hygienic quality of eggs is still not completely under control despite many efforts from authorities, the poultry and egg industry, researchers, and the pharmaceutical and feed industries. The risk of contamination by *Salmonella* is a real threat for human health and remains a major concern for the egg producing and the egg product manufacturing industries. Under healthy breeding conditions, the egg content is generally sterile just after laying. However, it can be contaminated by a diversified microbiota containing food spoilage microorganisms and sometimes pathogenic bacteria. Eggs can be contaminated externally, on the eggshell, and internally, that is, during its formation (Baron and Jan 2011). In the case of *Salmonella*, eggs may be contaminated during their formation in the ovary or in the oviduct of infected hens (vertical contamination). However, this type of contamination is sporadic. The exogenic contamination (sometimes called horizontal contamination) of eggs is by far more frequent than the vertical contamination for overall bacteria and corresponds to the contamination of the eggshell surface and subsequent penetration of microbes into the egg. Horizontal contamination occurs after laying, when the eggshells are in contact with hen's faecal microorganisms or with other microorganisms present in the farm environment, or, downstream, in the environment of the conditioning centres. Contamination may also occur during egg production or transport and/or packaging in farm or in the conditioning centre, either by the environment or from one egg to another. The presence of cracks on eggs increases the risk of internal contamination.

This chapter offers a comprehensive overview of the natural defences of egg white and of key parameters controlling hygienic quality and bacterial growth in egg white. In Section 2, a repertoire of major and minor egg white proteins with antibacterial activities is described, with particular emphasis on their mode of action and their relative importance. Section 3 discusses the physicochemical and environmental conditions (temperature, pH, viscosity, carbon dioxide (CO_2), ions) influencing the bacterial growth and the activity of antibacterial proteins. Section 4 then covers hen-associated factors influencing the natural defences of the egg white. Finally, Sections 5 and 6 explore the physicochemical and antibacterial properties of egg white observed during egg storage and incubation, and highlight potential applications and innovative methods to maintain or increase egg white antimicrobial activities.

2 Egg white antimicrobial molecules

Egg white consists of about 10% of proteins (giving a concentration of approximately 100 mg/mL). At least 150 different proteins are present in the egg white (Mann and Mann 2011); however, the ten most abundant proteins represent about 95% of the total proteins and the remaining 5% is composed of a large number of minor proteins. A number of macromolecules, such as lysozyme (which lyses Gram-positive bacteria), ovotransferrin (which deprives bacteria of iron) and protease inhibitors have all been commonly cited to explain the ability of egg white to prevent bacterial growth. These molecules are abundant in the egg white, but some minor components recently revealed by high-throughput approaches might also play a role in the defence against bacterial contaminations. The effect of most of these compounds is mediated according to three main mechanisms: bacterial membrane disruption, chelation of essential nutrients and inhibition of microbial enzymes. It is noteworthy that some antibacterial proteins can exhibit several mechanisms of action and that synergistic activities between these compounds or interfering effects might occur in egg white, given the diversity of egg white proteins and its peculiar physicochemical properties. A variety of antimicrobial molecules in egg white is discussed in this section, with attention given to both mode of action and relative importance.

2.1 Lysozyme

Lysozyme, present in many biological fluids, is an important component of the non-specific defence mechanisms. It is responsible for different types of bactericidal activity. Lysozyme (3.4% of egg white proteins) is a basic protein (isoelectric point is 10.7) of 14,307 daltons (129 amino acid residues), a molecular weight which is low compared with most proteins and enzymes. The structure of lysozyme consists of two domains: one is predominantly composed of helical structures, whereas the other domain contains β-sheets. Its amino acid sequence is cross-linked by four disulphide bridges; the enzymatic activity of lysozyme is lost when more than two disulphide bridges are destroyed.

One of the major antimicrobial mechanisms of lysozyme, and the best known, is the degradation of the glycosidic (1–4) β-linkage between the *N*-acetylglucosamine and the *N*-acetylmuramic acid of the peptidoglycan layer in the bacterial cell walls (muramidase activity). The peptidoglycan is a strong, woven mesh that maintains the cell's shape and allows the diffusion of solutes and other types of molecules via large openings in the mesh. The cell wall determines the shape of the cell and protects it from osmotic lysis. Without a wall or when it is attacked by lysozyme or other antimicrobial agents, the cell would swell and burst. Bacteria that are susceptible to the enzymatic action of lysozyme are not lysed when they are in an osmotically balanced medium. Cell death occurs by the lytic action of lysozyme on peptidoglycan only when in low-osmotic-strength media, or when the rate of the synthesis and polymerization process for new peptidoglycan formation is slower than the lysozyme-catalysed degradation.

Such a lysozyme mechanism is unfortunately, limited to certain Gram-positive bacteria. The peptidoglycan of some Gram-positive bacteria is indeed resistant to hydrolysis by lysozyme because of chemical modifications. This is the case for *Staphylococcus aureus*, which carries an *O*-acetylation on the C-6 hydroxyl group of the *N*-acetylmuramyl residues (Clarke and Dupont 1992), and for *Bacillus subtilis*, which has a de-*N*-acetylation on the *N*-acetylglucosamine group (Atrih et al. 1999). Peptidoglycan of pathogenic species is often modified, allowing them to resist host defences (Vollmer 2008). Lysozyme is very basic and therefore positively charged at physiological conditions and can bind to more or less abundant negatively charged teichoic acids depending on the species, preventing access to its peptidoglycan substrate.

Invasive Gram-negative bacteria are resistant to the enzymatic action of lysozyme. The cell walls of Gram-negative bacteria have a more complicated structure, which is constituted by a thin layer of peptidoglycan surrounded by the outer membrane (lipid membrane containing lipopolysaccharide (LPS) and lipoproteins). The peptidoglycan layer in Gram-negative bacteria is protected from the lytic action of lysozyme because the outermost surface of the bacteria functions as a permeability barrier. Lysozyme is trapped and inactivated by surface LPS of the outer membrane of these bacteria (Ohno and Morrison 1989). More broadly, in addition to the protection provided by their outer membrane, Gram-negative bacteria have developed mechanisms of resistance to hydrolysis by lysozyme: the production of lysozyme inhibitors. A gene product named 'Ivy' for 'Inhibitor of vertebrate lysozyme' has been shown to express a strong anti-lysozyme activity on *Escherichia coli* (Monchois et al. 2001). This product could form a complex with lysozyme resulting in the occlusion of the lysozyme active site by a loop protruding from the Ivy molecule (Abergel et al. 2007). A study showed that inactivation of Ivy gene in *E. coli* renders this microorganism more susceptible to the antimicrobial activity of egg white compared to the wild strain (Deckers et al. 2008). The Ivy gene was identified in the genomes of many Gram-negative bacteria and many proteins from the same family have been reported since then. Considering the widespread occurrence and effectiveness of lysozymes as antibacterial agents, it is not surprising that bacteria have in turn evolved mechanisms to resist.

In the last two decades, evidence has been accumulated for the existence of alternative mechanisms of bactericidal action of lysozyme, which are independent of its enzymatic activity.

In the avian egg, the rise in the pH of the egg white begins once the egg is laid (cf Section 5 'Variability of physicochemical and antibacterial properties of egg white during egg storage'). The pH increases with time from 7.6 to 9.5 (Burley and Vadehra 1979). Ibrahim et al. (1994) showed the absence of lytic activity of lysozyme at pH 9 (pH of the egg a few days after laying); therefore, the lysozyme could quickly lose its lytic activity after laying. At this range of pH, lysozyme is enzymatically inactive but substantially bactericidal (Wang and Shelef 1991). When hen eggs are stored, an irreversible dimeric form of lysozyme is generated and involves the active site. Therefore, it is still not certain that the enzymatic activity of lysozyme makes a large contribution to the defence against infections in eggs stored for a long time or stored under inappropriate conditions (favouring increase in pH).

Another indication is that lysozyme, denatured by heat or by reducing agents, such as dithiothreitol (DTT) which should impair its enzymatic activity, still has bactericidal properties against some bacteria (During et al. 1999). More convincing evidence was provided by Ibrahim et al. (2001), who employed site-directed mutagenesis to inactivate the enzymatic site. This mutation rendered the enzyme completely catalytically inactive but did not affect the bactericidal properties of lysozyme against *B. subtilis* and *Staphylococcus aureus*. Moreover, some have reported that peptides derived from lysozyme and devoid of the enzymatic activity can disrupt membranes (Masschalck et al. 2002). The correlation between these disturbances and cell death strongly suggests that the cytoplasmic membrane is the primary target of lysozyme. The membrane disruptive activity seems dependent on the energy status of the membrane, which is consistent with the effect observed for other cationic molecules.

Lysozyme possesses antibacterial activities against some Gram-negative bacteria such as *E. coli*, presumably via mechanisms involving the disruption of DNA/RNA synthesis or permeabilization of the bacterial membrane (Masschalck and Michiels 2003; Pellegrini et al. 1992, 2000). Lysozyme rapidly increases the permeability of the outer membrane of *E. coli* by inducing large size pore formation (Derde et al. 2013). The antibacterial effect against *E. coli* and Gram-negative bacteria does not seem to rely on the tridimensional structure of lysozyme but likely involves the cationic character and the numerous positive charges, which may help lysozyme to interact with the negatively charged bacterial membranes. The denaturation of lysozyme by dry heating (7 days, 80°C), probably resulting from the increased hydrophobicity and exposition of positive charges, can potentiate its effects on *E. coli* membrane (Derde et al. 2014). The insertion of lysozyme *in vitro* into the LPS monolayer requires the polysaccharide moiety of LPS and leads to a slight reorganization of the LPS monolayer (Derde et al. 2015b). These modifications might initiate the formation of pores in the outer membrane. Disruption of the inner membrane can also be induced by lysozyme but without any perforation (Derde et al. 2013). *In vitro* studies using a model of *E. coli* inner membrane demonstrated that lysozyme can interact specifically with the bacterial cytoplasmic membrane phospholipids by adsorbing on the lipid head groups resulting in lipid packing modifications and in lateral cohesion modifications between phospholipids (Derde et al. 2015a). This interaction likely occurs via electrostatic interactions.

Another mechanism of lysozyme antimicrobial activity was also attributed to the induction of autolysins in *B. subtilis* (Ibrahim et al. 2001), a phenomenon that had been observed earlier in oral streptococci exposed to lysozyme (Laible and Germaine 1985). Autolysins are peptidoglycan hydrolases that bacteria use to introduce breaks in the peptidoglycan chains at the point of cell wall growth but that, under particular conditions, are able to cause bacterial autolysis.

In conclusion, it is clear that lysozyme is involved in egg white antimicrobial activity *via* several different bactericidal mechanisms. However, the relative contribution of these mechanisms is not completely elucidated. The first and the well-known mechanism is based on the enzymatic peptidoglycan hydrolysis, which is characterized by spheroplast formation and possibly bacteriolysis in case of osmolarity changes. The second mechanism is non-lytic and is based on membrane disturbance. A third mechanism involves the induction of autolysins. Depending on the type of bacteria (composition and structure of the membranes and the cell wall, presence of a factor inhibiting lysozyme activity) and the environmental conditions that can affect the energization status of the bacterial membrane or the lysozyme structure, one of these mechanisms may predominate on the others and cause cell death or they may act in synergy.

2.2 Ovotransferrin

Ovotransferrin (or conalbumin) is the second most abundant protein of the egg white (12%) after ovalbumin (>50%). It is a glycoprotein of 77.7 kDa containing 15 disulphide bonds and possessing an isoelectric point of 6.1. Ovotransferrin is a protein chelating metal ion and belongs to the family of transferrins that have members in various animal fluids (blood transferrin, lactoferrin from milk). By creating an iron-deficient environment, it has a bacteriostatic effect on bacteria. Ovotransferrin contains two iron-binding sites. Like all transferrins, ovotransferrin is composed of two homologous lobes. Each lobe has the ability to bind ferric iron with extremely high affinities, in the 10^{20} M^{-1} range (Abdallah and Chahine 1999).

A large number of independent authors have found that ovotransferrin has a bacteriostatic effect on Gram-negative bacteria and that this effect is blocked when the protein is saturated with iron (Baron et al. 1997; Garibaldi 1970; Lock and Board 1992). The most likely hypothesis for this activity has been that ovotransferrin produces an iron-deficient environment. Iron is essential for all forms of life, as it participates in many major biological processes (Andrews et al. 2003). The iron content of egg white was generally reported to be insufficient to allow ovotransferrin saturation (Thapon and Bourgeois 1994).

It has been repeatedly shown that the bacteriostatic action of egg albumen against Gram-negative bacteria in general is reduced by the addition of iron in amounts sufficient to saturate ovotransferrin (Lock and Board 1992). Further studies (Baron et al. 1997; Garibaldi 1970) showed that the major role in preventing growth of Gram-negative spoilage bacteria in egg white is contributed by ovotransferrin and iron deficiency.

Many bacteria deposit intracellular reserves of iron within iron storage proteins. These iron stores can then be used to enhance growth when external iron supplies are restricted (Andrews et al. 2003). Moreover, certain bacteria can synthesize and secrete high-affinity extracellular ferric chelators called siderophores in response to iron restriction (Neilands 1995). Siderophores have a low molecular mass (<10,000 Da) which solubilizes exogenous iron, making it available for transport into cells. Pathogenic bacteria are able to counter the iron restriction imposed in biological fluids through the use of siderophores that can compete with iron-binding proteins for iron. Gram-negative bacteria take up ferri-siderophore complexes *via* specific outer membrane receptors in a process that is driven by the cytosolic membrane potential. Periplasmic binding proteins transport ferri-siderophores from the outer membrane receptors to cytoplasmic membrane ATP-binding cassette permease that, in turn, delivers the ferri-siderophores to the cytosol where the complexes are probably dissociated by reducing conditions. Gram-positive bacteria use binding protein dependent on the ATP-binding cassette. This binding protein is generally a lipoprotein linked to the external surface of the cytoplasmic membrane. However, it is possible for pathogens to acquire iron directly from iron-binding proteins by using receptor-mediated transport systems that are specific to host iron complexes (Andrews et al. 2003).

Several lines of evidence have suggested that transferrin, lactoferrin and ovotransferrin may have a direct effect on Gram-negative bacteria in addition to those resulting from nutritional deprivation of iron. While hydrophobic LPS/LPS or LPS/protein interaction contributes to maintenance of outer layer integrity, the presence of divalent cations within the membrane appears to be critical to stabilizing the strong negative charge of the LPS molecules. By binding these membrane-associated cations, metal-binding protein could have an effect on the membrane (Ellison et al. 1988). Several authors (Baron 1998; Valenti et al. 1986) suggested that ovotransferrin antimicrobial activity can result from a direct effect on the membranes of *Candida albicans* (fungus) and *Salmonella* Enteritidis. The interaction of ovotransferrin with the anionic outer membrane of Gram-negative bacteria can be hypothesized due to the cationic nature of a bactericidal region contained in the N-terminal moiety (N-lobe) of ovotransferrin (Leu109-Asp200). The 92-amino acid peptide (OTAP-92) derived from this region exhibits killing activities against *E. coli* through a membrane damage mechanism (Ibrahim et al. 2000). Aguilera et al. (2003) described a molecular mechanism common to all transferrins, which is permeation of the *E. coli* outer membrane to access the inner membrane where they cause permeation of ions in a selective manner. The consequence is an *in vivo* dissipation of $\Delta\Psi$ (transmembrane electrical potential) without a significant alteration of ΔpH (transmembrane pH gradient) and a decrease in the electrochemical potential from −198 mV to −56 mV. As a result of the uncoupling of the respiration-dependant energy production, bacteria enter bacteriostasis. Ovotransferrin also possesses bactericidal activities against Gram-positive bacteria, such as *Staphylococcus aureus* and *Bacillus cereus*. The killing effect on *Staphylococcus aureus* is mediated through the N-terminal moiety (N-lobe) of ovotransferrin and is not affected by iron saturation (Ibrahim et al. 1998). The bactericidal activities of egg white against the spore-forming bacteria *Bacillus cereus* have recently been attributed to ovotransferrin (Baron et al. 2014). A modification of the electrochemical potential of the *B. cereus* cytoplasmic membrane can be induced by ovotransferrin, suggesting it could act *via* a membrane disturbance mechanism. Interestingly, the bactericidal effect on *B. cereus* is maintained after the addition of iron under alkaline conditions (pH 9.3) but not at pH 7.3.

In conclusion, there is strong evidence that ovotransferrin possesses distinct mechanisms of bacteriostatic action against bacteria. The first is the well-known mechanism based on the chelation of iron that creates an iron-deprived environment for bacteria. The second is a direct interaction with the membrane resulting in the damage of the biological functions of bacterial cytoplasmic membrane. The two mechanisms are dependent on the concentration of iron in the medium: iron saturation of ovotransferrin eliminates the bacteriostatic effect.

2.3 Protease inhibitors

The egg white contains high amounts of active protease inhibitors, called antiproteases. Proteases are proteolytic enzymes catalysing the cleavage/degradation of peptidic chains and are involved in many biological processes in eukaryotes and prokaryotes. The low digestibility of raw egg white illustrates the abundance of egg white protease inhibitors, which are able to inhibit digestive enzymes. Although the reason for such a massive presence in egg white remains still unclear regarding the physiological function of the egg to support the development of an embryo, it is believed that they might be involved in the defence of the egg against bacterial contamination. Indeed, egg antiproteases could potentially inhibit proteases-secreting bacteria and/or have a direct bactericidal activity. Some bacteria are able to secrete extracellular proteases (e.g. *Pseudomonas aeruginosa, Staphylococcus aureus, Streptococcus, Porphyromonas gingivalis*, etc.) that are considered as virulent factors due to their role in the survival, growth or invasive potency of bacteria and playing critical role in the host–pathogen interactions. The proteolytic activity of these bacterial enzymes can be counteracted by at least four of the main representative active protease inhibitors in the egg white, namely ovomucoid, ovoinhibitor, ovostatin and cystatin.

Ovomucoid

Ovomucoid represents the third most abundant protein of the egg white (11%) after ovalbumin (>50%) and ovotransferrin (12%). This protein (28 kDa, pI 4.1) is highly glycosylated and contains 9 disulphide bonds. The glycosylations are composed of both sialyloligosaccharides and sulphated oligosaccharides, and constitute approximately 25% of ovomucoid. This glycoprotein has a protease-inhibiting activity (antitrypsin, antichymotrypsin) and contains three functional homologous kazal-like domains, each containing a reactive site. The antibacterial role of ovomucoid and its participation in the protection of egg white against bacterial contamination have not been clearly demonstrated. However, some bacterial proteases like subtilisins and *Streptomyces griseus* proteases A and B are inhibited by turkey ovomucoid (Ardelt and Laskowski 1985).

Ovoinhibitor

Ovoinhibitor (1.5% of total egg white proteins) is a 48 kDa glycoprotein possessing seven kazal-like domains and 21 disulphide bonds. Each domain contains an inhibitory site. Ovoinhibitor has been shown to inhibit several serine proteases such as trypsin, chymotrypsin and elastase. Subtilisin, a serine protease produced by *Bacillus* spp., is also inhibited suggesting that ovoinhibitor has antimicrobial activity, which has been recently revealed on *Bacillus thuringiensis* (Bourin et al. 2011).

Ovostatin

Ovostatin (or ovomacroglobulin) is a large size homotetrameric protein with four disulphide linked subunits (4 × 165 kDa) and belonging to the alpha2-macroglobulin family. This protein (0.5% of total egg white proteins) is known to bind and inhibit various proteases from different classes (serine-, cysteinyl-, metallo- and aspartyl-proteases) but with a better efficiency towards metalloproteases (collagenase, thermolysin, stromelysin). Ovostatin can be considered as an antimicrobial due to its ability of inhibiting proteases produced by several virulent bacteria inducing corneal tissue damages (*Serratia marcescens, Pseudomonas aeruginosa*) (Molla et al. 1987).

Cystatin

Cystatin is a non-glycosylated protein (12.7 kDa, pI 5.1) containing two disulphide bonds and representing 0.05% of total egg white proteins (Li-Chan et al. 1995). This antiprotease targets most cysteine proteases including ficin, papain, and cathepsins B, H, L. Cystatin exhibits antibacterial activity by preventing the growth of group A *Streptococcus* (Bjorck 1990), *Salmonella* Typhimurium (Nakai 2000) and *Porphyromonas gingivalis* (Blankenvoorde et al. 1996). Bactericidal activity has been observed for cystatin at low dose against *Acinetobacter lwoffii, E. coli, Oligella* spp. and *Pseudomonas aeruginosa* (Wesierska et al. 2005).

2.4 Vitamin-binding proteins

Avidin

Avidin is a cationic homotetrameric glycoprotein (68.3 kDa, pI 10) representing 0.05% of the total protein content of egg white (Li-Chan et al. 1995). This protein has a very high affinity for biotin (vitamin B8/H) with a dissociation constant of around 10^{-15} M (Green 1963), which is one of the strongest non-covalent interactions found in nature. Although its main role is probably related to the supply of the egg and the developing embryo with biotin, avidin can potentially inhibit the growth of biotin-requiring microorganisms. In addition, avidin has been found to bind to various Gram-negative and Gram-positive bacteria, including *E. coli, Klebsiella pneumonia, Serratia marcescens, Pseudomonas aeruginosa, Staphylococcus aureus* and *Staphylococcus epidermis* (Korpela et al. 1984).

Riboflavin-binding protein

Riboflavin-binding protein (RfBP) is a phosphoglycoprotein of ~32 kDa with a pI of about 4 and represents 0.8% of egg white proteins (Li-Chan et al. 1995). The apoprotein possesses a high affinity for the riboflavin (vitamin B2) with a dissociation constant of 1.3×10^{-9} M (Becvar and Palmer 1982). RfBP has a critical role in embryonic development (White and Merrill 1988). Present in the egg white of many avian species, the concentrations and the proportion of the apoform of RfBP can vary over a 10-fold range, which may reflect the possibility that RfBP in egg white has antimicrobial functions (White and Merrill 1988).

2.5 Defensins

Defensins are small (2–6 kDa) cysteine-rich cationic proteins involved in the host innate defence and are found in many living species (vertebrates, invertebrates, plants). These molecules are characterized by conserved cysteines involved in disulphide bonds making them extremely compact and stable. Most of these molecules possess a broad-spectrum activity directed against Gram-positive and Gram-negative bacteria, but also against fungi and viruses. Proteomic analysis of chicken egg white (Mann 2007) revealed the presence of two beta-defensin-related molecules: the avian beta-defensin 11 (AvBD11) and gallin.

AvBD11

AvBD11 (previously known as vitelline membrane outer layer protein 2 or VMO-2) is a long size beta-defensin (9.2 kDa) containing 12 cysteines involved in 6 disulphide bonds. Although the three-dimensional structure of AvBD11 has not yet been solved, its unusual long size and the fact that the number of cysteines/disulphide bonds is doubled compared to classic defensins suggest that AvBD11 is likely composed of two beta-defensin motifs. This cationic protein (theoretical pI 8.78) exhibits antimicrobial activities against *Salmonella* Enteritidis, *Salmonella* Typhimurium, *E. coli, Listeria monocytogenes* and *Staphylococcus aureus* (Herve-Grepinet et al. 2010). AvBD11 can bind to heparin, a negatively charged glycosaminoglycan used to purify the protein from egg vitelline membrane by affinity chromatography. In turn, heparin is also able to inhibit the antibacterial activity of AvBD11, which suggests that the heparin-binding site on the protein has a pivotal role in the mechanism of action (Guyot et al. 2016a).

Gallin (OvoDA1)

Gallin or OvoDA1 (4.7 kDa) is a member of ovodefensins, a sub-family of beta-defensins found in the egg white from different avian species (chicken, turkey, swan, duck) (Whenham et al. 2015). It is a cationic peptide (theoretical pI 9.22) containing 6 cysteines which form 3 disulphide bonds. The recombinant form of the peptide has been shown to inhibit *E. coli* (Gong et al. 2010). The antibacterial spectrum seems to be narrow as the synthetic gallin was found active against *E. coli* strains but not against *Salmonella enterica, Staphylococcus aureus* and *Listeria monocytogenes* (Herve et al. 2014). This observation contrasts with most of beta-defensins found in nature, which are generally active against a large variety of bacteria. The three-dimensional structure of synthetic gallin is organized as a β-defensin fold but with significant variations, which might be related to its apparent *E. coli* specificity or other yet unknown functions (Herve et al. 2014).

2.6 Other antimicrobial molecules

Tenp (Transiently Expressed in Neural Precursor/ Ovoglobulin G2)

Tenp (transiently expressed in neural precursor) or Ovoglobulin G2 (~49 kDa, theoretical pI 5.77) is a protein belonging to the BPI/LBP (bactericidal permeability increasing protein/LPS-binding protein)/PLUNC (palate, lung and nasal epithelium clone protein) family. Members of this family are able to bind bacterial LPS, a surface component of Gram-negative bacteria, and mediate alterations of the outer membrane and damage to the inner membrane of bacteria. These events lead to the inhibition of bacterial growth (Elsbach and Weiss 1998). By analogy with other members of this family, Tenp is supposed to be antibacterial but further investigations on this protein (purification, characterization) need to be performed to confirm the predicted function of Tenp. Tenp isolated from Emu egg white, however, exhibits antibacterial activities against the Gram-positive bacteria *Micrococcus luteus* and *B. subtilis*, but not against the Gram-negative bacteria *E. coli* and *Salmonella* Typhimurium (Maehashi et al. 2014). In spite of being structurally related, it seems, however, that Tenp and BPI have a different antibacterial spectrum.

Ovalbumin-related protein X

Ovalbumin-related protein X or OVAX is a glycosylated heparin-binding 45–50 kDa protein belonging to the ov-serpin family, which includes the major egg white protein, ovalbumin. The theoretical pI of the protein chain is 6.29. Serpins are serine protease inhibitors but, to date, no cognate proteases have been identified for the ovalbumin and related proteins. OVAX amount in egg white is estimated to be 100 times less concentrated than ovalbumin. In spite of high sequence identity between OVAX and ovalbumin, OVAX has recently been shown to be antimicrobial for two foodborne disease pathogens, *Listeria monocytogenes* and *Salmonella* Enteritidis, whereas ovalbumin has no antibacterial activity (Rehault-Godbert et al. 2013). A patent for the use of OVAX as an anti-*Listeria* agent has been deposited (Rehault-Godbert et al. 2011). Its activity may be related to the ability of OVAX, but not ovalbumin, to bind to the anionic surfaces/molecules such as the glycosaminoglycan heparin. Indeed, heparin, by blocking the heparin-binding site of OVAX, is able to inhibit anti-*Salmonella* properties of OVAX, presumably by competing with bacteria at the antibacterial site of OVAX. A cluster of cationic amino acid residues present at the surface of OVAX is supposed to interact with heparin and take part in the antibacterial mechanism (Rehault-Godbert et al. 2013).

VMO-1, Pleiotrophin and Beta-microseminoprotein-like

Three novel antibacterial proteins, present at low abundance in the egg white, have recently been identified using heparin-affinity chromatography, together with OVAX and AvBD11: VMO-1, pleiotrophin and beta-microseminoprotein-like (Guyot et al. 2016a). Antibacterial activities of these proteins were revealed after fractionation of egg white proteins using a gelose diffusion assay. VMO-1 is a cationic protein (18 kDa, pI 8.85) containing 4 disulphide bonds initially identified in the outer vitelline membrane of the hen egg. VMO-1 exhibits antibacterial activity against *Listeria monocytogenes* but not against *Salmonella* Enteritidis (Guyot et al. 2016a). It is likely, however, that this protein possesses other physiological roles in the egg, especially within the vitelline membrane where VMO-1 is abundant (Mann 2008). The tridimensional structure of VMO-1 is organized into a singular structural motif called beta-prism, also found in the domain II of the insecticidal delta-endotoxin (a pore-forming toxin produced by *Bacillus thuringiensis*) and in the plant jacalin-like lectin domain (Sankaranarayanan et al. 1996; Shimizu and Morikawa 1996). Beta-prism motifs are thought to interact with carbohydrates. The heparin-binding property of VMO-1 is partly involved in the anti-*Listeria* activity of VMO-1 but other mechanisms might act complementarily to this carbohydrate-binding property to destroy or inhibit the bacteria. Pleiotrophin (15.2 kDa, pI 9.6, 5 disulphide bonds) is a growth factor possessing a high affinity for heparin. The human homolog exhibits antibacterial activities against Gram-positive and Gram-negative bacterial strains *via* a membrane disruption mechanism (Svensson et al. 2010). Chicken pleiotrophin also possesses antibacterial activities, as demonstrated against *Listeria monocytogenes* and *Salmonella* Enteritidis (Guyot et al. 2016a). Beta-microseminoprotein-like (gi:513191195/LOC101750704) is a small size cationic cysteine-rich protein (9.9 kDa, pI 9.3) recently identified in the egg white by heparin affinity. This protein also exhibits antibacterial activities against *Listeria monocytogenes* and *Salmonella* Enteritidis (Guyot et al. 2016a). Human beta-microseminoprotein possesses a potent fungicidal activity against *Candida albicans* whereas no antibacterial activity was observed against *E. coli*, *S. agalactiae*, *S. pyogenes*, *S. aureus* and *E. faecalis* (Edstrom Hagerwall et al. 2012). The antibacterial activity of chicken pleiotrophin and beta-microseminoprotein-like is mediated via mechanisms involving their heparin-binding site(s) (Guyot et al. 2016a).

3 Environmental and physicochemical conditions controlling bacterial growth and dispersion in egg white

The bacterial growth in egg white and its antimicrobial activities under the control of its proteins and molecules are influenced by different physicochemical parameters, such as those related to environmental conditions (temperature) or the freshness of eggs (viscosity, pH). Other intrinsic factors such as CO_2 and divalent cations may also affect bacterial growth. Depending on these factors, the antimicrobial activity of egg white can be either bacteriostatic or bactericidal.

3.1 Temperature

It was clearly shown that temperature plays a key role on the growth and survival of bacteria in the egg white. For *Salmonella* Enteritidis, the major pathogen of eggs, the growth of *Salmonella* seems to be impossible at temperatures below 8°C (Ruzickova 1994; Schoeni et al. 1995). Moreover, a bactericidal effect is observed with the disappearance of the whole population at 4°C in a few days (Chen et al. 2005; Lock and Board 1992; Schoeni et al. 1995). Between 20°C and 30°C, a growth of one to four logarithmic units per mL, depending on temperature and incubation time, is observed (Baron and Jan 2011). For temperatures above 37°C, a bactericidal effect was observed depending on temperature and incubation time (Alabdeh et al. 2011; Clavijo et al. 2006; Kang et al. 2006). In agreement with previous studies showing a poor survival of *Salmonella* Enteritidis in forming eggs (Keller et al. 1995) or in egg white incubated at 42°C (Alabdeh et al. 2011), it is believed that hen's body temperature (around 40°C–42°C) enhances the antimicrobial activity of egg white proteins during the period of egg formation in the oviduct. This effect at 42°C is in connection with the egg white as, in optimum medium, the temperature of 42°C is bacteriostatic but allows the survival of *Salmonella* Enteritidis.

Given the results obtained at low temperature, the storage of eggs at refrigeration temperature represents an effective way of preventing bacterial growth. Moreover, the refrigeration temperature reduces the loss of the vitelline membrane integrity and the access of bacteria to egg yolk.

3.2 Egg white pH

After laying, the loss of CO_2 changes the pH value from 7.6 to about 9.5 in a few days (cf Section 5 'Variability of physicochemical and antibacterial properties of egg white during egg storage') . This increase in pH value has consequences on bacteria and on egg white molecules. The maintenance of intracellular pH around its optimum value (7.6–7.8) is essential for many biological functions, particularly for bacterial enzyme activities and status of the membranes. The alkaline pH mainly affects the respiratory status of the cells that are trying to preserve the proton motive power and their electrochemical potential. To do this, bacteria accelerate the importation of protons and block their export. They also suppress the energy cost systems such as those associated with flagellar biosynthesis and chemotaxis (Maurer et al. 2005). It has indeed been shown that *Salmonella* Enteritidis lacks flagella at alkaline pH (Baron 1998), which may limit its motility and further access to nutrients.

Concerning the effect of alkaline pH on egg white proteins, it has been suggested that alkaline pH may promote the antimicrobial activity of ovotransferrin by interfering with bacterial iron metabolism, preventing bacteria from obtaining sufficient iron for growth in the egg white, an environment already deficient in iron (Tranter and Board 1984). A higher activity of ovotransferrin was recorded at alkaline pH than at acidic pH, due to an acceleration of iron release under acidic conditions and a slowdown under alkaline conditions (Halbrooks et al. 2005; Okamoto et al. 2004). The antibacterial activity of lysozyme is also pH-dependent. Levashov et al. (2010) showed that the lysozyme activity was increased under alkaline pH; however, according to others (Ibrahim et al. 2001), the lysozyme peptidoglycan lytic activity is inhibited by alkaline pH without change in its other bactericidal activities. The optimal pH of lysozyme for bacterial cell lysis seems to be strongly dependent on the species of bacteria tested (Levashov et al. 2016). Optimal rates of cell lysis can be observed at pH 6.4 for *Alcaligenes faecalis*; pH 7.0 for *Lactobacillus acidophilus* and *Streptococcus agalactiae*; and pH 8.7 for *Bacillus megaterium*, *Proteus vulgaris*, *Pseudomonas aeruginosa* and *Staphylococcus haemolyticus*. These variations might also be related to the different mechanisms of action of lysozyme, dependent or independent of the intrinsic enzymatic (muramidase) activity. Concerning the other antimicrobials contained in egg white, the influence of pH changes on the antibacterial activity of these molecules remains largely unexplored.

The role of alkaline pH on the antimicrobial activity of egg white has been highlighted by several studies (Kang et al. 2006; Messens et al. 2004; Tranter and Board 1984). These studies show that at pH greater than 8.8, the egg white shows bacteriostatic or bactericidal effect in contrast to pH 7.5 or 8 that allow bacterial slight growth or bacteriostatic activity.

3.3 Egg white viscosity

The viscosity of the bacterial environment is a physical parameter affecting the motility of flagellated bacteria (Schneider and Doetsch 1974). The egg white is a viscous fluid composed of three distinct layers surrounding the egg yolk: the outer thin egg white, the thick egg white and the inner thin egg white, respectively, from the exterior to the interior of the egg. The viscosity of egg white probably makes difficult the motility of bacteria and the access to nutrients required for bacterial growth. Owing to its high viscosity, the thick egg white may be considered as a physical barrier, like the eggshell and membranes. The decrease in egg white viscosity observed during egg storage may help bacteria to spread in the egg. Some authors have shown that the growth of *Salmonella* is favoured at 20°C for eggs stored beyond 3 weeks (Duboccage et al. 2001; Messens et al. 2004) probably due to the decrease in viscosity that would favour either the passage of nutrients from yolk to white or migration of bacteria to the yolk (Chen et al. 2005; Humphrey and Whitehead 1993). The viscosity level of the thick egg white can be estimated by the measure of albumen height and Haugh units (HU). These quality parameters are routinely used to determine the degree of freshness of eggs. The HU score is calculated from the following formula: $HU = 100 * (\log (h - 1.7w^{0.37} + 7.57)$ where h is the albumen height in millimetres and w represents the egg weight in grams.

The egg white viscosity is attributed to the presence of ovomucin and more particularly to the complex it forms with lysozyme. Ovomucin is a sulphated, high molecular weight, highly glycosylated protein belonging to the family of mucins, a group of proteins found in the numerous secretions of epithelial tissues and involved in the gel-like properties of secretions in which they are present. The egg white is a biological fluid secreted by the tubular gland and epithelial cells of the magnum to protect the egg yolk, and numerous studies have suggested that ovomucin has an important role in the gel-like structure of the egg white, although its function in the avian egg and during embryo development remains still unclear. Its interaction with lysozyme at neutral pH is thought to alter antibacterial properties of lysozyme. Mucins have been shown to inhibit the muramidase activity of lysozyme (Park et al. 2006) and alter antibacterial activities of numerous antibacterial proteins/peptides. So it is likely that ovomucin may have similar effects on egg white antimicrobial molecules, and that the dissociation of lysozyme from ovomucin at alkaline pH observed during the egg white thinning could release additional antibacterial activities. Thick and thin egg whites, which are characterized by different ovomucin composition and/or properties, may also display distinct behaviours in terms of antibacterial activities. For instance, it has been shown that the growth of *Salmonella* Typhi is favoured in the thick egg white compared to the thin egg white (Yadav and Vadehra 1977). This may be due to the different structure of ovomucin between these two fractions as the addition of thick egg white ovomucin to the thin egg white has been shown to interfere with the antibacterial activity of thin egg white (Yadav and Vadehra 1977).

The egg white viscosity depends on several factors: the laying hen line, the fraction of egg white involved (thin vs. thick egg white), time and storage conditions. Thus, the viscosity of egg white decreases during storage of eggs (Cf Section 5 'Variability of physicochemical and antibacterial properties of egg white during egg storage'), the white gradually liquefies and loses its gel structure. This is typically explained by the dissociation of the complex ovomucin–lysozyme due to the increase in pH. This leads to the weakening of the electrostatic bonds of the complex. Although the egg white thinning probably favours bacterial spreading in the egg, this effect may be counterbalanced at least in part by the loss of flagella of bacteria due to alkaline pH.

3.4 Other parameters

The egg forms in the oviduct in a CO_2-rich environment. As a consequence, the egg white from freshly laid eggs contains dissolved and combined CO_2, which progressively escapes through the eggshell pores to the external milieu during egg storage (cf Section 5 'Variability of physicochemical and antibacterial properties of egg white during egg storage'). CO_2 is able to inhibit the bacterial growth of aerobic microorganisms by affecting the lag phase and the growth rate. This effect has been observed for several spoilage bacteria (Devlieghere and Debevere 2000; Gill and Tan 1980). A greater effect of CO_2 is generally observed for

Gram-negative bacteria by comparison to Gram-positive bacteria, which are mostly facultative or strict anaerobes. Strict anaerobes seem not to be affected by CO_2. CO_2 was found to inhibit the growth of Enterobacteriaceae, including *Salmonella* (Phillips 1996). However, a recent study suggests that *Salmonella* Enteritidis, when inoculated in egg white, may be resistant to its inhibiting effect (Gurtler and Conner 2009). Interestingly, besides its direct effect on bacteria, CO_2 can also potentially interact with protein activities, as demonstrated for lysozyme whose lytic activity is modulated by CO_2 (Banerjee et al. 2011).

Ionic composition of the egg white is also of particular interest. The absence of free iron (due to large amounts of ovotransferrin) makes the egg white quite unfavourable to the growth of iron-requiring bacteria (cf Section 2 'Egg white antimicrobial molecules', Subsection 2.2 'Ovotransferrin'). Egg white also contains divalent cations such as magnesium and calcium with potential effect on the efficiency of egg antibacterial components. Divalent cations (even at subphysiologic levels) have been shown to affect the activity of mammalian cationic antimicrobial peptides, as previously demonstrated for LL-37 (Deslouches et al. 2005). They are able to bridge LPSs on bacterial surfaces, suggesting that they may serve as competitive inhibitors of cationic antimicrobial molecules. Interestingly, Mg^{2+} and Ca^{2+} were found to inhibit the potent bactericidal activity of a heat-denatured form of egg white lysozyme (without muramidase activity) against *Staphyloccocus aureus* and *E. coli* K12 (Ibrahim et al. 1997). The effect of divalent cations on native egg white antibacterial proteins remains to be further investigated.

4 Variability of egg white natural defences induced by hen-associated factors

Egg white proteins are produced by the magnum, a specific segment of the oviduct. Most of egg white proteins (ovalbumin, ovotransferrin, ovomucoid, lysozyme) are synthesized within the tubular gland cells (Palmiter and Gutman 1972) and secreted into the lumen, but epithelial goblet cells (also named mucous cells) are thought to produce and secrete some egg white components, in particular avidin (Kami and Yasuda 1984) and ovomucin (Wyburn et al. 1970). The examination of magnum cells by electron microscopy reveals the presence of secretory granules, containing proteins, in tubular gland cells and epithelial goblet cells, but not in epithelial ciliated cells (Wyburn et al. 1970). Cells producing egg white proteins are absent before the sexual maturation of birds. In pullets, the formation of the oviduct is driven by steroid hormones during sexual maturation. The differentiation of the immature chick oviduct can be induced by oestrogen injections (primary stimulation) leading to the formation oestrogen-responsive tubular gland cells and progestin-responsiveness (Tuohimaa et al. 1989). In sexually mature hens (around 16 weeks of age), the synthesis of egg white proteins is under the control of sex steroid hormones (oestrogen, testosterone, progesterone) (Palmiter 1972). Insulin and glucocorticoids (dexamethasone) were also found to regulate the expression of egg white proteins (Hager et al. 1980; Kato et al. 1988b). In oestrogen-primed cells, avidin production is stimulated by progesterone (Chan et al. 1973; O'malley and Mcguire 1968), but a progesterone-independent synthesis can also be induced in response to inflammatory stimuli (Tuohimaa et al. 1989). The synthesis of OVAX, AvBD11, VMO-1 and pleiotrophin, four minor egg white proteins with antibacterial activities, seems also to be oestrogen-dependent, like the major egg white components. The treatment of chicks with an agonist of oestrogen receptor, diethylstilbestrol (synthetic non-steroidal oestrogen), is indeed able to induce the expression of OVAX, AvBD11, VMO-1 and pleiotrophin at mRNA and protein levels in the glandular and luminal epithelial cells of the magnum (Lee et al. 2012; Lim et al. 2011, 2013; Lim and Song 2015). The expression of AvBD11, VMO-1 and pleiotrophin is likely to be post-transcriptionally regulated by microRNAs (Lee et al. 2012; Lim et al. 2013; Lim and Song 2015). The protein composition of the egg white is known to be very stable, at least for the major constituents, whereas the amounts of minor proteins, in contrast, are probably more prone to variations. Several parameters related to the physiology and genetics of the hen can induce variations in the physicochemical properties and biological activities of the egg white and will be developed in the next paragraphs.

4.1 Age and laying cycle

The age of birds is known to negatively affect the quality of egg white. The increasing egg weight observed during the laying cycle is associated with a decline in the relative proportion of egg white and a slight reduction in the amount of dry matter (mostly composed of egg white proteins) (Sauveur 1988; Silversides and Scott 2001), concomitantly with the increase of egg weight and egg yolk proportion. In parallel, the thick egg white undergoes a progressive deterioration of its gel-like properties, which is characterized by a decreased albumen height (Silversides and Budgell 2004; Silversides and Scott 2001). It is noteworthy, however, that albumen height and HU score can be restored after a period of moulting (Donalson et al. 2005). A variability of biological activities can be observed depending on the hen age. Trziszka et al. (2004) compared lysozyme and cystatin activities in the egg white produced by 30–70-week-old hens and found that these activities were the highest with hens aged 40–50 weeks. In the interval of 35–50 weeks, a peak of egg white antitrypsin and antichymotrypsin activities (being mostly represented by ovomucoid and ovoinhibitor) was observed in eggs laid by 45-week-old hens (Swierczewska et al. 2005). The expression of some egg white proteins seems to be altered by the age of animals. In particular, a reduction of avidin synthesis resulting from post-transcriptional regulation occurs in the quail oviduct with age (Arendes et al. 1980).

4.2 Genetics

Differences between strains have been observed for several egg white parameters. Scott and Silversides (2000) found that the albumen height from ISA-White was higher compared to that from ISA-Brown hens, but no differences were observed between the two strains for pH. Hen line effects on egg white pH and height were observed in another study by Silversides and Budgell (2004) conducted on Brown Leghorn, ISA-Brown and Babcock hens. A comparative proteomic study (2D gel electrophoresis combined with LC-MS/MS analysis) conducted on white-shell and brown-shell eggs (produced by hen lines of Single Comb White Leghorns and Rhode Island Reds, respectively) revealed significant differences for at least 25 protein spots (10 proteins) including ovotransferrin, ovoinhibitor, riboflavin-binding protein and OVAX (Wang et al. 2012). Some egg white quality traits, such as albumen height and HU scores, which may be associated with the level of protection in eggs, are known to be heritable traits (Washburn 1979), whereas the concentrations in total proteins, lysozyme and ovotransferrin were shown to exhibit very low heritabilities (<0.1) (Sellier et al. 2007). Heritabilities for the antibacterial effects against *Staphylococcus aureus* (0.16) and *Salmonella* Enteritidis (0.11) are higher than those determined for the concentration of antimicrobial proteins (Sellier et al. 2007), but these values remain too low for practical applications in animal selection.

4.3 Nutrition

Very few dietary factors are known to alter egg white quality, especially the HU score and the albumen height, which are good indicators of quality of the thick egg white. Total protein and essential amino acid intakes have no significant effects on these parameters and are known to primarily affect the amount of secreted egg white (Sauveur 1988). Vanadium (contaminating element of phosphates) at 10 ppm can provoke a decrease in HU score after 4 weeks (Sell et al. 1982). In contrast, acidifying substances such as ammonium chloride (NH_4Cl) induce a spectacular increase in HU score in freshly laid eggs via an effect on pH and on calcium and magnesium contents (Sauveur 1976). In the eggs produced by hens fed with NH_4Cl-supplemented diet, HU score remains at a high value during storage for 28 days at 20°C, together with a pH maintained close to neutrality and with high and stable concentrations of calcium and magnesium. Dietary magnesium can increase the concentration of magnesium in the thick egg white and slow down the decrease in HU score and thick egg white amount observed during storage (Monsey et al. 1977). HU score and albumen height are weakly but significantly improved using a hen diet supplemented with garlic powder or garlic juice (Olobatoke et al. 2011; Lim et al. 2006; Mahmoud et al. 2010); however, this effect needs to be further confirmed as others found no significant effects of garlic on HU score and albumen index (Yalcin et al. 2006). The enrichment of the hen feed with mineral–humine preparation, n-3 polyunsaturated fatty acids (rapeseed meal) or antioxidants (hibiscus, vitamins A+E) mainly alters antipapain (cystatin) and antitrypsin activities in egg white while lysozyme activity seems to be weakly affected (Kopec et al. 2005). Omega-3-, vitamin- and lutein-enriched eggs, designed for their potential benefits in human health, can easily be produced by adding to the hen diet some specific products rich in omega-3 (e.g. fish oil, rapeseed, linseed or marine algae), lutein (e.g. alfalfa or yellow-corn meals) and vitamins (e.g. fortified vitamin premix). The effects of these hen diet modifications on egg white protein contents and activities is not well characterized; however, a comparative study using a proteomic approach (2D gel electrophoresis combined with LC-MS/MS) revealed differences in protein abundance for several protein spots between the enriched eggs and conventional eggs (Wang et al. 2012). Among the differential spots identified as antimicrobial proteins are ovotransferrin, lysozyme, ovomucoid, ovoinhibitor and riboflavin-binding protein.

4.4 Stress and disease

The egg white quality is slightly affected by bacterial diseases (cholera, typhoid), encephalomyelitis, Marek's disease and leucosis; however, Newcastle disease and infectious bronchitis have dramatic effects on the egg white quality (Sauveur 1988). The height of albumen can be reduced by half, and in some cases, it becomes completely liquid. Effects on the antibacterial potential of egg white remain unknown, but it is likely that the disintegration of the gel-like structure of the thick egg white could favour the mobility and dispersion of bacteria inside the eggs following contamination. The mechanism of this deterioration is thought to result from an increase in the water content and a decrease in the concentration of ovomucin and lysozyme in thick egg white, which are responsible for its gel-like properties (Butler et al. 1972). Several modifications of the oviductal epithelium can be observed in response to infectious bronchitis (reduction of the height of epithelial cells, modification of the shape of goblet cells, reduction or complete loss of cilia) (Ignjatovic and Sapats 2000), and it seems that the virus can induce a cessation of the synthesis of ovomucin by its action on goblet cells (Sauveur 1988). Stress may also result in the deterioration of egg white quality and antibacterial defences. In support of this, alterations in the abundance of some egg white proteins and a decrease in both HU score and albumen height were observed in a model of stress induced by dietary corticosterone (Kim and Choi 2014; Kim et al. 2015).

4.5 Immune stimulation

The stimulation of the immune system of hens can potentially modulate the antibacterial activities of egg white. In response to antigen administration or bacterial infection, some specific immunoglobulins produced by the hen are deposited in the forming egg yolk to protect the future chick during embryo development and after hatching. Therefore, it was hypothesized that the levels of antimicrobial proteins and activities could be also adjusted in egg white during egg formation, following the stimulation of hen immunity. Indeed, innate immune molecules such as antimicrobial peptides and proteins are usually upregulated during immune responses. The systemic administration of LPS (bacterial endotoxin) in hens stimulates the antibacterial properties of egg white proteins against *Staphylococcus aureus* (Bedrani et al. 2013a); however, the underlying mechanisms of this stimulation, which seem to be lysozyme-independent, remain to be elucidated. The magnitude of the stimulation potential of egg white antibacterial properties was explored in another experimental model using three extreme breeding conditions with increasing bacterial load: germ-free, specific pathogen-free and conventional hens (Bedrani et al. 2013b). In this experiment, antibacterial activities of egg white proteins against *Staphylococcus aureus* and *Streptococcus uberis* were increased in specific pathogen-free and in conventional groups by comparison to the germ-free group, which thus confirmed that egg white antibacterial activities could be reinforced by the microbial environment of hens and the level of the immune stimulation. With the exception of a tendency for a decreased antichymotrypsin activity in the germ-free group, lysozyme and antiprotease activities were not significantly altered by the breeding conditions. Altogether, these two studies demonstrated the feasibility of improving the egg white antibacterial defences by the stimulation of hen immunity. The molecule(s) involved in the upregulation of antibacterial activities still remains unknown. It is important to note that the stimulatory effects observed in these studies are moderate and might be restricted to a limited number of bacteria. The fact that the synthesis of egg white proteins (and antibacterial proteins) is mainly controlled by sex steroid hormones might explain at least in part the low magnitude of stimulation.

5 Environmental variability of physicochemical and antibacterial properties of egg white during egg storage

Shell eggs are generally consumed within days or weeks after laying. During this period, eggs are stored at various temperatures depending on where they are stored (laying farm, conditioning centre, food store, consumer's home, etc.) and might be subjected to bacterial contaminations. In the conditioning centre, the storage temperature is ideally between 5°C and 20°C (Buffet 2010). At the consumer's home, eggs are usually either refrigerated or stored at room temperature, if not immediately consumed. The storage of eggs between laying and consumption induces internal modifications, which are strongly influenced by temperature and duration of storage. Physicochemical modifications and antibacterial properties of the egg white during egg storage will be successively described in this section. Innovative methods for egg storage illustrated by the use of CO_2 will also be discussed in relation to the quality and safety of shell eggs.

During egg storage, albumen undergoes important modifications, the most dramatic change being the thinning of the thick white. These modifications are due to gas exchanges between the interior of the egg and the external milieu through the eggshell, and exchanges between the egg white and the egg yolk through the vitelline membrane (Thapon and Bourgeois 1994). Time and temperature are two factors having crucial roles on the evolution of egg white properties during storage.

Water and CO_2 are two egg white components leaking out of the egg through eggshell during egg storage. The water loss depends primarily on the integrity of the cuticle. Owing to the progressive degradation of the cuticle after the egg is laid, the water loss increases over time during egg storage and can be dramatically enhanced when eggs are washed. Water elimination through eggshell is affected by several parameters including storage temperature, relative humidity, storage time, eggshell surface and eggshell porosity.

The pH of the albumen depends on the equilibrium between dissolved CO_2, HCO_3^-, CO_3^{2-} and protein (Li-Chan et al. 1995). HCO_3^- and CO_3^{2-} concentrations are governed by the partial pressure of CO_2 in the egg environment, which is the main factor responsible for egg white pH variations. The freshly laid egg white is enriched in CO_2 (under dissolved and combined form) that makes the egg white pH close to neutrality. At oviposition, albumen contains dissolved CO_2 (4–5 mg/egg) and combined CO_2 present in bicarbonate (about 100 mg) (Sauveur 1988). During egg storage, CO_2 extricates and leaks through eggshell pores to the external milieu leading to the alkalinization of egg white. The rate of CO_2 leakage through the eggshell is a function of storage temperature (the higher the temperature the faster the rate of CO_2 leakage). The loss of CO_2 observed during egg storage has two remarkable consequences on the albumen properties: pH increase and egg white thinning, both being dependent on storage temperature. The egg white pH value varies from 7.6 at laying to 9.5 within a few days post-lay. The pH increase can be inhibited or limited by preventing CO_2 leakage out of the egg. The obstruction of pores (by using substances like oil, wax, silicates, etc.) in freshly laid eggs and the conservation of eggs in a CO_2-enriched atmosphere are methods that restrict pH increase in egg white. In this case, egg white pH relies on CO_2 content. Egg white thinning in stored eggs is characterized by a decrease in the viscosity of thick egg white. This phenomenon is accelerated by high storage temperatures. Several hypotheses have been proposed to explain the molecular mechanism of egg white thinning, including:

- Dissociation of lysozyme–ovomucin complexes resulting from electrostatic interactions at alkaline pH (Cotterill and Winter 1955). In thick egg white, a decrease in the ratio of lysozyme content between the gel fraction and the liquid fraction is observed during storage and egg white thinning and is associated with a decrease in lysozyme activity (Kato et al. 1978). The decrease in lysozyme activity not only depends on storage time but also on storage temperature and egg white pH. It is speculated that the dissociation of ovomucin–lysozyme complexes and decreased lysozyme activity might result from pH alkalinization and/or lysozyme denaturation.

- Dissociation of ovomucin components, in particular a solubilization of the β component, rich in glucids, which transfers to the thin albumen (Hayakawa et al. 1983). The fact that a portion of divalent cations (Ca^{2+} and Mg^{2+}) transfers from the albumen to the yolk during storage accelerates this phenomenon.
- Partial degradation of *O*-glycosidic bonds following the solubilization of β-ovomucin and release of carbohydrates (hexoses, hexosamine, sialic acid) by β-elimination at alkaline pH (Kato et al. 1979).

Owing to the difference of osmotic pressure between both sides of the vitelline membrane and to the lower hydration of yolk proteins, a transfer of water and minerals from the albumen to the yolk can be observed during egg storage. It is accompanied by sodium fluxes while potassium is less affected. This phenomenon affects primarily divalent cations such as calcium and magnesium. Their contents decrease rapidly in albumen and increase in the yolk. This loss in divalent cations is also supposed to contribute to the egg white thinning. A transfer of proteins through the vitelline membrane can also occur during storage. In this respect, increasing amounts of albumen proteins have been detected in egg yolk during egg storage at 37.8°C (Rehault-Godbert et al. 2014), but it is likely that yolk proteins can also diffuse into the egg white. The weakening of the vitelline membrane during the storage may facilitate these transfers.

During egg storage, antimicrobial proteins in egg white are challenged by the alkaline pH and storage conditions (time, temperature). Two proteomic analyses were performed on egg white proteins during storage to investigate the effects of duration and temperature of storage on the modifications of abundance of egg white proteins (Omana et al. 2011; Qiu et al. 2012b). The proteomic analysis of eggs stored for different periods of time (0, 10, 20, 30, 40 days) at room temperature suggested that ovotransferrin and ovoinhibitor were amongst the proteins showing significant changes during storage. Several spots corresponding to ovotransferrin BC and ovoinhibitor increased transiently but significantly in abundance after 10 or 20 days of storage at 22°C, while no changes in lysozyme spots were observed over a storage period of 40 days at the same temperature (Omana et al. 2011). A proteomic study conducted on eggs stored for 15 days at three different storage temperatures (4°C, 20°C, 37°C) revealed the formation of protein complexes in egg white during storage at 37°C (Qiu et al. 2012b). The abundance of a spot corresponding to an ovalbumin–lysozyme complex increased during storage, being highest at 20°C. The amounts of fragments of ovotransferrin (presumably degradation products) and a complexed form of ovotransferrin were significantly increased at 37°C. A protein spot corresponding to OVAX was stable during storage at 4°C but significantly decreased in abundance with increasing temperature. The abundance of a spot containing ovoinhibitor was found to be increased at 20°C, while another one was stable during storage at 4°C but decreased at 20°C and 37°C. Besides these proteomic studies, other studies investigated the changes of antimicrobial proteins during egg storage, as well as their activities. Egg white lysozyme content, assessed by HPLC, tends to decrease during egg storage, whereas lysozyme-specific activity measured at neutral pH does not seem to be changed during storage (Schäfer et al. 1999). The reduction of lysozyme content is similar between eggs stored at 5°C and 20°C but significantly accentuated after 6 weeks in eggs stored at 30°C. It is suggested that this decline may be due to the degradation of disulphide bonds at alkaline pH or to the formation of a lysozyme dimer which is devoid of lytic activity. In another study, a decrease of lysozyme activity by 4% to 17% has been observed in egg white following storage during 28 days at 15°C (Trziszka et al. 2004). In egg white, lysozyme undergoes a deamidation process on several amino acid residues during egg storage, leading to decreased activity (Kato et al. 1988a). As for lysozyme, a trend for a decrease in ovotransferrin content was observed in egg white during storage, regardless of the storage temperature (5°C, 20°C, 30°C) (Schäfer et al. 1999). Ovotransferrin activity (iron-binding capacity), however, was not altered. Besides lysozyme and ovotransferrin, antiprotease activities have also been investigated during storage but with inconsistent results depending on authors. A study demonstrated that egg white antichymotrypsin and antipapain activities were not altered by storage time or temperature, whereas antitrypsin activity which was stable at 4°C and 20°C deteriorated after 2 weeks of storage at 37°C (Rehault-Godbert et al. 2010). Another study showed that the inhibition capacity of cystatin, assessed using papain as target protease, tends to decrease by 5% to 12% after 4 weeks of storage at 15°C (Trziszka et al. 2004). Kopec et al. (2005) observed that storage at 15°C induces a reduction in antitrypsin activity after 2 weeks and a significant decline in the inhibitory activity of cystatin after 4 weeks.

Bacterial penetration and growth in egg white depends firstly on the egg integrity but also on the conditions of egg storage (time and temperature) (Baron and Jan 2011). The growth of *Salmonella* is favoured at 20°C for eggs not older than one week, as compared to eggs stored at this temperature for 2–3 weeks (Duboccage et al. 2001). Beyond three weeks, the growth may be stimulated. These favourable conditions for *Salmonella* multiplication are likely to be related to the alteration of the vitelline membrane and to the decrease in egg viscosity, both being due at least in part to the dissociation of the ovomucin–lysozyme complex at alkaline pH. Both factors favour the entrance of nutrients from egg yolk to egg white and the migration of *Salmonella* into the yolk (Chen et al. 2005; Humphrey and Whitehead 1993). Besides analyses of *Salmonella* growth in stored shell eggs, the evolution of the egg white antibacterial potential during egg storage has recently been studied using egg white samples isolated from shell eggs stored for different periods of time at variable temperature (Rehault-Godbert et al. 2010). In this study, the antibacterial potential of each egg white sample has been assessed by measuring *in vitro* the growth of *Salmonella* Enteritidis at 30°C for 3 days. Egg white antibacterial activity is significantly increased within a few days following oviposition. The activation of these anti-*Salmonella* defences is strongly influenced by not only storage time but also temperature. Indeed, antimicrobial properties in the egg whites sampled from eggs stored at 37°C are progressively increased between 0 and 5 days following oviposition; it is then progressively altered after two weeks of storage. This alteration might be partially explained by the degradation of ovotransferrin, which is the major anti-*Salmonella* agent in egg white. When the egg is stored at 20°C, the activation of egg white antimicrobial properties is more progressive, being maximal after ten days but then stabilizes up to one month after oviposition. These observations suggest that egg white thinning and/or increase in pH in egg white may be involved in the activation of anti-*Salmonella* properties of albumen during storage, at least during the first week of storage. Similar observations were made a long time ago by Sharp and Whitaker in 1927 with another Gram-negative bacteria, *Pseudomonas pyocyaneus*, isolated from a contaminated egg white. The authors suggested that the decreased ability of *P. pyocyaneus* to grow in the egg white of stored eggs compared to fresh egg white was due to egg white alkalinization (Sharp and Whitaker 1927). Considering the dual effect of high temperature during egg storage, favouring bacterial growth and the putative antibacterial activity of egg white, further experiments need to be conducted on shell eggs internally inoculated with *Salmonella* Enteritidis to find the conditions minimizing *Salmonella* growth during storage. One possibility could be to introduce a short period of higher temperature to stimulate the antibacterial activity of egg white during the first days of storage.

In conclusion, these observations altogether suggest that some conditions of egg storage can reinforce the antibacterial potential of the egg white, at least against *Salmonella*. This activation has been observed for *Salmonella* Enteritidis but it might possibly occur for other bacterial species. The underlying molecular mechanisms remain unclear, although egg white thinning and pH increase are likely to be involved. These observations underline the interest to applying temperature sequences. For instance, applying a moderate to high temperature to shell eggs for a short period (activation of anti-*Salmonella* defences) and then cool down eggs to keep both freshness and hygienic quality of egg white. A pulse of high temperature can be performed by pasteurization. This method is used in the United States to decontaminate shell eggs. However, the effect of such a treatment on the antibacterial activity of egg white and proteins remains to be investigated. Warm/hot temperatures applied to shell eggs can weaken the vitelline membrane, particularly in case of prolonged treatment. This degradation, also observed during storage at room temperature, may facilitate the migration of *Salmonella* from the egg white to the yolk. Refrigeration of shell eggs could stop or limit the heat-induced degradation of the vitelline membrane. In practice, it is noteworthy that freshly laid eggs nowadays are not refrigerated but stored under cool/mild temperature conditions within days before reaching the consumer's home. The refrigeration of shell eggs immediately after laying should be strongly encouraged because of its beneficial effect on the vitelline membrane integrity.

Some innovating methods based on the use of CO_2 can be employed to stabilize or improve the physicochemical properties of egg white during the storage of shell eggs. For instance, one way to control egg white properties during storage is the use of modified atmosphere packaging (MAP). Modified atmospheres are used to extend the shelf life of foods by inhibiting chemical, enzymatic and microbial spoilage. The gases used in most MAP applications include CO_2, O_2, N_2 or different combinations of these gases mainly used to preserve fruits and meat during storage. Physicochemical properties of eggs packed in air, 100% O_2, and 100% CO_2 and stored up to 45 days at 4, 20 and 36°C were evaluated during the RESCAPE European program (2006–9). The weight loss of eggs classically observed during egg storage was significantly decreased for all packed eggs. In normal storage conditions, the pH of albumen progressively increases up to 9.5 and the HUs decrease in relation to the egg white thinning, but the use of air and O_2 MAP stabilizes egg white pH and slows down the decrease in HUs when eggs are stored over a period of 28 days at 25°C (Rocculi et al. 2009). A more spectacular effect is observed when using CO_2 MAP, which improves the physicochemical egg white quality (decreased pH, stabilized HUs) (Rocculi et al. 2009). The rapid cooling of shell eggs using cryogenic CO_2 can also be used to improve the quality and safety of eggs. This method is patented (Curtis et al. 2000). Microbial counts were carried out in inoculated eggs submitted to cryogenic CO_2 cooling or traditional air cooling. Reduced *Salmonella* number was detected in the egg white of cryogenically CO_2 cooled eggs, in comparison to traditional air cooling. Interestingly, eggs rapidly cooled by CO_2 have not only a better hygienic quality but also a longer shelf life given its effect on HUs and pH.

6 Environmental variability of physicochemical and antibacterial properties of egg white during egg incubation

Under optimal conditions of temperature (37°C) and relative humidity (45–65% relative humidity), it takes 21 days for a chicken embryo to develop and hatch. Remarkable modifications occur in the egg during incubation. The egg white undergoes major changes in terms of pH, volume, water content, protein concentration and protein amounts. These rapid and important changes, occurring mainly during the first half of incubation, can potentially affect the natural defences of egg white and its ability to protect the embryo from bacterial contaminations.

In embryonated eggs, egg white pH increases during the first two days of incubation up to about 9.3–9.5 and then decreases progressively to neutrality (7–7.5) at day 12 (Cunningham 1974; Guyot et al. 2016b; Qiu et al. 2012a; Wang and Wu 2014). The egg white alkalinization observed at the very beginning of incubation (0–2 days) is similar to that observed in unfertilized eggs and results from CO_2 leakage through the eggshell pores. It is likely that these conditions can confer an efficient protection for the chicken embryo during this period. The environment of egg white under alkaline conditions is generally harsh for the survival of bacteria, and it enhances the iron-chelating activity of ovotransferrin (Tranter and Board 1984); however, it inhibits in counterpart the enzymatic activity of lysozyme (Chang and Carr 1971; Ibrahim et al. 1994, 2001; Smolelis and Hartsell 1952). Alkalinization is also associated with an alteration of the gel-like properties of the thick egg white during early incubation, as demonstrated by the measure of the albumen height (Fang et al. 2012a). In embryonated eggs, pH starts to decrease after the second day of incubation to be progressively neutralized, which probably results from the secretion of hydrogen ions by the blastoderm and the diffusion of the CO_2 produced by developing tissues (Dawes 1975), making the pH of the embryonic microenvironment optimal for embryonic development (Reijrink et al. 2008). This pH neutralization can favour the growth of some bacteria, being closer to their optimal pH, but it might also in turn improve the muramidase activity of lysozyme, the pH-optimum of which being around 5–7.

Egg white is a natural reservoir of water (approximately 88% of its weight). In embryonated eggs, a significant part of this water dissipates during incubation, as it is mostly utilized for the growth of embryo and for the supply of extra-embryonic fluids (amniotic and chorioallantoic fluids) and membranes. A part of water also dissipates by evaporation through the eggshell pores. These modifications result in the decrease of egg white volume and weight. At 10–12 days of incubation, the egg white is about 30% of that of non-incubated eggs (Carinci and Manzoli-Guidotti 1968). A threefold increase in the egg white protein concentration is also observed within 8–12 days of incubation (Guyot et al. 2016b). No further synthesis of egg white proteins occurs once the egg is laid. In fertile eggs, no substantial transfer of egg proteins to extra-embryonic fluids occurs during the first half of incubation (Carinci and Manzoli-Guidotti 1968; Romanoff and Romanoff 1933) and the composition of residual egg white proteins remains substantially unchanged during incubation, being thereby very similar to that before incubation (Liu et al. 2013, 2015). Therefore, this increase in egg white protein concentration is the direct consequence of the loss of water. The slight increase in egg white density observed from the 4th day of incubation (Meuer and Egbers 1990) likely results from these events. The progressive concentration of egg proteins might have repercussions on the bioactivities such as antibacterial activities.

Egg white proteins are strongly challenged by the temperature (37.7°C) and pH variations (rapid pH increase, alkaline conditions for several days, then a drop of 1.5–2.0 pH units) during incubation, with possible impacts on their stability and activity. A study by Cunningham revealed that lysozyme activity in fertile egg white rapidly decreases during incubation to reach about 20% of the initial activity after 17 days of incubation (Cunningham 1974). It was suggested that such a decline in lysozyme activity could result from the interaction of lysozyme with conalbumin (ovotransferrin), ovomucin or ovalbumins (Cunningham 1974). Other studies by Fang et al. showed that the specific activity of lysozyme decreases in fertilized egg white in the second day of incubation (Fang et al. 2012a,b). This observation may be in connection with the progressive alkalinization of egg white during the early stages of incubation. Reduction of lysozyme activity was also observed in unfertilized eggs maintained at 35°C, despite the high stability of egg white proteins (Feeney et al. 1952). In contrast, an increase in ovotransferrin activity and iron-binding capacity is observed concomitantly with the increase in egg white pH in fertilized egg white during early incubation (Fang et al. 2012a,b). It has been hypothesized that increased concentration results from the digestion of the vitelline membrane during the yolk sac development in embryonated eggs and the release of additional ovotransferrin in the egg white, as the vitelline membrane contains significant amounts of this protein (Mann 2008). Increased amounts and activity of ovotransferrin may also be explained by the decreasing amount of water in fertilized egg white. The assessment of antibacterial potential of egg white proteins using a turbidimetric method revealed that the inhibitory action of egg white proteins (at constant protein concentration) on *Listeria monocytogenes* and *Streptococcus uberis* is progressively altered during the first half of incubation, regardless of the presence or absence of a developing embryo (Guyot et al. 2016b). This antibacterial decline may result from a decreased lysozyme activity, as both bacteria are Gram-positive. The storage at high temperature for prolonged period is usually associated with the denaturation of proteins, resulting in the loss of activities (Koehler 1974). The apparent deterioration of antibacterial proteins in embryonated egg white might be counterbalanced, at least in part, by the increasing protein concentration associated with water loss. A higher concentration of egg white proteins indeed should theoretically be accompanied by higher antimicrobial activities. The assessment of antibacterial potential of total egg white inoculated with *Salmonella* Enteritidis using a colony counting method revealed that the antibacterial action of egg white is improved during early incubation: egg white activity is bacteriostatic during the first two days, then it becomes bactericidal at day 3 (Fang et al. 2012b). After 5 days, the antibacterial potential against *Salmonella* Enteritidis is then strongly altered.

Despite being highly challenged by the temperature and pH variations, the integrity of major egg white proteins is, in general, not dramatically compromised during the first part of incubation. This relative stability may be, at least in part, due to the presence of large amounts of protease inhibitors in the egg white. Several studies, however, showed that the incubation of fertilized eggs induced the formation of protein fragments and the modification in the amounts of specific proteins in fertilized egg white. Protein fragments with an apparent molecular weight of 30–35 kDa (as determined by SDS-PAGE) are generated during incubation, regardless of the presence or absence of an embryo (Guyot et al. 2016b). Similar fragments, identified as degradation products of ovotransferrin and ovalbumin, are also detected in unfertilized eggs stored at 37°C (Rehault-Godbert et al.

2010). The generation of these protein fragments is probably not dependent on pH but rather related to the temperature of incubation which might activate latent proteases such as aminopeptidase Ey, transmembrane protease serine 9, carboxypeptidase D and aminopeptidase A, previously identified in the egg white proteome (Mann and Mann 2011). The involvement of protease(s) in the cleavage of egg white proteins, however, remains to be demonstrated. Egg white is characterized by large amounts of protease inhibitors, which seem to be highly stable when challenged by temperature. In unfertilized eggs incubated at 37°C, no significant alteration of antitrypsin, antichymotrypsin and antipapain activities was observed within 12 days (Rehault-Godbert et al. 2010). Only antitrypsin activities were deteriorated after 2 weeks of incubation at 37°C. In fertilized eggs, the concentration of egg white cystatin gradually decreases up to day 8 of incubation and increases thereafter from day 10 to 14 (Golab et al. 2001). Another protease, pro-MMP2 (72 kDa gelatinase), has been identified in the egg white but its activity seems to decrease rapidly during early incubation in both fertilized and unfertilized eggs (Rehault-Godbert et al. 2008). Recent proteomic analyses of fertilized egg white proteins by 2D PAGE revealed changes in several protein spots during incubation (Liu et al. 2013, 2015; Qiu et al. 2012a; Wang and Wu 2014). Alterations during incubation are, in general, similar between fertilized and unfertilized eggs, which suggest that these changes likely result from incubation temperature rather than the presence of the embryo; however, some differences are observed between fertilized and unfertilized eggs for several protein spots corresponding to clusterin, ovoinhibitor, lysozyme and OVAY, suggesting pH-dependent modifications (Qiu et al. 2012a). In comparison to major egg white components, the stability of some proteins present in low abundance may be more susceptible to the incubation conditions. In this regard, the stability/integrity of beta-microseminoprotein-like (gi|513191195), a heparin-binding protein with antibacterial properties (Guyot et al. 2016a), is rapidly compromised during incubation in both embryonated and unfertilized eggs. As observed by SDS-PAGE, a protein band corresponding to this protein present in the egg white of freshly laid eggs or after a storage of three days at moderate temperature, disappears almost completely after 4 days of incubation (Guyot et al. 2016b). It is unclear whether this protein is degraded, complexed or transferred in another egg compartment during egg incubation but the loss of the native form of this protein may have repercussions on its biological functions including its antibacterial activities.

7 Conclusions

In conclusion, the egg white is a complex antimicrobial biological fluid. The most well-known antimicrobial proteins of egg white are lysozyme, ovotransferrin and protease inhibitors but several minor antimicrobial proteins have recently been characterized. It is likely that egg white contains still uncharacterized proteins or peptides involved in the bactericidal or bacteriostatic activities. The coexistence of a multitude of antibacterial compounds in the egg white with various spectra of antibacterial activities suggests that there may be interactions between these molecules or synergistic effects, which can explain why egg white is efficient against most environmental microbes. In addition to the presence of these molecules, physicochemical/environmental conditions are playing an important role in the bacterial growth and in the egg white antimicrobial activity. Extrinsic (temperature) and intrinsic (alkaline pH, viscosity, CO_2, ionic composition, etc.) factors can affect either bacteria or egg white molecules, leading to enhanced or decreased capacity of bacteria to grow and survive in egg white. Some physicochemical parameters and antibacterial activities can be modified during egg formation by hen-associated factors such as the feed, genetic strain, age, disease and the stimulation of hen immunity, but the mechanisms involved in such modifications are not well characterized. Even if it is difficult to know the part played by each factor and their interaction, egg storage conditions through pH and temperature strongly influence the potential of egg white antimicrobial activity and surely affect protein–protein interactions and their subsequent activity. Egg white undoubtedly has a variety of mechanisms allowing for the control of bacterial growth, depending on the type of bacteria and the environmental conditions. One of these mechanisms may predominate for a particular bacterium. In this regard, the iron deficiency due to ovotransferrin plays a major role in the egg white activity against *Salmonella* Enteritidis. Other bacteria are insensitive to this activity. For example, *Listeria* is mainly inhibited in the egg white by lysozyme. In addition, the alkalinization of egg white pH, which is an antimicrobial mechanism occurring during storage, allows the inhibition of a wide variety of bacteria. The egg white has a highly selective effect on microorganisms: the high incidence of *Salmonella* Enteritidis in the egg content could be explained by the fact that this bacterium has advantages for survival in egg white or for the colonization of the hen reproductive tract. Finally, it is noteworthy that internal egg antibacterial activities are not inert but evolve over time according to extrinsic environment factors (temperature, atmosphere). Although it adds further complexity to the current knowledge, the modulation of antibacterial properties during egg storage is a very interesting mechanism with promising potential to optimize hygienic quality of shell eggs. Understanding the mechanisms of egg white antibacterial defences with an emphasis on *Salmonella* Enteritidis is a prerequisite to develop innovative and efficient systems for safe eggs preservation.

8 Where to look for further information

Nys, Y., Bain, M. and Van Immerseel, F. (editors). Improving the Safety and Quality of Eggs and Egg Products (Volumes 1 & 2). Woodhead Publishing Limited, Cambridge, 2011.

Huopalahti, R., López-Fandiño, R., Anton, M. and Schade, R. (editors). Bioactive Egg Compounds. Springer-Verlag Berlin Heidelberg, 2007.

Stadelman, W.J. and Cotterill, O.J. (editors). Egg Science and Technology (4th Edition). Food Products Press, New York, 1995

9 References

Abdallah, F. B. and Chahine, J. M. (1999), Transferrins, the mechanism of iron release by ovotransferrin. *Eur. J. Biochem.* 263, 912–20.

Abergel, C., Monchois, V., Byrne, D., Chenivesse, S., Lembo, F., Lazzaroni, J. C. and Claverie, J. M. (2007), Structure and evolution of the Ivy protein family, unexpected lysozyme inhibitors in Gram-negative bacteria. *Proc. Natl. Acad. Sci. USA* 104, 6394–9.

Aguilera, O., Quiros, L. M. and Fierro, J. F. (2003), Transferrins selectively cause ion efflux through bacterial and artificial membranes. *FEBS Lett.* 548, 5–10.

Alabdeh, M., Lechevalier, V., Nau, F., Gautier, M., Cochet, M. F., Gonnet, F., Jan, S. and Baron, F. (2011), Role of incubation conditions and protein fraction on the antimicrobial activity of egg white against *Salmonella* Enteritidis and *Escherichia coli*. *J. Food Prot.* 74, 24–31.

Andrews, S. C., Robinson, A. K. and Rodriguez-Quinones, F. (2003), Bacterial iron homeostasis. *FEMS Microbiol. Rev.* 27, 215–37.

Ardelt, W. and Laskowski, M., Jr. (1985), Turkey ovomucoid third domain inhibits eight different serine proteinases of varied specificity on the same ... Leu18-Glu19 ... reactive site. *Biochemistry* 24, 5313–20.

Arendes, J., Zahn, R. K. and Muller, W. E. (1980), Age-dependent gene induction in quail oviduct. XI. Alterations on the post-transcriptional level (analytical aspect). *Mech. Ageing Dev.* 14, 49–57.

Atrih, A., Bacher, G., Allmaier, G., Williamson, M. P. and Foster, S. J. (1999), Analysis of peptidoglycan structure from vegetative cells of *Bacillus subtilis* 168 and role of PBP 5 in peptidoglycan maturation. *J. Bacteriol.* 181, 3956–66.

Banerjee, P., Keener, K. M. and Lukito, V. D. (2011), Influence of carbon dioxide on the activity of chicken egg white lysozyme. *Poult. Sci.* 90, 889–95.

Baron, F. (1998), Etude du comportement de *Salmonella* Enteritidis dans le blanc d'oeuf. Ecole Nationale Supérieure Agronomique de Rennes, Rennes.

Baron, F., Gautier, M. and Brule, G. (1997), Factors Involved in the Inhibition of Growth of *Salmonella* Enteritidis in Liquid Egg White. *J. Food Prot.* 60, 1318–23.

Baron, F. and Jan, S. (2011), Egg and egg product microbiology. In *Improving the Safety and Quality of Eggs and Egg Products: Egg Chemistry, Production and Consumption (Volume 1: Egg chemistry, production and consumption)*. Y. Nys, M. Bain and F. Van Immerseel eds. Woodhead Publishing Limited, Sawston, pp. 330–50.

Baron, F., Jan, S., Gonnet, F., Pasco, M., Jardin, J., Giudici, B., Gautier, M., Guerin-Dubiard, C. and Nau, F. (2014), Ovotransferrin plays a major role in the strong bactericidal effect of egg white against the *Bacillus cereus* group. *J. Food Prot.* 77, 955–62.

Becvar, J. and Palmer, G. (1982), The binding of flavin derivatives to the riboflavin-binding protein of egg white. A kinetic and thermodynamic study. *J. Biol. Chem.* 257, 5607–17.

Bedrani, L., Helloin, E., Guyot, N. and Nys, Y. (2013a), Systemic administration of lipopolysaccharide in laying hens stimulates antimicrobial properties of egg white against *Staphylococcus aureus*. *Vet. Immunol. Immunopathol.* 152, 225–36.

Bedrani, L., Helloin, E., Guyot, N., Rehault-Godbert, S. and Nys, Y. (2013b), Passive maternal exposure to environmental microbes selectively modulates the innate defences of chicken egg white by increasing some of its antibacterial activities. *BMC Microbiol.* 13, 128.

Bjorck, L. (1990), Proteinase inhibition, immunoglobulin-binding proteins and a novel antimicrobial principle. *Mol. Microbiol.* 4, 1439–42.

Blankenvoorde, M. F., Henskens, Y. M., van't Hof, W., Veerman, E. C. and Nieuw Amerongen, A. V. (1996), Inhibition of the growth and cysteine proteinase activity of *Porphyromonas gingivalis* by human salivary cystatin S and chicken cystatin. *Biol. Chem.* 377, 847–50.

Bourin, M., Gautron, J., Berges, M., Attucci, S., Le Blay, G., Labas, V., Nys, Y. and Rehault-Godbert, S. (2011), Antimicrobial potential of egg yolk ovoinhibitor, a multidomain Kazal-like inhibitor of chicken egg. *J. Agric. Food Chem.* 59, 12368–74.

Buffet, E. (2010), Conditionnement et emballage des oeufs de consommation. In *Science et technologie de l'oeuf. Volume 1: Production et qualité*. F. Nau, C. Guérin-Dubiard, F. Baron and J. L. Thapon eds. Tec&Doc Lavoisier, pp. 251–63.

Burley, R. W. and Vadehra, D. V. (1979), Chromatographic separation of the soluble proteins of hen's egg yolk: an analytical and preparative study. *Anal. Biochem.* 94, 53–9.

Butler, E. J., Curtis, M. J., Pearson, A. W. and McDougall, J. S. (1972), Effect of infectious bronchitis on the structure and composition of egg albumen. *J. Sci. Food Agric.* 23, 359–69.

Carinci, P. and Manzoli-Guidotti, L. (1968), Albumen absorption during chick embryogenesis. *J. Embryol. Exp. Morphol.* 20, 107–18.

Chan, L., Means, A. R. and O'Malley, B. W. (1973), Rates of induction of specific translatable messenger RNAs for ovalbumin and avidin by steroid hormones. *Proc. Natl. Acad. Sci. USA* 70, 1870–4.

Chang, K. Y. and Carr, C. W. (1971), Studies on the structure and function of lysozyme. I. The effect of pH and cation concentration of lysozyme activity. *Biochim. Biophys. Acta* 229, 496–503.

Chen, J., Shallo Thesmar, H. and Kerr, W. L. (2005), Outgrowth of Salmonellae and the physical property of albumen and vitelline membrane as influenced by egg storage conditions. *J. Food Prot.* 68, 2553–8.

Clarke, A. J. and Dupont, C. (1992), O-acetylated peptidoglycan: its occurrence, pathobiological significance, and biosynthesis. *Can. J. Microbiol.* 38, 85–91.

Clavijo, R. I., Loui, C., Andersen, G. L., Riley, L. W. and Lu, S. (2006), Identification of genes associated with survival of *Salmonella enterica* serovar Enteritidis in chicken egg albumen. *Appl. Environ. Microbiol.* 72, 1055–64.

Cotterill, O. J. and Winter, A. R. (1955), Egg White Lysozyme: 3. The Effect of pH on the Lysozyme-Ovomucin Interaction. *Poult. Sci.* 34, 679–86.

Cunningham, F. E. (1974), Changes in egg white during incubation of the fertile egg. *Poult. Sci.* 53, 1561–5.

Curtis, P., Ann, McBride, Anderson, K., Emil, Conner, D., Edward, Hughes, L., Ann and Keener, K., M. (2000), Use of CO2 cooling in treatment of poultry eggs. Pat. No. WO/2000/007453.

Dawes, C. M. (1975), Acid-base relationships within the avian egg. *Biol. Rev. Camb. Philos. Soc.* 50, 351–71.

Deckers, D., Vanlint, D., Callewaert, L., Aertsen, A. and Michiels, C. W. (2008), Role of the lysozyme inhibitor Ivy in growth or survival of *Escherichia coli* and *Pseudomonas aeruginosa* bacteria in hen egg white and in human saliva and breast milk. *Appl. Environ. Microbiol.* 74, 4434–9.

Derde, M., Guerin-Dubiard, C., Lechevalier, V., Cochet, M. F., Jan, S., Baron, F., Gautier, M., Vie, V. and Nau, F. (2014), Dry-heating of lysozyme increases its activity against *Escherichia coli* membranes. *J. Agric. Food Chem.* 62, 1692–700.

Derde, M., Lechevalier, V., Guerin-Dubiard, C., Cochet, M. F., Jan, S., Baron, F., Gautier, M., Vie, V. and Nau, F. (2013), Hen egg white lysozyme permeabilizes *Escherichia coli* outer and inner membranes. *J. Agric. Food Chem.* 61, 9922–9.

Derde, M., Nau, F., Guerin-Dubiard, C., Lechevalier, V., Paboeuf, G., Jan, S., Baron, F., Gautier, M. and Vie, V. (2015a), Native and dry-heated lysozyme interactions with membrane lipid monolayers: Lipid packing modifications of a phospholipid mixture, model of the *Escherichia coli* cytoplasmic membrane. *Biochim. Biophys. Acta* 1848, 1065–73.

Derde, M., Nau, F., Lechevalier, V., Guerin-Dubiard, C., Paboeuf, G., Jan, S., Baron, F., Gautier, M. and Vie, V. (2015b), Native lysozyme and dry-heated lysozyme interactions with membrane lipid monolayers: lateral reorganization of LPS monolayer, model of the *Escherichia coli* outer membrane. *Biochim. Biophys. Acta* 1848, 174–83.

Deslouches, B., Islam, K., Craigo, J. K., Paranjape, S. M., Montelaro, R. C. and Mietzner, T. A. (2005), Activity of the de novo engineered antimicrobial peptide WLBU2 against *Pseudomonas aeruginosa* in human serum and whole blood: implications for systemic applications. *Antimicrob. Agents Chemother.* 49, 3208–16.

Devlieghere, F. and Debevere, J. (2000), Influence of Dissolved Carbon Dioxide on the Growth of Spoilage Bacteria. *LWT Food Sci. Tech.* 33, 531–7.

Donalson, L. M., Kim, W. K., Woodward, C. L., Herrera, P., Kubena, L. F., Nisbet, D. J. and Ricke, S. C. (2005), Utilizing different ratios of alfalfa and layer ration for molt induction and performance in commercial laying hens. *Poult. Sci.* 84, 362–9.

Duboccage, L., Heyndrickx, M., Grijspeerdt, K. and Herman, L. (2001), Growth of *Salmonella* in egg white. *Meded Rijksuniv Gent Fak Landbouwkd Toegep Biol Wet* 66, 531–4.

During, K., Porsch, P., Mahn, A., Brinkmann, O. and Gieffers, W. (1999), The non-enzymatic microbicidal activity of lysozymes. *FEBS Lett.* 449, 93–100.

Edstrom Hagerwall, A. M., Rydengard, V., Fernlund, P., Morgelin, M., Baumgarten, M., Cole, A. M., Malmsten, M., Kragelund, B. B. and Sorensen, O. E. (2012), beta-Microseminoprotein endows post coital seminal plasma with potent candidacidal activity by a calcium- and pH-dependent mechanism. *PLoS Pathog.* 8, e1002625.

EFSA (2009), The community summary Report ontrends and sources of Zoonoses and Zoonic agents in the European Union in 2007. *The EFSA Journal* 223.

Ellison, R. T., 3rd, Giehl, T. J. and LaForce, F. M. (1988), Damage of the outer membrane of enteric gram-negative bacteria by lactoferrin and transferrin. *Infect. Immun.* 56, 2774–81.

Elsbach, P. and Weiss, J. (1998), Role of the bactericidal/permeability-increasing protein in host defence. *Curr. Opin. Immunol.* 10, 45–9.

Fang, J., Ma, M., Jin, Y., Qiu, N., Huang, Q., Sun, S., Geng, F. and Guo, L. (2012a), Liquefaction of albumen during the early incubational stages of the avian embryo and its impact on the antimicrobial activity of albumen. *J. Food Agric. Environ.* 10, 423–7.

Fang, J., Ma, M., Jin, Y., Qiu, N., Ren, G., Huang, X. and Wang, C. (2012b), Changes in the antimicrobial potential of egg albumen during the early stages of incubation and its impact on the growth and virulence response of *Salmonella* Enteritidis. *Ital. J. Anim. Sci.* 11, 92–7.

Feeney, R. E., Ducay, E. D., Silva, R. B. and MacDonnell, L. R. (1952), Chemistry of shell egg deterioration: the egg white proteins. *Poult. Sci.* 31, 639–47.

Fleming, A. (1922), On a remarkable bacteriolytic element found in tissues and secretions. *Proc. R. Soc. London. Series B, Containing Pap. Biol. Character* 93, 306–17.

Garibaldi, J. A. (1970), Role of microbial iron transport compounds in bacterial spoilage of eggs. *Appl. Microbiol.* 20, 558–60.

Gill, C. O. and Tan, K. H. (1980), Effect of carbon dioxide on growth of meat spoilage bacteria. *Appl. Environ. Microbiol.* 39, 317–19.

Golab, K., Gburek, J., Gawel, A. and Warwas, M. (2001), Changes in chicken egg white cystatin concentration and isoforms during embryogenesis. *Br. Poult. Sci.* 42, 394–8.

Gong, D., Wilson, P. W., Bain, M. M., McDade, K., Kalina, J., Herve-Grepinet, V., Nys, Y. and Dunn, I. C. (2010), Gallin; an antimicrobial peptide member of a new avian defensin family, the ovodefensins, has been subject to recent gene duplication. *BMC Immunol.* 11, 12.

Green, N. M. (1963), Avidin. 1. The Use of (14-C)Biotin for Kinetic Studies and for Assay. *Biochem. J.* 89, 585–91.

Gurtler, J. B. and Conner, D. E. (2009), Survival and growth of *Salmonella* Enteritidis in liquid egg products varying by temperature, product composition, and carbon dioxide concentration. *Foodborne Pathog. Dis.* 6, 561–7.

Guyot, N., Labas, V., Harichaux, G., Chesse, M., Poirier, J. C., Nys, Y. and Rehault-Godbert, S. (2016a), Proteomic analysis of egg white heparin-binding proteins: towards the identification of natural antibacterial molecules. *Sci. Rep.* 6, 27974.

Guyot, N., Rehault-Godbert, S., Slugocki, C., Harichaux, G., Labas, V., Helloin, E. and Nys, Y. (2016b), Characterization of egg white antibacterial properties during the first half of incubation: a comparative study between embryonated and unfertilized eggs. *Poult. Sci.* 95(12):2956-2970.

Hager, L. J., McKnight, G. S. and Palmiter, R. D. (1980), Glucocorticoid induction of egg white mRNAs in chick oviduct. *J. Biol. Chem.* 255, 7796–7800.

Halbrooks, P. J., Giannetti, A. M., Klein, J. S., Bjorkman, P. J., Larouche, J. R., Smith, V. C., MacGillivray, R. T., Everse, S. J. and Mason, A. B. (2005), Composition of pH-sensitive triad in C-lobe of human serum transferrin. Comparison to sequences of ovotransferrin and lactoferrin provides insight into functional differences in iron release. *Biochemistry* 44, 15451–60.

Hayakawa, S., Kondo, H. and Nakamura, R. (1983), Effect of beta-Ovomucin on the solubility of alpha-Ovomucin and further inspection of the structure of Ovomucin complex in thick egg white. *Agric. Biol. Chem.* 47, 815–20.

Herve-Grepinet, V., Rehault-Godbert, S., Labas, V., Magallon, T., Derache, C., Lavergne, M., Gautron, J., Lalmanach, A. C. and Nys, Y. (2010), Purification and characterization of avian beta-defensin 11, an antimicrobial peptide of the hen egg. *Antimicrob. Agents Chemother.* 54, 4401–9.

Herve, V., Meudal, H., Labas, V., Rehault-Godbert, S., Gautron, J., Berges, M., Guyot, N., Delmas, A. F., Nys, Y. and Landon, C. (2014), Three-dimensional NMR structure of Hen Egg Gallin (Chicken Ovodefensin) reveals a new variation of the beta-defensin fold. *J. Biol. Chem.* 289, 7211–20.

Humphrey, T. J. and Whitehead, A. (1993), Egg age and the growth of *Salmonella* Enteritidis PT4 in egg contents. *Epidemiol. Infect.* 111, 209–19.
Ibrahim, H. R., Higashiguchi, S., Sugimoto, Y. and Aoki, T. (1997), Role of divalent cations in the novel bactericidal activity of the partially unfolded Lysozyme. *J. Agric. Food Chem.* 45, 89–94.
Ibrahim, H. R., Iwamori, E., Sugimoto, Y. and Aoki, T. (1998), Identification of a distinct antibacterial domain within the N-lobe of ovotransferrin. *Biochim. Biophys. Acta* 1401, 289–303.
Ibrahim, H. R., Matsuzaki, T. and Aoki, T. (2001), Genetic evidence that antibacterial activity of lysozyme is independent of its catalytic function. *FEBS Lett.* 506, 27–32.
Ibrahim, H. R., Sugimoto, Y. and Aoki, T. (2000), Ovotransferrin antimicrobial peptide (OTAP-92) kills bacteria through a membrane damage mechanism. *Biochim. Biophys. Acta* 1523, 196–205.
Ibrahim, H. R., Yamada, M., Matsushita, K., Kobayashi, K. and Kato, A. (1994), Enhanced bactericidal action of lysozyme to *Escherichia coli* by inserting a hydrophobic pentapeptide into its C terminus. *J. Biol. Chem.* 269, 5059–63.
Ignjatovic, J. and Sapats, S. (2000), Avian infectious bronchitis virus. *Rev. Sci. Tech.* 19, 493–508.
Kami, K. and Yasuda, K. (1984), Immunoelectron microscopical demonstration of endogenous avidin in secretory granules of the hen oviduct mucosa: a preliminary study. *Histochem J* 16, 835–42.
Kang, H., Loui, C., Clavijo, R. I., Riley, L. W., and Lu, S. (2006), Survival characteristics of *Salmonella enterica* serovar Enteritidis in chicken egg albumen. *Epidemiol. Infect.* 134, 967–76.
Kato, A., Ogino, K., Kuramoto, Y. and Kobayashi, K. (1979), Degradation of the O-glycosidically linked carbohydrate units of ovomucin during egg white thinning. *J. Food Sci.* 44, 1341–4.
Kato, A., Shibata, M., Yamaoka, H. and Kobayashi, K. (1988a), Deamidation of Lysozyme during the storage of egg white. *Agric. Biol. Chem.* 52, 1973–8.
Kato, A., Wakinaga, T., Matsudomi, N. and Kobayashi, K. (1978), Changes in Lysozyme during egg white thinning. *Agric. Biol. Chem.* 42, 175–6.
Kato, S., Ito, S., Miura, Y., Noguchi, T. and Naito, H. (1988b), The effects of estrogen, insulin and dexamethasone on the synthesis and secretion of egg white proteins in primary cultured oviduct cells of laying Japanese quail (Coturnix coturnix japonica). *Comp. Biochem. Physiol. B* 89, 79–83.
Keller, L. H., Benson, C. E., Krotec, K. and Eckroade, R. J. (1995), *Salmonella* Enteritidis colonization of the reproductive tract and forming and freshly laid eggs of chickens. *Infect. Immun.* 63, 2443–9.
Kim, J. and Choi, Y. H. (2014), Differential abundance of egg white proteins in laying hens treated with corticosterone. *J. Agric. Food Chem.* 62, 12346–59.
Kim, Y.-H., Kim, J., Yoon, H.-S. and Choi, Y.-H. (2015), Effects of Dietary Corticosterone on Yolk Colors and Eggshell Quality in Laying Hens. *Asian Australas. J. Anim. Sci.* 28, 840–6.
Koehler, H. (1974), Physicochemical appraisal of changes in egg white during storage. *J. Agric. Food Chem.* 22, 288–93.
Korpela, J., Salonen, E.-M., Kuusela, P., Sarvas, M. and Vaheri, A. (1984), Binding of avidin to bacteria and to the outer membrane porin of *Escherichia coli. FEMS Microbiol. Lett.* 22, 3–10.
Laible, N. J. and Germaine, G. R. (1985), Bactericidal activity of human lysozyme, muramidase-inactive lysozyme, and cationic polypeptides against *Streptococcus sanguis* and *Streptococcus faecalis*: inhibition by chitin oligosaccharides. *Infect. Immun.* 48, 720–8.
Laschtschenko, P. (1909), Uber die keimtötende und entwicklungshemmende Wirkung von Hühnereiweiß. *Zeitschrift für Hygiene und Infektionskrankheiten* 64, 419–27.
Lee, J. Y., Jeong, W., Lim, W., Kim, J., Bazer, F. W., Han, J. Y. and Song, G. (2012), Chicken pleiotrophin: regulation of tissue specific expression by estrogen in the oviduct and distinct expression pattern in the ovarian carcinomas. *PLoS One* 7, e34215.
Levashov, P. A., Ovchinnikova, E. D., Morozova, O. A., Matolygina, D. A., Osipova, H. E., Cherdyntseva, T. A., Savin, S. S., Zakharova, G. S., Alekseeva, A. A., Belogurova, N. G., Smirnov, S. A., Tishkov, V. I. and Levashov, A. V. (2016), Human Interleukin-2 and hen egg white Lysozyme: screening for Bacteriolytic activity against various bacterial cells. *Acta Naturae* 8, 98–102.
Levashov, P. A., Sedov, S. A., Shipovskov, S., Belogurova, N. G. and Levashov, A. V. (2010), Quantitative turbidimetric assay of enzymatic gram-negative bacteria lysis. *Anal. Chem.* 82, 2161–3.

Li-Chan, E. C. Y., Powrie, W. D. and Nakai, S. (1995), 6 – The chemistry of eggs and egg products. *Egg Science and Technology (4th Edition)*. W. J. Stadelman and O. J. Cotterill eds. Food Products Press, New York, pp. 105–75.

Lim, W., Jeong, W., Kim, J., Yoshimura, Y., Bazer, F. W., Han, J. Y. and Song, G. (2013), Expression and regulation of beta-defensin 11 in the oviduct in response to estrogen and in ovarian tumors of chickens. *Mol. Cell Endocrinol.* 366, 1–8.

Lim, W., Kim, J. H., Ahn, S. E., Jeong, W., Kim, J., Bazer, F. W., Han, J. Y. and Song, G. (2011), Avian SERPINB11 gene: characteristics, tissue-specific expression, and regulation of expression by estrogen. *Biol. Reprod.* 85, 1260–8.

Lim, W. and Song, G. (2015), Differential expression of vitelline membrane outer layer protein 1: hormonal regulation of expression in the oviduct and in ovarian carcinomas from laying hens. *Mol. Cell Endocrinol.* 399, 250–8.

Liu, Y., Qiu, N. and Ma, M. (2013), Comparative proteomic analysis of hen egg white proteins during early phase of embryonic development by combinatorial peptide ligand library and matrix-assisted laser desorption ionization-time of flight. *Poult. Sci.* 92, 1897–1904.

Liu, Y., Qiu, N. and Ma, M. (2015), Comparative proteomic analysis of egg white proteins during the rapid embryonic growth period by combinatorial peptide ligand libraries. *Poult. Sci.* 94, 2495–2505.

Lock, J. L. and Board, R. G. (1992), Persistence of contamination of hens' egg albumen in vitro with *Salmonella* serotypes. *Epidemiol. Infect.* 108, 389–96.

Maehashi, K., Ueda, M., Matano, M., Takeuchi, J., Uchino, M., Kashiwagi, Y. and Watanabe, T. (2014), Biochemical and functional characterization of transiently expressed in neural precursor (TENP) protein in emu egg white. *J. Agric. Food Chem.* 62, 5156–62.

Mann, K. (2007), The chicken egg white proteome. *Proteomics* 7, 3558–68.

Mann, K. (2008), Proteomic analysis of the chicken egg vitelline membrane. *Proteomics* 8, 2322–32.

Mann, K. and Mann, M. (2011), In-depth analysis of the chicken egg white proteome using an LTQ Orbitrap Velos. *Proteome Sci.* 9, 7.

Masschalck, B., Deckers, D. and Michiels, C. W. (2002), Lytic and nonlytic mechanism of inactivation of gram-positive bacteria by lysozyme under atmospheric and high hydrostatic pressure. *J. Food Prot.* 65, 1916–23.

Masschalck, B. and Michiels, C. W. (2003), Antimicrobial properties of lysozyme in relation to foodborne vegetative bacteria. *Crit. Rev. Microbiol.* 29, 191–214.

Maurer, L. M., Yohannes, E., Bondurant, S. S., Radmacher, M. and Slonczewski, J. L. (2005), pH regulates genes for flagellar motility, catabolism, and oxidative stress in *Escherichia coli* K-12. *J. Bacteriol.* 187, 304–19.

Messens, W., Duboccage, L., Grijspeerdt, K., Heyndrickx, M. and Herman, L. (2004), Growth of *Salmonella* serovars in hens' egg albumen as affected by storage prior to inoculation. *Food Microbiol.* 21, 25–32.

Meuer, H.-J. and Egbers, C. (1990), Changes in density and viscosity of chicken egg albumen and yolk during incubation. *J. Exp. Zool.* 255, 16–21.

Molla, A., Matsumura, Y., Yamamoto, T., Okamura, R. and Maeda, H. (1987), Pathogenic capacity of proteases from *Serratia marcescens* and *Pseudomonas aeruginosa* and their suppression by chicken egg white ovomacroglobulin. *Infect. Immun.* 55, 2509–17.

Monchois, V., Abergel, C., Sturgis, J., Jeudy, S. and Claverie, J. M. (2001), *Escherichia coli* ykfE ORFan gene encodes a potent inhibitor of C-type lysozyme. *J. Biol. Chem.* 276, 18437–41.

Nakai, S. (2000), Molecular modification of egg proteins for functional improvement. *Egg Nutrition and Biotechnology*. J. S. Sim, S. Nakai and W. Guenter eds. CAB International, Wallingford, UK, pp. 205–17.

Neilands, J. B. (1995), Siderophores: structure and function of microbial iron transport compounds. *J. Biol. Chem.* 270, 26723–6.

O'Malley, B. W. and McGuire, W. L. (1968), Studies on the mechanism of action of progesterone in regulation of the synthesis of specific protein. *J. Clin. Invest.* 47, 654–64.

Ohno, N. and Morrison, D. C. (1989), Effects of lipopolysaccharide chemotype structure on binding and inactivation of hen egg lysozyme. *Eur. J. Biochem.* 186, 621–7.

Okamoto, I., Mizutani, K. and Hirose, M. (2004), Iron-binding process in the amino- and carboxyl-terminal lobes of ovotransferrin: quantitative studies utilizing single Fe^{3+}-binding mutants. *Biochemistry* 43, 11118–25.

Omana, D. A., Liang, Y., Kav, N. N. and Wu, J. (2011), Proteomic analysis of egg white proteins during storage. *Proteomics* 11, 144–53.

Palmiter, R. D. (1972), Regulation of protein synthesis in chick oviduct. I. Independent regulation of ovalbumin, conalbumin, ovomucoid, and lysozyme induction. *J. Biol. Chem.* 247, 6450–61.

Palmiter, R. D. and Gutman, G. A. (1972), Fluorescent antibody localization of ovalbumin, conalbumin, ovomucoid, and lysozyme in chick oviduct magnum. *J. Biological. Chem.* 247, 6459–61.

Park, W. K., Chung, J. W., Kim, Y. K., Chung, S. C. and Kho, H. S. (2006), Influences of animal mucins on lysozyme activity in solution and on hydroxyapatite surfaces. *Arch. Oral. Biol.* 51, 861–9.

Pellegrini, A., Thomas, U., von Fellenberg, R. and Wild, P. (1992), Bactericidal activities of lysozyme and aprotinin against Gram-negative and Gram-positive bacteria related to their basic character. *J. Appl. Bacteriol.* 72, 180–7.

Pellegrini, A., Thomas, U., Wild, P., Schraner, E. and von Fellenberg, R. (2000), Effect of lysozyme or modified lysozyme fragments on DNA and RNA synthesis and membrane permeability of *Escherichia coli. Microbiol. Res.* 155, 69–77.

Phillips, C. A. (1996), Review: Modified Atmosphere Packaging and its effects on the microbiological quality and safety of produce. *Int. J. Food Sci. Technol.* 31, 463–79.

Qiu, N., Ma, M., Cai, Z., Jin, Y., Huang, X., Huang, Q. and Sun, S. (2012a), Proteomic analysis of egg white proteins during the early phase of embryonic development. *J. Proteomics* 75, 1895–1905.

Qiu, N., Ma, M., Zhao, L., Liu, W., Li, Y. and Mine, Y. (2012b), Comparative proteomic analysis of egg white proteins under various storage temperatures. *J. Agric. Food Chem.* 60, 7746–53.

Rehault-Godbert, S., Baron, F., Mignon-Grasteau, S., Labas, V., Gautier, M., Hincke, M. T. and Nys, Y. (2010), Effect of temperature and time of storage on protein stability and anti-*Salmonella* activity of egg white. *J. Food Prot.* 73, 1604–12.

Rehault-Godbert, S., Gautron, J., Labas, V., Belghazi, M. and Nys, Y. (2008), Identification and characterization of the precursor of chicken matrix metalloprotease 2 (pro-MMP-2) in hen egg. *J. Agric. Food Chem.* 56, 6294–6303.

Rehault-Godbert, S., Labas, V., Helloin, E., Herve-Grepinet, V., Slugocki, C., Berges, M., Bourin, M. C., Brionne, A., Poirier, J. C., Gautron, J., Coste, F. and Nys, Y. (2013), Ovalbumin-related protein X Is a Heparin-binding Ov-Serpin exhibiting antimicrobial activities. *J. Biol. Chem.* 288, 17285–95.

Rehault-Godbert, S., Mann, K., Bourin, M., Brionne, A. and Nys, Y. (2014), Effect of embryonic development on the chicken egg yolk plasma proteome after 12 days of incubation. *J. Agric. Food Chem.* 62, 2531–40.

Rehault-Godbert, S., Nys, Y., Gautron, J., Labas, V., Helloin, E. and Slugocki, C. (2011), Fraction of proteins and peptides derived from egg white and protein derived from egg white and use thereof as anti-*Listeria* agent. Pat. No. WO/2011/151407.

Reijrink, I. A. M., Meijerhof, R., Kemp, B. and van den Brand, H. (2008), The chicken embryo and its micro environment during egg storage and early incubation. *Worlds Poult. Sci. J.* 64, 581–98.

Rocculi, P., Tylewicz, U., Pękosławska, A., Romani, S., Sirri, F., Siracusa, V. and Rosa, M. D. (2009), MAP storage of shell hen eggs, Part 1: effect on physico-chemical characteristics of the fresh product. *LWT Food Sci. Technol.* 42, 758–62.

Romanoff, A. L. and Romanoff, A. J. (1933), Gross assimilation of yolk and albumen in the development of the egg of gallus domesticus. *Anat. Rec.* 55, 271–8.

Ruzickova, V. (1994), Growth and survival of *Salmonella* Enteritidis in selected egg foods. *Vet. Med. (Praha)* 39, 187–95.

Sankaranarayanan, R., Sekar, K., Banerjee, R., Sharma, V., Surolia, A. and Vijayan, M. (1996), A novel mode of carbohydrate recognition in jacalin, a Moraceae plant lectin with a beta-prism fold. *Nat. Struct. Biol.* 3, 596–603.

Sauveur, B. (1988), *Reproduction des volailles et production d'oeufs*. INRA, Paris.
Schäfer, A., Drewes, W. and Schwägele, F. (1999), Effect of storage temperature and time on egg white protein. *Food/Nahrung* 43, 86–9.
Schneider, W. R. and Doetsch, R. N. (1974), Effect of viscosity on bacterial motility. *J. Bacteriol.* 117, 696–701.
Schoeni, J. L., Glass, K. A., McDermott, J. L. and Wong, A. C. (1995), Growth and penetration of *Salmonella* Enteritidis, *Salmonella* Heidelberg and *Salmonella* Typhimurium in eggs. *Int. J. Food Microbiol.* 24, 385–96.
Scott, T. A. and Silversides, F. G. (2000), The effect of storage and strain of hen on egg quality. *Poult. Sci.* 79, 1725–9.
Sellier, N., Vidal, M. L., Baron, F., Michel, J., Gautron, J., Protais, M., Beaumont, C., Gautier, M. and Nys, Y. (2007), Estimations of repeatability and heritability of egg albumen antimicrobial activity and of lysozyme and ovotransferrin concentrations. *Br. Poult. Sci.* 48, 559–66.
Sharp, P. F. and Whitaker, R. (1927), The relation of the hydrogen ion concentration of egg white to its germicidal action. *J. Bacteriol.* 14, 17–46.
Shimizu, T. and Morikawa, K. (1996), The beta-prism: a new folding motif. *Trends Biochem. Sci.* 21, 3–6.
Silversides, F. G. and Budgell, K. (2004), The relationships among measures of egg albumen height, pH, and whipping volume. *Poult. Sci.* 83, 1619–23.
Silversides, F. G. and Scott, T. A. (2001), Effect of storage and layer age on quality of eggs from two lines of hens. *Poult. Sci.* 80, 1240–5.
Smolelis, A. N. and Hartsell, S. E. (1952), Factors affecting the lytic activity of lysozyme. *J. Bacteriol.* 63, 665–74.
Svensson, S. L., Pasupuleti, M., Walse, B., Malmsten, M., Morgelin, M., Sjogren, C., Olin, A. I., Collin, M., Schmidtchen, A., Palmer, R. and Egesten, A. (2010), Midkine and pleiotrophin have bactericidal properties: preserved antibacterial activity in a family of heparin-binding growth factors during evolution. *J. Biol. Chem.* 285, 16105–15.
Swierczewska, E., Skiba, T., Sokolowska, A., Noworyta-Glowacka, J., Kopec, W., Koeniowska, M. and Bobak, L. (2005), Egg white biologically active proteins activity in relation to laying hen's age, in XI th European Symposium on the Quality of Eggs and Egg Products, Doorwerth (NED).
Thapon, J. L. and Bourgeois, C. M. (1994), Présentation générale de l'oeuf. *L'oeuf et les ovoproduits*. J. L. Thapon, and C. M. Bourgeois eds. Tec&Doc Lavoisier, Paris, pp. 2–108.
Tranter, H. S. and Board, R. G. (1984), The influence of incubation temperature and pH on the antimicrobial properties of hen egg albumen. *J. Appl. Bacteriol.* 56, 53–61.
Trziszka, T., Saleh, Y., Kopec, W., Wojciechowska-Smardz, I. and Oziemblowski, M. (2004), Changes in the activity of lysozyme and cystatin depending on the age of layers and egg treatment during processing. *Archiv für Geflügelkunde* 68, 275–9.
Tuohimaa, P., Joensuu, T., Isola, J., Keinanen, R., Kunnas, T., Niemela, A., Pekki, A., Wallen, M., Ylikomi, T. and Kulomaa, M. (1989), Development of progestin-specific response in the chicken oviduct. *Int. J. Dev. Biol.* 33, 125–34.
Valenti, P., Visca, P., Antonini, G. and Orsi, N. (1986), Interaction between lactoferrin and ovotransferrin and *Candida* cells. *FEMS Microbiol. Lett.* 33, 271–5.
Vollmer, W. (2008), Structural variation in the glycan strands of bacterial peptidoglycan. *FEMS Microbiol. Rev.* 32, 287–306.
Wang, C. H. I. and Shelef, L. A. (1991), Factors contributing to antilisterial effects of raw egg albumen. *J. Food Sci.* 56, 1251–4.
Wang, J., Liang, Y., Omana, D. A., Kav, N. N. and Wu, J. (2012), Proteomics analysis of egg white proteins from different egg varieties. *J. Agric. Food Chem.* 60, 272–82.
Wang, J. and Wu, J. (2014), Proteomic analysis of fertilized egg white during early incubation. *EuPA Open Proteomics* 2, 38–59.
Washburn, K. W. (1979), Genetic variation in the chemical composition of the egg. *Poult. Sci.* 58, 529–35.
Wesierska, E., Saleh, Y., Trziszka, T., Kopec, W., Siewinski, M. and Korzekwa, K. (2005), Antimicrobial activity of chicken egg white cystatin. *World J. Microbiol. Biotechnol.* 21, 59–64.

Whenham, N., Lu, T. C., Maidin, M. B., Wilson, P. W., Bain, M. M., Stevenson, M. L., Stevens, M. P., Bedford, M. R. and Dunn, I. C. (2015), Ovodefensins, an oviduct-specific antimicrobial gene family, have evolved in birds and reptiles to protect the egg by both sequence and intra-six-cysteine sequence motif spacing. *Biol. Reprod.* 92, 154.
White, H. B., 3rd, and Merrill, A. H., Jr. (1988), Riboflavin-binding proteins. *Annu. Rev. Nutr.* 8, 279–99.
Wurtz, R. (1890), De l'action bactericide du blanc d'oeuf. *C. R. Seances Soc. Biol. Fil.*
Wyburn, G. M., Johnston, H. S., Draper, M. H. and Davidson, M. F. (1970), The fine structure of the infundibulum and magnum of the oviduct of *Gallus domesticus*. *Q. J. Exp. Physiol. Cogn. Med. Sci.* 55, 213–32.
Yadav, N. K. and Vadehra, D. V. (1977), Mechanism of egg white resistance to bacterial growth. *J. Food Sci.* 42, 97–9.

Chapter 8

The effects of laying hen housing systems on egg safety and quality

Deana R. Jones, US Department of Agriculture, Agricultural Research Service, USA

1 Introduction

1.1 Consumer demands: egg quality, product price and laying hen well-being

Transitions in laying hen management and housing systems have constantly occurred throughout the history of commercial egg production. Around the world, there has been a rapid shift in hen housing requirements since the turn of the current century. In most cases, the changes in hen housing requirements have been due primarily to human social influence with consideration for hen well-being and industry sustainability. While consumers often agree with the social concept of housing system shifts, they do not feel the price difference compared with conventional housed eggs is warranted since quality is not noticeably greater (Hidalgo et al., 2008; Andersen, 2011).

Egg producers, equipment manufacturers and regulatory entities are working diligently to keep up with the rapid demand for less-intensive hen housing systems. The scientific community continues to generate research findings on the impact of hen housing on not only the laying hens but also egg safety and quality. As producers have shifted to less-intensive housing systems, concerns of maintaining egg quality and controlling costs have been noted. Whereas efforts to assist with the transition and develop management strategies continue, often producers are making the best management decisions they can and learning through the process. The first flocks in new housing systems present

http://dx.doi.org/10.19103/AS.2016.0012.09

unique challenges as producers learn to manage and understand the new equipment and hen needs. The inconsistency in egg production and quality can also be seen in the research setting as institutions adjust to the same challenges. As such, current research on egg safety and quality associated with hen housing systems often conflicts with initial published findings.

1.2 Research into hen housing and egg safety and quality

The first big shift in hen housing mandates occurred in Europe with the requirement that conventional cages be out of use by 2012. This resulted in a large quantity of research pertaining to egg safety impacts, in particular that associated with *Salmonella* Enteritidis, and some egg quality assessments being generated by the European scientific community. As presented by Holt et al. (2011), it is difficult to directly compare hen housing research findings from Europe to that of the United States egg industry. There are many differences between how eggs are produced in the two geographic areas including housing equipment, hen genetics, hen nutrition/feed formulation, flock size, egg handling and refrigeration. All of these factors can directly impact egg safety and quality. This chapter will include research from Europe, North America and other geographic areas. It is important for the reader to understand that such findings will often be contradictory.

Egg safety and quality research is conducted both on research farms and at local or commercial egg production facilities. Readers should be mindful that research farms allow for the control of set parameters to determine the impact on egg safety and quality. When egg production facilities (whether small or commercial farms) are utilized in research, they allow for 'real world' examination of issues. An inherent lack of control over the flock or management decisions can impact the scientific conclusions that can be drawn since field locations are in the business of producing income from the eggs, not generating research. Both controlled research and field sampling have important roles in understanding the outcome of hen housing systems on egg safety and quality and will be presented in this chapter.

2 Indicator organisms in monitoring egg safety

Safe food is an important component of human health. The safety of eggs can be compromised through contamination by pathogens or chemicals, and so researchers must explore how changes to hen housing and other management techniques impact these and related factors. In this section, the importance of monitoring indicator organisms is discussed. Section 3 will focus on *Salmonella* spp., which are the most prevalent pathogens found in eggs. Section 4 then reviews other pathogens, including *Campylobacter* spp., and the issue of chemical contamination.

Indicator organisms serve the role of providing insight into the general microbiological status of a product, process or production system. Total aerobic organisms, as well as coliforms or *Enterobacteriaceae* and yeasts and moulds, are assessed in egg production and processing. Total aerobic organisms (or aerobes) are a general group of bacteria that grow in the presence of oxygen. They are enumerated on non-selective agar and are an indication of a general level of cleanliness.

Enterobacteriaceae are a family of organisms that contain many of the human pathogens. These organisms are microaerophilic, meaning they can grow in the presence of limited oxygen. Coliforms are a classification of organisms within *Enterobacteriaceae* that are generally of faecal origin. For these reasons, *Enterobacteriaceae* or coliforms are often enumerated to assess the degree of faecal contribution to microbial populations.

Enumerating indicator organisms provides general microbial information, but will not provide information on specific pathogens such as *Salmonella, Campylobacter, Listeria* or *Escherichia coli*. Targeted cultural techniques are utilized to ascertain the presence of specific organisms.

A variety of studies have considered total aerobic population levels associated with eggs from a wide array of housing systems. Each study is unique in the housing systems compared, the samples collected and the handling of samples, making it difficult to directly compare and contrast results. General trends can be gleaned and findings are often contradictory.

De Reu et al. (2009) conducted shell rinses of eggs (floor eggs were excluded) from enriched cage, aviary and barn (without outdoor access) systems and determined that, biologically, the level of total aerobes was similar across housing systems and there were no differences in levels of *Enterobacteriaceae*. Previously, these authors (De Reu et al., 2006) sampled eggs from commercial conventional cage, organic aviary (with outdoor access) and barn housing systems in Belgium. They determined there were no biologically different levels of aerobes from eggs in the production systems. Eggs laid outside of the system (floor eggs) had much higher bacterial levels. In a subsequent study conducted on French egg laying farms, conventional cage, enriched cage and floor (barn) production systems were compared for total aerobes on the eggshell (Huneau-Salaün et al., 2010). Floor hens produced eggs with higher levels of aerobes on the shell surface compared with conventional cage hens, but these levels were within 0.5 log cfu (colony-forming units) and not biologically important. Furthermore, when nest run eggs were tested in Bavaria, Germany, the same magnitude of total aerobes was found on eggs among all housing systems examined.

Conversely, Mallet et al. (2006) compared two system styles of each of enriched and conventional cage systems. The authors reported one style of the enriched cages assessed resulted in dirtier eggs than the conventional cage systems. Overall, aerobic counts were greater for the enriched cages compared with the conventional cages. Furthermore, eggs laid outside of the nest box in the enriched systems had the highest microbial levels detected during the study. In 2013, Jones and Anderson compared conventional cage, cage-free barn and free-range (unrestricted outdoor access) systems. They found total aerobes to be greatest on conventional cage shells and the same eggs had the lowest levels of *Enterobacteriaceae*.

In another study, Jones et al. (2011) compared conventional cage and free-range production systems. Coliforms were lowest for conventional cage shell emulsions throughout the life of the flock. Similarly, Singh et al. (2009) assessed conventional cage and barn production and found coliforms to be lowest on conventional cage eggs compared with both nest box and floor eggs from the cage-free barn system. In a comparison of enriched cage, aviary and barn (without outdoor access), De Reu et al. (2009) found no difference in *Enterobacteriaceae* levels of shell rinses from the three housing systems. Floor eggs were not included in the sample sets for the study.

Eggs laid outside of the nest box in extensive systems are often classified differently for microbiological assessments. In particular, when hens have access to litter (aviary, cage-free, barn, etc.) or outdoors (free-range, some cage-free, organic, etc.), these eggs are referred to as 'floor eggs'. Floor eggs have been identified as microbiological concerns by many researchers including De Reu, et al., 2006; Jones and Anderson, 2013; and Jones et al., 2015. Floor eggs pose a microbiological concern for many reasons. Eggs laid outside of the nest box in extensive systems are more often laid in the litter or forage areas. In most instances, these eggs must be detected and hand collected. This makes it difficult to ensure timely collection and limited exposure to the litter or outdoors. Eggs laid in outdoor forage areas are exposed to all the natural flora, elements, insects and wildlife. De Reu et al. (2006) reported greater than 6.3 log cfu/egg total aerobes on floor eggs. Jones and Anderson (2013) detected the highest levels of *Enterobacteriaceae* on cage-free floor eggs when conventional cage, cage-free barn and free-range production systems were compared. Furthermore, Jones et al. (2015) detected the greatest level of total aerobes in cage-free aviary floor eggs when commercial conventional cage, enriched colony cage and cage-free aviary production systems were compared.

Flock management has an impact on environmental and egg microbiology (De Reu et al., 2009; Jones et al., 2011, 2015; Jones and Anderson, 2013). Dust has been identified as a major contributor to environmental and egg microbiology (De Reu et al., 2005, 2006; Huneau-Salaün et al., 2010; Kone et al., 2013; Jones et al., 2015), and the high surface area of dust provides more opportunity for microorganisms to attach. Dust easily moves through the air, especially in instances of high-velocity ventilation. De Reu et al. (2005, 2006) reported a correlation between initial egg bacterial levels and dust levels in housing systems. Jones and Anderson (2013) detected greater levels of total aerobes on the shells of eggs from conventional cage compared with cage-free barn and free-range production systems, but the authors also noted that comparatively the eggs from the conventional cage system were covered with dust as opposed to the eggs from the other housing systems. The higher levels of total aerobes enumerated from commercial cage-free aviary eggs (Jones et al., 2015) was attributed to the high levels of particulate matter detected in that production environment (Zhao et al., 2015). The same study also identified airborne particulate matter in the cage-free aviary housing system as a human health concern (Mitchell et al., 2015).

Laying hen genetics influence egg microbiology (Singh et al., 2009; Jones and Anderson, 2013). Jones and Anderson (2013) found that some breeds of laying hens produced eggs with lower microbial counts in certain housing systems. Singh et al. (2009) further found that laying hen genetics influenced nest box usage, which corresponded with egg microbiological differences. Nest box management also influences egg microbiology. Jones et al. (2011) reported free-range nest box swab coliform levels of 2–5.5 log cfu/mL and total aerobe levels of 3.2–6.1 log cfu/mL. In another study, Jones and Anderson (2013) detected elevated levels of *Enterobacteriaceae* in free-range and cage-free barn nest box swabs. In both of these studies, stagnant nest boxes (manual egg collection) were utilized. The authors indicate hens frequently roosted in the nest boxes since access was not limited. Both the cleaning and management of nest boxes can enhance environmental and egg microbiology. Additionally, Jones et al. (2015) reported high levels of coliforms associated with enriched colony cage scratch pads. The authors note the deposition of feed (high surface area) onto the artificial turf mat to encourage foraging behaviour also led to defecation and a build-up of feed and faecal matter on the scratch pad. Consideration should be made as to scratch pad material and design to reduce environmental microbial levels and potential egg safety concerns. Furthermore, Singh et al. (2009) found that egg coliform levels increased with hen age when comparing four genetic strains of laying hens in conventional cage and cage-free barn production systems.

3 *Salmonella* spp.

Salmonella is considered a primary food-borne pathogen for poultry and poultry products. While all species of *Salmonella* are human pathogens, *Salmonella* Enteritidis is most often listed as the predominant food-borne pathogen for eggs. As *Salmonella* spp. testing and detection have progressed, additional species have been identified as food safety concerns. With the swift shift in commercial hen housing systems, the impact of systems on *Salmonella* spp. transfer and persistence is not fully understood.

Challenge studies are integral to understanding the role of housing systems on bird health, egg contamination and pathogen transfer among hens. These studies must be conducted in at least animal biological safety level 2 (ABSL-2) animal containment areas due to the intentional contamination of the hen or environment with a pathogen. Challenge studies allow for controlled comparisons and are the best means to understand the potential transfer of pathogens in housing systems as it is not prudent or practical to introduce pathogens into commercial housing systems. It is expensive to maintain ABSL-2 or greater hen housing areas and conduct challenge studies. As such, there is a small group of researchers worldwide who conduct such comparisons. A chronological presentation of *Salmonella* spp. challenge studies examining the role of hen housing will be presented.

In 2005, Gast et al. introduced *Salmonella* Enteritidis and Heidelberg control strains and variant strains into specific pathogen-free (SPF) hens in conventional cages. Variant strains were strains of *S.* Enteritidis and Heidelberg, which had been cultured from faeces of inoculated hens and were naturally occurring variants of the control strain utilized as the inoculum. While this particular study did not compare housing systems, it is very important to the discussion of *Salmonella*, laying hens and the production environment in that the variant strains were shown to have a longer period of shedding (26.7 vs 17.5 days) and greater rate of egg contamination (6.4 vs 3.3%). Therefore, persistent *Salmonella* in the production environment may be more invasive than the original organism detected in the housing environment.

De Vylder et al. (2009) utilized *S.* Enteritidis to orally challenge hens in conventional cage, enriched cage and aviary housing systems. The prevalence of *S.* Enteritidis in ceca, spleen, liver, ovary and oviduct was assessed. The authors reported that less-intensive (enriched cage and aviary) housing systems, when compared to conventional cage, did not result in greater *S.* Enteritidis contamination. Additionally, there was no difference in detection among the organs cultured.

In 2011, De Vylder et al. inoculated floor hens with *S.* Enteritidis and three weeks later moved hens to conventional cage, enriched cage, aviary and cage-free barn housing systems. The authors found a greater occurrence of bird-to-bird transfer of *S.* Enteritidis in aviary and cage-free barn housing systems compared with conventional cage. Egg contamination was increased for the aviary housing system compared with conventional cage or cage-free barn systems.

Gast et al. (2013) challenged conventional cage and enriched cage hens with *S.* Enteritidis. Ceca, liver, spleen, ovary and oviduct were cultured to determine degree of hen contamination associated with each housing system. The ceca were most often contaminated with *S.* Enteritidis and there was no difference associated with housing system. Otherwise, the prevalence of liver, spleen, ovary and oviduct *S.* Enteritidis infection was significantly greater in conventional cage hens.

In 2014, Gast et al. again challenged conventional cage and enriched cage hens with *Salmonella* Enteritidis. Eggs were cultured throughout 5–25 days post-inoculation. The authors found a similar rate of contamination between the two housing systems. In a subsequent study (Gast et al., 2015), two phage types of *Salmonella* Enteritidis were utilized as inoculums for conventional cage and enriched cage hens. There was a significantly greater rate of *S.* Enteritidis shedding in conventional cage versus enriched cage hens. Furthermore, enriched cage hens shed *S.* Enteritidis for 8 weeks compared with conventional cage hens (10 weeks).

Stocking density in the enriched cage was compared with conventional cage controls in a *S.* Enteritidis challenge study (Gast et al., 2016). Enriched hens were housed at 648 cm^2 and 973 cm^2, whereas conventional cage hens were stocked at 648 cm^2. Enriched cage hens housed more intensely (648 cm^2) had significantly greater prevalence of *S.* Enteritidis in the liver and ovary compared with 973 cm^2 enriched cage hens. The enriched cage hens at 648 cm^2 had significantly lower spleen contamination compared with conventional cage hens at the same stocking density. The authors determined that housing and stocking density combine to influence *Salmonella* susceptibility.

Several groups of researchers have conducted reviews of published literature or sampled a survey of commercial farms to assess the risk factors for *Salmonella* spp. detection or infection. In preparation for the European Union mandated housing shift, Van Hoorebeke et al. (2011) conducted a review of published literature to assess the risk of the housing conversion from conventional cage to less-intensive systems on *Salmonella* prevalence. The authors determined that shifting from conventional cage to enriched colony cages and non-cage systems would not increase the incidence of *Salmonella* detection. The risk factors determined to a have greater impact than housing system on *Salmonella* prevalence were farm and flock size, stocking density, hen physiological stress and carry-over infections due to pests and housing hygiene. Denagamage et al. (2015) compared published data through fall (autumn) 2014 involving surveys and *Salmonella* spp. incidence. Thirteen published studies were included in this comparison. The review explains a metadata analysis of the information presented in the thirteen selected publications.

Snow et al. (2010) assessed the risk for *Salmonella* Enteritidis on United Kingdom egg farms between 2004 and 5. The identified risk factors were flock sizes >30 000 hens and frequent rodent sightings. Risks of infection were reduced with vaccination programmes, utilizing foot dips with brushes; washing and disinfecting after depopulation; clean parking away from barns; independent feed manufacturing, at least 1 km between farm locations; in-line farms and egg packaging; and the presence of cats and dogs on the farm. This survey was conducted before the European Union mandated shift from conventional cage housing. Additionally, the existence of an effective pest management plan would be a better option than having cats and dogs present on the farm, as both can be reservoirs of *Salmonella* spp. In fact, the US Food and Drug Administration rule entitled 'Prevention of *Salmonella* Enteritidis in shell eggs during production, storage, and transportation' (FDA, 2009) explicitly forbids cats on egg laying farms falling under the rule due to the potential transfer of *Salmonella* Enteritidis to the laying house from cats.

Carrique-Mas et al. (2009) summarized the findings of 264 *Salmonella* spp. detection incidents in Great Britain between 1998 and 2007 (before the shift in European Union housing requirements). Faecal and dust samples were assessed and an incident was defined as *Salmonella* being detected in a production house. The stated 264 incidents took place in 152 production houses. Van Hoorebeke and colleagues (2010b) conducted a survey of 292 laying flocks in Europe one month before depopulation. This study was also conducted before the European Union mandated housing shift. Twenty-nine of the flocks (approximately 10%) were positive for at least one species of *Salmonella*. Risk factors for *Salmonella* Enteritidis and Typhimurium detection included housing in conventional cages, absence of dry cleaning among flocks and sampling in winter.

Huneau-Salaün et al. (2009) collected faecal and dust samples from 519 French laying flocks at the end of lay in 2004 and 2005. The authors determined 17.9% of flocks were positive for *Salmonella* spp. Rousi et al. (2010) also conducted faecal and dust sampling in 92 laying flocks in Southern Greece (no time frame provided) and determined 64% of houses were positive for at least one species of *Salmonella*. Both groups of researchers reported higher rates of *Salmonella* detection in dust samples compared with fresh faecal samples. Galis and colleagues (2013) found dust was a major contributor to the long persistence of *Salmonella* in poultry houses. Dewaele et al. (2012) sampled Belgian egg production farms that had previously tested positive for *Salmonella* Enteritidis. After three cycles of production flocks and corresponding depopulation cleaning and disinfection in the identified houses, *S.* Enteritidis was still detected but at rates lower than initially found. Cleaning and disinfection of housing has been identified as one of the best means to reduce pathogens on farms (Doyle and Erickson, 2012).

The natural occurrence of *Salmonella* spp. on farms of various housing systems has been investigated by many groups of researchers since the early 2000s. The outcomes often conflict, making it difficult to understand fully the impact of hen housing on *Salmonella*. Unlike challenge studies which control parameters so that direct comparisons can be made, field or natural occurrence studies present 'real-world' findings. Many of these studies take place on egg production farms that are managed to produce a commodity at a profit. The variable nature of management strategies utilized at egg production farms can lead to conflicting results among research studies. As was the case with indicator organisms, the initial research comparing housing systems and the prevalence of *Salmonella* was conducted in Europe as preparation for the mandated European Union housing shift. Another wave of research followed in the United States as state and federal regulations concerning *Salmonella* status of flocks and hen housing requirements were published. Subsequently, both European Union and US commercial studies assessing *Salmonella* prevalence ceased due to regulatory implications of additional *Salmonella* testing at egg production farms.

A Belgium baseline study on *Salmonella* prevalence in the egg production house was reported in 2008 (Namata et al.). Dust and faecal samples were collected from cage and barn/free-range houses. The authors identified cage production as a higher risk of *Salmonella* infection than the barn/free-range houses. Huneau-Salaün et al. (2009) also reported a significantly greater *Salmonella* incidence in cage versus floor production flocks in France. Pieskus et al. (2008) tested faeces, dust and water on eight laying hen farms in Lithuania to determine *Salmonella* spp. prevalence. Conventional cage, enriched cage and aviary production systems were represented in the study. The authors reported that housing systems did not significantly impact the prevalence of *Salmonella* detected in the production environment.

Cage and non-cage (floor, free-range and organic) production systems in Belgium were compared by Van Hoorebeke and colleagues (2010a) for the incidence of *Salmonella* and it was determined that housing system did not significantly impact the detection. The authors reported previous *Salmonella* detection on the farm and age of the production system did significantly impact the prevalence of *Salmonella* in the production environment. A study of flocks in Belgium, Denmark and Germany (total of 41 flocks) housed in cage (conventional or enriched) and non-cage (barn, free-range, or organic) was conducted to determine the prevalence of *Salmonella*. All birds were negative for *Salmonella* before placement on a farm. A total of 10/41 flocks were found to be positive for *Salmonella* with no differences found between cage and non-cage housing systems.

In the United States, Green et al. (2009) conducted a study comparing *Salmonella* infection of hens from two styles of conventional cage systems (manure belt and high-rise) and cage-free barn housing. Small intestines and ceca homogenates were formed. The authors found no difference in *Salmonella* infection due to hen housing system. Jones et al. (2012) compared conventional cage and free-range housing systems and found no difference in the prevalence of *Salmonella* spp. associated with egg or environmental samples from the systems. A commercial comparison of conventional cage, enriched colony cage and cage-free aviary housing systems and *Salmonella* prevalence associated with egg and environmental microbiology has been conducted (Jones et al., 2015). The commercial production system comparison was conducted without concern for regulatory action as the research was conducted at an in-line egg products facility. Eggs produced solely for further processed egg products in the United States are not required to meet *Salmonella* Enteritidis testing requirements associated with shell egg production (FDA, 2009), which allowed researchers to assess pathogens in the flocks. All flocks were shedding *Salmonella* spp. at a high rate (89–100% prevalence in manure scraper blade swabs). No differences were found in the prevalence of *Salmonella* spp. detected in environmental swabs or eggshell pools from the three housing systems.

Jones and colleagues (2016) continued the commercial comparison study by assessing *Salmonella* infection of 120 hens from each housing system at flock termination. Adrenal gland, spleen, ceca, ovary and oviduct were cultured from each hen. Even though flocks were shedding *Salmonella* at high rates, only 7% of all 360 hens tested had at least one positive organ. There was a significant difference in the percentage of hens from each system that were *Salmonella* positive – conventional cage, 13.33%; enriched colony cage, 5%; and cage-free aviary, 3.33%. No *Salmonella* was detected in the 360 spleens sampled. The only positive ovaries were found in enriched colony cage hens (2.5%). A significantly greater prevalence of *Salmonella* was detected in the ceca of conventional cage (12.5%) hens compared with enriched colony cage (2.5%) and cage-free aviary (2.5%).

The two predominant strains detected in the US commercial comparison study were *Salmonella* Braenderup and Kentucky (Jones et al., 2016). The strains were consistently isolated from environmental, egg and organ samples across the three housing systems. *S.* Braenderup and Kentucky were equally distributed in manure scraper blade swabs across the housing systems. The equitable rate of shedding did not transfer to environmental and eggshell pools isolates, which were primarily *S.* Braenderup. No *Salmonella* Enteritidis, Typhimurium or Heidelberg were detected. In Europe, *Salmonella*

Enteritidis is often the primary isolate from egg and environmental samples associated with egg production (Pieskus et al., 2008; Carrique-Mas et al., 2009; Huneau-Salaün et al., 2009; Rousi et al., 2010; Schulz et al., 2011). However, *S.* Typhimurium has also been detected with some frequency (Pieskus et al., 2008; Huneau-Salaün et al., 2009) when housing system comparisons have been conducted.

4 Other pathogens and chemical contaminants

4.1 *Campylobacter* spp.

Campylobacter spp. are a major food safety concern and often detected in poultry production and processing. The predominant information on *Campylobacter* spp. and poultry has focused on broilers. As housing systems have evolved and laying hens are more readily able to interact with faeces and the outdoors, researchers have begun to consider the impact of housing on *Campylobacter* prevalence and infection.

In Finland, Sulonen et al. (2007) sampled fresh faecal droppings and eggs from 36 organic laying farms in autumn 2003 and spring 2004. Finnish organic farming consisted of hens in barns on litter with outdoor access. During the autumn sampling, 84% of farms were positive for *Campylobacter*, whereas 76% were positive during the spring sampling period. Throughout the study, 1 of the 360 eggshell samples was positive for *Campylobacter*. *Campylobacter jejuni* was the predominant isolate with a few farms being positive for *Campylobacter coli*. Both organisms are capable of causing human illness.

Jones et al. (2012) also reported a low level of eggshell *Campylobacter* spp. contamination when conventional cage and free-range production were compared. Free-range nest box swabs collected during the study had a greater prevalence of *Campylobacter* than free-range grass samples or conventional cage wire swabs. The authors noted that stagnant nest boxes were utilized and that faecal matter was often visible in the nest boxes during swabbing.

During the comparison of commercial conventional cage, enriched colony cage and cage-free aviary housing systems reported by Jones et al. (2015), the dry manure belt in the conventional cage system impacted the ability to detect *Campylobacter*. It was known that the conventional cage hens were shedding *Campylobacter* due to the detection of the organism in conventional cage system wire swabs, but manure scraper blade swabs from the system resulted in no *Campylobacter* detection. This was not the case for the manure scraper blade swabs from the enriched colony cage or cage-free aviary systems. The greatest prevalence of environmental *Campylobacter* was detected in cage-free aviary forage drag swabs and enriched colony cage scratch pads. A low level of *Campylobacter* prevalence was seen for all types of eggshell pools collected and was not significantly impacted by housing system.

A subsequent publication provided insight into the *Campylobacter* status of the laying hens from each of the housing systems at depopulation and speciation of the isolates collected throughout the study. Conventional cage hens (95%) were infected with *Campylobacter* spp. at a significantly greater prevalence than cage-free aviary (85%) hens. Across the 360 hens sampled from the three housing systems, 88% of ceca were *Campylobacter* positive with a significantly greater proportion being conventional (92.5%) and enriched colony cage (90%) hens compared with cage-free aviary (80.83%). Conversely, cage-free aviary ovaries were infected (9.17%) at a significantly greater rate than conventional (4.17%) or enriched colony cage (1.67%) hens. There was no difference in *Campylobacter* detection in the reproductive tracts of hens from the three housing systems. *Campylobacter jejuni* was the primary isolate from environmental swabs, whereas *C. coli* was the predominant isolate from manure scraper blades. The few positive eggshell pools (12 total) were principally *C. jejuni*. The positive organs collected from all three housing systems were equally identified as *Campylobacter coli* and *jejuni*.

Green et al. (2009) compared conventional cage with manure belt, high-rise conventional cage (with manure pit), and barn production systems for the prevalence of *Campylobacter* in the small intestines and ceca. During the winter sampling period, 80% of cage-free barn hens were positive for *Campylobacter* spp., significantly greater than the 49.4% of conventional cage hens. Conversely, during the summer sample period 27.5% of barn hens and 45.7% of conventional cage hens were similarly positive for *Campylobacter*. Within the conventional cage housing systems, high-rise hens were infected with *Campylobacter* at a lower rate than manure belt system hens. *Campylobacter coli* was the predominant species isolated in the autumn, while *C. jejuni* was most frequently isolated in summer.

Schwaiger et al. (2008) sampled organic and conventional cage laying farms in Germany. There was not a significant difference in the rate of *Campylobacter* detection in cloacal swabs from the two housing system (36.3% organic vs 29.3% conventional cage). When isolates were combined, *Campylobacter jejuni* was the primary isolate (82.8%), followed by *C. coli* (16.4%). The authors also report variable differences in *Campylobacter* resistance to certain antimicrobials. There was not a clear trend in antimicrobial resistance due to housing system.

4.2 Other pathogens

Listeria, especially *Listeria monocytogenes*, is an important food-borne pathogen. Particularly unique characteristics of *Listeria* are the ability to grow in refrigerated conditions and the difficulty to rid an environment of the organism after it begins to grow. *Listeria* has been detected in the shell egg and egg products processing environment (Leasor and Foegeding, 1989; Farber et al., 1992; Jones et al., 2006; Jones and Musgrove, 2007, 2008) but information as to on-farm incidence or impact of housing systems on *Listeria* prevalence is limited. Aury and colleagues (2011) examined the risk factors associated with *Listeria monocytogenes* prevalence in 84 flocks of caged laying hens in France. *L. monocytogenes* was detected in faecal samples of 30.9% of flocks tested. The authors reported that belted manure removal decreased the presence of *L. monocytogenes*. The presence of pets on the farm increased the risk of *L. monocytogenes*. The prevalence of *Listeria* spp. in the production environment and associated with eggs (shell pools and contents pools) of conventional cage and free-range housing systems was determined (Jones et al., 2012). *Listeria* spp. were detected in 7 out of 595 total environmental, eggshell and egg content samples collected from both housing systems. There was no difference in *Listeria* detection due to housing system. Both *Listeria monocytogenes* and *innocua* were isolated in the study.

Schwaiger et al. (2008) reported the antimicrobial resistance of *Salmonella* spp., *Campylobacter* spp. and *Escherichia coli* from conventional cage and organic laying farms in Germany. Overall, they reported antimicrobial resistance of gram-negative organisms to be greater for conventional cage than organic housing systems. The specific antibiotics to which organisms were resistant were variable among housing systems for each monitored population. The authors surmise that the management practices required under German organic egg production standards resulted in a lower degree of antimicrobial resistance.

In 2011, Harisberger and colleagues compared antimicrobial resistance of *Escherichia coli*, *Enterococcus faecalis* and *Enterococcus faecium* collected from faecal swabs of Swiss laying hens. The hens were housed in a variety of housing systems (organic, 14; free range, 33; barn outdoor access, 42; barn indoor, 10). A wide profile of antimicrobials was utilized in assessments. The authors report rare use of antimicrobials during the life cycle of the flocks monitored, yet antimicrobial resistance was frequently detected. There was no correlation between housing system and antimicrobial resistance. The authors surmise that hen housing and management practices do not solely influence antimicrobial resistance of organisms.

A research group in Spain sampled retail eggs from a variety of housing systems (Alvarez-Fernández et al., 2012). The types of eggs sampled were quail (conventional cage) and chicken (conventional cage, barn, free-range, organic and domestic production – which was not defined). Twenty *Escherichia coli* isolates from each type of egg source (total of 120) were tested for resistance from 12 predominant antimicrobials utilized in veterinary and human health care. Antimicrobial resistance was greatest in conventional cage and barn production, followed by free-range, organic and domestic breeding.

4.3 Chemical contamination

The contamination of eggs with dioxins, pesticides and heavy metals due to exposure from housing system conditions has been reviewed by Holt et al. (2011). Free-range, organic and barns with outdoor access are all systems allowing hens to interact with forage areas or outdoors. Hens are therefore exposed to soil contaminants, pests and plants through exploratory behaviour and feeding activities.

Kijlstra et al. (2007) examined the impact of flock size on dioxin concentration of eggs from hens with outdoor access (under organic management). Earthworm or soil concentration of dioxins did not correlate with egg dioxin concentration. There was a significant correlation between flock size and egg dioxin and polychlorinated biphenyl (PCB) levels. The authors surmise the correlation is due to the larger flocks spending more time outdoors. One of the farms sampled restricted outdoor access time resulting in a decrease in egg dioxin content. While the contamination of the soil and soil organisms with dioxins is an important factor in egg dioxin concentration, the behaviour of the hens is also important.

With the urban farming culture encouraging the use of laying hens in sustainability plans, a group examined the level of lead contamination in eggs from urban farming in New York City. In the urban landscape, it is difficult to ascertain the breadth of chemical contamination of soil. Urban farms or cooperatives may also bring soil into an area in preparation for farming further muddling the understanding of soil history. Spliethoff et al. (2014) reported urban farmed egg lead concentration was significantly correlated with soil lead levels. The authors also found that hen dietary calcium supplementation reduced egg lead levels. More work is needed to understand the mechanisms of the reduction.

The authors also note that mitigation strategies are needed for urban egg production to control egg lead contamination.

5 Egg quality

Consumers draw conclusions about housing systems in part based on the quality of the eggs from housing systems available in retail. Extensive housing systems often have higher production costs, resulting in more expensive eggs. A consumer has an expectation of enhanced quality with the price premium. Regulatory agencies want to ensure regulations for egg quality equitably assess quality across housing systems. Review articles have been published on the impacts of hen housing on egg quality (Sossidou and Elson, 2009; Holt et al., 2011; Rakonjac et al., 2014). In many instances, studies conflict on the outcomes of egg quality due to housing. Karcher et al. (2015) illustrate the complexity of comparing hen production and egg quality among housing systems and the factors which can significantly influence outcomes.

5.1 Shell quality

In 2004, Van Den Brand et al. (2004) compared hens housed in conventional cage and free-range systems. Significant hen age and housing system interactions were found for egg and eggshell weights, indicating different responses in egg quality as hens age in various housing systems. Free-range eggs were consistently broader than caged eggs. Wang et al. (2009) conversely found no difference in egg shape index for hens housed in conventional cage and free-range systems. Van Den Brand et al. (2004) also determined conventional cage hens began laying eggs three weeks earlier than free-range hens. The authors summarize that it was more difficult to maintain egg quality in the free-range system compared with conventional cage. Floor hens (barn) were found to produce heavier eggs compared with conventional cage hens in another study (Singh et al., 2009). Guesdon and Faure (2004) compared conventional and enriched cage systems and found no difference in egg weight.

Shell quality is an important since the shell is the first line of defence against external microbial contamination and is also the 'container' for the egg contents through retail and storage. A variety of methods can be utilized to assess shell quality making it important to discuss general trends in findings compared with absolute values. Vits et al. (2005) compared two strains of laying hens in three different styles of enriched cage systems. There were no differences in shell strength among the layer strains. There were also no differences in shell strength between the three styles of enriched systems. Conversely, Guesdon et al. (2006) compared two styles of each conventional and enriched cage systems and found shell strength to be impacted by enriched cage design. Wall and Tauson (2002) reported that egg saver wires and long nest box curtains reduced the percentage of cracked eggs in enriched cage systems. Guesdon et al. (2006) attribute the phenomenon of egg cracking in enriched cage systems to both increased use of nest boxes resulting in piling of eggs on the roll-out troughs and on egg laying from perches. In earlier research (Guesdon and Faure, 2004), enriched cage was also reported as having a higher percentage of cracked eggs compared with conventional cage. If only nest box eggs from the enriched cage system were considered in the analysis, there is no difference in broken eggs between the systems. Yet the same study and Valkonen et al. (2010) found no differences in shell strength between conventional and enriched cage systems. Similarly, Wang et al. (2009) detected no differences in shell strength or shell thickness between eggs from conventional cage and free-range production systems. Van Den Brand et al. (2004) found shell quality declined with hen age in conventional cage production, but remained consistent or increased in free-range production as hens aged.

In a study comparing Italian retail eggs (Hidalgo et al., 2008), shell strength was greatest for conventional cage compared with free-range, barn and organic production systems. In a United States survey of retail eggs (Jones et al., 2010), brown and white shell eggs from conventional cage, cage-free, and free-range systems were compared. Shell strength was similar within brown eggs and also within white eggs, irrespective of housing system.

Abrahamson and Tauson (1998) compared five production flocks in aviary housing and found shell strength to be variable among flocks. Nest box usage influenced shell cleanliness, which was also reported by Singh et al. (2009). Abrahamson and Tauson (1998) stated that the incidence of cracked eggs reported in extensive systems may not be accurate. The authors speculate that eggs laid outside nest boxes may be broken or consumed by hens preventing an accurate assessment of egg production and incidence of cracked eggs. This phenomenon could also explain the findings of Mertens et al. (2006) in a comparison of conventional cage, enriched cage, aviary and free-range housing systems. Shell strength was found to be greatest for aviary eggs and weakest for free-range eggs. The dynamic stiffness of the shell followed the same trend. Yet the authors report the percentage of cracked eggs to be greatest for conventional and enriched cage eggs, with aviary and free range having similar levels of egg breakage.

5.2 Interior quality

Interior egg quality has a direct impact on consumer perception and also egg performance in a food matrix. The most common assessment of interior egg quality is the Haugh unit (Haugh, 1937) or thick albumen height. When Haugh unit scores were compared for hens in a variety of enriched cage production systems (Vits et al., 2005), a significant difference was found between systems. The range of average Haugh unit scores in the study was 80.8–82.9, which is a tight range for Haugh unit scores and all fall within the United States egg grade status of Grade AA (Haugh unit ≥ 72; USDA, 2000). Singh et al. (2009) found albumen height to be greater for conventional cage eggs in comparison with those from free-range production. The authors speculate that lower albumen height values for eggs laid outside of the nest box in free-range production might be due to ammonia exposure.

In both the Italian (Hidalgo et al., 2008) and US (Jones et al., 2010) retail egg quality surveys, Haugh unit scores were considered. Hidalgo et al. (2008) found similar Haugh unit scores for cage, free-range and barn production systems; organic production had significantly lower Haugh unit scores. Jones et al. (2010) found the highest Haugh unit scores in conventional cage eggs, but also noted these eggs had been in retail for the shortest period of time based on carton pack dates. Hidalgo and colleagues (2008) also conducted the functional assessment of whipping capacity. Organic eggs had the greatest whipping capacity, while cage eggs had the lowest. This falls in line with the reported findings on Haugh unit scores due to the inverse relationship between Haugh unit scores and whipping capacity. When Haugh unit or albumen height values are higher, proteins in the thick albumen are generally more tightly bound. The pH of the albumen increases as an egg ages, resulting in loosened protein structures in the thick albumen, allowing a greater foam to be formed during whipping.

Singh et al. (2009) found greater component (albumen and yolk) weights in eggs from floor hens compared with conventional cage eggs. Yolk colour scores were also greater from the floor housing system. In enriched housing systems, Wall (2011) found egg quality to be greater in the lower stocking densities compared. Overall egg quality was reported as being similar between the conventional cage and free-range systems compared by Wang et al. (2009).

In the United States, egg quality standards (USDA, 2000) are based on eggs being within grade standards for the entire period of retail presentation (45 days for USDA-grade shield product; USDA, 2006). These standards were also developed for conventional cage egg production. Eggs in the United States may also be marketed without participating in the federal voluntary-grade shield programme and then must meet the standards for marketing eggs in each state, which in general is a maximum of 30 or 45 days in commerce. In 2014, Jones and colleagues published a comparison of egg quality decline during extended storage based on commercial conventional cage, enriched colony cage and cage-free aviary housing systems. A full profile of egg quality characteristics were assessed over 12 weeks of refrigerated storage (4°C): egg weight, shell dynamic stiffness, static compression shell strength, albumen height, Haugh unit, yolk index, and vitelline membrane strength and elasticity. The study found that egg quality declined at the same rate for the three housing systems and that US egg grading standards therefore equitably define quality of eggs across these commercial housing systems.

6 Summary

Egg safety and quality are influenced by flock management, nutrition, laying hen genetics, environment and housing conditions. The changes in hen housing requirements are occurring faster than research can be completed for informed management and regulatory decisions to be made. As such, there is a need for more comprehensive studies to garner how housing systems impact all aspects of food safety, product quality, production efficiency, and hen health and well-being.

In extensive systems, it is imperative to promote egg laying in nest boxes. When hens have access to forage, litter or the outdoors, eggs laid in these locations should not be utilized in the human food supply. Furthermore, when hens have access to these areas, producers and consumers should be mindful of potential chemical contaminants in eggs produced. All housing systems have the ability to be managed in a safe and economically feasible manner. Research, extension and outreach needs to focus on housing system design, flock management and egg handling to enhance product safety and quality.

7 Future trends in research

There is a vast amount of research needed to understand the impact of commercial and small-scale hen housing systems on egg safety and quality. In an era of dwindling research funding, it is imperative for collaborative, multidisciplinary research projects to be conducted. Not only do projects such as the Coalition for Sustainable Egg Supply (Swanson et al., 2015) provide large amounts of pertinent information, the interrelation of multiple disciplines allows for a holistic approach to the research problem. This project examined the commercial housing of hens in conventional cage, enriched colony cage, and cage-free aviary and the resulting impacts on hen health and well-being, food safety and product quality, environment, worker health and economics. At this point, at least twenty peer-reviewed publications have been generated on the flocks. Furthermore, numerous outreach and scientific presentations, as well as popular press and media interviews have resulted.

International teams of experts are needed to tackle the issue of hen housing and food safety and product quality. As housing system manufacturers and animal health companies expand the options available to the world egg industry, independent research is necessary to assist with maximizing the efficiency of egg production while ensuring product safety and quality.

8 References

Abrahamsson, P. and Tauson, R. (1998), 'Performance and egg quality of laying hens in an aviary system', *J. Appl. Poult. Res.*, 7(3), 225–32.

Alvarez-Fernández, E., Dominguez-Rodríguez, J., Capita, R. and Alonso-Calleja, C. (2012), 'Influence of housing systems on microbial load and antimicrobial resistance patterns of *Escherichia coli* isolates from eggs produced for human consumption', *J. Food Prot.*, 75(5), 847–53.

Andersen, L. (2011), 'Animal welfare and eggs – Cheap talk or money on the counter?', *J. Ag. Econ.*, 62(3), 565–84.

Aury, K., Le Bouquin, S., Toquin, M., Huneau-Salaün, A., Le Nôtre, Y., Allain, V., Petetin, I., Fravalo, P. and Chemaly, M. (2011), 'Risk factors for *Listeria monocytogenes* contamination in French laying hens and broiler flocks', *Prev. Vet. Med.*, 98(4), 271–8.

Carrique-Mas, J. J., Breslin, M., Snow, L., McLaren, I., Sayers, A. R. and Davies, R. H. (2009), 'Persistence and clearance of different *Salmonella* serovars in buildings housing laying hens', *Epidemiol. Infect.*, 137(6), 837–46.

De Reu, K., Grijspeerdt, K., Heyndrickx, M., Uyttendaele, M., Debevere, J. and Herman, L. (2006), 'Bacterial shell contamination in the egg collection chains of different housing systems for laying hens', *Brit. Poult. Sci.*, 47(2), 163–72.

De Reu, K., Grijspeerdt, K., Heyndrickx, M., Zoons, J., De Baere, K., Uyttendaele, M., Debevere, J. and Herman, L. (2005), 'Bacterial eggshell contamination in conventional cages, furnished cages, and aviary housing systems for laying hens', *Brit. Poult. Sci.*, 46(2), 149–55.

De Reu, K., Rodenburg, T. B., Grijspeerdt, K., Messens, W., Heyndrickx, M., Tuyttens, F. A. M., Sonck, B., Zoons, J. and Herman, L. (2009), 'Bacteriological contamination, dirt, and cracks of eggshells in furnished cages and noncage systems for laying hens: An international on-farm comparison', *Poult. Sci.*, 88(11), 2442–8.

De Vylder, J., Dewulf, J., Van Hoorebeke, S., Pasmans, F., Haesebrouck, F., Ducatelle, R. and Van Immerseel, F. (2011), 'Horizontal transmission of *Salmonella* Enteritidis in groups of experimentally infected laying hens housed in different housing systems', *Poult. Sci.*, 90, 1391–6.

De Vylder, J., Van Horrebeke, S., Ducatelle, R., Pasmans, F., Haesebrouck, F., Dewulf, J. and Van Immerseel, F. (2009), 'Effect of the housing system on shedding and colonization of gut and internal organs of laying hens with *Salmonella* Enteritidis', *Poult. Sci.*, 88(12), 2491–5.

Denagamage, T., Jayarao, B., Patterson, P., Wallner-Pendleton, E. and Kariyawasam, S. (2015), 'Risk factors associated with *Salmonella* in laying hen farms: Systematic review of observational studies', *Avian Dis.*, 59(2), 291–302.

Dewaele, I., Van Meirhaeghe, H., Rasschaert, G., Vanrobaeys, M., De Graef, E., Herman, L., Ducatelle, R., Heyndrickx, J. and De Reu, K. (2012), 'Persistent *Salmonella* Enteritidis environmental contamination on layer farms in the context of an implemented national control programme with obligatory vaccination', *Poult. Sci.*, 91(2), 282–91.

Doyle, M. P. and Erickson, M. C. (2012), 'Opportunities for mitigating pathogen contamination during on-farm food production', *Intern. J. Food Microbiol.*, 152, 54–74.

Farber, J. M., Daley, E. and Coates, F. (1992), 'Presence of *Listeria* spp. in whole eggs and wash water samples from Ontario and Quebec', *Food Res. Int.*, 25, 143–5.

FDA. (2009), 'Prevention of *Salmonella* Enteritidis in shell eggs during production, storage, and transportation; Final rule', 21 CFR parts 16 and 118, https://www.gpo.gov/fdsys/pkg/FR-2009-07-09/pdf/E9-16119.pdf (accessed 1 July 2016).

Galis, A. M., Marcq, C., Marlier, D., Portetelle, D., Van, I., Beckers, Y. and Théwis, A. (2013), 'Control of *Salmonella* contamination of shell eggs – preharvest and postharvest methods: A review', *Comprehen. Rev. Food Sci. Food Safety*, 12(2), 155–82.

Gast, R. K., Guard-Bouldin, J. and Holt, P. S. (2005), 'The relationship between the duration of fecal shedding and the production of contaminated eggs by laying hens infected with strains of *Salmonella* Enteritids and *Salmonella* Heidelberg', *Avian Dis.*, 49(3), 382–6.

Gast, R. K., Guraya, R., Jones, D. R. and Anderson, K. E. (2013), 'Colonization of internal organs by *Salmonella* Enteritidis in experimentally infected laying hens housed in conventional or enriched cages', *Poult. Sci.*, 92, 468–73.

Gast, R. K., Guraya, R., Jones, D. R. and Anderson, K. E. (2014), 'Contamination of eggs by *Salmonella* Enteritidis in experimentally infected laying hens housed in conventional or enriched cages', *Poult. Sci.*, 93(3), 728–33.

Gast, R. K., Guraya, R., Jones, D. R. and Anderson, K. E. (2015), 'Persistence of fecal shedding of *Salmonella* Enteritidis by experimentally infected laying hens housed in conventional or enriched cages', *Poult. Sci.*, 94, 1650–6.

Gast, R. K., Guraya, R., Jones, D. R., Anderson, K. E. and Karcher, D. M. (2016), 'Colonization of internal organs by *Salmonella* Enteritidis in experimentally infected laying hens housed in enriched colony cages at different stocking densities', *Poult. Sci.*, 95, 1363–9.

Green, A. R., Wesley, I., Trampel, D. W. and Xin, H. (2009), 'Air quality and bird health status in three types of commercial egg layer houses', *J. Appl. Poult. Res.*, 18(3), 605–21.

Guesdon, V., Ahmed, A. M. H., Mallet, S., Faure, J. M. and Nys, Y. (2006), 'Effects of beak trimming and cage design on laying hen performance and egg quality', *Brit. Poult. Sci.*, 47(1), 1–12.

Guesdon, V. and Faure, J. M. (2004), 'Laying performance and egg quality in hens kept in standard and furnished cages', *Anim. Res.*, 53(1), 45–57.

Harisberger, M., Gobeli, S., Hoop, R., Dewulf, J., Perreten, V. and Regula, G. (2011), 'Antimicrobial resistance in Swiss laying hens, prevalence and risk factors', *Zoonoses Public Health*, 58(6), 377–87.

Haugh, R. R. (1937), 'The Haugh unit for measuring egg quality', *US Egg Poult. Mag.*, 43, 552–5, 572–3.

Hidalgo, A., Rossi, M., Clerici, F. and Ratti, S. (2008), 'A market study on the quality characteristics of eggs from different housing systems', *Food Chem.*, 106(3), 1031–8.

Holt, P. S., Davies, R. H., Dewulf, J., Gast, R. K., Huwe, J. K., Jones, D. R., Waltman, D. and Willian, K. R. (2011), 'The impact of different housing systems on egg safety and quality', *Poult. Sci.*, 90(1), 251–62.

Huneau-Salaün, A., Chemaly, M., Le Bouquin, S., Lalande, F., Petetin, I., Rouxel, S., Michel, V., Philippe, F. and Nicolas, R. (2009), 'Risk factors for *Salmonella enterica* subsp. *Enterica* contamination in 519 French laying hen flocks at the end of the laying period', *Prev. Vet. Med.*, 89(1–2), 51–8.

Huneau-Salaün, A., Michel, V., Huonnic, D., Balaine, L. and Le Bouquin, S. (2010), 'Factors influencing bacterial eggshell contamination in conventional cages, furnished cages and free-range systems for laying hens under commercial conditions', *Brit. Poult. Sci.*, 51(2), 163–9.

Jones, D. R. and Anderson, K. E. (2013), 'Housing system and laying hen strain impacts on egg microbiology', *Poult. Sci.*, 92, 2221–5.
Jones, D. R., Anderson, K. E. and Guard, J. Y. (2012), 'Prevalence of coliforms, *Salmonella*, *Listeria*, and *Campylobacter* associated with eggs and the environment of conventional cage and free-range egg production', *Poult. Sci.*, 91, 1195–202.
Jones, D. R., Anderson, K. E. and Musgrove, M. T. (2011), 'Comparison of environmental and egg microbiology associated with conventional and free-range laying hen management', *Poult. Sci.*, 90, 2063–8.
Jones, D. R., Cox, N. A., Guard, J., Fedorka-Cray, P. J., Buhr, R. J., Gast, R. K., Abdo, Z., Rigsby, L. L., Plumblee, J. R., Karcher, D. M., Robison, C. I., Blatchford, R. A. and Makagon, M. M. (2015), 'Microbiological impact of three commercial laying hen housing systems', *Poult. Sci.*, 94, 544–51.
Jones, D. R., Guard, J., Gast, R. K., Buhr, R. J., Fedorka-Cray, P. J., Abdo, Z., Plumblee, J. R., Bourassa, D. V., Cox, N. A., Rigsby, L. L., Robison, C. I., Regmi, P. and Karcher, D. M. (2016), 'Influence of commercial laying hen housing systems on the incidence and identification of *Salmonella* and *Campylobacter*', *Poult. Sci.*, 95, 1116–24.
Jones, D. R., Karcher, D. M. and Abdo, Z. (2014), 'Effect of a commercial housing system on egg quality during extended storage', *Poult. Sci.*, 93, 1282–8.
Jones, D. R. and Musgrove, M. T. (2007), 'Pathogen prevalence and microbial levels associated with restricted shell eggs', *J. Food Prot.*, 70(9), 2004–7.
Jones, D. R. and Musgrove, M. T. (2008), 'Assessment of microbial contaminants present on vacuum loaders in shell egg processing facilities', *J. Food Safety*, 28(3), 346–54.
Jones, D. R., Musgrove, M. T., Anderson, K. E. and Thesmar, H. S. (2010), 'Physical quality and composition of retail shell eggs', *Poult. Sci.*, 89(3), 582–7.
Jones, D. R., Musgrove, M. T., Caudill, A. B. and Curtis, P. A. (2006), 'Frequency of *Salmonella*, *Campylobacter*, *Listeria*, and *Enterobacteriaceae* detection in commercially cool water-washed shell eggs', *J. Food Safety*, 26, 264–74.
Karcher, D. M., Jones, D. R., Abdo, Z., Zhao, Y., Shepherd, T. A. and Xin, H. (2015), 'Impact of commercial housing systems and nutrient and energy intake on laying hen performance and egg quality parameters', *Poult. Sci.*, 94(3), 485–501.
Kijlstra, A., Traag, W. A. and Hoogenboom, L. A. P. (2007), 'Effect of flock size on dioxin levels in eggs from chickens kept outside', *Poult. Sci.*, 86(9), 2042–8.
Kone, A. Z., Jan, S., Le Marechal, C., Grosset, N., Gautier, M., Puterflam, J. and Baron, F. (2013), 'Identifying risk factors for eggshell contamination by *Bacillus cereus* group bacteria in French laying farms', *Brit. Poult. Sci.*, 54(3), 298–305.
Leasor, S. B. and Foegeding, P. M. (1989), '*Listeria* species in commercially broken raw liquid whole eggs', *J. Food Prot.*, 52, 777–80.
Mallet, S., Guesdon, V., Ahmed, A. M. H. and Nys, Y. (2006), 'Comparison of eggshell hygiene in tow housing systems: Standard and furnished cages', *Brit. Poult. Sci.*, 47(1), 30–5.
Mertens, K., Bamelis, F., Kemps, B., Kamers, B., Verhoelst, E., De Ketelaere, B., Bain, M., Decuypere, E. and De Baerdemaeker, J. (2006), 'Monitoring of eggshell breakage and eggshell strength in different production chains of consumption eggs', *Poult. Sci.*, 85(9), 1670–7.
Mitchell, D., Arteaga, V., Armitage, T., Mitloehner, F., Tancredi, D., Kenyon, N. and Schenker, M. (2015), 'Cage versus noncage laying-hen housings: Worker respiratory health', *J. Agromed.*, 20(3), 256–64.
Namata, H., Méroc, E., Aerts, M., Faes, C., Abrahantes, J. C., Imberechts, H. and Mintiens, K. (2008), '*Salmonella* in Belgian laying hens: An identification of risk factors', *Prev. Vet. Med.*, 83(3–4), 323–36.
Pieskus, J., Kazeniauskas, E., Butrimaite-Ambrozeviciene, C., Stanevicius, Z. and Mauricas, M. (2008), '*Salmonella* incidence in broiler and laying hens with the different housing systems', *J. Poult. Sci.*, 45, 227–31.
Rokonjac, S., Bogosavljević-Bošković, S., Pavlovski, Z., Škrbić, Z., Dosković, V., Petrović, M. D. and Petričević, V. (2014), 'Laying hen rearing systems: A review of chemical composition and hygienic conditions of eggs', *World's Poult. Sci.*, 70, 93–104.

Rousi, V., Madouvalou, S., Pasiotou, M., Diamantis, X. and Burriel, A. R. (2010), 'Investigating *Salmonella* serotypes colonizing laying hen housing across Southern Greece: Implications to public health', *J. Anim. Vet. Adv.*, 9(4), 841–3.

Schulz, J., Van Hoorebeke, S., Hald, B., Hartung, J., Van Immerseel, F., Radtke, I., Kabell, S and Dewulf, J. (2011), 'The dynamics of *Salmonella* occurrence in commercial laying hen flocks throughout a laying period', *Avian Pathol.*, 40(3), 243–8.

Schwaiger, K., Schmied, E. M. V. and Bauer, J. (2008), 'Comparative analysis of antibiotic resistance characteristics of gram-negative bacteria isolated from laying hens and eggs in conventional and organic keeping systems in Bavaria, Germany', *Zoonoses Public Health*, 55, 331–41.

Singh, R., Cheng, K. M. and Silversides, F. G. (2009), 'Production performance and egg quality of four strains of laying hens kept in conventional cages and floor pens', *Poult. Sci.*, 88(2), 256–64.

Snow, L. C., Davies, R. H., Christiansen, K. H., Carrique-Mas, J. J., Cook, A. J. C. and Evans, S. J. (2010), 'Investigation of risk factors for *Salmonella* on commercial egg-laying farms in Great Britain, 2004–2005', *Vet. Rec.*, 166(19), 579–86.

Sossidou, E. N. and Elson, H. A. (2009), 'Hens' welfare to egg quality: A European perspective', *World's Poult. Sci.*, 65, 709–18.

Spliethoff, H. M., Mitchell, R. G., Ribaudo, L. N., Taylor, O., Shayler, H. A., Greene, V. and Oglesby, D. (2014), 'Lead in New York City community garden chicken eggs: Influential factors and health implications', *Environ. Geochem. Health*, 36(4), 633–49.

Sulonen, J., Kärenlampi, R., Holma, U. and Hänninen, M. L. (2007), '*Campylobacter* in Finnish organic laying hens in autumn 2003 and spring 2004', *Poult. Sci.*, 86(6), 1223–8.

Swanson, J. C., Mench, J. A. and Karcher, D. (2015), 'The Coalition for Sustainable Egg Supply project: An introduction', *Poult. Sci.*, 94(3), 473–4.

USDA. (2000), 'United States standards, grades, and weight classes for shell eggs', AMS 56, https://www.ams.usda.gov/sites/default/files/media/Shell_Egg_Standard%5B1%5D.pdf (accessed 1 July 2016).

USDA. (2006), 'Regulations governing the voluntary grading of shell eggs', 7 CFR part 56, https://www.gpo.gov/fdsys/pkg/CFR-2006-title7-vol3/xml/CFR-2006-title7-vol3-part56.xml (accessed 1 July 2016).

Valkonen, E., Venäläinen, E., Rossow, L. and Valaja, J. (2010), 'Effects of calcium diet supplements on egg strength in conventional and furnished cages, and effects of 2 different nest floor materials', *Poult. Sci.*, 89(11), 2307–16.

Van Den Brand, H., Parmentier, H. K. and Kemp, B. (2004), 'Effects of housing system (outdoor vs cages) and age of laying hens on egg characteristics', *Brit. Poult. Sci.*, 45(6), 745–52.

Van Hoorebeke, S., Van Immerseel, F., De Vylder, J., Ducatelle, R., Haesebrouck, F., Pasmans, F., de Kruif, A. and Dewulf, J. (2010a), 'The age of production system and previous *Salmonella* infections on-farm are risk factors for low-level *Salmonella* infections in laying hen flocks', *Poult. Sci.*, 89(6), 1315–19.

Van Hoorebeke, S., Van Immerseel, F., Haesebrouck, F., Ducatelle, R. and Dewulf, J. (2011), 'The influence of the housing system on *Salmonella* infections in laying hens: A review', *Zoonoses Public Health*, 58(5), 304–11.

Van Hoorebeke, S., Van Immerseel, F., Schulz, J., Hartung, J., Harisberger, M., Barco, L., Ricci, A., Theodoropoulos, G., Xylouri, E., De Vylder, J., Ducatelle, R., Haesebrouck, F., Pasmans, F., de Kruif, A. and Dewulf, J. (2010b), 'Determination of the within and between flock prevalence and identification of risk factors for *Salmonella* infections in laying hen flocks housed in conventional and alternative systems', *Prev. Vet. Med.*, 94(1–2), 94–100.

Vits, A., Weitzenbürger, D., Hamann, H. and Distl, O. (2005), 'Production, egg quality, bone strength, claw length, and keel bone deformities of laying hens housed in furnished cages with different group sizes', *Poult. Sci.*, 84(10), 1511–19.

Wall, H. (2011), 'Production performance and proportion of nest eggs in layer hybrids housed in different designs of furnished cages', *Poult. Sci.*, 90(10), 2153–61.

Wall, H. and Tauson, R. (2002), 'Egg quality in furnished cages for laying hens – effects of crack reduction measures and hybrid', *Poult. Sci.*, 81(3), 340–8.
Wang, X. L., Zheng, J. X., Ning, J. H., Qu, L. J., Xu, G. Y. and Yang, N. (2009), 'Laying performance and egg quality of blue-shelled layers as affected by different housing systems', *Poult. Sci.*, 88(7), 1485–92.
Zhao, Y., Shepherd, T. A., Li, H. and Xin, H. (2015), 'Environmental assessment of three egg production systems – Part 1: Monitoring system and indoor air quality', *Poult. Sci.*, 94, 518–33.

Chapter 9

Egg washing to ensure product safety

Margaret Sexton, Primary Industries and Regions, South Australia (PIRSA), Australia

1 Introduction

Eggs are a unique animal food product with structures that pose specific problems for washing, especially in large commercial operations. However, a good understanding of the important parameters and problems for anyone washing eggs, from small manual egg washing operations to large highly mechanized systems, will ensure safer outcomes.

The pitfalls of poor egg washing can be seen in the fact that egg washing is not permitted for table eggs in the European Union, no doubt due to historical reports of increased egg spoilage being linked to poor egg washing (Hutchison et al., 2004). Elsewhere in the world, it is common practice to wash eggs and has been carried out in those countries in order to maximize the recovery rate of table eggs that are visually contaminated with faeces, dirt, litter, yolk, albumen and/or blood. With the push towards more free-range egg production, the proportion of eggs requiring washing is increasing, particularly with the use of nest pads that retain faeces and the problem of eggs being laid on the floor.

The aim of the washing process is to make the egg visually clean without damaging the shell (cracking) and also to ensure that the bacterial load on the shell is significantly reduced. It is paramount that the washing solution is not drawn into the egg. However, it must be understood that washing and sanitizing, even when implemented correctly, does not guarantee that there are no microorganisms in the shell or membranes or already inside the egg due to the conditions to which the egg has been exposed prior to washing.

http://dx.doi.org/10.19103/AS.2016.0012.10

This includes the contamination to which the egg is exposed from the time it is laid, the effectiveness of the cuticle (2.1), and the thermodynamic properties of the egg (2.6), which together can draw bacteria into the egg and, thus are then not accessible to the cleaning and sanitising agents. (Sparks and Board, 1985).

Effective egg washing requires an understanding of how the eggs are produced and processed, the food safety risks that can occur throughout production and how processing and handling can affect the quality and safety of the product. It also requires an understanding of the principles and problems of cleaning (and cleaning systems) and how the two interact.

2 Egg properties relevant to the washing process

2.1 Cuticle

It is a glycoprotein that plugs the majority of the pores of the shell and is the only effective measure to prevent microorganisms from entering the pores because the cracks on its surface are significantly less than 1 μm and are smaller than most bacteria (Sparks, 2016). However, the amount of glycoprotein produced varies considerably within a flock (and therefore a batch of eggs). As cleaning chemicals break up and remove protein, it also leads to some removal of the cuticle. Damage to the cuticle may also be sustained as the egg is laid, because the cuticle can take approximately three minutes to harden, during which time bacteria may enter the shell structure (Sparks, 1987; Sparks and Broad, 1985).

2.2 Porous shell

It is largely composed of calcium carbonate on an organic matrix and is covered with thousands of very small pores 15–65 μm in diameter. These pores make the surface extremely difficult to clean and are large enough to allow the passage of microorganisms, which are typically 1–5 μm, but very difficult to remove. Some of the calcium carbonate can be removed during the washing process and thus add calcium ions to the solution turning it into 'hard' water.

2.3 Membranes – inner, outer and limiting, beneath the shell

These are a matrix of fibres, but the inter-fibre space, even the more densely packed inner limiting membrane, is significantly greater in size than the dimensions of most bacteria (Sparks, 1985) and poses no barrier in preventing the movement of bacteria into the egg contents particularly if aided by the presence of moisture and at a temperature that is lower than the egg contents. The membranes may also provide protection for the microorganisms from the adverse conditions in the albumen.

2.4 Cracks

These are areas of the shell that may be very hard to detect but where the shell structure has separated, therefore allowing the washing solution direct access to the membranes underneath the shell. If the membranes rupture, the contents of the egg will leak into the wash solution, thereby greatly increasing the soil load because of the high protein and fat content.

2.5 Contents of the egg: albumen (protein) and yolk (fat)

The albumen contains biochemicals that have several protective mechanisms to prevent bacterial growth, particularly one that binds iron, a necessary element for the growth of bacteria, such as *Salmonella*. However, iron is readily available in the yolk, and *Salmonella* growth in the yolk can be very rapid at room temperature. Due to the presence of protein in the albumen, a suitable cleaning chemical at the correct concentration needs to be used to break up the protein structure. The high concentration of protein in the albumen means that the chemical can be depleted very quickly if a lot of albumen is present (such as with cracked, leaking eggs). The yolk requires a chemical that will emulsify the fats and suspend them in solution without causing a foam.

2.6 Thermodynamic properties of the egg

If the internal egg temperature is higher than the surrounding temperature, the contents will start to cool, contract and set up a negative pressure that draws anything on the surface of the egg into the internal contents, particularly when moisture is present (Hutchison et al., 2003). The temperature of the egg is 41°C when laid (matching the internal body temperature of the hen), and as the ambient temperature is usually lower, the egg begins to cool down immediately.

3 General cleaning principles

3.1 Time

The duration of the cleaning process requires sufficient time for the cleaning action to occur, which includes penetration of the soils by the cleaning chemicals and removal of the soils from the surfaces being cleaned.

3.2 Temperature

The cleaning chemical needs to be used in its optimum temperature range to maximize its effectiveness. However, it is important to understand how the optimum temperature for washing may affect the structure being washed.

3.3 Agitation

Agitation assists the washing solution to penetrate the soils, and it should cover all of the surfaces that need to be cleaned (e.g. hand scrubbing, high-pressure water, mechanical brushes).

3.4 Cleaning chemical

The chemical(s) selected must take into account the following:

- the properties of the surfaces being cleaned (e.g. porous, non-porous, reactive, non-reactive);
- the types of soils (e.g. protein, fat) being removed (requiring that all the soils are held in solution and easily rinsed off);
- the environment in which the washing is being carried out (e.g. enclosed or open system);
- the application method of the solution;
- the final disposal of the solution.

The water that is used to make up the chemical must be potable with the absence of any pathogenic microorganisms and organic matter. Suitable levels of minerals must also be considered, as some types of minerals may interfere with the activity of the chemical.

3.5 Chemical concentration

It is important that the concentration of the active chemicals is at an effective level at the surface of the structure that is to be cleaned (and sanitized), allowing a sufficient period of time for the chemical to carry out its activity at the operating temperature.

3.6 Chemical application

There are several methods for delivering the chemical to the required surface (e.g. foaming, high/low-pressure spraying, immersion), so it is important that the chemical selected is appropriate for the application method. There are also different ways of dosing the chemical – for example, powdered products by weight or volume and liquid products by weight or volume. This can be achieved by mechanical dosing using pumps controlled by conductivity probes, pulse dosing (by time) or flow meters.

3.7 Rinse

This step is required to remove the cleaning solution and any remaining soils.

3.8 Sanitizer

The sanitizer used needs to be capable of significantly decreasing the level of target microorganisms on the entire surface. It may be required to provide a complete covering of the active chemical for a period of time after application to ensure that the surface being sanitized is protected from recontamination.

3.9 Drying

Drying is a step that may be crucial to protect the surfaces from further contamination, especially porous surfaces. However, if the washing and sanitizing processes have not been effective, the use of forced air drying may result in aerosoling of organic matter and microorganisms into the environment, which may then cause recontamination.

3.10 Monitoring, process control and verification

The controls in each of the abovementioned categories must be monitored at frequent, regular intervals throughout the cleaning process to ensure that all the crucial parameters are maintained within acceptable levels. Monitoring checks need to be quick and effective, and adjustments need to be made immediately to ensure that the system is kept within the acceptable control limits. These assessments verify that the process achieves the required outcome(s) during production.

3.11 Validation

It is a process that provides evidence that the system is effective. It examines the purpose of the system, obtains evidence that the objective is achieved and usually involves scientifically approved methods. It is required initially to prove that the system is effective and any subsequent changes require revalidation of the system.

4 Cleaning principles applied to egg washing: time, temperature, agitation and use of chemicals

Given the unique parameters required to successfully wash eggs, it is important that these cleaning principles and their interactions are understood in order to apply them to egg washing. In the next three sections, the application of each cleaning principle to egg washing will be discussed.

4.1 Time

Time is a crucial aspect of egg washing. For all cleaning systems, there is a minimum cleaning time that depends on temperature, agitation and chemical concentration. These factors in return are determined by the types of chemicals used. If the cleaning period is too long, the temperature of the internal egg contents will rise to the level of the wash solution and increase the risk of facilitating the movement of surface fluid and bacteria into the egg.

Most commercial mechanized washing machines have an egg contact time within the washer of approximately 30 seconds. With such a contact time, there is an approximate 1°C rise in internal egg temperature. If the internal egg temperature is 15°C and the wash solution temperature at the egg surface is 45–47°C, this would ensure that the internal egg temperatures are easily kept below normal air-conditioned room temperatures (22–25°C) assisting in preventing the drawing of the wash solution into the contents of the egg, as described in Section 2.6.

With short washing times, all cleaning and sanitizing parameters must be optimized in order to achieve a significant improvement in both visual and microbial cleanliness.

Care must be taken with small manual operations, where eggs are submerged in a wash solution. Eggs should be submerged for a short period of time so that the internal egg temperatures do not rise to the temperature of the wash water.

4.2 Temperature

Temperature is another important parameter when washing eggs. The temperature of the wash solution must always be greater than that of the internal egg temperature, to prevent the solution being drawn into the egg due to negative pressure.

It is advantageous to have a processing system that ensures uniform internal temperatures of eggs, and this is usually achieved by placing fresh eggs into a controlled cooling environment. If eggs are conveyed directly from the sheds to the washing process, it is important to measure the internal egg temperatures arriving at the washing machine. The temperatures could vary from 41°C to the ambient shed temperature, or may even be affected by outdoor temperatures if they are conveyed a considerable distance. Regardless of the system, it is critical to measure the internal temperature of a range of eggs at the point of entry into the washing system in order to ensure that the correct wash solution temperatures are being used. These checks should be carried out throughout the year to monitor any fluctuations that may arise due to changing environmental conditions or other procedures.

It is important that the solution temperature is measured very close to where it contacts the egg surface.

Washing machines often have a thermometer located in either the solution inflow line or the recycle holding tank, but this may not accurately reflect the temperature of the solution at the cleaning surface because there is often a considerable distance between the spray nozzles and the eggs with brushes in between. In machines with rotating roller brushes, the temperature of the solution can drop up to 10°C from the nozzle to the egg surface. This may not only affect the performance of the cleaning chemical solution but also come perilously close to the internal temperature of the eggs. Eggs washed directly from the shed pose a greater hazard, because the internal egg temperature is 41°C when laid (and may remain close to that temperature if the eggs are being conveyed distances from the sheds and exposed to high ambient temperatures in hot countries).

Conversely, the opposite problem can develop if the temperature of the shell is reduced too low. For example, during winter, when the outdoor ambient temperatures can quickly lower the shell temperature, the temperature differential between the shell and the wash solution is too great, cracks in the shells may appear (Hutchison et al., 2001). This can lead to excessive wastage of table quality eggs or contamination of the eggs if the cracks are undetected. Wesley and Beane (1967) reported a significant increase in the number of thermal cracks if the temperature differential was greater than 10°C and although the general recommendation now is no more than 25–28°C temperature differential from shell to water temperature, it is likely that many factors influence the threshold level at which cracks begin to appear. This problem can be easily overcome by ensuring that a prewash solution is used at a lower temperature than the wash solution, and the shell is warmed up in stages to reduce cracking. This occurs in many commercial machines, where the overflow from the wash solution is used as the prewash. However, it is still critical to check that the temperature of the prewash solution is not less than the internal temperature of the eggs.

In sum, it is crucial to measure the temperature of the solutions in all sections of the washer at the surface of the egg, the internal egg temperatures and the temperature of the surface of the egg prior to entry into the washing machine.

Most washing machines aim to have a gradient of increasing temperatures throughout the process, ending with a higher temperature in the sanitizing stage to assist in the drying process (Hutchison et al., 2003).

This temperature differential applies equally to eggs that are washed manually, particularly with regard to the temperature of a rinse spray.

4.3 Agitation

As the washing time is minimal, agitation is an essential part of washing eggs, to improve the effectiveness of the washing solution at the interface between the egg and the soils.

Several methods of agitation are used in egg washing. Most commercial machines have rotating rollers with depressions, which cup each egg and rotate the egg so that all surfaces are exposed to the wash solution sprayed from above. In machines without brushes, spray jets are used to provide agitation, which assists the penetration and action of the wash solution. It is critical that the nozzles do not become blocked and the jets are correctly aligned to adequately cover all surfaces of the egg with adequate chemical. The pressure of the sprays is also important – if the pressure is too little the spray will be ineffective; if it is too much, the spray may damage the eggs. The distance from the spray jets to the eggs determines the actual pressure on the eggs and the effectiveness of the shape of the spray pattern.

Two types of brush actions are used in egg washing: rotating soft dense brushes that need to be close enough to the eggs to lap over the whole of the egg shape, and horizontal reciprocating brushes that are cut to the shape of the egg, contacting the ends, and applying an even pressure over the shape of the eggs. The height of the brushes must be adjustable, and the brushes need to be flexible to accommodate small eggs and not put excessive pressure on large eggs that may damage the shell. The brushes also need to be regularly examined for wear and adjusted or replaced to ensure optimal agitation. Sprays must not be blocked and must cover the width of the brushes to ensure that adequate chemicals are being applied to all eggs.

The cleanliness of the brushes is another indicator of the effectiveness of the washing solution.

In small manual operations where the eggs are submerged in the wash solution, agitation provided by bubbling air through the solution rather than by circulation of the solution is more effective. The eggs are best kept in a basket to limit their movement (and to minimize the risk of cracking).

4.4 Cleaning chemicals[1]

The correct selection of chemicals to clean egg surfaces is important due to the short contact time for cleaning and the difficulty of cleaning egg surfaces. To achieve this, the following properties of chemicals are required:

- They penetrate soils quickly (a surfactant).
- They break up protein such as albumen, blood and faeces.
- They emulsify fats from the yolk.

1. Please note that although egg shells are not a food as such, the chemicals used for washing and sanitizing egg shells should be 'recognized as generally safe' and in some countries may require an approval rating or process.

- They prevent minerals (such as calcium from the egg shells or minerals from hard water) from interfering with the detergent action (requiring a sequestrant) or from depositing and blocking the spray jets (requiring a chelating action).
- They hold all the above in suspension in the solution.
- They are free rinsing (leaving no residues).

Egg-washing chemicals must be effective in the temperature range of 40–60°C due to the reasons mentioned in Section 4.2.

Alkaline cleaning compounds are less aggressive on the egg shell than acid-cleaning compounds at their effective concentrations.

If the wash solution is reused the combination of chemicals in the solution also needs to be able to prevent the growth of *Salmonella* and other bacteria and help prevent the transfer of these organisms through the washing system (by being free rinsing). Chlorine is often used in these solutions and is an easy parameter to measure but requires the activity of all the other chemicals to be effective against *Salmonella* in these conditions. And thus if chlorine is being measured and is the first chemical to be depleted (due to its reaction with organic matter) then this builds a safety factor into the monitoring system.

4.5 Chemical concentration and application

To remain effective, chemicals need to be kept at their working concentration, and this requires the constant application of additional chemicals. This becomes difficult in many washing machines because the wash solution is recycled. The chemical concentration in the recycled water is dependent on the soil load coming in on the eggs, its consequent level in the solution and the rate of usage of the active chemicals. If the eggs are broken and the contents leak into the wash solution, the chemicals are deactivated in the recycled solution because the protein content of the albumen is very high. The best outcome is achieved if a new washing solution is applied directly to the eggs. The level of active chemical needs to be measured frequently to ensure that the required concentration level is maintained and the chemical use remains effective.

It is paramount that the measurement of the chemical is one that indicates the active levels of the cleaning chemical in the washing solution. Measurements such as pH and conductivity may be misleading as to the effectiveness of the cleaning activity. For example, a moderately buffered solution, diluted by half, can retain the same pH, and yet all the active chemicals are halved in strength.

Chemical dosing pumps that are controlled by a conductivity probe in the recycled solution can give misleading readings due to the build-up of salts (calcium) and soils in the solution. Volumetric dosing of the input water requires an accurate measuring of the flow rates with the chemical concentrations to establish effective dosing.

Recycling washing systems with no continual dosing system are very difficult to manage. Such systems must be closely checked and constantly dosed manually, and depending on the soil load, the solution must be regularly dumped and refilled.

Excessive foaming may be a problem associated with egg washing machines and can be due to the following causes:

- incorrect use of a foaming detergent for a spray application;
- use of some types of caustic-based detergents (e.g. sodium hydroxide) which may react with the fat from yolks to produce soap, known as a saponification reaction;

- levels of chemical actives, which are insufficient to break down protein in a peptizing reaction, resulting in a stable protein foam.

A solution for excessive foaming is to add defoamers, but ultimately, the root cause must be identified and corrected.

4.6 Sanitizer

The sanitizer used on the eggs needs to be effective at the temperature required for use and ideally needs to remain active on the eggs throughout the drying process, and right through the rest of the grading process to packing with the ability to reduce the target microorganisms, particularly *Salmonella*. The equipment in contact with the eggs after washing can become contaminated from the contents of broken eggs, and *Salmonella*-contaminated yolk material (from either contaminated shells or faecal material that has not been removed from the eggs in the washing process) will multiply and may contaminate more eggs. After washing, eggs may have increased susceptibility to contamination because the protective cuticle has been damaged, and the only antimicrobial control will be the activity of the sanitizer on the egg. (All egg contact surfaces should be effectively cleaned and sanitized after each production period to remove contamination).

The sanitizing solution must be applied to the entire egg surface. If the sanitizer is applied after the rinsing of the wash solution, it is also important to ensure that there is no over-spray from the rinse as this will dilute the concentration of the sanitizing solution on the egg. The temperature of the sanitizer solution at the surface of the egg can be up to 60°C (as it is applied for such a short duration) to facilitate drying.

5 Cleaning principles applied to egg washing: rinsing, drying and decontamination of equipment

5.1 Rinsing

There is no point in attempting to wash eggs unless the solution that contains all the soils and microorganisms is effectively rinsed off the surface of the egg. Potable water must be used for rinsing. A sanitizer can also be used in the rinse solution, provided that the sanitizer is compatible with the chemicals in the wash solution. The temperature of the rinse water should be slightly higher than that of the wash solution (at the surface of the egg) to ensure that the temperature gradients are maintained.

In some machines, there is insufficient separation or protection of the eggs in the rinse section from over-spray from the wash section, thus negating the effect of the rinse. This problem can be easily rectified by the insertion of a barrier in the washing unit to segregate the rinse section. If a machine does not have a rinse section, rinse sprays can be placed above the eggs just outside the spray cabinet.

If the rinse water drains into the recycled washing solution, it dilutes the chemicals in the washing solution. Dosing must be adjusted to maintain the concentration of the active ingredients in the washing solution.

In small manual operations, it is also important to rinse eggs effectively after washing in order to remove the wash solution of soils and chemicals. This can be achieved by hand-spraying the eggs, provided that all surfaces are adequately covered with potable water at a temperature higher than the washing solution.

5.2 Drying

Drying of eggs is usually carried out by the use of blowers, and some machines stop the rotation of the eggs for a period to enhance the effects of evaporation. However, Hutchison et al. (2003) states that it is difficult to dry eggs effectively in the short time available, due to the volume, speed and dryness of air required and the porosity of the shell.

However, drying is an important step because if the temperature of the moisture on the egg falls below the internal temperature of the egg, the moisture can then facilitate the movement of bacteria into the egg due to negative temperature differentials. This scenario is most likely to happen when eggs are washed straight from the shed environment (because of their potentially high internal temperatures that continue to cool through the packing process). This can be mitigated by the use of hot air in the driers to counteract the wind chill factor. Drawing air from outside, filtering it to remove microorganisms and dehumidifying it are also advantageous. Cleaning and sanitizing the air ducts regularly will prevent bacterial contamination.

5.3 Decontamination of equipment

All equipment coming into contact with eggs must be adequately cleaned and sanitized after each production period. Equipment that frequently contact the broken egg material should be cleaned more often – equipment such as suction cups and/or egg rotators that transfer the eggs from the incoming egg trays onto the washing machine conveyors. *Salmonella* will grow exponentially at room temperatures after three hours if lodged in these cups with yolk material. Therefore, ideally these should be removed every two hours and replaced with cleaned, sanitized and dried suction cup sets and rotators cleaned and sanitized. Equipment prior to egg washing is likely to be more contaminated. Egg trays used for collection from the sheds must be cleaned effectively, sanitized and dried before reuse, or if cardboard trays are used, they need to be discarded after a single use to prevent cross-contamination.

Unfortunately, much of the egg-processing equipment has not been designed to be cleaned with high-volume or high-pressure washing solutions and must be hand-cleaned and sanitized carefully with low-volume sprays after each production period. A sanitizer with residual activity should be used.

The washing machines must also be cleaned and sanitized after each production period and may require periodic descaling.

6 Monitoring, process control, verification and validation

6.1 Monitoring, process control and verification

The consequences of egg washing systems falling outside safe operating parameters are severe and can lead to spoilage and/or food safety risks. A washing system should only be used if it is carefully and regularly monitored, and appropriate adjustments are

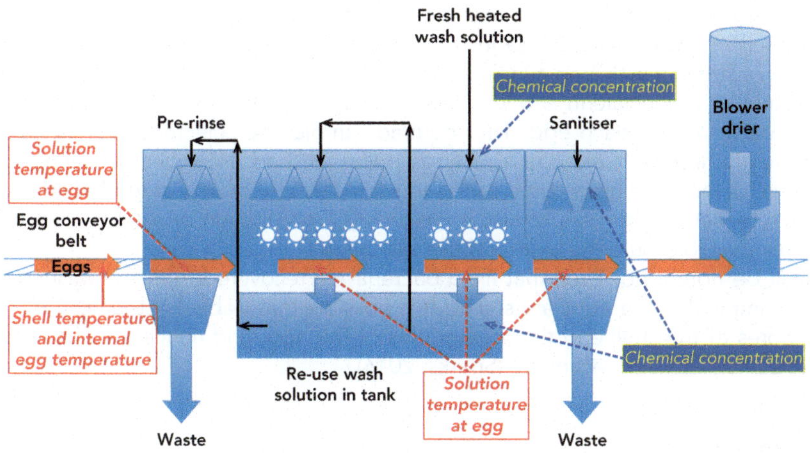

Figure 1 Schematic diagram of an egg washing machine showing the points at which temperatures and chemical concentrations need to be monitored.

made to maintain the system within safe operating boundaries (Fig. 1). Should these parameters be broken, immediate action must be taken to carefully assess the resultant risks and a decision made as to the disposition of the eggs likely to be affected. As a result, regular checks must be made to identify trends towards being out of specification. The use of alarms, particularly on solution temperatures, is a very useful tool. During operation, checks of solution temperatures, chemical concentrations and equipment (such as nozzle operation) are vital, especially with the use of recycled solutions. If in situ temperature probes are not close to the egg surfaces, the difference between the actual egg surface and the probe can be measured and adjusted, for example, the probe may need to read 55°C to achieve 45°C on the egg shell. Brush wear (where applicable) should be monitored and corrected (if required) at the end of each production period. It must be mentioned that the success for controlling these types of operations lies in the ability of the operator to easily and effectively make changes to the equipment.

Effective egg washing and sanitation relies on the accuracy of the controlling equipment and the training and instruction of the operator.

The checks carried out as described above can assist in verifying that the process is under control. Probes must be cleaned regularly and calibrated to ensure that they read correctly.

6.2 Validation

It is vital that validation is carried out for any egg washing system and should be done at least annually. However, validation may be required in winter and summer, and should be done when any changes are made to the system.

The effectiveness of an egg washing system can be validated by using a method that enumerates the bacterial population on egg shells prior to the washer and then after the washer. Statistical analysis can then be applied to determine whether the bacterial load has been significantly reduced by washing. A simple method uses 10 ml of sterile rinsate such as peptone water and massages each individual egg gently in the solution for two minutes in a sterile plastic bag. One millilitre of rinsate for total viable count (TVC) is plated onto Petrifilm (or equivalent) and incubated for two days at 35°C. This gives a count per millilitre of rinsate for each egg. The required sample size selected must be calculated to sufficiently represent the batch of eggs being washed. However, the sample size is often small if the eggs have come from the same farm, flock and date of collection, and there is a reasonable degree of visual uniformity. Consultation of a statistician is highly recommended to ensure the vigour of the validation.

It must be noted, however, that if no bacteria are recovered from the washed egg, it does not imply that the eggs are sterile. It is possible that viable bacteria can be lodged in the pores of the shell and the membranes of the egg, and hence are not able to be removed by this testing technique (Sparks, 2016).

7 Summary

The egg washing process should remove all the extraneous organic matter from the exterior surface of the shell, as well as *Salmonella*, which may have formed biofilms on the eggs' surface. However, egg washing should not damage the shell and preferably minimize the removal of the cuticle. To this end, great attention must be paid to the operation and maintenance of the washing system and process control.

It must be understood that the handling of eggs prior to washing may lodge food poisoning and food spoilage organisms deep in the shell pores and in the membranes of the egg, where they are protected from the washing and sanitizing processes.

It is crucial to ensure that any liquid coming into contact with the shell must be maintained at a higher temperature than the internal egg temperature as this will help to prevent the movement of bacteria (and solution) into the egg contents (Sparks, 2016).

8 Where to look for further information

Further information regarding washing equipment can be obtained from the commercial manufacturers of egg washing and grading machines. Information regarding chemicals and their application, use and effectiveness can be obtained from commercial chemical companies producing the chemicals specifically for the washing and sanitizing of table eggs.

The FSA (Food Standards Agency) have a report titled 'A review of published literature concerning commercial egg washing with particular emphasis on the control of Salmonella.' which is available through the following website and provides some very comprehensive information. http://tna.europarchive.org/20131001174833/http://www.food.gov.uk/science/research/foodborneillness/eggsresearch/b15programme/b15projects/b03017/.

The EFSA (European Food safety Authority) also published a document on egg washing as follows: Opinion of the Scientific Panel on Biological Hazards on the request from the Commission related to the Microbiological Risks on Washing of Table Eggs. *The EFSA Journal* (2005), 269, 1–39. http://www.efsa.europa.eu/en/scdocs/doc/269.pdf.

The USDA (United States Department of Agriculture), Egg Safety and Quality Research website has up-to-date information available regarding publications and research to be found at http://www.ars.usda.gov/main/site_main.htm?modecode=60-40-10-25.

A useful book titled *Improving the safety and quality of eggs and egg products Vol 2: Egg safety and nutritional quality*, Edited by Filip Van Immerseel, Yves Nys and Maureen Bain, Woodhead Publishing Ltd, 2011, can be found at http://store.elsevier.com/Improving-the-Safety-and-Quality-of-Eggs-and-Egg-Products/isbn-9781845697549/.

9 References

Hutchison, M., Gittins, J., Walker, A., Sparks, N. and Moore, A. 2001. A review of published literature concerning commercial egg washing with particular emphasis on the control of *Salmonella*. *Report for the Food Standards Agency.*

Hutchison, M., Gittins, J., Walker, A., Moore, A. Burton, C. and Sparks, N. 2003. Washing table eggs: a review of the scientific and engineering issues. *Worlds' Poultry Science Journal*. 59: 233–48.

Hutchison, M., Gittins, J., Walker, A., Sparks, N., Humphrey, T., Burton, C. and Moore, A. 2004. An assessment of the microbiological risks involved with egg washing under commercial conditions. *Journal of Food Protection*. 67: 4–11.

Sparks, N. and Board, R. 1985. *Australian Veterinary Journal*. 62: 169–70.

Sparks, N. 1987.The hen's eggshell: a resistance network. PhD., Bath University.

Sparks, N. 2016. Working with the egg structure to minimise salmonellosis. *Proceedings of the 27th Annual Australian Poultry Science Symposium*, 101–8.

Wesley, R. L. and Beane, W. L. 1967. The effect of various wash water and internal egg temperatures on the number of thermal cracks of oiled and un-oiled eggs. *Poultry Science*. 46: 1336.

Chapter 10

New developments in packaging of eggs to improve safety and quality

Pietro Rocculi, University of Bologna, Italy

1 Introduction

Various quality attributes of the albumen and yolk are lost as the egg ages (Stadelman and Cotterill, 1995). This happens because eggs are breathable material and allow moisture and CO_2 to permeate through the shell (Caner, 2005). The gradual evaporation of water causes a decrease in density (initially approx. 1.086 g/cm^3; the daily reduction coefficient is about 0.0017 g/cm^3) and air cell enlargement. The diffusion of CO_2, which starts soon after the egg is laid, causes a sharp rise in pH, especially in egg white (Belitz and Grosch, 1999). The pH of the albumen of a newly laid egg is between 7.5 and 8.8 and that of the yolk approximately 6 (Keener et al., 2000; Moran, 1937). As soon as the egg is laid, the pH of the albumen begins to increase as a result of CO_2 release by diffusion (Li et al., 1985; Fabbri et al., 2011). It is also important to note that the buffering capacity of the egg is minimal between pH 7.5 and 8.5 and during storage, the pH increases at a temperature-dependent rate to about 9.8 (Keener et al., 2000). In general, the rate of egg deterioration is closely related to the loss of CO_2 through the shell and to the alkalinity increase of the egg contents (Overfield, 1982). In addition, during storage, the viscosity of the egg white and yolk drops. The change of several properties, such as egg white whipping behaviour and foam stability, is critical for egg processing (Belitz and Grosch, 1999). In the EU, except for a few countries, for fresh egg preservation, stabilizing treatments such as those traditionally used in the United States by the food industry for sanitation (washing with aqueous solution of chemical agents) and microbial growth inhibition (refrigeration) are not permitted (Regulation (CE) no. 1028/2006).

Given these egg characteristics, the correct choice of packaging system is a key factor in order to maintain high-quality characteristics of the product during storage, both in terms

http://dx.doi.org/10.19103/AS.2016.0012.12

of type of packaging, gas permeability and interaction with water vapour released from the egg during storage.

The most used packaging systems for shell eggs all over the world present a very high permeability to air and vapour because of open/not sealed enclosure and/or the kind of material used such as moulded cartons with different percentages of paper pulp (from about 15% to 90%), expanded polystyrene or polyvinyl chloride (PVC).

Although the use of a low-barrier packaging material is not optimal in terms of minimization of the weight loss and of egg freshness, this system reduces the possibility of water condensation inside the packaging or on the shell surface, avoiding serious safety problems bound to the increased water activity promoted by water condensation.

The use of modified atmosphere packaging (MAP) could be a useful alternative to the traditional treatments commonly used for food stabilization, and it agrees with EU restrictions for fresh egg preservation. MAP consists of the modification of the atmospheric composition in the package headspace of a packed product.

Despite MAP with a high level of CO_2 having a high potential to improve the maintenance of freshness of the egg and to permit the maintenance/modulation of the technological properties of egg albumen, the possible condensation of water on the shell surface caused by the high barrier of the packaging cannot assure product safety.

A possible packaging solution in order to have the beneficial effects of CO_2 while avoiding water condensation is the combination of MAP with a water absorber/regulator. This could be done easily and with low cost just by choosing the right type of moulded carton.

2 Shell egg packaging in low- and high-barrier materials

The most used packaging for shell eggs all over the world presents a very high permeability to air and vapour because of the kind of material and/or not sealed enclosure, such as support of moulded carton with different percentages of paper pulp (from about 15% to 90%), expanded polystyrene or PVC (Koelkebeck et al., 2001).

The most important mass transfer phenomena characterizing fresh eggs during storage are the loss of CO_2 and water by transpiration (Caner, 2005), as shown in Fig. 1 for high-permeability packaging.

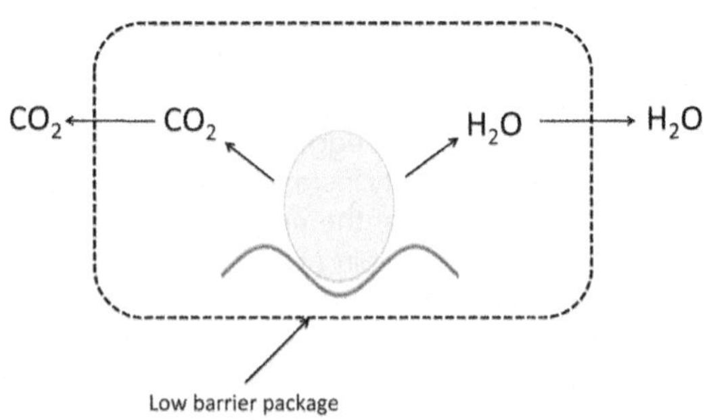

Figure 1 Schematic representation of mass transfer phenomena of fresh shell egg packed in low-barrier packaging material during storage.

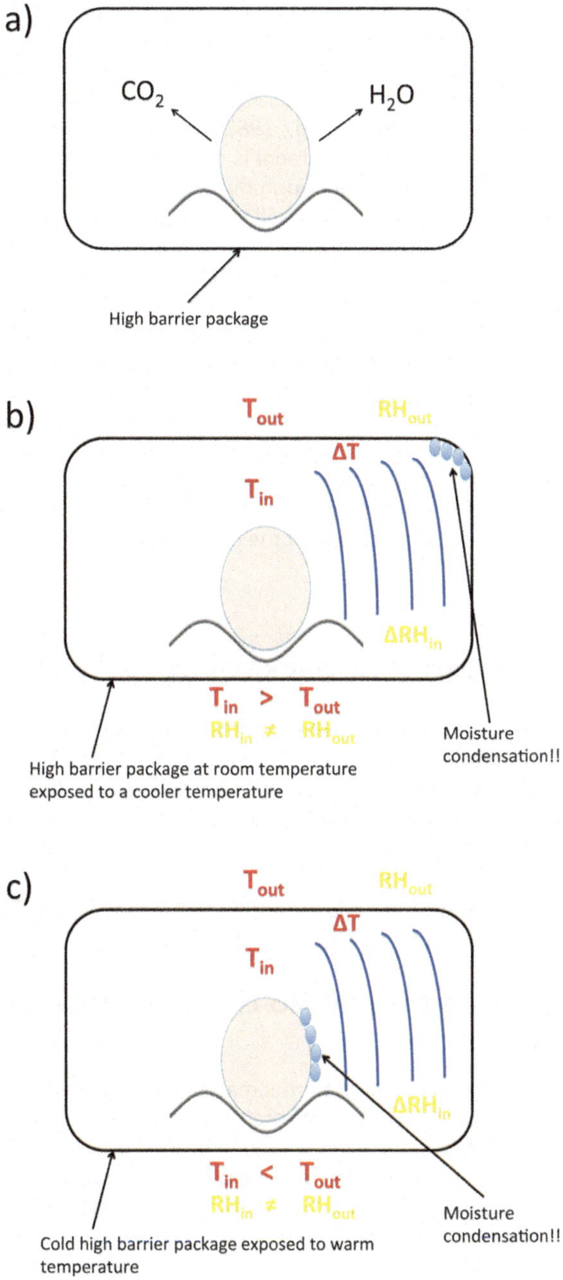

Figure 2 Schematic representation of mass transfer phenomena of fresh shell egg packed in high-barrier packaging material during storage (a) and when the package is exposed to lower (b) or higher (c) temperature.

During 28 days of storage at 25°C, fresh shell eggs packed in low-barrier material had a weight loss ranging from 6% to about 10% of the initial weight (Li et al., 1985; Bhale et al., 2003; Caner, 2005; Rocculi et al., 2009). The CO_2 loss through the shell promotes an increase in pH, particularly of the albumen, reaching a value of about 10 at the end of the storage period (Keener et al., 2000; Li et al., 1985; Rocculi et al., 2009, 2011).

The use of a low-barrier packaging material is not optimal in terms of minimization of the weight loss and of egg freshness, although this system reduces the possibility of water condensation inside the packaging or on shell surface, avoiding serious safety problems bound to the increase of water activity promoted by water condensation.

Even if the use of high-barrier packaging could promote a better maintenance of egg freshness and initial weight (Rocculi et al., 2009); in this condition, CO_2 and water vapour are released into the package headspace until a dynamic equilibrium with egg is reached (Fig. 2a). Unfortunately, the normal fluctuation of the storage temperature associated with the commercial life of the product (e.g. during distribution, sale, storage by the consumer) could promote water condensation on the internal surface of the packaging material (when the environmental temperature is lower than the egg temperature, for example when the package is placed in the refrigerator by the final consumer (Fig. 2b) or on the shell surface when it is removed from the refrigerator (Fig. 2c). When condensation occurs, egg safety could be seriously compromised because of the natural microbial load of the shell and the effect of increased water activity.

3 Modified atmosphere packaging (MAP) of shell eggs

The use of MAP could be a useful alternative to the traditional treatments commonly used for food stabilization, and it complies with EU requirements for fresh egg preservation. MAP consists of modification of the atmospheric composition in the package headspace of a packed product. It is a widely used food preservation technique, which minimally affects fresh product characteristics, resulting in consumer perception as a natural and additive-free technique (Rocculi et al., 2009).

In the following paragraphs, the most important effects of MAP with different gases on the freshness and on quality, stability and functional properties of shell eggs are reviewed.

3.1 Effect of MAP on the physico-chemical properties of shell egg

In comparison with eggs packed in high-barrier material in air or in MA, the greatest decline of interior egg quality was observed for the sample that was not packed, which showed the highest weight loss, pH increase and Haugh unit decrease. As mentioned, packaging eggs in high-barrier material reduces the egg weight loss during storage and limits Haugh unit decrease and pH increase. The substitution of the air in the package headspace with 100% N_2 did not provide any benefit in terms of physico-chemical characteristics of the product (Rocculi et al., 2009).

Eggs packed in 100% CO_2 maintained the initial values of Haugh unit for the whole storage period and the albumen pH of this sample was 1.5–2 points lower compared with the not packed product until the end of storage. This finding is of great importance from a technological point of view, considering that thermal coagulation and foaming ability of egg albumen are strongly related to its pH, as reported in the following paragraph. Nevertheless, using CO_2 atmosphere, the normally light yellow colour of the white progressively deepened during storage and the eggshell gradually developed a slight powdery surface (Rocculi et al., 2009).

3.2 Effect of MAP on the functional properties of the albumen

From a technological point of view, the egg may be considered a multifunctional ingredient because it can perform several functions by means of its coagulating, foaming, emulsifying, colouring, flavouring, anti-crystallizing and nutritional properties (Yang and Baldwin, 1995). Products of egg albumen are important food commodities because they are used in a vast number of food formulations (Liang and Kristinsson, 2007).

Heat coagulation is one of the most important functional properties of egg white (Hermansson, 1979). This phenomenon is a multistage process that plays a key role in determining the rheological and textural properties of products such as cakes, egg-based creams, confectionery and surimi seafood (Raikos et al., 2007). Coagulation is described as the random aggregation of already-denaturated protein molecules (Hermansson, 1979). Hydrogen bonds are broken, polypeptide chains are uncoiled and reactive groups are then exposed to denaturation (Powrie and Nakai, 1985).

In addition to coagulation, egg albumen proteins are responsible for different functional properties such as foaming, whipping and gelling properties (Hsieh and Regenstein, 1989). The foam properties are determined by the ability to rapidly adsorb at the air–liquid interface during whipping or bubbling and to form a cohesive viscoelastic film by way of intermolecular interactions (Mine, 1995).

Whereas a model interface can establish equilibrium for hours or days, a foam made with egg white proteins starts changing within minutes of formation (Foegeding et al., 2006). The excellent foaming capacity of egg white proteins and the stability of the resulting foams (even when subjected to heating) are applied in the food industry in the preparation of, for example, cakes and meringues (Van der Plancken et al., 2007). For these products, conditions become more complex because stability must be maintained during additional formulation and heat processing.

Rocculi et al. (2011) compared the functional properties (gel hardness, foam drainage and meringue crispness) of albumen of non-packed hen table eggs with those of eggs packed in high-barrier material with three different atmospheric conditions (air, 100% N_2 and 100% CO_2) during 28 days of storage at 25°C.

The results showed that the hardness of coagulated albumen exhibited a strict correlation with raw albumen pH, and 100% CO_2 MAP caused the formation of a soft and puffy coagulum (Fig. 3) with very low hardness for the entire storage period as a consequence of a rapid and marked pH reduction and the expansion of solubilized CO_2 during heating. The foam stability decreased during storage for all samples, but eggs packed in 100% CO_2 showed a more stable foam than fresh eggs until 15 days, reaching values similar to those of the other samples only at the end of the storage period.

In addition to the stabilizing effect on albumen foam, packing eggs in CO_2 improved meringue crispness compared with the other packaging conditions. The final product presented a more uniform and finer structure, showing that the CO_2 effect persisted after the formulation and processing needed for meringue preparation.

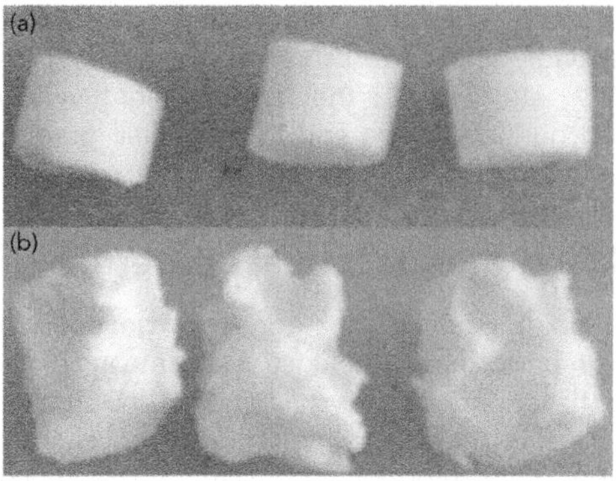

Figure 3 Samples of coagulated albumen of shell eggs packed in air (a) and 100% CO_2 MAP (b) after 3 days of storage at 25°C.

3.3 Effect of MAP on pathogen and spoilage bacteria

Despite the CO_2 MAP potentiality to improve the freshness maintenance of the egg and to maintain/modulate the technological properties of egg albumen, the possible condensation of water on the shell surface caused by the high barrier of the packaging cannot assure the safety of the product, as reported in the previous section.

Pasquali et al. (2012) monitored the effects of air, 100% CO_2 and 100% O_2 MAP on the survival of experimentally inoculated pathogen bacteria (*Salmonella Enteritidis, Escherichia coli* and *Listeria monocytogenes*) as well as on spoilage bacteria (total aerobic mesophilic bacteria) on table eggs during 30 days of storage at 4, 25 and 37°C using the colony count method. In general, temperatures rather than the different MAP played a major role in influencing bacterial survival. In particular, the lowest microbial loads were registered at 4°C for *E. coli* and spoilage bacteria, whereas 37°C was the best storage temperature to avoid the development of the psychrotropic microorganism *L. monocytogenes* regardless of the gas used. A total of 100% CO_2 packaging, in association with a low storage temperature (4°C), had a significant positive effect in reducing Salmonella loads. For eggs inoculated with *L. monocytogenes* and stored at 4°C as well as on eggs containing only spoilage bacteria and stored at 25°C, 100% CO_2 resulted in the best results in comparison with air and O_2. One hundred per cent CO_2 packaging showed no negative effect on pathogen survival compared to air. Pasquali and co-authors suggested control of RH within the package in order to limit bacteria growth/survival.

4 Case study: combination of CO_2 MAP and active packaging with humidity regulator for shell eggs storage

Active packaging is the inclusion of subsidiary constituents into the packaging material or the package headspace with the intent of enhancing the performance of the packaging system (Hernandez and Giacin, 1998).

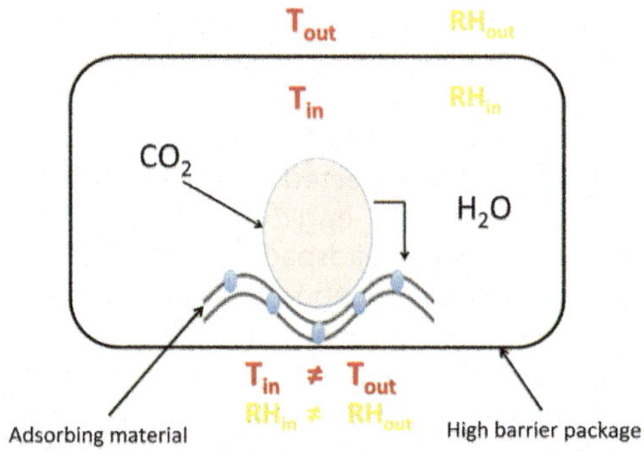

Figure 4 Schematic representation of mass transfer phenomena of fresh shell egg packed in MAP with 100% CO_2 with humidity regulator during storage.

A possible packaging solution in order to have the beneficial effects of CO_2 while avoiding possible safety risks promoted by water condensation on the shell surface can be achieved by the combination of MAP with high CO_2 level and an active packaging with humidity absorber/regulator.

This permits the simultaneous adsorption of CO_2 from the egg and of water vapour in the adsorbing packaging system (Fig. 4).

In order to test this solution, a preliminary research has been performed by our research group, testing the effect of the different packaging conditions on the physico-chemical and microbiological aspects of fresh eggs. In this study, eggs were packed with high-barrier packaging material in the following conditions, during 28 days of storage at 25°C: carton support, not packed (Control); carton support, packed in air (Air–C); plastic support, packed in air (Air–P); carton support, packed in 100% CO_2 (CO_2–C); plastic support, packed in 100% CO_2 (CO_2–P); carton support, packed in 100% CO_2 with silica gel as humidity regulator (CO_2–Abs–C) and plastic support, packed in 100% CO_2 with silica gel as humidity regulator (CO_2–Abs–P).

As expected, with increased storage time, the sample that was not packed showed increased weight loss, reaching values of about 6.5% of the initial weight at the end of the storage period (around 3.5–4 g for each egg). These results are in agreement with those of previous investigations (Rocculi et al., 2009, 2011). Small differences in weight loss among studies may be due to storage conditions, temperature, egg size and shell porosity. In contrast, all packed samples showed a similar and very small weight loss until the end of the storage period. Samples packed in air and in CO_2 with plastic supports showed the lowest weight loss, losing only 0.5% of the initial weight after 28 days of the experiment. The weight loss of samples Air–C and CO_2–C were higher compared with Air–P and CO_2–P. The absorbing effect of the moulded carton was evident. The Air–C group dehydration kinetic was very similar with that of the CO_2–Abs–P group, showing that, in this situation, the adsorbing effect of the carton support was comparable to that of the water absorber used (silica gel). Eggs packed with carton supports and water absorber showed the highest weight loss, reaching values of about 2% at the end of the experiment.

In terms of pH, the albumen of control sample showed an increasing trend from the beginning to the end of storage caused by CO_2 loss through the shell (Fig. 5a). Starting from fresh eggs with an average pH value of 8.6, the albumen of the control sample reached values of about 10 after 28 days of storage. The albumen of samples packed in air with both plastic and carton supports showed quite constant values of pH during storage, while the use of 100% CO_2 was responsible for a rapid and marked pH decrease (of about two points) as a consequence of CO_2 solubilization in the albumen, confirming results of previous investigations. As expected, this effect was evident also on the yolk. The absorption of the CO_2 gas by the egg contents takes place until equilibrium is established between the surrounding headspace and the CO_2 concentration within the egg. In particular, the equilibrium between egg white and CO_2 resembles the one between blood serum and this gas where the proteins react to the addition of acid by reversible combination with H^+. Observing the pH behaviour of CO_2-packed samples, it seems that this condition was reached in our experiments after 4–7 days of storage. Our results are in agreement with those obtained in previous studies, which demonstrated that the albumen pH is bound to the partial pressure of CO_2 in the surrounding atmosphere and ranges roughly from 9.7 in air to 6.5 in 100% CO_2 (Rocculi et al., 2009).

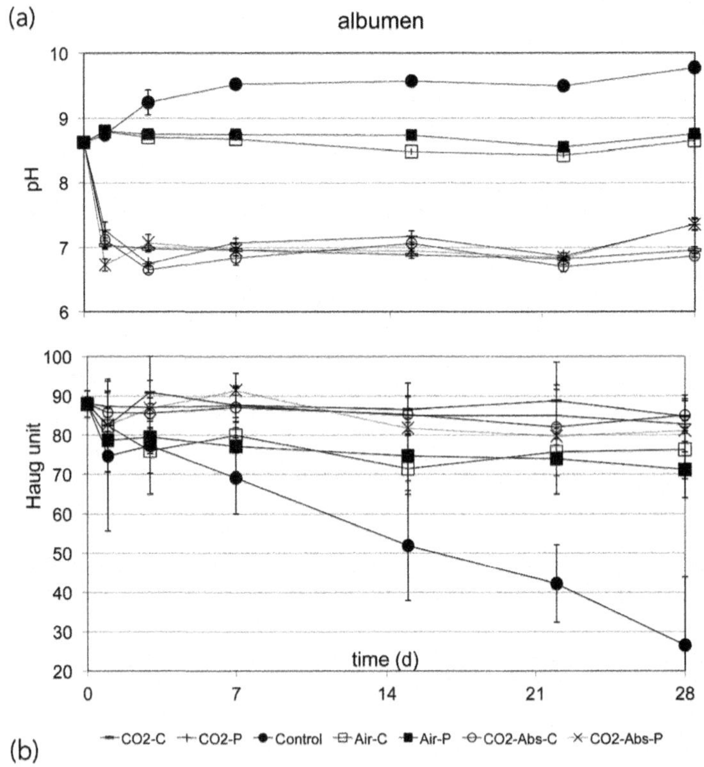

Figure 5 Albumen pH (a) and HU (b) values of differently packed eggs during 28 days of storage at 25°C (control: carton support, not packed; Air–C: carton support, packed in air; Air–P: plastic support, packed in air; CO_2–C: carton support, packed in 100% CO_2; CO_2–P: plastic support, packed in 100% CO_2; CO_2–Abs–C: carton support, packaged in 100% CO_2 with silica gel as humidity regulator and CO_2–Abs–P: plastic support, packaged in 100% CO_2 with silica gel as humidity regulator).

The results of albumen luminosity (L*) detected by a *tristimulus* colorimeter showed that MAP storage of eggs in 100% CO_2 promotes a significant decrease of this parameter, while L* decrease for the other samples were smaller and very similar. In general, results of colour of CO_2 samples indicate a shift of the albumen to a darker yellow and cloudy colour that, in our opinion, can be associated with the typical albumen colour of very fresh eggs, due presumably to the reduced swelling of the mucin fibres (Moran, 1937).

The average HU value of the fresh eggs used for our experiments (Fig. 5b) was about 88 corresponding to that of a fresh, good-quality egg. As expected, this value decreased rapidly for the control sample. HU reduction is due to decreased thick albumen height. Multiple hypotheses for this phenomenon have been proposed, including the breakdown of the ovomucin–lysozyme complex, decreasing carbohydrate content of ovomucin during storage and increasing pH. The exact mechanism of this albumen quality decrease with time is currently being investigated. All the tested packaging systems better preserved eggs in terms of HU compared to the control. Samples packed in air (Air–C and Air–P) showed a very similar decreasing trend and they maintained higher values for HU for the entire storage period. All samples packed in CO_2 maintained HU values similar to those of the fresh eggs for the whole storage period, confirming the beneficial effect of CO_2 for fresh egg maintenance.

As far as microbial growth is concerned (Table 1), control samples maintained the initial value of microbial load for the entire storage period, with no evidence of mould proliferation. In contrast, for both samples packed with plastic support (air and CO_2), microbial load reached values higher than 10^5 UFC g^{-1} after 15 days of storage and moulds growth was observed after just 8 days. Compared to these samples, Air–C and CO_2–C showed lower values of microbial growth and moulds were visible only on shell samples of eggs stored for 28 days, probably because of the water absorption performed by the carton supports. Confirming this statement, in samples with water absorbers, microbial growth was very low, similar to that of the samples that were not packed, and mould growth was never detectable. In general, for all the samples, the packaging with CO_2 did not promote any additional benefit in terms of microbial growth inhibition.

The water-absorbing capacity of different types of moulded carton in very high RH conditions (c.a. 98%) has been assessed (Fig. 6); the different samples evidenced different

Table 1 Total microbial load values detected on egg shell during 28 days of storage at 25°C

	Total microbial load* (Log CFU g^{-1})						
Time (d)	Control	Air–P	Air–C	CO_2–P	CO_2–C	CO_2–Abs–P	CO_2–Abs–C
0	3.7 (±0.7)	3.7 (±0.7)	3.7 (±0.7)	3.7 (±0.7)	3.7 (±0.7)	3.7 (±0.7)	3.7 (±0.7)
7	3.0 (±0.3)	5.5‡ (±0.3)	3.8 (±0.7)	3.7‡ (±0.7)	3.3 (±0.5)	3.1 (±0.4)	3.0 (±0.1)
15	2.7 (+0.2)	5.8‡ (±0.5)	5.0 (±0.4)	5.5‡ (±0.3)	4.1 (±0.8)	2.3 (±0.2)	2.9 (±0.6)
28	2.1 (±0.5)	5.6‡ (±0.4)	4.7 (±0.2)	5.2‡ (±0.1)	4.7‡ (±0.8)	3.0 (±1.2)	2.0 (±0.5)

*Average ± SD.
‡Visible moulds on the shell surface.

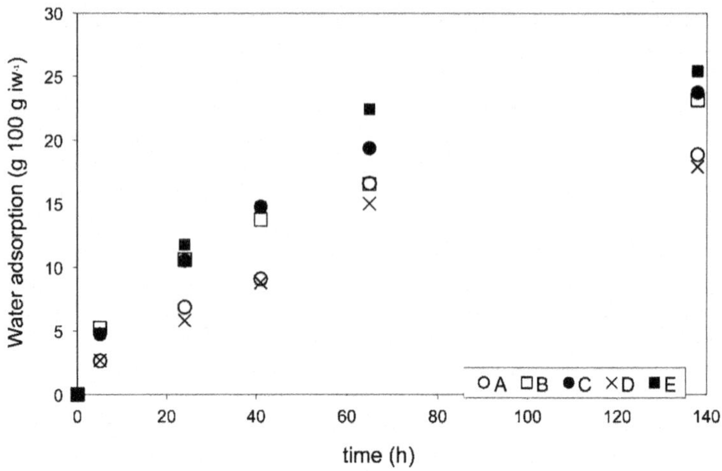

Figure 6 Adsorbing capacity at 98% RH of different kinds of moulded carton for egg packaging.

values of water adsorption, probably because they were made with different percentages of paper pulp (according to Koelkebeck et al. (2001), they can range from about 15% to 90%) and/or processing technologies.

From a practical point of view, for egg MAP, the kind/quantity of the moulded carton support has to be chosen considering its absorbing capacity; this could avoid the use of a specific humidity regulator, such as salt pouches inside the packaging, that could be perceived negatively by the consumer and increase the cost of the final product.

5 Future trends in research

The application of the innovative packaging system proposed can ameliorate the loss of quality characteristics of eggs during storage and maintain their functionality particularly for the production of albumen-based food products. Commercially, CO_2 MAP with water absorber could permit the diversification of the egg constituents characteristics during storage, providing to the market shell eggs with tailored functional properties connected to the quality of specific derived products (e.g. 'special fresh shell eggs for meringue preparation'). This application could give an added value to shell eggs, both in terms of producer profits and consumer needs.

It is important to underline that in the European Union eggs are marketed at room temperature, whereas in the United States eggs are kept at 7.2°C after washing, grading and packing. The higher storage temperature promotes a faster loss of egg freshness and of the technological properties of its constituents. Under these circumstances, the application of the proposed innovative packaging system represents a protective system against disruption of the cold chain or temperature abuse or both while promoting the improvement of the quality maintenance of the product during transport, retail and domestic storage. To better understand egg constituent modifications linked to CO_2 absorption, further investigations are currently underway in our laboratory to model CO_2

distribution in shell eggs at different CO_2 partial pressures and storage temperatures and to fully investigate the consequent albumen and yolk chemical changes and the modification of their functional effect in different formulated foods.

6 Where to look for further information

International research projects

RESCAPE—*Reducing Egg Susceptibility to Contaminations in Avian Production in Europe*. Sixth Framework Programme, Priority 5, Food Quality and Safety. Contract no. FOOD-CT-2006-036018, Specific Targeted Research Project. Periodo: 2006-2009. Responsabile: Prof. Nys Y. Partner: UNIBO.

Research centres

Department of Agri-Food Science and Technology, *Alma Mater Studiorum*, University of Bologna, Campus of Food Science, Piazza Goidanich 60, 47023 Cesena (FC), Italy.
Website: www.distal.unibo.it
Inter-Departmental Centre for Agri-Food Industrial Research (CIRI Agroalimentare), University of Bologna, piazza Goidanich, 60, I-47521 Cesena (FC), Italy.
Website: http://www.agroalimentare.unibo.it/ciri-agroalimentare

7 References

Belitz, H. D. and Grosch, W. (1999), *Food Chemistry*, New York: Springer-Verlag.

Bhale, S., No, H. K., Prinyawiwatkul, W., Farr, A. J., Nadarajah, K. and Meyers, S. P. (2003), 'Chitosan coating improves shelf life of eggs', *J. Food Sci.*, 68, 2378–83.

Caner, C. (2005), 'The effect of edible eggshell coatings on egg quality and consumer perception', *J. Sci. Food Agric.*, 85, 1897–902.

Fabbri, A, Cevoli C., Cocci, E. and Rocculi, P. (2011), 'Determination of CO_2 mass diffusivity of egg components by finite element model inversion', *Food Res. Int.*, 44, 204–8.

Foegeding, E. A., Luck, P. J. and Davis, J. P. (2006), 'Factors determining the physical properties of protein foams', *Food Hydrocoll.*, 20, 284–92.

Hermansson, A. M. (1979), *Functionality and Protein Structure*, in E. A. Pour (ed.), ACS Symposium Series No. 92. American Chemical Society, Washington, DC.

Hernandez, R. J. and Giacin, J. R. (1998), 'Factors affecting permeation, sorption, and migration processes in package-product systems', in I. A. Taub and R. P. Singh (eds), *Food Storage and Stability*, CRC Press, Boca Raton, FL, chap. 10.

Hsieh, Y. L. and Regenstein, J. M. (1989), 'Texture changes on heating spray-dried egg white', *J. Food Sci.*, 54, 1206–8.

Keener, K. M., Lacrosse, J. D., Curtis, P. A., Anderson, K. E. and Farkas, B. E. (2000), 'The influence of rapid air cooling and carbon dioxide cooling and subsequent storage in air and carbon dioxide on shell egg quality', *Poult. Sci.*, 79, 1067–71.

Koelkebeck, K. W., Bell, D. D., Carey, J. B. K., Anderson, K. E. and Darre, M. J. (2001), 'Egg marketing in national supermarkets: products, packaging, and prices – Part 3', *Poult. Sci.*, 80, 396–400.

Liang, Y. and Kristinsson, H. G. (2007), 'Structural and foaming properties of egg albumen subjected to different pH treatments in the presence of calcium ions', *Food Res. Int.* 40, 668–78.

Li, L. Y., Lai, C. C. and Gilbert, S. G. (1985), 'Keeping quality of eggs packaged in acrylonitrile pouches', *J. Food Proc. Pres.* 9, 179–87.

Mine, Y. (1995), 'Recent advances in the understanding of egg white protein functionality', *Trends Food Sci. Technol.*, 6, 225–32.

Moran, T. (1937), 'Gas storage of eggs', *J. Soc. Ind. Chem.* 56, 96T.

Overfield, N. D. (1982), *Quality Testing of Eggs. Reference Book 428*, Ministry of Agricultural, Fisheries and Food, London.

Pasquali, F., Manfreda, G., Olivi, P., Rocculi, P., Sirri, F. and Meluzzi, A. (2012), 'Modified-atmosphere packaging of hen table eggs: effects on pathogen and spoilage bacteria', *Poult. Sci.*, 91, 3253–9.

Powrie, W. D. and Nakai, S. (1985), *Food Chemistry*, in O. R. Fennema (ed.), Marcel Dekker, New York, NY.

Raikos, V., Campbell, L. and Euston, S. R. (2007), 'Effects of sucrose and sodium chloride on foaming properties of egg white proteins', *Food Res. Int.*, 40, 347–55.

Rocculi, P., Cocci, E., Sirri, F., Cevoli, C., Romani, S. and Dalla Rosa, M. (2011), 'Modified atmosphere packaging of hen table eggs: effects on functional properties of albumen', *Poult. Sci.*, 90, 1791–8.

Rocculi, P., Tylewicz, U., Pekoslawska, A., Romani, S., Sirri, F., Siracusa, V. and Dalla Rosa, M. (2009), 'MAP storage of shell hen eggs, Part 1: effect on physico-chemical characteristics of the fresh product', *LWT-Food Sci. Technol.*, 42, 758–62.

Stadelman, W. J. and Cotterill, O. J. (1995), *Egg Science and Technology*, Food Products Press, New York.

Van der Plancken, I., Van Loey, A. and Hendricks, M. (2007), 'Effect of moisture content during dry-heating on selected physicochemical and functional properties of dried egg white', *J. Agric. Food Chem.*, 55, 127–35.

Yang, S. and Baldwin, R. E. (1995), *Egg Science and Technology*, 4th ed., in W. J. Stadelman and O. J. Cotterill (eds), The Haworth Press, Binghamton, NY.

Part 3

Sensory and nutritional quality

Chapter 11

Egg quality: consumer preferences and measurement techniques

Bart De Ketelaere, Katholieke Universiteit Leuven, Belgium; Koen De Reu, Institute for Agricultural and Fisheries Research (ILVO), Belgium; and Steven Vermeir, Katholieke Universiteit Leuven, Belgium

1 Introduction

This chapter deals with egg quality – how consumers perceive egg quality, and how instrumental techniques offer the possibility of measuring those quality characteristics that are important to them. The structure of this chapter is as follows. In the first section, the perception of egg quality by consumers is discussed, as well as their preferences. In the subsequent sections, instrumental techniques able to assess those quality attributes in an objective way are discussed. We focus on techniques that are non-destructive and fast, since they can be used online on commercial grading machines that sort up to 200000 eggs an hour. Besides, as there are several recent papers and books on egg quality and the measurement thereof, the focus of this chapter is on recent developments in instrumental techniques. The use of instrumental techniques is split up according to their application. The first section deals with the eggshell quality, both from a mechanical point of view (presence of cracks, strength) and from its visual appearance (dirt, colour). The next section gives an overview of techniques that can be used to determine the internal quality of the egg. This embraces a multitude of aspects such as the freshness which relates to the viscosity of the albumen, odour and air cell height. The last section on instrumental techniques measuring egg quality is devoted to the detection of anomalies in the egg, such as blood spots and bacterial infection. In our concluding section, we map consumer preferences to current state-of-the-art sensing technologies and define some future perspectives for research.

http://dx.doi.org/10.19103/AS.2016.0012.13

2 Consumer perceptions of egg quality and consumer preferences for eggs

2.1 General consumer surveys

The physical appearance and packaging of an egg make the first impression upon the consumer. If the product does not meet perceived expectations, consumer confidence diminishes (Jones and Musgrove, 2005). The most readily perceived basic quality characteristics valued by the consumer include visual attributes and flavour, typically associated with egg freshness for example, a thick, cohesive egg white and a firm yolk that does not easily rupture (Burley and Johnson, 2013; USDA-Agricultural Marketing Service, 2000). In a consumer survey of 1999 with the objective of validating and ranking food quality attributes proposed by the European Consumer Association (BEUC), consumers from Spain and Germany indicated safety, hygiene and freshness as the most important quality factors, followed by ethical aspects, animal feeding and nutritional value (Hernandez et al., 2000; Hernandez, 2005). Other surveys conducted in France, Germany, Italy, UK, Spain, Poland and Greece found that consumers defined egg quality more through tangible characteristics of the egg; most especially shell strength, albumen consistency and intense yolk colour (Hernandez et al., 2000). Differences between countries were found. Shell strength was considered as more important (26–65% of responses) compared to shell cleanliness (17–30% of responses). Concerning the albumen, the consensus of all country responses was that albumen consistency and clarity characterize quality albumen. Haugh unit (HU) is the standard for quantifying interior egg quality and freshness (see Section 4). An earlier study carried out in England (Fearne and Lavelle, 1996b) pointed out that freshness, no cracks and good value for money were the most important factors taken into consideration by consumers when purchasing eggs. The most important attributes in the formation of consumer preferences towards eggs found by Mesias et al. (2011) were price (30.1%), rearing conditions (19.7%), feed (19.2%) and the choice between different egg sizes (14.4%). Results of a pilot survey in five European countries found that freshness, clear labelling and free range are attributes that can be considered as principal warranties of safety (Hansstein, 2011). Based on focus group discussions and questionnaires, Jibir et al. (2012) found quality attributes in order of importance: egg size, shell cleanliness, shell colour uniformity in tray and shell texture. The respondents, students of agriculture in Nigeria, expected clean eggs to signify hygienic, freshness and safety, while dark-coloured eggs are expected to be more nutritious and to have a stronger shell.

When exploring the drivers of purchase intent and the US consumer satisfaction of hen eggs and quality attributes, Wardy et al. (2014a) found that freshness, absence of cracks, sale price and label stating 'packaging/best-before-date' were key attributes affecting the purchase intent of the consumer. Must-be attributes were also shell cleanliness, USDA-certified farm eggs, egg grade, secure packaging and availability. Finally, egg size was also classified as an attractive attribute. Yolk colour, shell colour, shell thickness, organic egg and nutrient-fortified egg were considered as less important.

2.2 Yolk colour preferences

When offering egg samples with different yolk colour, a majority of the surveyed persons preferred the darkest of the yolk colours presented (corresponding to colour 14 in the

Roche Fan) (Hernandez et al., 2000; Hernandez, 2005). Similarly, in the study by Bejaei et al. (2011), most respondents in British Columbia (Canada) indicated yolk colour as an important quality parameter. Most of the respondents (67.3%) preferred darker yolk colour, whereas 18.8% were not sure which yolk colour they preferred and 13.9% did not like darker yolk colour. In addition, a relationship existed between the preference of dark yolk colour and the type of egg that consumers used. Consumers of white regular eggs preferred lighter yolk colour, but consumers of brown regular eggs preferred darker yolk colour. The respondents who consumed free-range eggs preferred darker yolk colour. Consumers of organic eggs did not appreciate a lighter yolk colour. Of particular interest were the reasons provided for a darker yolk preference. Of all the respondents, 43.5% stated that their preference was because eggs with darker yolks taste better; 35.9% indicated that eggs with darker yolk colour have better visual effects; 28.3% believed that hens producing eggs with darker yolks are better fed; 10.5% thought these hens are raised under better animal welfare conditions, 20.9% believed that eggs with darker yolks have better nutritional value; 18.9% thought darker yolks provide better health and 12.8% believed that eggs with darker yolks are locally produced. Although many respondents in the study by Bejaei et al. (2011) indicated that darker yolk colour results in better taste, results of a blind taste test of Fearne and Lavelle (1996a) indicated that consumers were not able to distinguish any significant difference in the taste of different types of eggs and yolk colours. Similarly, Bejaei et al. (2011) indicated that the Canadian respondents reported that brown regular and specialty eggs are tastier than white regular eggs, while in a blind test by Fearne and Lavelle (1996a), the consumers could not make a distinction between the taste of regular and specialty eggs. In a Turkish study by Mizrak et al. (2012), again the majority of households (81.2%) preferred dark-coloured egg yolks, while 8.74% preferred light-coloured ones. The most important factor in the Turkish household's desire for dark-coloured yolks was the belief that eggs with such yolks are more delicious (50.7%) and have a high nutritional value (37.4%); the rate of those who preferred them from a visual perspective was only 10.3%. Jacob et al. (2000) also stressed the importance of the yolk colour as quality parameter. The group found that consumer preferences especially for yolk colour were very subjective and varied widely from country to country. Moreover, besides the yolk colour being a key aspect in egg quality, the composition of the colour, texture, firmness and smell of the yolk are also important (Coutts and Wilson, 2007; Hernandez et al., 2009).

2.3 Eggshell colour preferences

The complex nature of eggshell colour is still under investigation, and the fact that certain hens lay brown eggs, others lay white and some lay even blue-shelled indicate the high heritability of shell colour (Froster et al., 1996; Francesh et al., 1997; Zhang et al., 2005). Studies showed that the brown colouration of the shell is an important shell quality parameter and has a positive influence on consumers' preferences (Samiullah et al., 2015). In Europe and Asia, many markets sell brown eggs. Among those markets, variation exists in preference for intensity of shell colour (Arthur and O'Sullivan, 2005). In Japan, uniform dark brown eggs are most valued, whereas in other markets, a light-tinted brown egg is more acceptable. Uniformity of shell colour among eggs in the same package appears to be an important characteristic as well. According to consumers' opinion in Canada, brown eggs also have higher nutritional value than white regular eggs (Bejaei et al., 2015). In a study on the acceptance of brown-shelled eggs in a white-shelled egg market in the United States, white eggs were preferred over brown eggs; however, brown eggs gained in acceptance but did not exceed white as white eggs are considered to be more nutritious or flavourful and fresher. Panellists also preferred the appearance of light-tinted brown eggs to that of darker brown eggs (Johnston et al., 2011). In the study by Mizrak et al. (2012), 34% of the Turkish consumers were found to prefer white-shelled eggs, while 30% preferred brown-shelled eggs, and 35% of the consumers did not pay attention to the colour of the shell. However, a great proportion (70%) of those who stated a preference for white eggs insisted on their choice.

While studying the possibilities of edible eggshell coatings on the quality of eggs during storage (weight loss, pH, HU, yolk index and yolk and shell colour), different authors found that shell colour is a critical factor affecting the acceptance of eggs (Wardy et al., 2014b; Caner, 2005). Surface smoothness and glossiness of eggshell increase purchase intent from 61% to 72–77% (Wardy et al., 2014b).

2.4 Egg size preferences

In the study by Bett et al. (2013) on consumers' willingness to buy indigenous eggs, it was shown that besides eggshell and yolk colour, egg size is also an important factor. Consumer preferences on egg size were also studied by Mizrak et al. (2012). More than half of the respondents (58.6%) stated a preference for medium eggs and more than one-third (36%) for larger eggs, whereas only a small proportion (3%) stated a preference for small eggs. In the study by Grijspeedt et al. (1999), it was concluded that the majority of eggs sold in Belgium were medium-sized ones (43%).

2.5 Other egg quality preferences

Thickness of the eggshell is considered as important to the consumer. More than half (61%) of the Turkish households preferred eggs with shells having a moderate thickness, while 24.3% opted for thick-shelled eggs and 14.8% preferred thin-shelled ones (Mizrak et al., 2012). On the other hand, poor shell quality will result in a greater number of cracks, which will result in a lowering of consumer confidence. The study by Mizrak et al. (2012) also showed that egg packaging influenced the decisions of 67.6% of the consumers. The same study also showed that consumers (64.4%) paid most attention to the production date. Remarkably, 72.4% of the families were not knowledgeable about organic eggs. In a consumer survey in Belgium in 1998, it was found that 88% of the consumers bought brown eggs and, comparable to the previous mentioned study (Mizrak et al., 2012), the majority of the people claimed to notice the shelf life (75.2%), which is important in the framework of risk analyses (Grijspeedt et al., 1999). Also, the absence of blood spots and meat spots on the yolk are considered as important quality aspects (De Ketelaere et al., 2004a). In some cultures, because of religious laws, the consumption of blood-spotted eggs is even unacceptable (Johnston et al., 2011).

2.6 Consumer perceptions of animal welfare concerns

Sales of eggs that are differentiated from conventional eggs by the circumstances used for raising hens have increased during the last decade. Studies on consumer perceptions and preferences on this changing habit of egg consumption are also available. Animal welfare is becoming increasingly important. European laws have recognized that farm animals can feel, experience and suffer. These laws are now the basis for animal welfare standards in many countries. For eggs, this means that consumers may base their decisions on whether eggs are coming from hens in cages or in non-cage systems. In the study by Heng et al. (2013), most US respondents perceived caged housing or other conventional management practices as reducing hens' welfare and were willing to pay a premium for eggs produced in cage-free and in other non-conventional production systems. Respect for animal welfare, in the mind of the consumer, is not only a quality but also a safety cue. As several food scares have shown, safety concerns can lead to a rapid decrease in the level of consumption for certain foods, seriously affecting producers (Hansstein, 2011). For a certain segment of egg consumers, animal welfare is not only important, but even fundamental for their choice (Grunert, 2005). Animal welfare is linked to sustainable production. Consumers prefer to buy free-range eggs because they feel they are doing something good and valuable (not only for the environment, but also for themselves) (Hansstein, 2011). Using models, Bejaei et al. (2015) found that consumers of regular eggs were less educated, older, price sensitive, major chain store buyers and store flyer users. They also had a lesser awareness about the different types of eggs and less concern regarding animal welfare issues. On the other hand, most of the non-caged egg consumers were less concerned about the price, had higher awareness about the different types of table eggs, purchased their eggs from local/organic grocery stores, farm gates or farmers markets and were more concerned about care and feeding of hens than consumers of other egg types. The same Canadian research group also believed that most consumers felt that free-run, free-range and organic eggs had higher nutritional value than regular eggs (Bejaei et al., 2011). Andersen (2011) found that people were willing to pay a higher premium for organic eggs, which was attributed to consumers perceiving organic eggs as healthier and to their familiarity with the organic label.

2.7 Consumer perceptions of eggs and health

The human health benefits offered by food with nutritionally added properties are also a consumer interest. Among the different products delivering essential nutrients to the body, an egg has arguably a special place, being a rich and balanced source of essential amino and fatty acids as well some minerals and vitamins. Advances in technology can improve the nutritional quality of eggs, resulting in functional foods or more popular 'health foods'. The benefits offered by those 'enriched eggs' responds the consumers' worries for their health. Omega-3-enhanced eggs (which also have increased levels of vitamin E) and enhancing levels of anti-oxidants and selenium are some examples. A conjoint analysis of consumer preferences (Mesias et al., 2011) found that only a certain consumer segment is willing to pay a premium for functional foodstuffs such as omega-3-enhanced eggs, namely, the pro-organic consumer segment. These consumers also have a medium income level and the highest per capita egg consumption. Canadian consumers were also shown to be willing to pay a premium for omega-3 eggs (Asselin, 2005). In the study by Bejaei et al. (2011), consumers indeed chose nutrient-enhanced eggs because of their nutritional value and because they were considered a healthy choice.

Purchase of eggs is also affected by their cholesterol content, perceived by consumers as a risk for coronary heart disease. However, there are also some critical voices, as so far only a limited number of studies have specifically explored the relation between egg consumption and coronary heart disease (Hansstein, 2011). Some recent studies have revealed that the cholesterol in some foods, such as eggs, does not contribute seriously to the cholesterol levels in human blood (Mizrak et al., 2012). Further research in this area is needed, especially on how egg consumption influences the overall cholesterol ratio (Gray and Griffin, 2009).

3 Instrumental techniques for measuring eggshell quality

The first impression of eggs when bought is dictated by the quality of the shell. Absence of cracks and cleanliness are the essential aspects. In addition, the quality of the shell is also determined by irregularities in shape, and calcium deposits and so-called body checks are undesirable features for the consumer. From the point of view of egg production, the strength of the shell is also an essential aspect, since it is related to the probability of cracking of the egg during transportation from farm to fork. We divide those aspects into two classes, (1) the mechanical and (2) the visual quality of the shell. The mechanical quality of the shell embraces aspects such as the presence of cracks and the strength of the shell. The visual appreciation relates to the colour, cleanliness and anomalies.

3.1 Mechanical shell quality

The mechanical quality of the egg is an essential part for many reasons. While research has shown that microorganisms preferably penetrate the shell through cracks (Widdicombe et al., 2009), it is also known that the force an egg can withstand is related to the quality of its material and the thickness of the shell.

Detecting cracks in eggs is classically performed by using a human candling operation, which includes a bright light source to illuminate the egg and to enhance the contrast between crack and sound shell, and a trained, human observer who judges the egg in the presence of cracks. This operation has proven to be valuable, both for off-line candling and for in-line mass-candling of eggs, but due to the increasing throughput of modern egg grading machines (grading up to 200 000 eggs per hour), alternatives for candling are researched. The main directions are through (1) computer vision, mimicking the human candler and (2) mechanical techniques that include exciting the egg and measuring its response to it. A combination of both approaches was investigated by Pan et al. (2011).

Numerous studies have shown the possibilities of computer vision to replace human inspection. Although some studies report acceptable results (e.g. Omid et al., 2013), many studies also report difficulties in detecting small cracks ('hairline cracks') and false rejects that occur because natural translucent spots and stripes are falsely considered as cracks. In several studies (Lawrence et al., 2009; Jones et al., 2010, 2011; Li et al., 2012), a set-up that combines repeated imaging of a batch of eggs when subjected to a dynamic pressure has been proposed. As such, an image is taken when the crack is 'closed', as well as when the crack is 'opened' under the modified pressure. The difference between these two images is used to find cracks. Lawrence et al. (2009) report an overall accuracy of 99.6% with only 0.3% false positives, compared to 94.2% for human graders with 1.2% false positives. The same authors also investigated the potential increase in crack size due to the manipulation, but found no significant influence on the potential increase in bacterial contamination (Jones et al., 2011).

Crack detection can also be assessed by using a mechanical technique. In most cases, a small impactor excites the egg, and the response of the egg after being impacted is recorded using a microphone, an accelerometer or a piezo sensor (Cho et al., 2000; De Ketelaere et al., 2000; Sun et al., 2013). When the egg is intact, multiple excitations around the eggs' equator produce similar responses, and the similarity between those responses can be used as an indicator for cracks. This forms the basis of one commercial system for off-line grading using a low-mass impactor and a microphone to capture the eggs' response (http://www.columbus-tester.com, an acoustic egg tester, Fig. 1). Rather than analysing the eggs' response, the behaviour of the impactor itself can be investigated. In a commercial system developed by the Dutch company MOBA b.v. (http://www.moba.nl), the number of rebounds of an impactor on the eggs' surface is used as a basis for crack detection (Moayeri, 1996). Close to a crack, the local stiffness of the egg is decreased so that the number of rebounds will be lower when compared to an intact region. Scanning the egg surface in this way can reveal the presence of cracks, and provides a clear indication of the severity and location, which is less straightforward in the case of acoustic impulse response measurements. Jin et al. (2015) used another implementation of a mechanical detector. They used the egg itself as the impactor while it was rolling on a 'step' surface. Each time the egg hit a step the vibration produced was acquired by using a microphone,

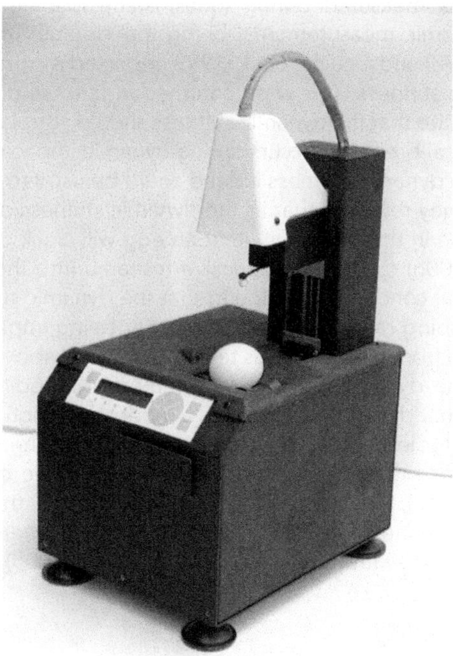

Figure 1 The Columbus Egg Tester, a device that can measure mechanical shell properties such as the presence of cracks as well as the strength expressed as 'dynamic stiffness' (http://www.columbus-tester.com).

and the signals were fed to a data processing unit. The authors reported a crack detection rate of 90% but allowed a 10% false reject which is unacceptably high for practical use.

When it comes to shell strength, quasi-static compression of the egg between two parallel plates is the classical method. By recording the force-deformation curve until breakage, several characteristics can be derived. First, the maximal force is often referred to as the breaking strength – it is a 'direct' way of assessing the strength. Obviously, it is destructive and can be measured only once. This value can also be obtained by using a quasi-static puncture test during which a needle is forced through the shell and the force required to do so is measured (Widdicombe et al., 2009).

The slope of the force-deformation curve before failure of the shell is an estimate of the strength as well, and is called static stiffness (Coucke et al., 1999). The static stiffness can be measured by applying only a limited load on the egg so that it becomes a non-destructive method that can be repeated along the equator of the egg. This way of measuring is called 'indirect', since it is not related to the ultimate failure characteristic of it (Hamilton, 1982). Research has shown that the relation between the static stiffness and the breaking strength is only moderate (De Ketelaere et al., 2004). Due to the moderate correlation between the static stiffness and the breaking strength and the fact that it is time-consuming, extensive research was performed into non-destructive methods that can estimate the shell strength in a faster and better way. Instead of looking at the static stiffness, Coucke (1998) and De Ketelaere et al. (2002) defined a dynamic stiffness; whereas the static stiffness was measured during (quasi-)static loading, the dynamic stiffness was derived from dynamic measurements, being the vibration of the egg after impact excitation. Coucke (1998) and Coucke et al. (1999) reported a correlation of 0.71 between the static and dynamic stiffness. This was confirmed in later studies (De Ketelaere et al., 2003). These authors state that the dynamic stiffness should correlate better with breakage in practice since practical breakage occurs during dynamic impacts on the egg. Bain et al. (2006) showed that the dynamic stiffness indeed could be used to predict the failure of an egg during packing. They revealed that as the dynamic stiffness decreases <1500 kN/m, there is a rapid increase in the probability that an egg will crack during routine handling. Also, Mertens et al. (2006) conducted extensive research into the probability of an egg breaking under practical conditions and found that the dynamic stiffness was a key factor, together with the damping ratio of the vibration after being impacted. Both parameters can be readily obtained from the Columbus tester mentioned above. Those findings would motivate the use of the dynamic stiffness in commercial breeding programmes, and is supported by the fact that dynamic stiffness has a moderately high heritability (Dunn et al., 2005). These authors studied a pedigree population of laying hens which had previously undergone selection for breaking strength and found that the dynamic stiffness had a moderate heritability (0.53), indicating that genetic selection for this parameter is feasible. Furthermore, dynamic stiffness showed a high genetic and phenotypic correlation with eggshell breaking strength, with values of around 0.5 (Bain, 2005).

3.2 Visual shell quality

Since human candling is also the classical operation for removing soiled eggs, replacing this operation by computer vision is the logical step towards automation, and a report from Elster and Goodrum (1991) was one of the first attempts in this direction. Since that period, camera technology, as well as computer power, has improved drastically, and novel approaches, both in the way the analysis was performed and in the way the digital

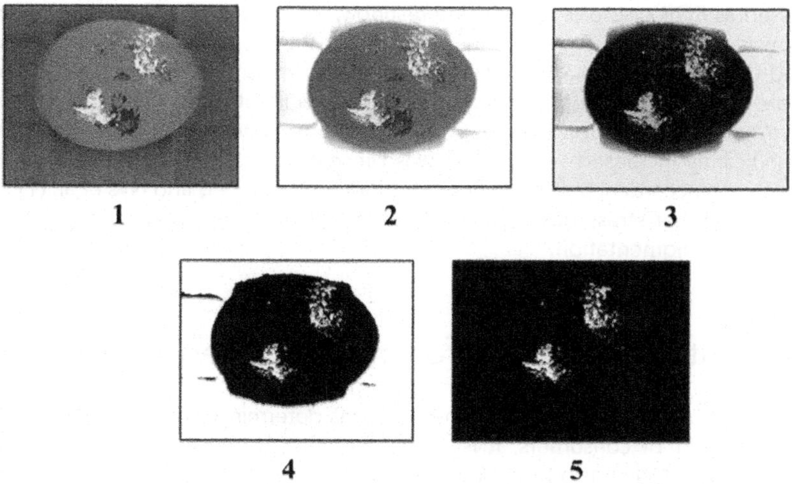

Figure 2 Computer vision for dirt detection on egg. This figure shows an example for the detection of white stains caused by uric acid and embraces five steps: (1) original image; (2) image after extraction of the blue colour plane; (3) image after adjustment of brightness and contrast properties of the image; (4) image after setting a threshold and (5) image after applying the image mask and a filter.

Source: Mertens et al. (2005).

images were taken, have been proposed. In 2005, Mertens et al. showed that different types of dirt can be detected and classified, together with the quantification of the amount of each dirt type on the shell (Fig. 2). More recently, Omid et al. (2013) reported on an expert system for egg grading based on computer vision and artificial intelligence. Instead of using red-green-blue (RGB) images, they found that transformation to the hue–saturation–value (HSV) colour space was more informative. Since the development of hardware and software, computer vision is applied in modern egg graders for automated mass-candling.

A new direction of research is the use of hyperspectral imaging instead of classical colour imaging. In hyperspectral imaging, from each pixel in the image a spectrum is obtained, rather than only an intensity value (eventually the RGB intensity values). During hyperspectral imaging, advanced statistical tools allow for determining the most discriminating wavelengths for detection of the desired flaw, and a practical solution that focuses on those wavelengths (e.g. through optical filters or dedicated illumination) can be built, typically providing images with a better contrast (Lunadei et al., 2012).

As mentioned in Section 2, the colour of the shell is an important characteristic for many consumers and is mainly determined by the pigment protoporphyrin in the shell. Samiullah and Roberts (2013) found that the majority of protoporphyrin was located in the calcareous part of the eggshell (80–87%), while only a minority is present within the cuticle (13–20%) and is thus visible to the human eye. Because of this interest, different measurement techniques for defining eggshell colour have been developed, and classical reflectometry is the most used of them. With this method, the egg is illuminated with a well-defined light source, which is calibrated against a reference tile and the reflected light is captured by using an optical sensor (Wei and Bitgood, 1989). The captured light is then expressed as RGB or L*a*b* values. Interestingly, Mertens et al. (2010) showed that shell colour was a subtle indicator of stress situations in the barn. Bamelis (2003) defined the so-called transmission colour value (TCV), which is based on the transmission of light at two wavelengths: 643 nm, a dominant absorption wavelength of protoporphyrin, and 610 nm as a normalizing factor. During a well-designed experiment, it was found that the TCV significantly increased when stress was present, indicating that the colour was brighter, which is in accordance with the findings of Mills et al. (1991) and Nys et al. (1991), all of whom stated that stress results in egg whitening as a consequence of premature termination of shell pigmentation.

4 Instrumental techniques for measuring egg freshness

Whereas it is the outer characteristics of the egg that determine to a significant extent the buying behaviour of consumers, it is the interior that is consumed, also requiring high standards. One of the important aspects is the freshness of the egg, which is mainly expressed by two quality traits: the viscosity of the albumen and the air cell size. We will describe in short the mechanisms behind those changes and will elaborate on sensor technologies that are capable of determining. As an extra parameter, we will discuss sensor technologies that are able to analyse odour as an indicator for egg freshness.

4.1 Albumen viscosity

Kemps et al. (2006) reviewed some mechanisms behind the thinning of the albumen during storage, a process that is complex and not fully understood. Robinson and Monsey (1972) mentioned that the pH increase during storage led to a destabilization of the ovomucin–lysozyme complex, which subsequently gave rise to albumen thinning. The pH of an egg at oviposition has a value around 7.6 and increases to a value of about 9.5 during storage, after which it levels off or even slightly decreases after a long storage period (Lapao et al., 1999). Dam (1971), on the other hand, suggested that changes in ovomucin might be sufficient to explain thinning of the thick albumen. Donovan et al. (1972) suggested that alkaline hydrolysis of the disulphide bonds of ovomucin was responsible for the thinning.

The most widely used destructive method to assess the albumen quality is through the measurement of the HU (Haugh, 1937). The HU is based on the weight of the intact egg and the height of the albumen when carefully broken on to a smooth, flat surface. For a fresh egg, the yolk is in a central position surrounded by the thick albumen. When a stale egg is broken, the yolk is often displaced to one side and the surrounding thick albumen becomes thinner, resulting in a large area of watery albumen. This results in a decreased albumen height, which consequently leads to a decreased HU. Figure 3 shows the decrease in HU during storage at different temperatures (Kemps, 2006). Since the pH changes during storage as mentioned in the previous paragraph, it is a sensible indicator as well. It is most often measured in the thick albumen. Sharp and Powell (1930) defined the yolk index as a measure for egg freshness. The method was later simplified by Funk (1948) and involved measuring the height and the width of the yolk without removing it from the albumen. Alternatively, Hildago et al. (1995) demonstrated that furosine, as an indicator of the initial steps of a Maillard reaction, is a good index of shell egg freshness.

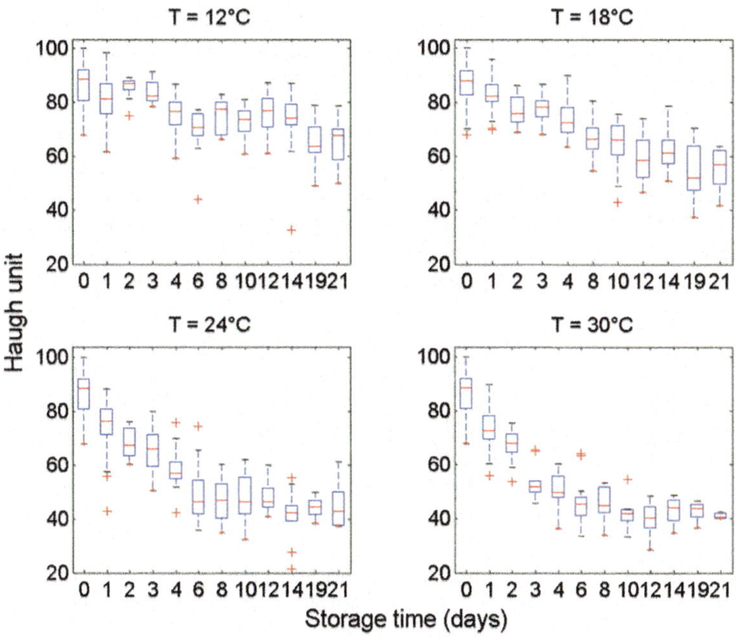

Figure 3 Decrease in HU during storage at different temperatures.
Source: Kemps (2006).

Just as for the mechanical properties of the shell, it is important for the assessment of albumen quality to move from destructive, time-consuming methods towards methods that are non-destructive and fast, so that they can be used for online applications on large egg-sorting machines. Of all studied non-destructive methods, visible (VIS) and near-infrared (NIR) spectroscopy clearly stands out as the most researched method for assessing albumen freshness. The use of VIS-NIR spectroscopy for the determination of egg freshness was first studied by Norris (1996). He studied the effect of storage of the egg on its optical properties and noticed changes in the spectral data that occurred over several hours just after laying. Later, Schmilovitch et al. (2002) measured NIR transmission spectra of eggs. They stated that, during the days after lay, the size of the air cell, the weight loss and pH could be predicted accurately. However, the high correlation coefficients refer to group means rather than to individual eggs and thus need cautious interpretation. In 2006, Kemps et al. showed that NIR spectroscopy provides quantitative information about egg freshness. They used partial least squares (PLS) models to link the non-destructive spectral measurements to both the HU and the pH, resulting in moderate results. The research group observed that the models obtained for both parameters were strikingly similar, indicating that the HU and pH have the same physicochemical background. Since then, several authors have worked further on the use of VIS-NIR spectroscopy for albumen quality assessment, mostly with acceptable to good results (Liu et al., 2007; Zhao et al., 2010; Lin et al., 2011; Abdel-Nour et al., 2011; Aboonajmi and Najafabadi, 2014). As is typical with VIS-NIR spectroscopy, the way of calibrating, validating and testing the models is a crucial aspect. Kemps et al. (2010) provided a cautious note on the results from literature by showing that, for the same dataset, results were completely different when changing the calibration strategy. To date, the use of NIR spectroscopy for accurate assessment of albumen viscosity is still under debate.

Besides VIS-NIR spectroscopy, other techniques have been reported to be effective in assessing the albumen quality. Laghi et al. (2005) showed that the proton NMR relaxation time was significantly influenced by storage, mainly resulting from an increase in proton exchange rate due to the pH increase attributed to the loss of carbon dioxide through the eggshell. Yongwei et al. (2009) showed that an electronic nose was able to distinguish eggs of different storage times under cool and room-temperature storage. These authors reported that the HU could be predicted with a standard error of prediction of 3.74 units and a correlation coefficient of 0.91. Ragni et al. (2010) used a so-called waveguide technique for non-destructive determination of egg quality parameters. They reported that the range between 10.5 and 11.5 GHz could be used to assess the HU, yolk index and the air cell height with an $R^2 > 0.85$. Aboonajmi et al. (2014) used ultrasound to assess the freshness of the egg. Their proposed method could predict the HU and other quality attributes by computing the ultrasound phase velocity within the egg material. Very recently, Soltani and Omid (2015) showed that the dielectric properties in the radio frequency range, when coupled with appropriate data analysis techniques, were indicators of egg freshness.

In order to prolong shelf life for fruit, use of modified atmosphere has become common. For eggs, however, this is still in its infancy but might gain attention in the coming years. Karoui et al. (2008) showed that increased CO_2 levels during storage slowed down the degradation of albumen quality as expressed by HU. Yanagisawa et al. (2010) investigated changes in HU, weakening of vitelline membranes, and the generation of volatiles of eggs stored under modified atmosphere conditions. Their results show that the technique of super chilling and high carbon dioxide concentration enabled eggs to be most effectively stored for up to 90 days. Rocculi et al. (2011) compared several technological properties (gel hardness, foam drainage, and meringue crispness) of albumen of non-packed eggs with those of eggs that were packed in high-barrier plastic pouches with three modified atmosphere packaging (MAP) conditions (air, 100% N_2 and 100% CO_2) during 28 days of storage at 25°C. These researchers also found that MAP, especially those with high CO_2 level, slowed down quality deterioration during storage.

4.2 Air cell size

Several authors (Cepero et al., 1995; Rossi et al., 1995) reported that the size of the air cell can be used as an indication of egg freshness. The loss of water during storage directly relates to the increase in air cell. The height of the air cell is affected mainly by egg weight, shell characteristics, temperature and relative humidity of the storage environment. In the United States, the size of the air cell divides the eggs into different quality categories AA, A and B (USDA-Agricultural Marketing Service, 2000). Air cell height is the only quantitative egg freshness parameter considered by the European Union regulation (Commission Regulation (EC) No 589/2008). Theoretically, a grade A egg at packaging has to keep the characteristics of its grade (air cell height <6 mm) until its expiry date. However, the strong dependence of air cell height upon relative humidity and temperature makes it difficult to guarantee egg quality without a strict control of these variables throughout the egg marketing cycle. Furthermore, these values are not completely reliable as the initial height of the air space varies substantially.

When candling the egg, the size and position of the air cell can easily be visualized and used for grading. However, translating this operation to automated sorting is less straightforward. Wang et al. (2009) used computer vision to extract different aspects from the imaged egg, including the air cell. Aboonjami and Najafabadi (2014) showed that the air cell height could be determined by using NIR spectroscopy, yielding models with an R^2 of 0.86. As for HU, the use of NIR spectroscopy for air cell size needed to be approached with some reservations.

4.3 Odour as a freshness indicator

Odour is used frequently and intuitively by consumers to judge the freshness of food products. The human nose can be considered a very sensitive sensory device in food quality control when a limited number of samples needs to be analysed. However, when it comes to the analysis of larger number of samples in an industrial context, this 'device' fails due to the poor repeatability, low processing speed and inter-personal inconsistencies in threshold detection levels. Therefore, reference technologies to measure aroma components including specific gas chromatography (GC) technologies, in combination with selective extraction and isolation procedures, are required. Specific and extensive procedures are required for the quantification of volatile sulphur components (VSC), considered as characteristic flavour and off-flavour components in food (McGorrin, 2011). A study, using GC technology, revealed the presence of volatile sulphur aroma components (e.g. dimethyl sulphide, dimethyl disulphide, dimethyl trisulphide) and ethanol, associated with egg freshness (Brown et al., 1986). GC-based methods analyse the headspace of samples accurately under very controlled conditions. Typically, they are time consuming and, hence, used off-line for laboratory control on a limited number of samples. Although modern portable GC devices (such as Torion T-9 GC/MS, http://torion.com/products/torion.html and Syft, http://www.syft.com/about-syft-technologies/) are commercially available nowadays and allow for online monitoring of aroma components in industrial processes, their potential with respect to quality measurements on individual egg samples remains unclear.

Therefore, several analytical approaches are suggested in the recent literature to determine the egg freshness in a fast, non-destructive way, showing intrinsically the potential to be implemented in an industrial context. These analytical approaches can be differentiated on the basis of the specificity of the sensor detection mechanism. Sensor systems can be either developed for the selective quantification of one (or a limited number of) component(s) or for the measurement of a large number of non-selective interactions between the different aroma components with an array of non-specific sensors. Whereas the first approach results directly in concentration numbers of individual components, the second ('E-nose') approach requires multivariate statistical techniques to correlate the sensor output with the envisaged research or quality control objectives.

With respect to the first class, an extensive review of Pandey et al. (2012) provides very detailed information on the different measurement technologies for the detection of hydrogen sulphide, which is responsible for the characteristic 'rotten eggs' odour. In their review, the authors summarize the advantages and disadvantages of metal oxide, electrochemical, surface acoustic wave (SAW), conducting polymer and optical sensor technologies. Their general conclusion reads that optical sensor technologies outperform the other technologies when it comes to response time and detection limit. Nevertheless, most of the discussed technologies were still under development in a laboratory environment, with no real validation in an industrial context. Within this context, laser photo-acoustic spectroscopy possibly forms an interesting alternative technology, with reported detection limits below the ppm-level (Wu et al., 2015).

The second class of sensor technologies involves the group of E-nose technologies, whose potential was frequently investigated in a research context for a wide range of food quality and safety monitoring objectives (Sliwinska et al., 2014). As a result of the E-nose analysis, specific digital signal patterns are obtained and further processed by multivariate statistical methodologies to obtain semi-quantitative or qualitative information on the analysed samples. A schematic representation of this approach is shown in Fig. 4. In the framework of egg freshness measurements, some reports are published indicating the potential of E-nose technologies towards this application. In an initial study by Dutta et al. (2003), an E-nose technology, composed of an array of four sensors, was used to classify egg samples with and without holes, in three different freshness categories (within specified 'use by date', 11–20 days after use by date and 21–40 days after use by date), based on the signal patterns from a large dataset of analysed egg samples. Depending on the applied methodology, classification accuracies between 65% and 95% were retrieved for eggs without holes, indicating the potential of this technology to analyse egg freshness non-destructively. In a second study (Yongwei et al., 2009), an E-nose composed of an array of eight sensors was used to assess egg freshness. These researchers looked more in detail into the effect of storage (at room temperature vs. at 4°C) of egg samples, stored for up to five different weeks, on the performance of the technology. The study indicated that differentiation of eggs in different storage classes using the E-nose technology became more difficult, especially for eggs that were stored at 4°C and originating from week 0, 1 and 2. In the same study, a model was developed to predict the HU and yolk coefficient from the E-nose output signals of all the different egg samples. Relative errors of 5.37% and 2.89% were reported, which is encouraging. In a third study (Liu and Tu, 2012), a sensor array composed of ten individual sensors was used to classify egg samples based on their freshness and in close relation to the total volatile basic nitrogen (TVB-N) content of the

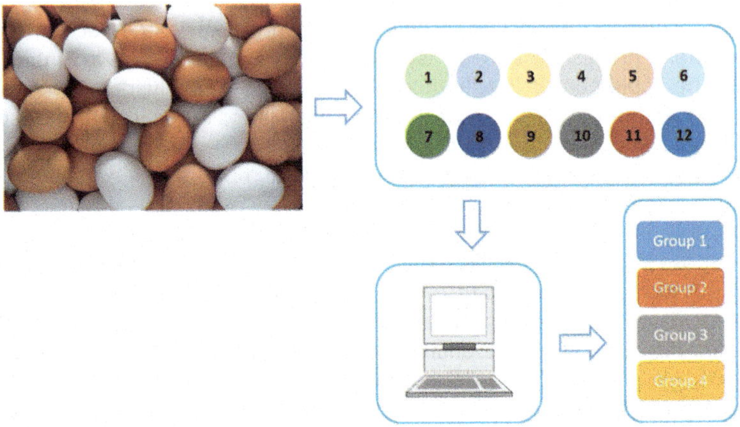

Figure 4 Schematic representation of an electronic-nose device, composed of a sensor array of 12 (non-specific) sensors. After data processing and analysis, the original sample set can be divided into several subgroups, based on the interactions of the volatile components with the E-nose system.

egg samples. This parameter increases during storage. Eggs were stored for a total period of 72 days and distinct patterns in the E-nose signal output were observed for the different classes. In addition, the elaborated prediction model, based on principal component analysis (PCA) and a support vector regression model (SVM), allowed the estimation of the TVB-N content in the different samples. These studies clearly indicate the potential of the E-nose technology, but it is important to mention that the total analysis time per sample remains too long (from 30 min to 1 h) for many industrial practices. Besides, sensors based on E-nose systems are associated with limited stability and repeatability (drift) and are strongly influenced by interferential parameters (such as humidity), requiring further research.

As an alternative to the aforementioned approach, recent studies have examined the potential of colorimetric sensor arrays for food quality and safety measurements, including poultry products (Salinas et al., 2012; Chen et al., 2014). These colorimetric sensor arrays consist of a large number of dyes (interacting with odour components from the headspace), which are printed on a paper. This paper is analysed by a traditional flatbed scanner before and after interaction with the sample to monitor the change in colour intensities. Multivariate statistical techniques are indispensable here as well to convert the signal patterns into useful qualitative or semi-quantitative information. Compared to the traditional E-nose technology, the applied methodology is much more user-friendly, eliminating the need for extensive calibration steps. Typical analysis times for this technology range from a few minutes to 15 min, but this depends strongly on the set-up used (dynamic vs. static head-space condition) and on the sensitivity of the printed dyes towards the odour-specific components.

5 Instrumental techniques for measuring other quality attributes

5.1 Detection of anomalies such as blood spots

Inclusions in the egg can be detected by optical methods. One of the most important inclusions that are not desired by the consumer is blood, which is often present as a dense blood spot located close to the yolk of the egg. The optically active pigment of blood is haemoglobin and has very specific absorption peaks in the visible region of the electromagnetic spectrum (at 415, 541 and 577 nm; Brant et al., 1953), giving it its characteristic red colour. For detection purposes, only the last two wavelengths can be used since the shell of the egg absorbs all transmitted light lower than 500 nm. To correct for other egg characteristics such as the thickness and colour of the shell and the size of the egg, most often the transmission of light at 577 nm is standardized by dividing it by a reference transmission value, depending on the authors, between 585 and 610 nm (Gielen et al., 1979). The ratio between these two transmission values is called the 'blood value' and is used as a rejection criterion. According to Brant et al. (1953), large bloodspots and bloody white can be detected with this technique, but smaller bloodspots remain hidden.

The blood value has been used in commercial graders for many years and seem to be able to reject eggs including blood. However, the detection rate is largely influenced by the amount of blood, and even more by whether it concerns a dense blood spot, or the presence of diffused blood in the albumen. Dense blood spots are particularly difficult to detect. Another complicating factor is the colour of the shell, which is mainly determined by the pigment protoporphyrin as mentioned earlier. Protoporphyrin has a molecular structure that is very similar to haemoglobin, giving it a very similar light absorption spectrum. Besides the main absorption peak around 643 nm, protoporphyrin has also a clear absorption peak at 589 nm, which is close to the main absorption peak of haemoglobin (577 nm), thus complicating the detection.

In a recent study, Chen et al. (2015) used visible absorbance spectroscopy in the range between 200 and 1100 nm and showed that analysis of the absorbance spectra in this range yielded detection rates of >95%, although their results need to be considered with care since they based themselves on artificial blood eggs (blood was injected using a syringe), for which the probability of diffusive blood is higher than typically observed for natural blood containing eggs.

5.2 Bacterial contamination

Although food safety aspects of egg products are not only limited to the detection and quantification of *Salmonella* sp., emphasis in scientific research is strongly on the detection of bacteria from this genus. Other bacteria that can be present on the egg shells are *Campylobacter jejuni*, *Listeria monocytogenes* and *Yersinia enterocolitica* (De Reu et al., 2008). This strong focus on *Salmonella* sp. is due to the fact that these microorganisms can be considered as important causative agents for food-related intestinal infections and diseases and that these outbreaks are often associated with eggs or egg-related products, specifically caused by the presence of *Salmonella* serotype Enteritidis (Malorny et al., 2007). Because of the fact that the classical microbiological culture methods for the detection of those pathogens are time-consuming and laborious, a significant number of research papers have been published that all describe novel methodologies to detect the presence of these pathogens in a much faster way, allowing the initiation of appropriate measures to limit the magnitude of the contamination. For this overview, we highlight shortly the new molecular methods and also the potential of innovative biosensor technologies within this context.

Molecular methods include protein- or nucleic acid-based assays, which allow for very sensitive, specific, cost- and labour-effective detection of the unwanted pathogens in food samples. Besides, those methods are devised to be fast so that they allow for industrial applications. For the detection and characterization of *Salmonella* sp. in poultry and poultry products, a recent review by Park et al. (2014) describes in much detail the progress made with these methods and accurately discusses the advantages and disadvantages of the different methods for the detection of the pathogen. The different methods that are reported frequently in literature are methods based on (1) enzyme-linked immunosorbent assays (ELISA), (2) methods using polymerase chain reaction (PCR), (3) DNA microarray technology and (4) next-generation sequencing (NGS) technologies. Among the methods based on PCR, different approaches can be distinguished, such as single, multiplex or real-time quantitative PCR (qPCR) methods. The availability of these methods allows companies to develop elaborated sampling plans to improve efficiency and to reduce the number of analyses (Pasquali et al., 2014).

Biosensor technologies form an interesting new class of sensor technologies. In general, the detection of pathogenic microorganisms in food products has been of great interest for the scientific biosensor community. The main advantage of biosensors is that they have the potential to execute measurements on-site in a cost-effective and convenient way.

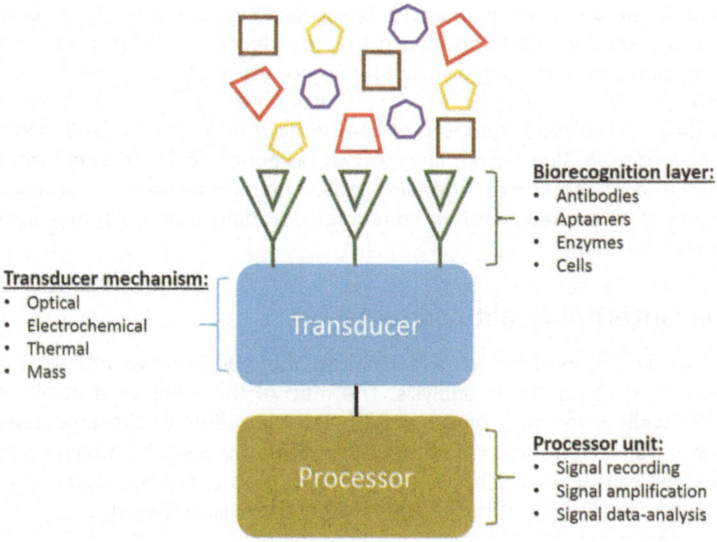

Figure 5 Schematic representation of the different components of a biosensor.

Compared to traditional sensor systems, biosensors are autonomous devices, combining biological entities (such as enzymes, antibodies and nucleic acid structures) with optical and electronic components. These devices are able to detect a wide variety of target components (including low molecular weight molecules, mycotoxins, proteins, nucleic acid structures and cellular structures). The integrated biological entity (and its specificity towards the target component) greatly influences the overall specificity of the biosensor and must be chosen with great care to obtain reliable results from the developed biosensor device. A schematic representation of a biosensor is shown in Fig. 5.

For the detection of the presence of pathogens on egg shells, Chai et al. (2012) described a method for rapid and sensitive detection of *Salmonella typhimurium*. Their method is based on a wireless magneto-elastic biosensor technology. Small magnetic particles are coated with virus-type particles (phages), selectively designed for binding to *S. typhimurium* cells. When placed in a varying magnetic field, these magnetic particles exhibit a different vibration behaviour, depending on whether these particles are free or bound to other physical entities (such as microorganisms). This analysis can be executed in a time period of 30 min. Initial experiments with this method resulted in detection limits of 1.6×10^2 CFU/cm^2 when clean eggs were inoculated with different concentrations of *S. typhimurium* pathogens.

5.3 Origin/fraud detection

Given the increased interest of consumers in animal welfare and the origin of the eggs they buy (whether they come from hens housed in cages or free range), research into this direction has been initiated as well. Gregory et al. (2005) developed a methodology for testing whether eggs that were sold as either 'barn' or 'free range' were laid under cage conditions. These authors analysed more than 11 000 eggs from cage as well as from barn and free-range systems. Their approach is based on illuminating the eggs with ultraviolet (UV) light, and capturing the fluorescence that is induced by the protoporphyrin and/or dust from the cage systems using a computer vision system. The prevalence of fluorescent white double parallel lines was found to be a distinguishing feature for eggs laid on wire floors in cages. The authors state that, if five or more eggs are found in a sample of 90 eggs that have double fluorescent lines, it can be concluded with very high probability that the batch contains some cage-laid eggs. This is interesting, because nowadays large amounts of eggs are sorted on commercial grading machines, thus increasing the potential to detect fraud.

5.4 Other internal quality aspects

Kuchida et al. (1999) performed research into the non-invasive measurement of the yolk:albumen ratio using image analysis. The ratio of the number of pixels of light and dark parts as well as the red, green and blue components of the image was used as image analysis traits. Correlation coefficients between the yolk:albumen ratio and analysis traits were significant from which the authors concluded that the image analysis method accurately predicted the yolk:albumen ratio without breaking the egg.

In the consumer perception study above, yolk colour was mentioned to be an important aspect. Traditionally, colour is assessed by using a colour fan, such as the Roche or DSM Yolk colour fan, but this approach is destructive and subjective, although guidelines for use of the fan aim to minimize errors. Different analytical methods are available to quantitatively identify colour through measuring the carotenoid content, since it is the main determining factor. The two standard analytical methods are spectrophotometric methods described by the Association of Official Analytical Chemists (AOAC) and high-performance liquid chromatography (HPLC). The disadvantage of both methods, however, is that they are time-consuming and require well-trained technical operators. In order to overcome those disadvantages, in a recent study, Islam and Schweigert (2015) compared a novel technique with the above-mentioned ones. This newly developed method claims to be laboratory-independent and fast and is commercially available (iCheck, BioAnalyt GmbH, Germany). The study, based on a sample of 80 eggs, shows that carotenoid values obtained by the reference methods AOAC and HPLC are highly correlated with the iCheck method with a coefficient of determination of 0.99 and 0.94 for iCheck versus AOAC and iCheck versus HPLC, respectively (both $p < 0.001$). Based on these results, the authors conclude that the novel rapid and portable iCheck method is a valid and effective tool to determine total carotenoid of egg yolk under laboratory-independent conditions with minimally trained personal. It remains to be said, however, that the method is destructive and thus cannot be used to screen eggs online.

6 Conclusions and future perspectives

From the above it is clear that classical, external quality aspects – shell cleanliness and colour, egg size and absence of cracks are still important factors for many consumers. For those quality aspects, technology has reached full maturity: computer vision is used for colour, size and cleanliness evaluation, whereas mechanical techniques are used for crack detection. Those methods are commercially available for online grading on large sorting lines that grade up to 200 000 eggs per hour.

When looking at the internal quality, the absence of blood is an important issue, together with yolk colour. For blood detection, spectroscopic methods have been used for many years and provide reasonable results. For yolk colour, however, there is no fast and non-destructive method that has hit the commercial market, and additional research is required. Albumen viscosity, often expressed as HU, is extensively investigated in the recent scientific literature since it is influenced by the freshness of the eggs, but it remains unclear what spectroscopic methods, especially in the VIS-NIR region, can offer. Reports vary in their conclusions, and it seems that often indirect measures are obtained. Indeed, much of the spectroscopic information comes from the shell of the egg and changes in the properties of the shell are picked up as well. We feel additional research is required in this field, and additional techniques might be required for an accurate assessment of the albumen quality. One alternative way of determining the freshness is through the measurement of the air cell size, but also here, additional research is needed. The odour of the egg can be used as a freshness indicator, and E-noses have shown potential in this direction, although they exhibit undesired behaviour such as drift and lack of specificity. A recent trend in sensor technology for measuring quality aspects of biological products is the introduction of sensor fusion approaches, combining different technologies in order to increase classification accuracy.

One direction that is clear from recent advances in sensor technology is that they have become more specific and more sensitive. Biosensors are a good example: they can be tailored towards the detection of a very specific molecule and are often small and fast enough for industrial application. It is believed that biosensors will gain momentum in the years to come, and will be able to solve some of the bottlenecks that remain to date.

7 Where to look for further information

The Knowledge Centre of the Poultry Site is a good collection of different aspects related to poultry and egg production. Especially the section 'Egg Quality Guide' is a good starting point for those interested in egg quality as a whole. The Guide can be found at: http://www.thepoultrysite.com/publications/1/egg-quality-handbook/. The book from which the information comes can be bought at: http://www.5mbooks.com/optimum-egg-quality.html.

Another relevant source of information is the World's Poultry Science Association (WPSA – http://www.wpsa.com/) that aims to be a 'unique organization that strives to advance knowledge and understanding of all aspects of poultry science and the poultry industry'. Their publication 'The World's Poultry Science Journal'. It features review papers of all topics related to Poultry Science.

The field of sensor technologies is very broad. Two books that cover a large range of food applications are:

- High Throughput Screening for Food Safety Assessment, 1st Edition. ISBN: 9780857098016.
- Instrumentation and Sensors for the Food Industry, 2nd Edition. ISBN: 9781855735606.

8 References

Abdel-Nour, N., Ngadi, M., Prasher, S. and Karimi, Y. (2011). Prediction of egg freshness and albumen quality using visible/near infrared spectroscopy. *Food Bioprocess Tech.* 4: 731–6.

Aboonajmi, M. and Najafabadi, T. A. (2014). Prediction of poultry egg freshness using Vis-Nir spectroscopy with maximum likelihood method. *Int. J. Food Prop.* 17: 2166–76.

Aboonajmi, M., Setarehdan, S. K., Akram, A., Nishizu, T. and Kondo, N. (2014). Prediction of poultry egg freshness using ultrasound. *Int. J. Food Prop.* 17: 1889–99.

Andersen, L. M. (2011). Animal welfare and eggs – cheap talk or money on the counter? *J. Agric. Econ.* 62: 565–84.

Arthur, J. A. and O'Sullivan, N. (2005). Breeding chickens to meet egg quality needs. *International Hatchery Practice* 19: 7–9.

Asselin, A. M. (2005). Eggcentric behavior: consumers characteristics that demonstrate greater willingness to pay for functionality. *Am. J. Agric. Econ.* 87: 1339–44.

Bain, M. M. (2005). Recent advances in the assessment of eggshell quality and their future application. *Worlds Poultry Sci. J.* 61(2): 268–77.

Bain, M. M., Dunn, L. C., Wilson, P. W., Joseph, N., De Ketelaere, B., De Baerdemaeker, J. and Waddington, D. (2006). Probability of an egg cracking during packing can be predicted using a simple non-destructive acoustic test. *Br. Poult. Sci.* 47(4): 462–9.

Bamelis, F. R. (2003). Non-invasive assessment of eggshell conductance and different developmental stages during incubation of eggs. PhD. diss., Katholieke Universiteit Leuven, Heverlee, Belgium.

Bejaei, M., Wiseman, K. and Cheng, K. M. (2011). Influences of demographic characteristics, attitudes, and preferences of consumers on table egg consumption in British Columbia. *Canada. Poult. Sci.* 90: 1088–95.

Bejaei, M., Wiseman, K. and Cheng, K. M. (2015). Developing logistic regression models using purchase attributes and demographics to predict the probability of purchases of regular and specialty eggs. *British Poult. Sci.* 56(4): 425–35.

Bett, H. K., Peters K. J., Nwankwo, U. M. and Bokelmann, W. (2013). Estimating consumer preferences and willingness to pay for the under utilized indigenous chicken products. *Food Policy* 41: 218–25.

Brant, A. W., Norris, K. H. and Chin, G. (1953). A spectrophotometric method for detecting blood in white shelled eggs. *Poult Sci.* 32: 357–63.

Brown, M. L., Holbrook, D. M., Hoerning, E. F., Legendre, M. G. and Angelo, A. S. J. T. (1986). Volatile indicators of deterioration in liquid egg products. *Poult. Sci.* 65: 1925–33.

Burley H. K. and Johnson C. L. (2013). Market survey of quality and freshness of eggs produced under an enhanced hen nutrition and egg production program. *J. Appl. Poult. Res.* 22: 929–33.

Caner C. (2005). The effect of edible eggshell coating on egg quality and consumer perception. *J. Sci. Food. Agr.* 85: 1897–902.

Cepero R., Alfonso, M., Arnaiz, A., Alvaro, J. R., Elia, L. and Enfedaque, A. (1995). Effects of transport and storage conditions on the commercial quality of eggs. *Proceedings of the VI European Symposium on the Quality of Eggs and Egg Products*, Zaragoza, 155–70.

Chai, Y., Li, S., Horikawa, S., Park, M.-K., Vodyanoy, V. and Chin, B. A. (2012). Rapid and sensitive detection of salmonella typhimurium on eggshells by using wireless biosensors. *J. Food Prot.* 75: 631–6.

Chen, M., Zhang, L. and Xu, H. (2015). On-line detection of blood spot introduced into brown-shell eggs using visible absorbance spectroscopy. *Biosyst. Eng.* 131: 95–101.

Chen, Q., Hui, Z., Zhao, J. and Ouyang, Q. (2014). Evaluation of chicken freshness using a low-cost colorimetric sensor array with AdaBoost-OLDA classification algorithm. *LWT – Food Sci. Technol.* 57: 502–7.

Cho, H. K., Choi, W. K. and Paek, J. H. (2000). Detection of surface cracks in shell eggs by acoustic impulse method. *Trans ASAE* 43(6): 1921–6.

Coucke, P. (1998). Assessment of some physical egg quality parameters based on vibration analysis. PhD. thesis, Katholieke Universiteit Leuven, Belgium.

Coucke, P., Dewil, E., Decuypere, E. and De Baerdemaeker, J. (1999). Measuring the mechanical stiffness of an eggshell using resonant frequency analysis. *Br. Poult. Sci.* 40: 227–32.

Coutts, J. A. and Wilson, G. C. (2007). Optimum egg quality – a practical approach. In S. Fernandez, E. Rosales, G. Weber and J. Hernandez (Eds), 5M Enterprises Ltd., Queensland, Australia.

Dam R. (1971). In vitro studies on the lysozyme–ovomucin complex. *Poult. Sci.* 50: 1824–31.

De Ketelaere, B., Coucke, P. and De Baerdemaeker, J. (2000). Eggshell crack detection based on acoustic resonance frequency analysis. *J. Agric. Eng. Res.* 76(2): 157–63.

De Ketelaere, B., Bamelis, F., Kemps, B., Decuypere, E. and De Baerdemaeker, J. (2004). Nondestructive measurements of the egg quality. *Worlds Poult. Sci. J.* 60(3): 289–302.

De Ketelaere, B., Govaerts, T., Coucke, P., Dewil, E., Decuypere, E. and De Baerdemaeker, J. (2002). Measuring the eggshell strength of 6 different genetic lines of laying hens: techniques and comparisons. *Br. Poult. Sci.* 43: 238–44.

De Ketelaere, B., Vanhoutte, H. and De Baerdemaeker, J. (2003). Parameter estimation and multivariable model building for the non-destructive, on-line determination of the eggshell strength. *J. Sound Vib.* 266(3): 699–709.

De Reu, K., Messens, W., Heyndrickx, M., Rodenburg, B., Uyttendaele, M. and Herman, L. (2008). Bacterial contamination of table eggs and the influence of housing systems. *Worlds Poult. Sci. J.* 64: 5–19.

Donovan, J. W., Davis, J. G. and Wiele, M. B. (1972). Viscosimetric studies of alkaline degradation of ovomucin. *J. Agric. Food Chem.* 20: 223–8.

Dunn, I. C., Bain, M., Edmond, A., Wilson, P. W., Joseph, N., Solomon, S., De Ketelaere, B., De Baerdemaeker, J., Preisinger, R. and Waddington, D. (2005). Heritability and genetic correlation of measurements derived from acoustic resonance frequency analysis; a novel method of determining egg shell quality in domestic hens. *Br. Poult. Sci.* 46(3): 280–6.

Dutta, R., Hines, E. L., Gardner, J. W., Udrea, D. D. and Boilot, P. (2003). Non-destructive egg freshness determination: an electronic nose based approach. *Meas Sci. Technol.* 14: 190–8.

Elster, R. T. and Goodrum, J. W. (1991). Detection of cracks in eggs using machine vision. *Trans ASAE.* 30(1): 307–12.

Fearne, A. and Lavelle, D. (1996a). Segmenting the UK egg market: results of a survey of consumer attitudes and perceptions. *Br. Food J.* 98(1): 7–12.

Fearne, A. and Lavelle, D. (1996b). Perceptions of food 'quality' and the power of marketing communication: results of consumer research on a branded-egg concept. *J. Prod. Brand Manage.* 5: 29–42.

Francesh, A., Estany, J., Alfonso, L. and Iglesias, M. (1997). Genetic parameters for egg number, egg weight, and shell color in three Catalan poultry breeds. *Poult. Sci.* 76: 1627–31.

Froster, A., Jaenecke, D., Wittmann, M., Flock, D. K. and Kreuzer, M. (1996). Use of photometrically determined shell color parameters as selection criteria for marketable brown shelled eggs. *Arch Geflugelkd* 60: 1–6.

Funk, E. M. (1948). The relation of yolk index determined in natural position to the yolk index as determined after separating the yolk from the albumen. *Poult. Sci.* 27: 367.

Gray, J. and Griffin, B. (2009). Eggs and dietary cholesterol – dispelling the myth. *Nutr. Bull* 34: 66–77.

Gielen, R. M. A. M., De Jong, L. P. and Kerkvliet, H. M. M. (1979). Electrooptical blood-Spot detection in intact eggs. *IEEE Trans. Instrum. Meas.* 28: 177–83.

Gregory, N. G., Gepp, M. J. and Babidge, P. J. (2005). Method for checking label accuracy in barn and free range eggs. *J. Sci. Food Agr.* 85: 1421–6.

Grijspeedt, K., Herman, L. and Billiet, L. (1999). A microbiological safety consumer survey on household use of eggs in Belgium. *J. Food Saf.* 19: 249–61.

Grunert, K. G. (2005). Food quality and safety: consumer perception and demand. *Eur. Rev. Agric. Econ.* 32: 469–78.

Hamilton, R. M. G. (1982). Methods and factors that affect the measurement of eggshell quality. *Poult. Sci.* 61: 2022–39.

Hansstein, F. (2011). Profiling the egg consumer: attitudes, perceptions and behaviours. In Y. Nys, M. Bain and F. Van Immerseel (Eds), *Improving the Safety and the Quality of Eggs and Egg*

Products, Volume 1: Egg Chemistry, Production and Consumption, 39–61, Cambridge, UK, Woodhead Publishing Limited. ISBN 978-1-84569-754-9.

Haugh R. R. (1937). The Haugh Unit for measuring egg quality. *U.S. Egg Poultry Magazine* 43: 552–5.

Heng, Y., Peterson, H. H. and Li, X. (2013). Consumer attitudes toward farm-animal welfare: the case of laying hens. *J. Agric. Resour. Econ.* 38: 418–34.

Hernandez, J. M. (2005). European consumer surveys about egg quality: how to improve the nutritional value. *Proceeding XVIIth European Symposium on the Quality of Poultry Meat & XIth Symposium on the Quality of Eggs and Egg Products*, Proceedings CD, 245–50, Doorwerth, Nederland, 22–26 April 2005.

Hernandez, J. M., Hamelin, C., Hayakawa, T., Ibarra, B. and Souza, R. (2009). Adding value to egg products. *World Poultry* 25: 24–6.

Hernandez, J. M., Seehawer, J., Hamelin, C., Bruni, M. and Wakeman, W. (2000). Egg quality – the European consumer perception. *Roche Vitamins Europe Publ.*, 6–55.

Hildago, A., Rossi, M. and Pompei, C. (1995). Furosine as a freshness parameters of shell eggs. *J. Agric. Food Chem.* 43: 1673–7.

Islam, K. M. S. and Schweigert, F. J. (2015). Comparison of three spectrophotometric methods for analysis of egg yolk carotenoids. *Food Chem.* 172: 233–7.

Jacob, J. P., Miles, R. D. and Mather, F. B. (2000). Egg Quality. IFAS Extension, University of Florida, p. PS24.

Jibir, M., Garba, S. and Ribah, M. I. (2012). Relative importance of external quality attributes of shell eggs: a consumer preference approach. *Int. J. Appl. Agric. Apic. Res.* 8(1): 31–8.

Jin, C., Xie, L. and Ying, Y. (2015). Eggshell crack detection based on the time-domain acoustic signal of rolling eggs on a step-plate. *J. Food Eng.* 153: 53–62.

Johnston, N. P., Jeffereis, L. K., Rodriguez, B. and Johnston, D. E. (2011). Acceptance of brown-shelled eggs in a white shelled egg market. *Poult. Sci.* 90: 1074–9.

Jones, D. R., Lawrence, K. C., Yoon, S. C. and Heitschmidt, G. W. (2010). Modified pressure imaging for egg crack detection and resulting egg quality. *Poult. Sci.* 89: 761–5.

Jones, D. R., Lawrence, K. C., Yoon, S. C. and Heitschmidt, G. W. (2011). Salmonella contamination in shell eggs exposed to modified pressure imaging for microcrack detection. *Poult. Sci.* 90: 1616–19.

Jones, D. R. and Musgrove, M. T. (2005). Effects of extended storage on egg quality factors. *Poult. Sci.* 84: 1774–7.

Karoui, R., Nicolai, B. and De Baerdemaeker, J. (2008). Monitoring the egg freshness during storage under modified atmosphere by fluorescence spectroscopy. *Food Bioprocess Tech.* 1: 346–56.

Kemps, B. (2006). VIS/NIR spectroscopy for the assessment of internal egg quality. PhD. thesis, Katholieke Universiteit Leuven, Belgium.

Kemps, B., Mertens, K., Saeys, W., Darius, P., De Baerdemaeker, J. and De Ketelaere, B. (2010). How partial least squares validation procedures may lead to deceptive models. *J. Near Infrared Spec.* 18: 231–7.

Kemps, B. J., Bamelis, F. R., De Ketelaere, B., Mertens, K., Kamers, B., Tona, K., Decuypere, E. M. and De Baerdemaeker, J. G. (2006). VIS-NIR transmission for the assessment of internal egg quality. *J. Sci. Food Agric.*, 86(9): 1399–406.

Kuchida, K., Fukaya, M., Miyoshi, S., Suzuki, M. and Tsuruta, S. (1999). Nondestructive prediction method for yolk:albumen ratio in chicken eggs by computer image analysis. *Poult. Sci.* 78(6): 909–13.

Laghi, L., Cremonini, M. A., Placucci, G., Sykora, S., Wright, K. and Hills, B. (2005). A proton NMR relaxation study of hen egg quality. *Magn. Reson. Imaging* 23: 501–10.

Lapao, C., Gama, T. L. and Chaviero Soares, M. (1999). Effects of broiler breeder age and length of egg storage on albumen characteristics and hatchability. *Poult. Sci.* 78: 640–5.

Lawrence, K. C., Yoon, S. C., Jones, D. R., Heitschmidt, G. W., Park, B. and Windham, W. R. (2009). Modified pressure system for imaging egg cracks. *Trans ASABE* 52(3): 983–90.

Li, Y., Dhakal, S. and Peng, Y. (2012). A machine vision system for identification of micro-crack in egg shell. *J. Food Eng.* 109: 127–34.

Lin, H., Zhao, J., Sun, L., Chen, Q. and Zhou, F. (2011). Freshness measurement of eggs using near infrared (NIR) spectroscopy and multivariate data analysis. *Innov. Food Sci. Emerg. Technol.* 12: 182–6.

Liu, P. and Tu, K. (2012). Prediction of TVB-N content in eggs based on electronic nose. *Food Control* 23: 177–83.

Liu, Y., Ying, Y., Ouyang, A. and Li, Y. (2007). Measurement of internal quality in chicken eggs using visible transmittance spectroscopy technology. *Food Control* 18: 18–22.

Lunadei, L., Ruiz-Garcia, L., Bodria, L. and Guidetti, R. (2012). Automatic identification of defects on eggshell through a multispectral vision system. *Food Bioprocess Tech.* 5: 3042–50.

Malorny, B., Bunge, C. and Helmurth, R. (2007). A real-time PCR for the detection of Salmonella Enteritidis in poultry meat and consumption eggs. *J. Microbiol. Methods* 70: 245–51.

McGorrin, R. J. (2011). The significance of volatile sulfur compounds in food flavors: an overview. In M. C. Qian, X. Fan and K. Mahattanatawee (Eds), *Volatile Sulfur Compounds in Foods*, American Chemical Society, ISBN13:9780841226166. doi: 10.1021/bk-2011–1068, 3–31.

Mertens, K., Bamelis, F., Kemps, B., Kamers, B., Verhoelst, E., De Ketelaere, B., Bain, M., Decuypere, E. and De Baerdemaeker, J. (2006). Monitoring of egg shell breakage and egg shell strength in different production chains of consumption eggs. *Poult. Sci.* 85(9): 1670–7.

Mertens, K., De Ketelaere, B., Kamers, B., Bamelis, F. R., Kemps, B. J., Verhoelst, E. M., De Baerdemaeker, J. G. and Decuypere, E. M. (2005). Dirt detection in brown eggs by means of color computer vision. *Poult. Sci.* 84(10): 1653–9.

Mertens, K., Vaesen, I., Löffel, J., Kemps, B., Kamers, B., Perianu, C., Zoons, J., Darius, P., Decuypere, E., De Baerdemaeker, J. and De Ketelaere, B. (2010). The transmission color value: a novel egg quality measure for recording shell color used for monitoring the stress and health status of a brown layer flocks. *Poult. Sci.* 89(3): 609–17.

Mesias F., Martinez-Carrasco, F., Martinez, J. and Gaspar, P. (2011). Functional and organic eggs as an alternative to conventional production: a conjoint analysis of consumers' preferences. *J. Sci. Food Agric.* 91: 532–8.

Mills, A. D., Nys, Y., Gautron, J. and Zawadski, J. (1991). Whitening of brown shelled eggs: individual variation and relationships with age, fearfulness, oviposition interval and stress. *Br. Poult. Sci.* 32: 117–29.

Mizrak, C., Durmus, I., Kamanli, S., Erdogan Demirtas, S., Kalebasi, S., Karademir, E. and Dogu, M. (2012). Determination of egg consumption and consumer habits in Turkey. *Turkish J. Vet. Anim. Sci.* 36(6): 592–601.

Moayeri, H. (1996). Probe, device and method for testing eggs. US Patent 5,728,939.

Norris, K. H. (1996). History of NIR. *J. Near Infrared Spec.* 4: 31–7.

Nys, Y., Zawadzki, J., Gautron, J. and Mills, A. D. (1991). Whitening of brown-shelled eggs: mineral composition of uterine fluid and rate of protoporphyrin deposition. *Poult. Sci.* 70: 1236–45.

Omid, M., Soltani, M., Dehrouyeh, M. H., Mohtasebi, S. S. and Ahmadi, H. (2013). An expert egg grading system based on machine vision and artificial intelligence techniques. *J. Food Eng.* 118: 70–7.

Pan, L., Zhan, G., Tu, K., Tu, S. and Liu, P. (2011). Eggshell crack detection based on computer vision and acoustic response by means of back-propagation artificial neural network. *Eur. Food Res. Technol.* 233(3): 457–63.

Pandey, S. K., Kim, K.-H. and Tang, K. T. (2012). A review of sensor-based methods for monitoring hydrogen sulfide. *Trends Anal. Chem.* 32: 87–99.

Park, S. H., Aydin, M., Khatiwara, A., Dolan, M. C., Gilmore, D. F., Bouldin, J. F., Ahn, S. and Ricke, S. C. (2014). Current and emerging technologies for rapid detection and characterization of Salmonella in poultry and poultry products. *Food Microbiol.* 38: 250–62.

Pasquali, F., De Cesare, A., Valero, A., Olsen, J. E. and Manfreda, G. (2014). Improvement of sampling plans for Salmonella detection in pooled table eggs by use of real-time PCR. *Int. J. Food Microbiol.* 184: 31–4.

Ragni, L., Cevoli, C. and Berardinelli, A. (2010). A waveguide technique for non-destructive determination of egg quality parameters. *J. Food Eng.* 100: 343–8.

Robinson, D. S. and Monsey, J. B. (1972). Changes in the composition of ovomucin during liquefaction of thick egg white. *J. Food Sci. Agric.* 23: 29–38.

Rocculi, P., Cocci, E., Sirri, F., Cevoli, C., Romani, S. and Rosa, M. D. (2011). Modified atmosphere packaging of hen table eggs: effects on functional properties of albumen. *Poult. Sci.* 90: 1791–8.

Rossi, M., Hildago, A. and Pompei, C. (1995). Classical methods and new proposals for rapid evaluation of shell egg freshness. In *Proceedings of the VI European Symposium on the Quality of Eggs and Egg Products*, Zaragoza, Spain, 23–31.

Salinas, Y., Ros-Lis, J. V., Vivancos, J.-L., Martinez-Manez, R., Marcos, M. D., Aucejo, S., Herranz, N. and Lorente, I. (2012). Monitoring of chicken meat freshness by means of a colorimetric sensor array. *Analyst* 137: 3635–43.

Samiullah, S. and Roberts, J. R. (2013). The location of protoporphyrin in the eggshell of brown-shelled eggs. *Poult. Sci.* 92: 2783–8.

Samiullah, S., Roberts, J. R. and Chousalkar, K. (2015). Eggshell color in brown-egg laying hens – a review. *Poult. Sci.* 94: 2566–75.

Schmilovitch, Z., Hoffman, A., Egozi, H. and Klein, E. (2002). Determination of egg freshness by NNIRS (near-near infrared spectroscopy), presented at EurAgEng, Budapest, paper No. 02-AP-023.

Sharp, P. F. and Powell, C. K. (1930). Decrease in internal quality of hen's eggs during storage as by the yolk. *Ind. Eng. Chem. Res.* 22: 909–10.

Sliwinska, M., Wisniewska, P., Dymerski, T., Namiesnik, J. and Wardencki, W. (2014). Food analysis using artificial senses. *J. Agric. Food Chem.* 62: 1423–48.

Soltani, M. and Omid, M. (2015). Detection of poultry egg freshness by dielectric spectroscopy and machine learning techniques. *LWT – Food Sci. Technol.* 62(2): 1034–42.

USDA-Agricultural Marketing Service (2000). The Egg Grading Manual. Handbook 75: 17.

Wang, Q., Deng, X., Ren, Y., Ding, Y., Xiong, L., Ping, Z., Wen, Y. and Wang, S. (2009). Egg freshness detection based on digital image technology. *Sci. Res. Essays* 4(10): 1073–9.

Wardy, W., Mena, B, Nongtaodum, S., No, H. K. and Prinyawiwatkul, W. (2014a). Exploring the drivers of purchase intent and consumer satisfaction of chicken eggs using principal component analysis and the Kano model. *J. Sens. Stud.* 29: 463–73.

Wardy, W., Martinez, K. D. P., Xu, Z., No, H. K. and Prinyawiwatkul, W. (2014b). Viscosity changes of chitosan solution affect physico-functional properties and consumer perception of coated eggs during storage. *Food Sci. Technol.* 55: 67–73.

Wei, R. and Bitgood, J. J. (1989). A new objective measurement of eggshell color. 1. A test for potential usefulness of two color measuring devices. *Poult. Sci.* 69: 1775–80.

Widdicombe, J. P., Rycroft, A. N. and Gregory, N. G. (2009). Hazards with cracked eggs and their relationship to egg shell strength. *J. Sci. Food Agric.* 89(2): 201–5.

Wu, H., Dong, L., Zheng, H., Liu, X., Yin, X., Ma, W., Zhang, L., Yin, W., Jia, S. and Tittel, F. K. (2015). Enhanced near-infrared QEPAS sensor for sub-ppm level H2S detection by means of a fiber amplified 1582 nm DFB laser. *Sens. Actuators, B: Chemical* 221: 666–72.

Yanagisawa, T., Ariizumi, M., Shigematsu, Y., Kobayashi, H., Hasegawa, M. and Watanabe, K. (2010). Combination of super chilling and high carbon dioxide concentration techniques most effectively to preserve freshness of shell eggs during long-term storage. *J. Food Sci.* 75(1): 78–82.

Yongwei, W., Wang, J., Zhou, B. and Lu, Q. (2009). Monitoring storage time and quality attribute of egg based on electronic nose. *Anal. Chem. Acta.* 650: 183–8.

Zhang, L. C., Ning, Z. H., Xu, G. Y., Hou, Z. C. and Yang, N. (2005). Heritabilities and genetic and phenotypic correlations of egg quality traits in brown-egg dwarf layers. *Poult. Sci.* 84: 1209–13.

Zhao, J., Lin, H., Chen, Q., Huang, X., Sun, Z. and Zhou, F. (2010). Identification of egg's freshness using NIR and support vector data description. *J. Food Eng.* 98: 408–14.

Chapter 12

Determinants of egg appearance and colour

C. Hamelin, CCPA, France and F. Cisneros, DSM, Switzerland

1 Introduction

It is essential for the consumer to select those types of eggs that are safe, nutritious and tasty. In this chapter, we begin by giving an overview of consumer decisions about whether to eat eggs and if so, which eggs to purchase. This includes reviewing regional differences in preferences and purchasing decisions, and the significance of packaging. Subsequent sections focus on the three key parts of the egg: the shell, the albumen and the yolk. We discuss consumer preferences for each and then review the factors that determine the appearance and colour of these parts of the egg.

2 Consumer purchasing decisions: choosing to eat eggs and choosing which eggs to eat

When introducing new kinds of products, it is important that the consumer's point of view is always considered, so that the effect on consumer acceptance can be assessed. In the case of eggs, consumers first consider whether they should include eggs in their diets (i.e. choice of food ingredient). Once eggs are selected as a source of nutrients, the next decision-making would be in buying the right type of eggs. Factors to consider here include bird welfare, sustainability of production, pricing, housing, nutritive value

http://dx.doi.org/10.19103/AS.2016.0012.14

and antibiotic usage. The importance given to each factor depends on the market and target population. A third level of consumer consideration concerns attractiveness of individual egg product brands and packaging at purchase. Pack size, freshness and brand recognition play a role, as well as pricing and the perceived quality. This all means that the egg industry must compete at different levels with different competitors in order to be perceived as a good choice by the end consumer.

2.1 Eggs in the mind of the consumer

The perception of what an egg is, and when we should eat eggs, varies among different cultures. According to the International Egg Commission, egg consumption is highest in Mexico and Japan, with an annual per capita consumption of more than 300 eggs (IEC, 2014). In Mexico, eggs are perceived as a must for every breakfast, and price influences choice with only the poorest sectors of the economy (Cámara de diputados, 2015). According to Cruz (2012), the adjusted egg price in Mexico went down from 60 to 20 pesos/kg from 1962 to 2008, supporting the idea that a high egg intake today is also related to affordability. In Japan, where culture is very important, eggs are consumed mostly raw combined with rice or specialty dishes such as Tamago Kake Gohan (Nobuoka, 2015) or egg cakes.

In Brazil, the per capita egg consumption is low (i.e., 168; IEC, 2014) and it is even lower in households with children under 6 years old (dos Santos et al., 2009), even under conditions in which choice of nutritive foods is reduced. In India, on the other hand, with 47% of children underweight (The World Bank, 2015), there are cultural challenges to overcome before annual per capita egg consumption increases above 65 eggs. As eggs are not considered part of a vegetarian diet in India, they are not included in school meals across the country. Therefore, price is not considered a major determinant of purchasing decisions (Sastry, 2014). Another interesting case concerns the United States, where,

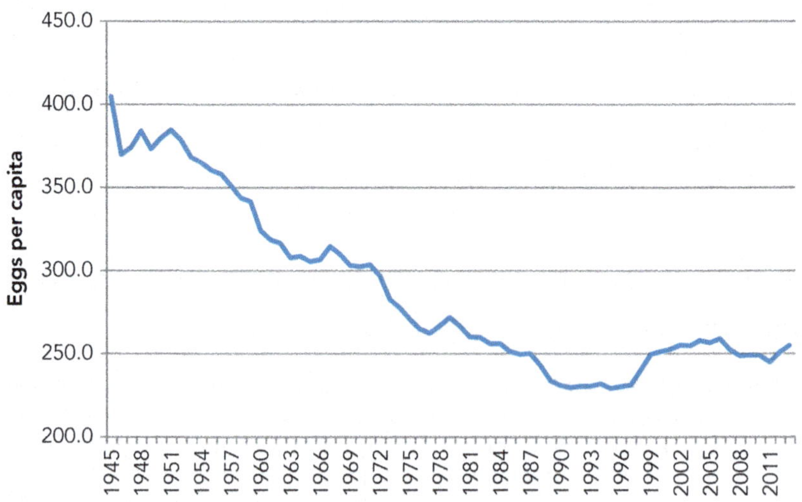

Figure 1 *Per capita* egg consumption USA (adapted from USDA).

in 1945, the annual per capita consumption was more than 400 (Fig. 1), a figure that dropped to under 230 by 1996. Fear of cholesterol, combined with the food guideline recommendations to reduce or limit egg consumption, explains partially this reduction, leading many in the country to replace eggs with cereals in their breakfast tables. However, the greater availability of beef, pork and chicken as competing protein sources for lunch and dinner also played a role. Only in the last few years has egg consumption in the United States started increasing, with a 10% rise recorded from 1998 to 2016 (American Egg Board, 2016), thanks to a change in the consumers' perception of eggs with acceptance as a more satisfying breakfast ingredient (Fallaize et al., 2013) and – in moderation – an important part of a healthy diet (US Department of Health and Human Services and US Department of Agriculture, 2015).

Thus we can see, as outlined by the examples of Mexico, Japan, Brazil, India and the United States, that the decision to include eggs in the daily diet may be cultural, but can also be related to economic and demographic factors with different variables playing important roles, depending on the country.

2.2 Factors affecting egg purchase decisions

Once somebody decides on eggs for their table, the next step is what kind of eggs. According to Caswell (1998), the quality attributes for food products can be classified as food safety, nutrition, value, package and processes. All of these can have more or less influence on the decision-making process and, depending on culture, country and demographics, can be classified as must-be quality, attractive quality, one-dimensional quality (price), indifferent or even reverse or questionable qualities, according to the Kano Model (Mikulic, 2007). In a study on egg consumer choices in the United States, Wardy et al. (2014) found price and freshness as the main one-dimensional attributes; and clean eggshell, freshness, secure packaging and absence of cracks as the main must-be qualities. The must-be attributes are qualifiers that, when present or absent, qualify that egg to be among the candidates for purchase. Therefore, according to this study, freshness, price, shell cleanliness, secure package and lack of cracks are the main attributes the consumer looks for in an egg.

In another study by Jibir (2013) in Nigeria, where eggs sell singly, size and shell cleanliness were identified as the main quality attributes in the mind of the consumer. The related expectations were:

- a bigger egg will be heavier and contain more edible material;
- a clean shell is related to a hygienic, fresh and safe egg.

In this particular case, price was not among the studied factors and, given the conditions of the country, it was expected that egg size would be the most important factor. In a different study by Parrott et al. (2013) in the United Kingdom, work by Harper Adams University identified production system and animal welfare as the most important factors in the purchasing decision in consumers' minds. However, 'best before' date (freshness) also ranked very highly. Similarly, Sossidou and Elson (2009) believed that, from a European perspective, welfare should be an integral part of any egg quality programme. In British Columbia, Canada, Bejaei et al. (2011) found that 23% of consumers preferred nutritionally enhanced eggs and that higher-income respondents consumed white cage eggs less regularly, but more free-range eggs than lower-income participants. Education level was also found to be important. In Malaysia, egg functionality and egg size were the most important characteristics when purchasing eggs (Hanis et al., 2013). Welfare or production systems were not even ranked or considered.

2.3 Regional differences

As previously demonstrated, when the consumer decides to buy eggs, cultural characteristics and demographics are essential considerations. Different countries consider different factors as important: in the United Kingdom, welfare; in Nigeria, egg size; in the United States, egg integrity; and, in Malaysia, enriched eggs (functional); of course, one should bear in mind that price also plays a role. But even in the same country, regional differences may be expected. In a study by Chang et al. (2010), quarterly egg sales were analysed for over 1985 stock keeping units across the entire United States for 20 quarters from 2004 and 2008. In addition, weekly data were obtained for two metropolitan areas: Dallas/Fort Worth, Texas, and San Francisco/Oakland, California. For the United States, about 95% of the eggs were sold as generic (conventional + none indicated production systems). This number is similar for Dallas, but 10 points lower for San Francisco (Table 1), where the debate for cage-free production has already taken place. Again, cultural and geographical differences should be considered when deciding what kind of eggs to be taken to the market. The cultural trends are very important.

2.4 Packaging and marketing

Packaging is important to protect fragile products such as eggs. According to the FAO (2003), the most important function of packaging is to protect the egg from microorganisms, natural predators, loss of moisture, tainting, extreme temperatures and possible crushing. The same publication distinguishes three kinds of packaging: baskets, cardboard fillers and retail-ready packages.

Baskets can be lined with natural, non-industrial, products such as straw or rice husks, and this is suitable for bulk selling and short-distance transportation. The filler cardboard tray (egg carton or egg box) is by far the most popular form of egg packaging in every country. It is used to transport from the farm to the packing station, and it is also a

Table 1 Percentage of fresh eggs sold by market

Production method	USA	San Francisco/Oakland	Dallas/Fort Worth
Cage-free	1.17	2.29	0.50
Conventional fresh	87.35	84.47	79.4
None indicated	7.43	2.34	17.5
None indicated + conventional*	94.8	86.81	96.9
Organic	1.02	4.43	0.41
Eggs sold (billion)	108	0.73	0.94

*Authors considered 'none' indicated conventional.

Source: Adapted from Chang et al. (2010).

display unit for bulk retailing in many parts of the world. Even in highly industrialized countries such as Germany, the egg tray is still popular in farmers' markets through to high-end supermarkets. The World Poultry Science Association (WPSA) food safety guide recommends using the recycled cardboard trays only once (WPSA, 2016). However, the author has seen cardboard trays being used many times over in countries such as Nigeria and India (Fig. 2). In addition (according to CFIA, 2016), there are other packaging materials for trays designed to transport eggs (plastic, fibreboard) plus wooden or plastic pallets and dividers and others that do not come in direct contact with the egg. The Canadian Food Inspection Agency (CFIA) states that, as a general rule, all packaging material should be clean, dry and protective, that is, able to protect eggs and not cause them any harm. Therefore, plastic trays have the advantage of being washable and reusable, although local regulations must be observed (CFIA, 2016).

Retail-ready packaging can come in many different materials, sizes and presentations. In supermarkets, packaging is related to marketing. According to OMAFRA (2016), food packaging must meet legal requirements, be compatible with the product, protect against environmental contamination, control the product's environment and protect from mechanical damage. It must be sanitary, tamper-proof, attractive, convenient, inexpensive, lightweight, environmentally sound, serve as retailing container, sell itself, identify the product and supply all required information. The same document states that there is no ideal packaging that covers all these different criteria, but that care should be taken to ensure that the most important ones are met, as well as all regulatory standards obeyed. Wardy et al. (2014) found that secure packaging was a must-be driver when purchasing eggs in Louisiana, USA. However, this was not considered an attractive attribute, meaning that an egg that was not securely packed was rejected by consumers. In another study, Parrott et al. (2013) found packaging as ranking 17 out of 18 factors of importance when buying eggs in the United Kingdom. A possible explanation for the low importance of packaging in these two markets is the high quality of eggs offered to the public in general. Thus, factors such as production system and welfare are considered more important and good packaging is regarded as self-evident. In contrast, Jibir (2013) found that, in Nigeria, the conventional 30-egg crate is the standard way to pack, transport and sell eggs (Fig. 2); attention is given to egg size and not to packaging because, in general, there is no secure egg packaging available in that market.

The decision on package size, materials, colours and wording (including claims) is directed by marketing requirements and depends on the country. Packaging can be very creative or very conventional. However, health and nutritional claims are highly regulated in many countries, such as the European Union (EU) (EFSA, 2006), the United States and Japan. However, at the moment, claims relating eggs with brainpower or long life can be seen in countries such as China and India. In a market study of 479 brands covering 20 different countries in Asia, Europe and America (Cisneros et al., 2014), 80% of the brands made no nutritional claims, while the most popular claim with the other brands concerned omega-3- and vitamin E-enriched eggs. On production systems, 73% of the brands were produced in cages, 8% from barn (aviary or cage-free) systems and 19% from free-range systems (including a third organic). Of course, organic and free-range systems claimed the highest prices (up to 300% more than regular eggs) and the consumers in many countries are willing to pay this premium.

Pack size is important. Earlier, eggs were sold by the dozen or in bulk. However, packs of 2, 4, 6, 7, 9, 10, 12 and 30 eggs are now found easily in almost any store around the world. In a study in the Romanian market, Hamelin and Stefanescu (2014) found consumers

Figure 2 Bulk eggs in a Nigerian market near Abuja, 2015. Note the cardboard filler that has been previously used and is soiled.

preferring to buy in 10 and 30 egg packs (39% and 34%, respectively), even where packs with 6, 20, 15 or 8 eggs were also offered.

3 The shell

3.1 Description of eggshell appearance and colour

Eggshell colour is determined mainly by the presence of the pigment protoporphyrin in the shell. Another pigment called 'biliverdin' also contributes to the shell colour of blue eggs. Brown and light-brown shells mainly contain protoporphyrin, which is almost completely lacking in white eggshells. Tinted shells are cream coloured (or light brown). A study with a Chinese strain of hen showed that biliverdin content was significantly higher in blue-shelled eggs than in brown-shelled eggs ($P < 0.01$). The data supported the belief that the avian shell gland is most likely to be the site of biosynthesis and secretion of both eggshell porphyrins and biliverdin (Zhao et al., 2006). The colour strongly depends on the breed of hen. Varieties originating from White Leghorn lay white eggs, while medium-type layers from Rhode Island Red, Plymouth Rock and New Hampshire hens produce brown eggs. Layers of tinted eggs are hybrids of the white and brown egg layers (Li et al., 2006). The heritability of shell colour is moderately high ($h^2 = 0.35–0.45$), which allows breeding companies to improve this trait through selection (Cavero et al., 2012).

Colour measurements can be carried out using equipment such as the Minolta Chromameter CR-300. The values of lightness (L*), redness (a*), yellowness (b*) and chroma (C*) of eggshells can be used to secure objective and quantitative measurements of eggshell colour. In practice, the blunt end of the egg is generally chosen for measuring eggshell colour. The a* value at this single point can be used to classify the three types of eggshell colour (white, blue, tinted/brown). However, the tinted and brown eggs cannot be distinguished accurately with a* value only (Li et al., 2006).

The two main quality defects occurring in eggshells are cracks and dirt stains. Cracks are known to be a factor for transmission of bacteria such as *Salmonella* inside the egg. The shell is thought to be contaminated in the oviduct, by faeces during expulsion of the egg, or when collecting or packing the egg. Dirt stains on eggs are caused mainly by faeces (black to light-brown stains), uric acid (white stains), yolk as a result of leakage from other eggs, and blood.

Reduction of egg defects by selection is of economic importance. Egg defects were evaluated for a fully pedigreed purebred brown-egg experimental layer line. Data from five generations (1 031 045 single egg records from 11 738 hens) were available (Wolc et al., 2012). The percentage of defective eggs (bloody, broken, with calcium deposits, dirty, double-yolked, misshapen, pee-wee, shell-less and soft-shelled eggs) was found to total 3.86. Most of these defects were related to the shells. High-production hens had a lower genetic liability for laying defective eggs. Greater egg weight and body weight were genetically associated with increased frequency of double yolks and, to a lesser extent, with shell quality defects. Each type of defect, except bloody eggs that had a low heritability, was genetically correlated with at least one production or quality trait. Shell thickness and shell ratio heritability were 0.743 and 0.579 in a White Leghorn strain flock in western India (Rath et al., 2015), showing that these traits can be improved by selection.

According to Sun et al. (2012), eggshell thickness was positively correlated with breaking strength (r = 0.4709; $P < 0.01$), and the uniformity of eggshell thickness was also positively correlated with breaking strength (r = 0.3405; $P < 0.01$). This study was run with 52-week-old Hyline brown hens and the uniformity was calculated from seven measurements from the blunt to the sharp end of the egg. Recorded shell thickness of the sharp end was found to be the same as average measurements for a complete egg. With Hisex brown laying hens in three periods during a laying cycle (20–24, 38–42 and 56–60 weeks of age), weight of the three components (yolk, albumen and shell) increased with hen age, but yolk weight increased at a faster rate and therefore made up a greater proportion of the egg than albumen or shell (Tumova and Ledvinka, 2009). In this study, some correlations were found between eggshell strength and shell thickness (0.47), and shell weight (0.40), showing that some other factors are responsible for shell resistance to damage. In both studies, shell thickness appeared to partly explain shell strength. The structural design of the eggshell significantly influences its mechanical properties.

Ultrastructure (the extent and disposition of major structural units within a polycrystalline calcium carbonate ceramic or calcite) and microstructure (the size of crystals, their shape and crystallographic orientation) are especially important. Weaker eggshells are formed by crystals of abnormal sizes (generally larger), shapes or orientation, which negatively affect their mechanical performance. A number of experimental observations support the role of eggshell matrix proteins (e.g. ovalbumin, ovocleidin-116, ovocalyxin-32) in determining the fabric of the shell and therefore its mechanical properties. The well-known reduction in shell strength in aged hens coincides with a change in the relative proportions of matrix proteins in the shell. Old hens lay eggs with shells consisting of larger crystals and highly oriented crystals with an unusual inverted cone shape. These changes in the eggshell crystal texture are associated with the modifications of the quantity and composition of organic matrix, which in turn controls the crystal growth. Forced moulting partially restores the shell strength and reverses the changes previously observed for matrix composition and microstructure of the shell (Rodríguez-Navarro et al., 2014). Because formation and structure of the *Gallus gallus* eggshell are still not fully understood, more research is needed to better understand the function of key components (Hincke et al., 2012).

3.2 Consumer preferences for eggshells

Shell appearance and colour are very important to the consumer, the preference being for a uniform white or brown egg without extreme variations (Schwägele, 2011). Also, the shell is expected to have a normal shape and texture as well as being clean and unbroken. Some countries, such as Australia, the United States, Canada and Japan, allow eggs to be washed and graded before being packed for retail sale, whereas the EU countries do not allow the washing of grade A eggs. Washing has the potential of damaging the cuticle, although manufacturers of modern egg washing equipment aim to reduce impact on the cuticle (Samiullah and Roberts, 2014).

White and brown are the most common eggshell colours found in the market place. In the United States, white shells are the most preferred. In Europe and Asia, many markets prefer brown eggshells. Among the latter markets, there is variance in the intensity of shell colour preferred (Arthur and O'Sullivan, 2005). In Japan, uniform dark brown eggs are the most valued, whereas in other markets a light-tinted brown egg is more acceptable. Uniformity of shell colour shade in the carton appears to be an important characteristic. One disadvantage of brown eggs noted by Arthur and O'Sullivan (2005) is that consumers prefer eggs without obvious blood and meat spots, and brown eggs have a higher incidence of such spots than white eggs. Within both Jewish and Muslim cultures, because of religious laws, the eating of blood-spotted eggs is unacceptable. In China, consumers generally prefer brown eggs to white eggs, while tinted and blue eggs have been developed in recent years. In a consumer preference survey run by DSM Nutritional Products in five cities in China (Beijing, Shanghai, Guangzhou, Jinan and Chengdu), a total of 1500 respondents said the most important criteria for shell quality were 'a smooth and clean eggshell', 'shining eggshell' and 'uniform eggshell colour' (DSM unpublished data, 2007). In India, a consumer preference survey conducted in 18 cities with 1922 respondents revealed that white shells were clearly the most preferred and 32% of people related the white colour with freshness (DSM unpublished data, 2010). In Brazil, there was less clear preference between brown (36%) and white (45%) shells. There, non-broken and clean shells were more important for consumers than shell colour. The survey was conducted in São Paulo with 160 consumers (Souza, 2008).

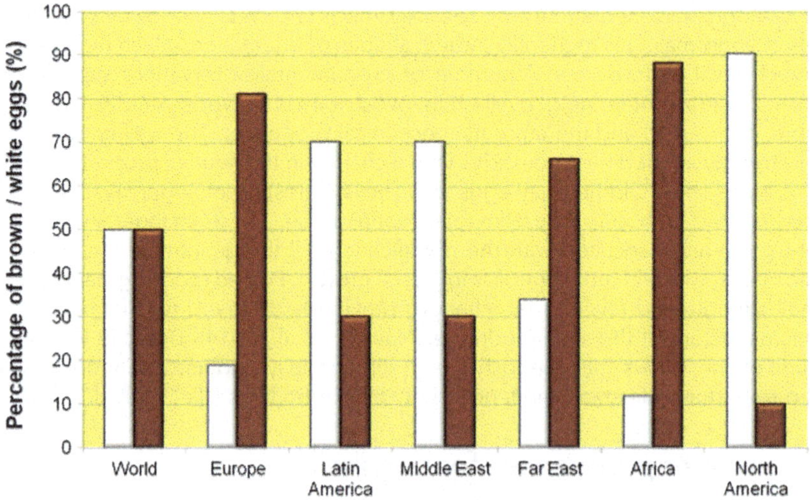

Figure 3 Brown and white eggs market share according to Cavero et al. (2012).

More preferred by African consumers are brown-shelled eggs (Cavero et al., 2012; Fig. 3). But one study in Accra, Ghana, showed that 55% of respondents were indifferent to eggshell colour, 32% preferred brown eggs and 13% preferred white eggs. This outcome disagreed with the general idea that shell colour is an important quality criterion for consumers. Cleanliness of eggs was regarded as much more important in this study (Ayim-Akonor and Akonor, 2014).

In Romania, 1073 interviews were conducted in six main cities. Eggshell cleanliness and resistance to damage were ranked as more important criteria than eggshell colour. Only brown eggs can be found in the Romanian market, which probably influenced the result (Hamelin and Stefanescu, 2014).

In the traditional white shell market of the United States, a survey (from the Utah region) looked at how the consumer viewed properties of eggs with regard to shell colour and at preferences for brown colour intensity (Johnston et al., 2011). Brown eggs represented a very small part of the market and were sold as specialty eggs. Panelists showed a preference ($P < 0.05$) for lighter brown shell colours. Of the panellists in this study who had encountered red or brown spots in eggs (84.6%), almost half responded by noting that they would discard such eggs rather than use them, indicating that such spots are considered major defects. For people who habitually buy brown eggs, this is probably not an issue. But for those who traditionally buy white eggs, switching to brown eggs might be somewhat problematical because the tendency would be to discard many of the eggs. This study highlighted how strong consumer preferences can be regarding shell colour.

Dirty eggs at retail level are normally rare in Western countries. However, they can be found in markets and supermarkets in some other countries (Fig. 4). Dirty eggs are normally downgraded before packing. In a survey run with eggs bought from Argentinian supermarkets, the percentage of dirty shells was 1.12% in white shell eggs (3475 eggs were checked) and 1.74% in brown shell eggs (total 1949 eggs) (Soria et al., 2013). Here, the white eggs were lighter in weight (59.3 ± 6.1 g) than the brown eggs (61.0 ± 6.5 g), a fact which is generally known.

Figure 4 Dirty eggs can still be found in retail settings in many countries.

3.3 Factors influencing eggshell appearance and colour

Eggshell colour is influenced by stress, disease and the age of the laying hen. As the hen gets older, egg weight increases, while shell strength, shell thickness, albumen level and Haugh unit decrease (Silversides and Scott, 2001).

Moulting is a practice that has been used by the commercial egg industry to rejuvenate flocks by extending the laying cycle and restoring egg quality. The main objective of forced moulting is to stop egg production. The hens then enter a non-reproductive state. The eggshell strength, as well as the red chromaticity and the yellow chromaticity, was significantly increased in eggs from moulted Lohmann brown hens during the post-moult period (Aygun, 2013). Stress can negatively affect shell quality. Hens were relocated from one type of housing (floor pens) to another (cages). Shell resistance was decreased after relocation. The incidence of coated, white banded and slab-sided eggs increased for several days after relocation. The yolk colour also reduced in intensity. During the final period, changes were less critical, suggesting that the birds had settled down after the process of relocation (Leary et al., 1997).

Stressful conditions from a high-temperature environment are a major concern to egg producers, particularly in summer or in tropical climates (Oguntunji and Alabi, 2010). A decrease in shell strength and thickness with thermally stressed hens can be attributed to lower feed intake of calcium, reduction in free ionized calcium in the blood and diversion of more blood to the skin for cooling, thereby reducing blood flow and calcium supply to the uterus. Environmental stress can lead to a reduction in the bird's defence mechanisms or to suppressed immune response. A series of studies show consistent benefits from feeding additional vitamin E to laying hens experiencing heat stress. During heat stress, vitamin C supply from endogenous biosynthesis can be inadequate for meeting all physiological requirements. Heat stress tends to lead to deterioration of all kinds of eggshell quality parameters including egg weight, specific gravity, breaking strength and shell thickness. Vitamin C (200 mg/kg) and/or vitamin E (250 mg/kg) supplementation in diets can prevent decreases in eggshell quality and tibia bone strength by alleviating stressful effects from high environmental temperature in broiler breeder hens (Chung et al., 2005).

Odabasi et al. (2007) found that shell colour intensity decreased with age, but this could have also been attributable to an increase in egg size without an accompanying increase in pigmentation. The lighter-coloured eggshells (higher L*) also had a reduced redness hue (a*). This means that once corrected for egg weight, very little change in eggshell pigmentation was observed.

Samiullah et al. (2015) listed the influential factors on the deposition of pigments in brown shell eggs as: housing system, hen age, hen strain, diet, stressors and certain diseases such as infectious bronchitis. However, hen nutrition is not the main factor affecting egg colour variation. With the increase in free-range commercial farming in different parts of the world, there is interest in investigating the factors causing pale shell colour and how to improve brown eggshell colour. It has been reported that eggs from free-range flocks may be lighter in colour than those from cage systems (Samiullah et al., 2014; Sekeroglu et al., 2010). Hen age has an influence because eggshell colour generally gets lighter as the hen ages. When laying hens are stressed, they might retain their egg in the shell gland beyond the normal time of laying and this can result in the deposition of extra-cuticular calcium, which makes brown eggs appear paler. Walker and Hughes (1996) found that the design of the nest box could affect the proportion of eggs laid inside the nest. This affected whether birds experience stress before and during oviposition.

Cases of brown free-range flocks producing a high incidence of white shell eggs were reported in the United Kingdom (Gittins, 2006), particularly during summer months. White-shelled eggs represent a loss of revenue to producers because they are downgraded. Health issues were detected with high titre levels for variant infectious bronchitis strains. Feather loss and reddening of exposed skin was found in problem flocks. It was noticed that if free-range flocks were confined indoors, then shell colour gradually improved after three days. Deterioration in shell colour subsequently occurred when the same flock reverted to free-range.

As the total pigment deposition is affected by stress or disease, Mertens et al. (2010) suggested that colour of brown eggs could be used to monitor the health of the flock, which is directly related to flock productivity. The transmission colour value (TCV) method as visible near-infrared transmission spectroscopy was considered a more informative measure than reflectometry for monitoring a change in shell colour.

In the European Union, housing systems have been changed recently by law (banning of the conventional cages systems by 1 January 2012, Council Directive 1999/74/EU). Many studies were carried out to evaluate effects concerning performance and egg quality. Egg quality depends very much on nest and egg collection systems, and it is not possible to say that one housing system is good or bad. An increase in dirty or cracked eggs has been reported in new alternative systems such as aviaries or barn systems, or free-range systems where there is the possibility of contact with litter. There was a necessary adjustment at packer level to detect faults and a general consensus emerged that more faults and variable quality had to be expected from new systems thought to provide better welfare for hens (Englmaierová et al., 2014). However, when a low proportion of eggs are laid on the floor, and the use of nest and egg collection systems are properly managed, dirty and cracked eggs can be limited to the same level as in conventional systems. Cepero and Hernándiz (2015) concluded their review saying that, if the management and feeding conditions were adequate, no relevant consequences were found in the majority of scientific studies, although in non-cage systems a greater variability in several commercial egg quality parameters is often observed. In recent years, there have been many improvements in design and management of both furnished cages and non-cage housing, aimed towards solving some problems such as increased incidence of dirty and cracked eggs.

While hen nutrition does not influence shell colour, it is a major determinant of shell strength and an important topic for the feed industry. Broken eggs cannot be sold as class A eggs. Bouvarel and Nys (2014) conducted a detailed review of the topic. Adjusting calcium level in the diet according to hen age is a crucial factor in shell strength as well

Table 2 Major determinants of brown shell colour

Disease (e.g. infectious bronchitis)
Environmental stress (e.g. high temperature, design of the nest box, hen relocation)
Ageing
Egg size
Hen strain
Moulting
Housing (e.g. free-range flocks)

Table 3 Major nutritional and feeding determinants of shell strength

Calcium level, carbonate particle size
Fat level
Phosphorus level, phytase level
Vitamin D level
Vitamin C
Sodium bicarbonate
Copper, zinc, manganese, selenium
Feed particle size

as in the prevention of bone demineralization. Part of the calcium should be delivered as coarse calcium that can be stored and released in a targeted way when shell formation occurs, mainly during the night. Calcium retention in the gut can be decreased by the formation of soap, especially with saturated fatty acids. Use of vegetable oils and a minimum of linoleic acid are current practices. Too much calcium or phosphorus above the needs of the bird are not beneficial and can also negatively influence shell quality. The use of phytase is of great importance because phytase improves not only the digestibility of phosphorus, but also that of calcium, trace minerals and amino acids. Vitamin D_3 is known to be a major determinant in bone mineralization and shell quality. Vitamin D_3 metabolites regulate homeostasis of calcium and phosphorous, that is, increase absorption of calcium and phosphorous from the small intestine, as well as affect the re-absorption and secretion of these minerals in the renal tubules, and increase uptake of minerals by bones. Hepatic problems such as fatty liver can decrease plasma levels of vitamin D3 metabolites and affect shell quality. Vitamin 25-hydroxy-D_3 has a higher biological activity and has been used for many years. Vitamin C promotes shell quality in high-temperature weather conditions. Too high levels of chloride can decrease shell quality. The use of sodium bicarbonate as a partial replacement of sodium chloride is useful in achieving good shell quality. Supplementation of copper, zinc, manganese and selenium is necessary for shell formation. The use of organic sources of trace minerals as a substitute for mineral sources could help to secure eggshell strength (Solomon and Bain, 2012), although other studies have been inconclusive in this context. Organic sources are also better bioavailable sources and lower dosages can be used to reduce losses in the environment, while retaining eggshell quality (Saldanha et al., 2009).

The effect of feed technology (milling method, thermal treatment) and presentation (pellet, meal, particle size) on egg quality has not been widely studied. Due to the fact that hens have a preference for larger feed particles, the particle size distribution may affect egg quality, as nutrients may not be equally ingested or effectively utilized.

4 The albumen

Albumen is one of the two main components of the internal egg. It is important not just from the point of view of what brand of eggs to choose, but also in consideration of eggs as part of everyday ingredients for baking, regional dishes and desserts. A good functional albumen is one of the best allies in the kitchen. Albumen constitutes about 60% of the whole egg weight and comprises 88% water and 10% proteins (Sugino et al., 1997). The proteins in the albumen have a very important protective role (Table 3), but at the same time are important in determining quality and functionality of the egg white.

For the food processing industry, the most important characteristics of egg albumen quality are related to gelling, foaming and other rheological characteristics. Hammershøj et al. (2001) found that the percentage of egg rejection for poor gelling in industrially boiled eggs can reach up to 60% for 60-week-old hens, compared to just 20% in the same flock at 33 weeks of age; similar trends were found for other farms in the same study. As the hen ages, there is a strong reduction of albumen dry matter and therefore protein content. Differences according to hen strain were significant, but not as important as hen age. In a different study, Alamprese et al. (2012) also found a strong effect of hen age on gel structure and loss of liquid consistency. However, foaming capability was not affected as much. In the same work, the effect of hen housing was also studied and, although significant differences were found, they were of minor magnitude compared with hen age.

On the other hand, for the family carer, albumen is important from two different standpoints: transparency (able to demonstrate absence of foreign particles and bloody whites) and consistency. The absence of meat and/or blood particles is very important for the appeal of the egg product (Hamelin and Stefanescu, 2014). Deaton et al. (1984) found a 2.5% incidence of eggs with blood spots >3.2 mm. However, this incidence varies due to influence of genetics, housing, age and nutrition (Jeffrey, 1945). Blood spots in albumen can become a hazard because they contribute to the survival of *Salmonella* in infected eggs (Smith and Musgrove, 2008) and therefore their importance goes beyond mere cosmetics. However, the vast majority of commercial eggs are free of this condition, thanks to the use of various candling methods. On the other hand, an opaque albumen in candling can also be caused by blood or bacterial infection. Therefore, transparency defined as the absence of non-albumen materials such as blood, meat or signs of microbial growth is a very important qualifier if the particular egg is to be eaten and this is therefore the most basic and important sign of quality for albumen.

Albumen consistency is given by the quaternary configuration of the protein ovomucin (responsible for the gelling properties of egg white), and the composition and configurations are affected by hen age and by egg storage conditions. However, some

Table 4 Composition and role of the major egg white proteins

Protein	Albumen (%, dry mass basis)	Characteristics
Ovoalbumin	54	Phosphoglycoprotein
Ovotransferrin	12	Binds metallic ions
Ovomucoid	11	Inhibits trypsin
Ovomucin	3.5	Staloprotein, viscous
Lysozyme	3.4	Bacteriolytic
Avidin	0.05	Biotin-binding

Source: Adapted from Mine (1995).

Table 5 Haugh unit preference by Australian customers

Respondents (N = 1617)	Haugh unit 3	Haugh unit 33	Haugh unit 71	Haugh unit 84	Haugh unit 92
First preference	21	103	244	131	118
Second preference	45	70	157	229	116
Final preference	5	4	1	2	3

Source: Adapted from Kelleway (2008).

other aspects such as hen nutrition, housing, disease and toxicity need to be considered as well (Miles and Henry, 2004; Alamprese et al., 2014; Honkatukia, et al., 2013).

The consistency of the albumen is usually measured as Haugh unit (HU), that is, based on the height of thick albumen immediately surrounding the yolk adjusted by egg weight (Haugh, 1933). The lower the value, the more watery the appearance of the egg white. The HU value is influenced by egg storage time and temperature, hen breed and even housing conditions and therefore it is not necessarily a measurement of freshness, but of the consistency of the albumen (Coutts et al., 2007).

In a unique study in Australia cited by Kellaway (2008), eggs with HUs from 3 to 92 were presented to consumers in order to study their preference. The consumers' preference was towards a mid-range albumen consistency with an HU at around 71. However, because it gave an impression of freshness, a thick albumen was significantly more preferred in this study than one with a runny, flat consistency. On the other hand, an egg presenting very high HU value proves very difficult to peel when hard-boiled.

5 The yolk

Yolk appearance is mainly characterized by yolk colour. Yolk colour can be determined by using the DSM YolkFan™ (DSM, 2016), a subjective method, and expressed in the 1–16 scale, or by a colorimeter (MiniScan XE HunterLab, Minolta colorimeter CR-300). This is an objective method with results expressed in the HunterLab scale (L, lightness; a, redness and b, yellowness). Both methods are well correlated: YolkFan = 11.5 + 0.0169 L* + 0.376 a* – 0.0687 b* with R^2=85% ($P < 0.01$) (Hamelin and Thoby, 2015). Because the YolkFan is more economical and easy to use, it is often the preferred method with shell eggs and egg products. A third method based on spectrophotometry used in particular in Italy for pasta eggs (AOAC, 1973) determines the carotenoid content of the yolk expressed as beta-carotene equivalent. The same method is available as a fast and portable spectrophotometer calibrated for egg products (Schweigert et al., 2010). It gives a quantitative evaluation of the carotenoid content in egg powders or liquids. It is perfectly correlated with the equivalent standardized method carried out in the laboratory (R^2 = 97%).

5.1 Consumer preferences for egg yolk

Few studies have reported any information regarding consumer preferences concerning egg yolk. Yolk colour is one of the main criteria for assessing quality of eggs. Since 1997,

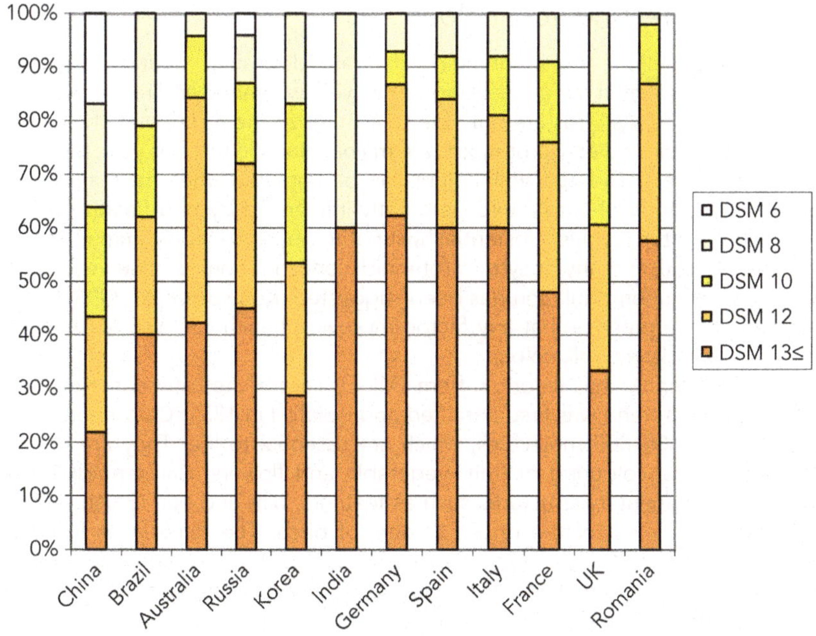

Figure 5 Consumer preference for yolk colour according to consumer surveys run in different countries from 1997 to 2014, per cent consumers that preferred colour 6, 8, 10, 12 or more than 13 as evaluated by the DSM YolkFan.

studies have been conducted in various countries by DSM on how consumers perceive egg quality (Fig. 5). Part of the work was dedicated to questions regarding yolk colour preference. Most consumers in the surveys expressed preference for the yolk colours higher than 12 presented in samples of eggs with yolks corresponding to YolkFan 8, 10, 12 and 14 (see Fig. 5). In the last survey, conducted in six cities in Romania, 57% of consumers preferred colour 14 and 29% preferred colour 12. The yolk colour was said to be the second most important criterion in egg quality after the freshness (Hamelin and Stefanescu, 2014). In an egg-quality survey run in Saint-Brieuc, France (Hamelin and Thoby, 2015), the average yolk colour was 11.6 ± 1.4, mottled eggs represented 3.3% of all eggs and yolk colour spots 5.8%. The homogeneity of yolk colour in eggs within a single retail pack improved significantly when colour intensity increased. Cage and high-value (Label Rouge) eggs presented the highest YolkFan value when compared to organic and free-range eggs. Contrary to this, the housing system was not influential in a survey conducted in the cities of Spain and Portugal (Hamelin and Martinez-Aleson, 2014). The average colour of the yolk was 12.4 ± 1.2. Also, the homogeneity of the colour inside the pack (six eggs) improved when the colour intensity increased. Taking a global view (36 cities in 3 continents), the yolk colour was 10.3 ± 2.62, lower for cage eggs (10.0) than barn eggs (10.7) and highest in free-range eggs (11). Vast differences in yolk colour were detected among the different countries (Cisneros et al., 2014).

5.2 Factors affecting yolk colour

Yolk colour comes from the carotenoids that are deposited during egg formation. Carotenoids are present in the animal kingdom in yellow, orange and red and are typically associated with birds. As birds are not able to synthesize them *de novo*, the carotenoids originate from the bird's diet. Carotenoids are mainly found in the egg yolk, the bird's skin, fat, liver and feathers. The egg yolk colour can be impaired when hens are challenged by disease. Deposition of carotenoids is greatly influenced by bird health status. Since these challenges often involve substantial host pathology and chronic immune responses, an increased bursal and thymocyte carotenoid concentration is observed. A role for carotenoids in immunomodulation has been suggested (Koutsos et al., 2003; Surai et al., 2003) based on the premise that any factor improving health can also help achieve the best and most consistent yolk colour.

The word 'carotenoid' has its origin from *Daucus carota*, the Latin name of the carrot, from which beta-carotene was first extracted and isolated in 1831. Carotenoids comprise a family of some 600–800 molecules, which are essential to life. They are responsible, in particular, for the colouring of fruit, vegetable and flowers. Carotenoids have been identified in integumentary structures such as feathers, bills and skin (Brush, 1981). They are responsible for most of the bright colours of birds. The differences in carotenoid profiles are determined by the processes responsible for their absorption, transport and metabolic changes. The range of carotenoids found in birds is rather broad.

In poultry production, there are two components in the pigmentation process. The first refers to the saturation phase and involves deposition of yellow carotenoids to create the yellow base. This yellow base is very important to ensure good colour consistency. The addition of red carotenoids is called colour phase and changes the hue towards more orange. For table eggs, quantities of yellow and red carotenoids can be added depending on the market requirement. Feed supplementation should be carried out for 10 days minimum to achieve a constant level of yolk colour. Higher concentrations of yellow carotenoids in the feed are necessary to fulfil the requirement for some applications such as pasta production (Sirri et al., 2007), so that colour can resist the aggressive processing of feed ingredients required in achieving the final products.

In modern poultry production, carotenoids come from the following ingredients (Table 6): yellow corn and derivatives (corn gluten, dried distillers corn solubles), alfalfa and derivatives (concentrate), nature-identical (canthaxanthin, apo-ester, citranaxanthin) or carotenoids are extracted from marigold flowers (tagetes) or from peppers. Lutein and zeaxanthin are yellow carotenoids found in corn and its by-products and in marigold products. Extracts from peppers can provide red carotenoids, mainly capsanthin and capsorubin. However, they also contain lutein and zeaxanthin (Soto and Hernandez, 2001; Huyghebaert, 2008), although the percentage of carotenoids of low efficiency (beta-carotene, violaxanthin, cryptoxanthin) is almost one-third of the total carotenoid content. Apo-ester (or apo-carotene-ester), canthaxanthin and citranaxanthin are commercially available sources of nature-identical carotenoids (Amaya et al., 2014). Due to its higher pigmentation efficacy compared to citranaxanthin, canthaxanthin is the preferred red xanthophyll in poultry farming (Breithaupt, 2007). The synthetic products are usually considered as more standardized and stable than extracted products. In the study by Sirri et al. (2007), the content of carotenoids in feed samples showed very little variation with respect to table values for apo-ester compared with marigold extracts, which were 30% lower than the expected levels. This result is partly due to storage losses. Esterification and saponification usually help to produce more stable tagetes and paprika products, as does spray drying protection. There is a wide variation in the quality of the coating. Gadient et al. (2010) checked the stability of various red carotenoids in premixes. After 3 months of storage, recovery rates varied from 66 to 92.1% (canthaxanthin products), 75.6% (citranaxanthin product) and 38.5% (capsanthin from a paprika extract). The quality must be checked carefully when comparing commercial products. Magnin et al. (2009b) reviewed the stability of 10 tagetes and 7 paprika extracts from the market and found high variability in recovery of carotenoids according to the storage time in premixes. Carotenoid stability depends on their isomeric forms. *Trans*-configuration is the most common form and is more efficiently deposited in the yolk compared with *cis*-configuration, which can be produced when damage from heat, light or oxygen occurs (Leigh Hadden et al., 1999). Good coating can prevent such damage. Some research (Fredriksson et al., 2006, Kotrbacek et al., 2013) looked at the effect of algal products in the hen's diet. Lutein and zeaxanthin proved to contribute to yolk colour, but few products are commercially available and use in practice is low.

A Greek study (Sinanoglou et al., 2011) looked at the carotenoid content of egg yolks from five different avian species (ostrich, turkey, quail, duck and geese) under free-range conditions with free access to grazing grasses and to a commercial corn-based diet. The high-performance liquid chromatography (HPLC) analysis of carotenoids from egg yolk lipids revealed the presence of six carotenoids: lutein, zeaxanthin and canthaxanthin in the highest quantities; neolutein, cryptoxanthin and beta-carotene in smaller amounts. The carotenoid levels in the yolk were mainly species dependent. However, housing conditions and diet preference (e.g. grass, grains, seeds and insects) were also considered as factors. A German survey (Schlatterer and Breithaupt, 2006) revealed the presence of six carotenoids in egg yolks: lutein, zeaxanthin and canthaxanthin in the highest amounts followed by citranaxanthin, apo-carotene-ester and cryptoxanthin. Experiments with boiled eggs proved that apo-carotene-ester had the highest stability after boiling, whereas lutein suffered the most degradation (loss of 19%). Organic eggs contained only lutein, zeaxanthin and cryptoxanthin as the specifications for organic production do not

Table 6 List of ingredients, sources of carotenoids used in modern poultry production. NA = ingredient not available in the market; Nature-identical = produced from synthesis; Nature-extracts = substances extracted from plants (tagetes, paprika) with solvents and then saponified

	Natural source of carotenoids and their metabolites	Use in poultry feed
Apo-carotenal	Grass, alfalfa, green vegetables, citrus fruits	NA
Apo-carotene-ester		Nature-identical
Canthaxanthin	Cantarella mushroom, algae, insects	Nature-identical
Citranaxanthin	Citrus fruits	Nature-identical
Cryptoxanthin	Fruits, pumpkin, yellow maize, algae, peach and clams	NA
Lutein, zeaxanthin	Yellow maize, alfalfa, petals of yellow flowers, green vegetables, fruits, algae	Nature-extracts
Capsanthin	Paprika	Nature-extracts

allow any synthetic carotenoid in feed. When hens have access to grass, as in free-range systems, it is well known that yolk colour can be enhanced (Dvorak et al., 2011). However, this contribution can vary according to season.

Carotenoid content in the yolk can be assessed using the HPLC analysis and the deposition rate of individual carotenoids can also be calculated. In the study by Hammershoj et al. (2010), alpha- and beta-carotene deposition rates in egg yolk were below 1%, whereas deposition of lutein and zeaxanthin varied from 19 to 27%. The birds mainly accumulate oxycarotenoids. Beta-carotene does not contribute significantly to pigmentation. Apo-ester is the best-deposited yellow carotenoid in egg yolk. Sirri et al. (2007) measured the transfer efficiency of apo-ester and marigold from the diet to the liquid egg as 50–5% and 16–18%, respectively. Fru et al. (2006) reported a deposition rate of about 60% for apo-ester. Canthaxanthin is the most efficiently deposited red carotenoid in egg yolk. In the study by Grashorn and Steinberg (2002), the deposition rate of canthaxanthin was 40%, independently of dietary supplementation level. However, canthaxanthin deposition success is also source dependent. Pigmentation was also significantly more stable with reference to egg cooking than the new products (Fru et al., 2007). Clasadonte et al. (2014) looked at the stability of canthaxanthin products during storage (3 months at each stage) of straight products, in the premix and in the feed. The lowest stability for a product resulted in availability of only 18% of the initial content, while the most stable product recorded 66% in this respect. Bioavailability of formulated products can be influenced by many factors, for example, matrix type and structure, formulation particle size, possibility of formation of non-absorbable complexes with matrix components and degradability of carrier. Research on lycopene supplementation showed that it is possible to accumulate lycopene in the yolk, thus enhancing the colour. Nevertheless, the deposition rate of lycopene was rather low, about 13%, which may explain why lycopene is not widely used for this application (Olson et al., 2008). Astaxanthin has a potential for yolk pigmentation but with transfer efficiency of not more than 15%, it is the carotenoid widely used in salmonid and crustacean aquaculture.

It has been established that esterified carotenoids have to be hydrolysed before their intestinal absorption and deposition in the yolk, mainly as free alcohol independently of the dietary form. Saponified xanthophylls can exhibit a higher pigmenting ability (Galobart et al., 2004, Lokaewmanee et al., 2011). This effect could be even more pronounced with young chicks.

The results of many trials indicate that 3 g of lutein/zeaxanthin in the feed is required to substitute 1 g of apo-ester and achieve similar yolk colour (Umar-Faruk et al., 2014; Sirri, 2007; Nys, 1999; Steinberg et al., 2000; Grashorn and Seehawer, 1999). Steinberg et al. (2000) found that increase in total xanthophylls in egg yolks and liquid eggs was not linear with increasing dietary inclusion levels of tagetes, while apo-ester was deposited in a linear way. The efficiency ratio was higher with high dietary inclusion levels (40 ppm apo-ester or 120 ppm tagetes) than the assumed ratio of 3:1. Apo-ester is the preferred yellow carotenoid when looking for high-colour products (>30 ppm beta-carotene equivalent). Because the carotenoid profile of several paprika extracts can vary, the efficiency for egg pigmentation was evaluated at 22–39% of canthaxanthin, which is a 2.6:1 to 4.5:1 ratio of efficiency (Huyguebaert, 2008). Magnin et al. (2009a) found a ratio of 1:3 between canthaxanthin and paprika extracts. The paprika extracts are not commonly used in egg production because of low efficiency. Citranaxanthin is less efficient than canthaxanthin and 1.56 ppm citranaxanthin was found to be equivalent to 1 ppm canthaxanthin (Huyguebaert, 1993).

Table 7 Main determinants of yolk appearance and colour

Carotenoid levels in the diet of hens
Hen's health, especially gut health
Supplement of carotenoids for minimum 10 days
Right balance of yellow and red carotenoids in the diet
Stability and efficiency of sources of carotenoids
Hen strain
Lipid supplementation and quality in the diet

While the composition in carotenoids in the diet and bird health are the most important determinants of yolk appearance, bird's genetics can also play a role. Hyline white birds had a lower deposition rate of yolk carotenoids than Lohmann brown birds (Sirri et al. 2007). Dekalb white hens had yolks of less intensive colour than ISA brown hens (Grobas et al. 2001). Finally, the oxycarotenoids present in the diet are lipid soluble, and it is well known that their absorption from the intestine is likely to be improved when fed in parallel with lipids.

6 Conclusion

Whereas Eggs can be seen as a very standard and predictable product, it appears that their appearance and color can be influenced by many factors. Most important factors are bird genetic, bird health, nutrition and feeding technics, feed quality, bird housing and management. Preference for egg appearance is local and is different according to the market. There is no global concensus or overview on the desirable egg appearance. Local surveys exist about egg preference from the consumers but there is no consolidation. This can be seen as a paradox when considering that there are only a few number of genetic companies supplying hens worlwide.

7 Where to look for further information

Information can be found at local or international associations of egg producers:

https://www.internationalegg.com/
http://www.aeb.org/ http://www.eggs.ca/
http://www.institutohuevo.com/
http://lesoeufs.fr/
http://uovoitaliano.it/assoavi/
http://www.eggs.org.au/
http://www.britisheggindustrycouncil.co.uk/Home/
http://www.laidinbritaineggs.co.uk/

There are a number of companies that produce the major hens used worldwide:

http://www.ltz.de/
http://www.isapoultry.com/
http://www.hyline.com/
http://www.novogen-layer.com/

A Good documented book was recently published: *Improving the safety and quality of eggs and egg products*, Volumes 1 and 2:

http://store.elsevier.com/Improving-the-Safety-and-Quality-of-Eggs-and-Egg-Products/isbn-9781845697549/

8 References

Alamprese, C., Casiraghi, E. and Rossi, M., 2012, 'Foaming, gelling and rheological properties of egg albumen as affected by the housing system and the age of the laying hens', *Int. J. Food Sci. Technol.*, 47: 1411–20.

Amaya, E., Becquet, P., Carne, S., Peris, S. and Miralles, P., 2014, 'Carotenoids in animal nutrition', FEFANA Publication ISBN 978-2-9601289-4-9, p. 37. Available at: http://www.fefana.org/sites/default/files/files/booklets/2014-12-04_booklet_carotenoids.pdf

American Egg Board, 2016, http://www.aeb.org/farmers-and-marketers/industry-overview (accessed 1 June 2016).

AOAC, 1973, 'AOAC final action method 17.002', *J. AOAC.*, 56: 272.

Arthur, J. A. and O'Sullivan, N., 2005, 'Breeding chickens to meet egg quality needs', *Int. Hatchery Practice*, 19(7): 7, 9.

Aureli, R., Jenn, P., Schierle, J. and Umar Faruk, M., 2014, 'Egg yolk pigmentation efficiency of apo-ester compared to two products with high zeaxanthin concentration', *19th European Symposium on Poultry Nutrition*, Potsdam, Germany, 26–29 August 2014, WPSA.

Aygun, A., 2013, 'Effects of force molting on eggshell colour, egg production and quality traits in laying hens', *Rev. Med. Vet.*, 164(2): 46–51.

Ayim-Akonor, M. and Akonor, P. T., 2014, 'Egg consumption: patterns, preferences and perceptions among consumers in Accra metropolitan area', *Int. Food Res. J.*, 21(4): 1457–63.

Baiao, N. C., Mendez, J., Mateos, J., Garcia M. and Mateos, G. G., 1999, 'Pigmenting efficacy of several oxycarotenoids on egg yolk', *J. Appl. Poult. Res.*, (8): 472–9.

Bejai, M., Wiseman, K. and Cheng, K. M., 2011, 'Influences of demographic characteristics, attitudes and preferences of consumers on table egg consumption in British Columbia, Canada', *Poult. Sci.*, 90: 1088.

Bouvarel, I. and Nys, Y., 2012, 'Optimizing egg mass and quality traits in modern laying hens through nutrition', *19th European Symposium on Poultry Nutrition*, Potsdam, Germany, 26–29 August 2014, WPSA.

Breithaupt, D. E., 2007, 'Modern application of xanthophylls in animal feeding – a review', *Trends Food Sci. Technol.*, 18: 501–6.

Brush, A. H., 1981, 'Carotenoids in wild and captive birds', in C. Bauernfeind (ed.), *Carotenoids as Colorants and Vitamin A Precursors*, Academic Press Inc., London Ltd., UK Edition, pp. 539–62.

Cámara de diputados, 2015, http://www.cefp.gob.mx/publicaciones/documento/2015/abril/cefp0072015.pdf (accessed 8 February 2016).

Caswell, J. A., 1998, 'Valuing the benefits and costs of improved food safety and nutrition', *Aus. J. Agric. Res. Econom.*, 42: 409.

Chang, J. B., Lusk, J. L. and Norwood, F. B., 2010, 'The price of happy hens: A hedonist analysis of retail egg prices', *J. Agric. Res. Econom.*, 35: 406.

Cavero, D., Schmutz, M., Icken, W. and Preisinger, R., 2012, 'Attractive eggshell color as a breeding goal', *Lohmann Information*, October 2012, 47(2): 15.

Cepero, R. and Hernándiz, A., 2015, 'Effects of housing systems for laying hens on egg quality and safety', *XXII European Symposium on the Quality of Poultry Meat and XVI European Symposium on the Quality of Eggs and Egg Products*, 10–13 May 2015, Nantes, France.

CFIA, 2016, '7.7.1. Packaging materials', http://www.inspection.gc.ca/food/eggs-and-egg-products/manuals/shell-egg/chapter-7/eng/1375307783346/1375307784283#a7.7.1 (accessed 2 June 2016).

Chang, J. B., Lusk, J. L. and Norwood, F. B., 2010, 'The price of happy hens: A hedonic analysis of retail egg prices', *J. Agric. Res. Econom.*, 35: 406.

Chung, M. K., Choi, J. H., Chung, Y. K. and Chee, K. M., 2005, 'Effects of dietary vitamins C and E on egg shell quality of broiler breeder hens exposed to heat stress', *Asian-Australasian J. Anim. Sci.*, 18(4): 545–51.

Cisneros, F., Hamelin, C., Miranda, J. F., Yao, D. and Hayakawa, T., 2014, 'Claim, value and Yolk color in 479 brands of eggs in twenty different countries', *Egg: Bioactive Treasure Proceedings Banff Egg Forum*, June 2014, Wroclaw.

Clasadonte, L., Gadient, M., Schierle, J. and Hamelin, C., 2014, 'Stability of red and yellow carotenoids in stored poultry premix', *XIVth European Poultry Conference*, 23–27 June 2014, Stavanger, Norway.

Conley, A., 2011, 'Consumer survey of egg preferences', University of Kentucky, https://www.uky.edu/fsic/SurveyeggPref.pdf.

Council Directive 1999/74/EC of 19 July 1999 laying down minimum standards for the protection of laying hens, Official Journal of the European Communities, 3 August 1999, L203, pp. 53–7.

Couts, J. A., Wilson, G. C., Fernández, S., Rosales, E., Weber, G. and Hernández, J. M., 2007, 'Optimum egg Quality: A practical approach', pp. 50–1, 5M publishers, Sheffield, GB. ISBN 0-9530150-6-8.

Cruz, J. S., 2012, 'El Mercado de huevo en Mexico, 1965–2010', Master Thesis. Colegio de Postgraduados. Mexico.

Deaton, J. W., Reece, F. N. and Thornberry, F. D., 1986, 'Atmospheric ammonia and incidence of blood spots in eggs', *Poult. Sci.*, 65: 1427–8.

Dos Santos, J. I., Schlindwein, M. M. and Scheurmann, G. N., 2009, 'Fatores determinantes do consume de ovos no Brasil', *Ver. De Economía Agrícola, São Paulo*. 56: 37.

DSM, 2015, 'Yolk colour fan' – A0015, HMB Printed in Switzerland (1/0802:9.3).

Dvořak, P., Suchý, P., Straková, E. and Kopřiva, V., 2011, 'Possibilities of enhancing the colour of egg yolk', *J. Sci. Food Agric.*, 92(4): 853–6.

EFSA, 2006, 'Addition of Vitamins and minerals', http://ec.europa.eu/food/safety/labelling_nutrition/vitamins_minerals/index_en.htm.

Englmaierova, M., Tumova, E., Charvatova, V. and Skrivan, M., 2014, 'Effects of laying hens housing system on laying performance, egg quality characteristics, and egg microbial contamination', *Czech J. Anim. Sci.*, 59(8): 345–52.

Fallaize, R., Wilson, L., Gray, J., Morgan LM. and Griffin, BA., 2013, 'Variation in the effects of three different breakfast meals on subjective satiety and subsequent intake of energy at lunch and evening meal', *Eur. J. Nutr.*, 52: 1353.

FAO, 2003, 'Egg marketing. A guide to the production and sale of eggs', FAO agricultural services bulletin 150, Roma, ftp://ftp.fao.org/docrep/fao/005/y4628E/y4628E00.pdf (accessed 1 June 2016).

Fredriksson, S., Elwinger, K. and Pickova, J., 2006, 'Fatty acid and carotenoid composition of egg yolk as an effect of microalgae addition to feed formula for laying hens', *Food Chem.*, 99(3): 530–7.

Fru, F., Philipps, P., Aureli, R. and Schierle, J., 2006, 'Comparing the efficiencies of a gelatine free apo-ester versus a gelatine apo-ester formulation in egg yolk pigmentation', Tagung Schweine- und Geflügelernährung, Halle (Saale), 28–30 November.

Fru, F., Phillips, P., Aureli, R. and Schierle, J., 2007, 'Comparing the yolk pigmentation efficacy of various red carotenoids and their in-yolk cooking stability', *XII European Symposium on the Quality of Eggs and Egg Products*, Prague.

Gadient, M., Schierle, J. and Hernandez, J. M., 2010, 'Stability of red carotenoids in stored poultry premix', *XIIIth European Poultry Conference*, August 2010, Tours.

Galobart, J., Sala, R., Rincon-Carroyo, X., Manzanilla, E. G., Villa, B. and Gasa, J., 2004, 'Egg yolk color as affected by saponification of different natural pigmenting sources', *J. Appl. Poult. Res.* 13: 328–34.

Gittins, J., 2006, 'When brown hens give white eggs', *Poult. World*, 160(12): 28.

Grashorn, M. A. and Seehawer, J., 1999, 'Use of apo-ester and tagetes extracts for yolk pigmentation in fresh and boiled eggs', *VIII European Symposium on the Quality of Eggs and Egg Products*, 19–23 September 1999, Bologna, Italy, pp. 163–6.

Grashorn, M. A. and Steinberg, W., 2002, 'Deposition rates of canthaxanthin in egg yolks', *Arch. Geflugelkd*, 66(6): 258–62.

Grobas, S., Mendez, J., Lazaro, R., De Blas, C. and Mateos, G. G., 2001, 'Influence of source and percentage of fat added to diet on performance and fatty acid composition of egg yolks of two strains of laying hens', *Poult. Sci.*, 80(8): 1171–9.

Hamelin, C. and Stefanescu, T., 2014, 'Egg yolk colour preference and its importance on the perception of egg quality in Romania', *XIVth European Poultry Conference*, 23–27 June 2014, Stavanger, Norway.

Hamelin, C. and Thoby, J. M., 2015, 'Claims, value and internal egg quality in table eggs from supermarkets in the area of Saint-Brieuc', *XVI European Symposium on the Quality of Eggs and Eggs Products and XXII European Symposium on the Quality of Poultry Meat*, 10–13 May 2015, Nantes-France.

Hamelin, C. and Stefanescu, T., 2014, 'Egg yolk colour preference and its importance on the perception of egg quality in Romania', *XIVth European Poultry Conference*, 23–27 June 2014, Stavanger, Norway.

Hamelin, C. and Martinez-Aleson, R., 2014, 'Estudio de huevos de gallina adquiridos en supermercados de distintas regiones en la península Ibérica: Etiquetado, precio y color de la yema', *59° Simposio Cientificode avicultura AECA/ WPSA*, 2–3 de Octubre 2014, Valencia Spain.

Hammershoj, M., Kidmose, U. and Steenfeldt, S., 2010, 'Deposition of carotenoids in egg yolk by short-term supplement of coloured carrot (Daucus carota) varieties as forage material for egg-laying hens', *J. Sci. Food. Agric.* 90: 1163–71.

Hammershøj, M., Larsen, L. B., Ipsen, R. H. and Qvist, K., 2001, 'Effect of hen egg production and protein composition on textural properties of egg albumen gels', *J. Texture Studies*, 32: 105–29.

Hanis, A., Izani, A. H., Shamsudin, M. N. and Shahrim, M., 2013, 'Consumer's preferences for eggs attributes in Malaysia: Evidence from conjoint survey', *Int. Food Res. J.*, 20(5): 2865–72.

Haugh, R. R., 1933, 'A new method for determining the quality of an egg', *U.S. Egg and Poultry Magazine*, 39: 27–49.

Hernandez, J. M., Hamelin, C., Hayakama, T., Ibarra, B. and Souza, R., 2009, 'Adding value to egg products', *World Poult.*, 25(5): 24, 26.

Hincke, M. T., Nys, Y., Gautron, J., Mann, K., Rodriguez-Navarro, A. B. and McKee M. D., 2012, 'The eggshell: structure, composition and mineralization', *Front. Biosci.*, 17: 1266–80, 1 January 2012.

Honkatukia, M., Tuiskula-Haavisto, M., Arango, J., Tabell, J., Schmutz, M. , Preisinger, R. and Vilkki, J., 2013. 'QTL mapping of egg albumen quality in egg layers', *Genet. Selec. Evol.*, 45: 31.

Huyghebaert, G., 2008, 'The utilisation of red oxy-carotenoids from paprika extracts for egg yolk pigmentation', *23rd World's Poultry Congress in Brisbane*, Australia.

Huyghebaert, G., 1993, 'The biological effectiveness of canthaxanthin and citranaxanthin in yellow-corn-alfalfa based diets for egg yolk pigmentation', *5th European Symposium on the Quality of Eggs and Egg Products*, Vol. II, 4–8 October 1993, Tours, pp. 348–56.

IEC, 2014, http://marketdata.internationalegg.com/2014/ (accessed 6 February 2016).

Jeffrey, F. P., 1945, 'Blood and meat spots in chicken eggs', *Poult. Sci.*, 24: 363–74.
Jibir, M., Isa, A. M., Garba, S., Jibrila, I. and Ibrahim, A., 2013, 'Value perception of shell egg characteristics as determined by size and colour', *J. Anim. Sci. Adv.*, 3: 457.
Johnston, N. P., Jefferies, L. K., Rodriguez, B. and Johnston, D. E., 2011, 'Acceptance of brown-shelled eggs in a white-shelled egg market', *Poult. Sci.*, 90(5): 1074–9.
Kellaway, J., 2008, 'The egg category: differentiate or die- a case study', *Proceedings IEC*, Shanghai, China.
Kotrbacek, V., Skrivan, M., Penkava, O., Hudeckova, P., Uhrikova, I. and Doubek, J., 2013, 'Retention of carotenoids in egg yolks of laying hens supplemented with heterotrophic Chlorella', *Czech J. Anim. Sci.*, 58(5): 193–200.
Koutsos, E. A., Calvert, C. C. and Klasing, K. C., 2003, 'The effect of an acute phase response on tissue carotenoid levels of growing chickens (Gallus gallus domesticus)', *Comp. Biochem. Physiol. A.*, 135: 635–46.
Leary, A., Roberts, J. R. and Ball, W., 1997, 'The effect of relocation from floor pens to cages on egg shell quality in laying hens', *Proceedings of Australian Poultry Science Symposium*, 9(9): 224.
Leigh Hadden, W., Watkins, R. H., Levy, L. W., Regalado, E., Rivadeneira, D. M., Van Breemen, R. B. and Schartz, S., 1999, 'Carotenoid composition of marigold (Tagetes erecta) flower extract used as nutritional supplement', *J. Agric. Food Chem.*, 47(10): 4189–94.
Li, X. Y., Xu, G. Y., Hou, Z. C., Zhao, R. and Yang, N., 2006, 'Variation of eggshell colour in different egg-type chickens', *Archiv fur Geflugelkunde*, 70(6): 278–82.
Lokaewmanee, K., Yamauchi, K., Komori, T. and Saito, K., 2011, 'Enhancement of yolk color in raw and boiled egg yolk with lutein from marigold flower meal and marigold flower extract', *J. Poult. Sci.*, 48(1): 25–32.
Magnin, M., Jeanmichel, P. and Mahieu, A., 2009a, 'Comparaison de l'efficacité de trois sources de pigments rouges naturels et de la canthaxanthine pour la coloration du jaune de l'œuf', *Huitièmes Journées de la Recherche Avicole*, 25–26 Mars 2009, St Malo, France, p. 20.
Magnin, M., Jeanmichel, P. and Mahieu, A., 2009b, 'Stabilité comparée de dix extraits de tagetes et de sept extraits de paprika, sources de pigments naturels jaunes et rouges pour les volailles', *Huitièmes Journées de la Recherche Avicole*, 25–26 Mars 2009, St Malo, France, p. 20.
Mertens, K., De Ketelaere, B., Kamers, B., Bamelis, F. R., Kemps, B. J., Verhoelst, E. M., De Baerdemaeker, J. G. and Decuypere, E. M., 2005, 'Dirt detection on brown eggs by means of color computer vision', *Poult. Sci.*, 84(10): 1653–9.
Mertens, K., Vaesen, I., Loffel, J. Kemps, B., Kamers, B., Perianu, C., Zoons, J., Darius, P., Decuypere, E., De Baerdemaeker, J. and De Ketelaere, B., 2010, 'The transmission color value: A novel egg quality measure for recording shell color used for monitoring the stress and health status of a brown layer flock', *Poult. Sci.* 89: 609–17.
Mikulić, J., 2007, 'The kano model–A review of its application in marketing research from 1984 to 2006', *Proceedings of the 1st International Conference Marketing Theory Challenges in Transitional Societies*, University of Maribor, pp. 87–96.
Miles, R. D. and Henry, P. R., 2004, 'Effect of time and storage conditions on albumen quality of eggs feed Vanadium', *J. Appl. Poult. Res.*, 13: 619.
Mine, Y., 1995, 'Recent advances in the understanding of egg white functionality', *Trends Food Sci. Technol.*, 6: 225.
Nobouka, S., 2015, http://marketdata.internationalegg.com/japan/ (accessed 8 February 2016).
Nys, Y., 1999, 'Evaluation de l'efficacité relative de xanthophylles végétales et de l'apocarotene ester en presence de canthaxanthine pour colorer le jaune d'oeuf', *3èmes Journées de la Recherche Avicole*, 23–25 March 1999, St Malo, pp. 423–6.
Odabasi, A. Z., Miles, R. D., Balaban, M. O. and Portier, K. M., 2007, 'Changes in brown eggshell color as the hen ages', *Poult. Sci.*, 86(2): 356–63.
Oguntunji, A. O. and Alabi, O. M., 2010, 'Influence of high environmental temperature on egg production and shell quality: A review', *World's Poult. Sci. J.*, 66(4): 739–50.

Olson, J. B., Ward, N. E. and Koutsos, E. A., 2008, 'Lycopene incorporation into egg yolk and effects on laying hen immune function', *Poult. Sci.*, 87(12): 2573–80.

OMAFRA, 2016, http://www.omafra.gov.on.ca/english/food/industry/food_proc_guide_html/chapter_7.htm'Packaging and labelling (accessed 3 March 2016).

Parrott, P., Walley, K. and Custance, P., 2013, 'Consumer defined dimensions of egg quality', *XV European Symposium on the Quality of Eggs and Egg Products*, September 2013, WPSA, Bergamo, Italy.

Rath, P. K., Mishra, P. K., Mallick, B. K. and Behura, N. C., 2015, 'Evaluation of different egg quality traits and interpretation of their mode of inheritance in White Leghorns', *Vet. World*, 8(4): 449–52.

Robert, F., Panheleux-Le Bastard, M., Hamelin, C. and Boulard, C., 2007, 'Effects of canthaxanthin supplementation in the ROSS breeder diet on oxidative stress of chicks', *WPSA European Symposium on Poultry Nutrition*, France

Rodríguez-Navarro, A. B., Dominguez-Gasca, N., Gonzalez-Segura, A., Gautron, J. and Nys, Y., 2014, 'Factors determining eggshell structure organization and its influence on eggshell properties', *XXII European Symposium on the Quality of Poultry Meat and XVI European Symposium on the Quality of Eggs and Egg Products*, 10–13 May 2015, Nantes, France

Saldanha, E. S. P. B., Garcia, E. A., Pizzolante, C. C., Faittarone, A. B. G., da Sechinato, A., Molino, A. B. and Laganá, C., 2009, 'Effect of organic mineral supplementation on the egg quality of semi-heavy layers in their second cycle of lay', *Brazilian J. Poult. Sci.*, October to December 2009, 11(4): 215–22.

Samiullah, S. and Roberts, J. R., 2014, 'The eggshell cuticle of the laying hen', *World's Poult. Sci. J.*, 70(4): 693–708.

Samiullah, S., Roberts, J. R. and Chousalkar, K., 2015, 'Eggshell color in brown-egg laying hens-a review', *Poultry Science*, 94(10): 2566–75.

Sastri, B. S. R., 2014, http://marketdata.internationalegg.com/india/ (accessed 5 February 2016).

Schlatterer, J. and Breithaupt, D., 2006, 'Xanthophylls in commercial egg yolks: Quantification and identification by HPLC and LC-(APCI)MS using a C30 Phase', *J. Agric. Food Chem.*, 54: 2267–73.

Schwägele, F. C., 2011, 'Egg quality assurance schemes and egg traceability', *Improving the Safety and the Quality of Eggs and Egg Products*, vol. 1, Woodhead Publishing, p. 65.

Schweigert, F. J., Hurtinne, A., Mothes, R. and Sschierle, J., 2010, 'Analysis of total carotenoids in egg yolk – A fast and laboratory-independent assay', *XIIIth European Poultry Conference*, Tours

Sekeroglu, A. and Altuntas, E., 2010, 'Effects of egg weight on egg quality characteristics', *J. Sci. Food Agric.*, 89: 379–83.

Silversides, F. G. and Scott, T. A., 2001, 'Effect of storage and layer age on quality of eggs from two lines of hens', *Poult. Sci.*, 80(8): 1240–5.

Sinanoglou, V. J., Strati, I. F. and Miniadis-Meimaroglou, S., 2011, 'Lipid, fatty acid and carotenoid content of edible egg yolks from avian species: A comparative study', *Food Chem.*, 124: 971–7.

Sirri, F., Iaffaldano, N., Minelli, G., Meluzzi, A. and Rosato, M. P., 2007, 'Comparative pigmentation efficiency of high dietary levels of apo-ester and marigold extract on quality traits of whole liquid egg of two strains of laying hens', *J. Appl. Poult. Res.*, 16: 429–37.

Smith, D. P. and Musgrove, MT., 2008, 'Effect of blood spots in table egg albumen on salmonella growth', *Poult. Sci.*, 87: 1659.

Solomon, S. E. and Bain, M. M., 2012, 'Structural and physical changes in the hen's eggshell in response to the inclusion of dietary organic minerals', *Br. Poult. Sci.*, 53(3): 343–50, June 2012.

Soria, M. A., Bueno, D. J. and Bernigaud, I. I. C., 2013, 'Comparison of quality parameters in hen's eggs according to egg shell color', *Int. J. Poult. Sci.*, 12(4): 224–34.

Sossidou, E. N. and Elson, H. A., 2009, 'Hens' welfare to egg quality: a European perspective', *Poult. Sci. J.*, 65: 709.

Soto-Salanova, M. and Hernandez, J. M., 2001, 'Effect of storage time on the stability of several carotenoid sources (paprika xanthophylls vs canthaxanthin) in premix and layer mash feed: difference between yolk pigmenting carotenoids and total xanthophylls', *13th European Symposium Poultry Nutrition*, October 2001, Blankenberge, Belgium.

Souza, R., 2008, 'Egg marketing in Brazil', *Zootechnica*, May 2008.

Steinberg, W., Kluenter, A. M., Schierle, J. and Blanch, A., 2000, 'Comparative pigmentation efficiency of two products containing either apo-ester or tagetes extracts in egg yolks and liquid eggs', *Arch.Gefluegelk.*, 64(4): 1–8.

Sugino, H., Nitoda, T. and Juneja, L. R., 1997, 'General chemical composition of hen eggs', in Yamamoto et al. (eds), *Hen Eggs: Their Basic and Applied Science*, CRC Press, Osaka, p. 19. ISBN 0-8493-4005-5.

Sun, C. J., Chen, S. R., Xu, G. Y., Liu, X. M. and Yang, N., 2012, 'Global variation and uniformity of eggshell thickness for chicken eggs', *Poult. Sci.*, 91(10): 2718–21.

Surai, A. P., Surai, P. F., Steinberg, W., Wakeman, W. G., Speake, B. K. and Sparks, N. H., 2003, 'Effect of canthaxanthin content of the maternal diet on the antioxidant system of the developing chick', *Br. Poult. Sci.*, 44(4): 612–9.

The World Bank, 2015, http://web.worldbank.org/WBSITE/EXTERNAL/COUNTRIES/SOUTHASIAEXT/0 contentMDK:20916955~pagePK:146736~piPK:146830~theSitePK:223547,00.html (accessed 8 February 2016).

Tůmová, E. and Ledvinka, Z. 2009, 'The effect of time of oviposition and age on egg weight, egg components weight and eggshell quality', *Archiv fur Geflugelkunde*, 73(2): 110–15.

Umar Faruk, M., Schierle, J., Aureli, R. and Cisneros, F., 2014, 'The pigmenting efficiency of two different yellow carotenoid products in egg yolk', *XIVth European Poultry Conference*, 23–27 June 2014, Stavanger, Norway.

US Department of Health and Human Services and US Department of Agriculture. 2015–2020, Dietary guidelines for Americans. 8th Ed. December 2015. http://www.cnpp.usda.gov/dietary-guidelines.

USDA, 2016, http://www.ers.usda.gov/data-products/food-availability-(per-capita)-data-system/.aspx (accessed 2 February 2016).

USDA, 2016b, http://www.fsis.usda.gov/wps/portal/fsis/topics/regulatory-compliance/labeling/Claims-Guidance (accessed 3 March 2016).

Walker, A. W. and Hughes, B. O., 1998, 'Egg shell colour is affected by laying cage design', *Br. Poult. Sci.*, 39(5): 696–9.

Wardy, W., Behanis, M., Nongtaodum, S., Kyoon No, H. and Prinyawiwatkul, W., 2014, 'Exploring the drivers of purchase intent and consumer satisfaction of chicken eggs using principal component analysis and the Kano model', *J. Sensory Studies*, 29: 463.

Wolc, A., Arango, J., Settar, P., O'Sullivan, N. P. , Olori, V. E., White, I. M. S., Hill, W. G. and Dekkers, J. C. M., 2012, 'Genetic parameters of egg defects and egg quality in layer chickens', *Poult. Sci.*, 91: 1292–8.

World Poultry Science Association, 2016, http://www.wpsa-foodsafety.com/index.php?item=119 (accessed 3 March 2016).

Zhao, R., Xu, G., Liu, Z., Li, J. and Yang, N., 2006, 'A study on eggshell pigmentation: Biliverdin in blue-shelled chickens', *Poult. Sci.*, 85(3): 546–9.

Chapter 13

Understanding and improving the shelf-life of eggs

Juliet R. Roberts, University of New England, Australia

1 Introduction

The shelf-life of food products is determined by several characteristics, including the appearance of the product, taste, sensory quality and microbiological safety. The changes occurring in food that cause loss of quality are physical, chemical, enzymatic and microbiological (Gould, 2000a). In the case of table eggs, the main determinants of shelf-life are the internal quality of the egg and the microbiological status of the egg and its contents. Many strategies are available for food preservation, but only some of these are suitable for table eggs (Gould, 2000a).

Avian eggs have unique structural properties. The avian egg possesses many intrinsic anti-microbial properties that are designed to ensure the safety of the developing chick. However, these same properties also protect the consumer of the egg. Many factors influence the internal quality and microbiological safety of eggs, including the strain of hen, the age of the flock, the microbiological status of the flock, the quality of the eggshells (ES), egg collection procedures, egg handling including washing and packaging, and any specific technologies that are employed to increase shelf-life.

Other chapters in the collection to which this chapter belongs 'Achieving Sustainable Production of Eggs' outline the composition and properties of the ES, egg albumen and egg yolk (EY). Eggs are regularly implicated in cases of food-borne illness (OzFoodNet Working Group, 2012). Therefore, the microbiological safety of eggs is of paramount importance. Other chapters in the collection describe the pathogenic bacteria affecting eggs, how these pathogens are transmitted into eggs and the process of sampling and

http://dx.doi.org/10.19103/AS.2016.0012.15

detection of pathogens in eggs. Other chapters outline ways of enhancing the natural defences of the egg and management of safety and quality during the egg production process, and evaluate the effect of the production system on egg safety and quality. Mechanisms for improving the safety of eggs by washing, sterilization and packaging are reviewed in other chapters. All of the factors reviewed in the above chapters have an impact on the shelf-life of table eggs.

This chapter reviews technologies for improving the shelf-life of table eggs. Section 1 outlines how shelf-life is defined in different parts of the world. Section 2 describes the properties of eggs that limit their shelf-life and the changes that occur in these properties during storage. The following sections review a range of technologies for extending shelf-life. Of primary importance is the handling and storage of eggs, with refrigeration being one of the most important factors. As eggs may take considerable time to cool down following packaging, rapid cooling has been investigated, although it is important that this process does not induce microcracks in the shells of the eggs. Pulsed visible light, electrolysed water, modified-atmosphere packaging, ultraviolet radiation, ionizing radiation, whole egg pasteurization and ultrasonic treatment have all been investigated, at least experimentally. Various coatings of the egg, such as oils and waxes, carbohydrate coatings such as those based on cellulose or chitosan, emulsions of oils and chitosans, proteins and synthetic polymers, have all been tested. Oils, both mineral and vegetable, have excellent moisture proofing properties, whereas chitosan coatings have anti-microbial properties, which makes the combination of these two substances promising for use in extending the shelf-life of eggs. The chapter concludes by discussing the commercial application of technologies for extending shelf-life of eggs and how different technologies can be combined for optimal effectiveness.

The official determination of shelf-life of table eggs varies among and within countries. For example, in the United States, the Department of Agriculture (USDA) provides the following guidelines for the handling of table eggs (Anon., 2015b). Expiration dates, the dates by which eggs must be sold, can be no longer than 30 days from the day the eggs were packed into the carton; 'use by' dates (also called 'best by' or 'best if used by'), the dates by which eggs should be used or discarded, 'may not exceed 45 days including the day the eggs were packed into the carton' (Anon., 2015b). The USDA recommends that eggs be stored in a refrigerator (40°F or 4.4°C) for 3–5 weeks and that eggs and products containing eggs be cooked thoroughly. The American Egg Board (Anon., 2015a) also provides guidelines on the handling, cooking and storage of eggs and egg products. Australian Egg Corporation (Anon., 2010a) published a voluntary code of practice for shell eggs in Australia. The advice given is that eggs be marked with a 'best before' date of a period of six weeks or less from the date of pack. The Food Standards Agency in the United Kingdom (Anon., 2015c) and the European Food Safety Authority state that eggs must be delivered to the consumer within a maximum of 21 days of laying (Anon., 2015d). In many countries, including Australia, eggs must be stamped to facilitate trace-back when problems arise with egg quality and safety.

2 Factors affecting egg internal quality

Commercial egg production inevitably requires that eggs be stored for a period of time, even though all attempts are made to minimize the time interval between when the egg is laid and when it is consumed. The hen's egg is nature's incubation chamber for a new chick and, therefore, its properties are directed to ensuring that a healthy chick is able to hatch from the egg. The domestic chicken's ancestor, the jungle fowl, would typically lay a clutch of eggs and then incubate the eggs all at once. Therefore, as discussed by Board (2000), the first laid egg of a clutch needs to have an anti-microbial system that is active for about 20 days at ambient temperature and this is reflected in the time frame of effective anti-microbiological defences of eggs.

Egg internal quality is measured both experimentally and commercially as albumen quality (albumen height, Haugh Unit, albumen viscosity) and yolk colour (Stadelman, 1995a). The presence of pores in the ES means that carbon dioxide and water vapour are lost from the egg contents during storage (Mueller, 1958, 1959; Reinke et al., 1973), causing a deterioration of egg internal quality. Another change that occurs with increased storage time is the osmotic movement of water from the albumen into the yolk. This causes the yolk to expand in volume and places pressure on the perivitelline membrane, making it more prone to rupture. At the time of lay, the yolk is at the centre of the egg, held in place by the paired chalazae. In a stale egg, the chalazae undergo changes in such a way that the yolk may become displaced and drift towards the uppermost region of the egg.

The effects of storage time and conditions on egg internal quality have been documented over many decades (Akyurek and Okur, 2009; Heath, 1977; Hidalgo et al., 1996; Jones and Musgrove, 2005; Kahraman-Dogan et al., 1994; Roberts, 2006; Scott and Silversides, 2000; Silversides and Scott, 2001; Williams, 1992). Jones and Musgrove (2005) measured changes in egg internal quality of eggs stored for 10 weeks at 4°C and 80% relative humidity (RH). Albumen height decreased from a mean of 7.05 at week 0 to 4.85 at week 10 and Haugh Unit decreased similarly from 82.59 at week 0 to 67.43 at week 10. However, vitelline strength did not change significantly during the 10-week storage period under storage conditions. Roberts (2006) stored eggs at three temperatures, 4°C, 10°C and 25°C, and measured albumen height and Haugh Units over a 10-week period. Albumen height and Haugh Unit decreased with increased storage time and this occurred more quickly at higher temperatures. A similar finding was reported by Akyurek and Okur (2009) using a smaller sample size of eggs from hens at 22 and 50 weeks of age, stored at 4°C or 20°C and 55% RH for 3, 7, 10 or 14 days. Both albumen height and Haugh Unit decreased with increased storage time and occurred more rapidly at 20°C than at 4°C. Albumen quality was also lower in eggs from the older hens. Yolk index (yolk height divided by yolk width) generally declined with increased storage time and decreased more rapidly at the higher storage temperature, although the results for storage of eggs from 50-week-old hens at 4°C showed no clear pattern. Air cell height increased and egg-specific gravity decreased with increased storage time and this occurred to a greater extent at the higher storage temperature.

The changes in egg albumen during storage have been studied at the level of the individual albumen components. Egg albumen constitutes about 59% of the total egg weight. In recent years, new technologies have facilitated the identification of new components in eggs, including albumen proteins (Gautron et al., 2011). Although only 13 egg white (EW) proteins had been identified up until 1989 (Li-Chan and Nakai, 1989), subsequent research, including that of Mann (2007) and D'Ambrosio et al. (2008), increased that number to 148. The loss of carbon dioxide from the egg after the egg is laid has a number of consequences (Stadelman, 1995b). The pH of the egg albumen increases and this leads to the thinning of the egg albumen, which can be measured as lower albumen height and Haugh Unit. A number of hypotheses have been proposed to explain these changes (reproduced from Nys and Guyot, 2012):

- a partial destruction of the lysozyme-ovomucin complex due to an inactivation of lysozyme which would lose 25% of its activity after 25 days of storage at 2°C (Feeney et al., 1952);
- a destruction of electrostatic bonds between $-NH^{3+}$ groups of lysyl residues and COO– groups of sialic acids of ovomucin when pH increases (Robinson and Monsey, 1972; Kato et al., 1975);
- a dissociation of ovomucin components, in particular a solubilization of the β component, rich in glucids, which transfers to the thin albumen. The fact that a part of the divalent cations (Ca^{++} and Mg^{++}) transfer from the albumen to the yolk during storage accelerates this phenomenon;
- a modification of the structure of ovomucin and especially that of β component (Robinson, 1972). This would lead to the transformation of the type I lysozyme-ovomucin complex found at pH7 to a type II complex at pH9;
- a partial degradation of *O*-glycosidic bonds following the solubilization of β-ovomucin and release of carbohydrates (hexoses, kexosamine, sialic acid) by β-elimination at alkaline pH (Kato et al., 1979).

The thinning of EW also results in changes to the functional properties of egg albumen, owing to the transformation of ovalbumin to *S*-ovalbumin. Nys and Guyot (2012) summarize the most common hypotheses for this transformation:

- a rearrangement of free SH and disulphide bonds (Smith and Back, 1968 a,b);
- modifications of charges at the surface of the molecule (Nakamura et al., 1980)
- a more compact structure of S-ovalbumin (Nakamura and Ishimaru, 1981).

In order to prevent the changes in egg internal quality and thus increase the shelf-life of eggs, it is necessary to reduce the loss of carbon dioxide and moisture from the egg. This can be best achieved by sealing the pores of the eggs with an edible substance and storing the eggs under refrigeration.

3 Factors affecting egg microbiological safety

The avian egg is a cleidoic egg, which is nature's incubation chamber for a new bird. Therefore, the egg must contain all the nutrients that the developing chick requires (except for oxygen gas that diffuses into the egg) to develop and hatch and also to support the chick for a short time following hatch (Board et al., 1994). In order to protect the developing chick from invading microorganisms, the egg contains many chemical and physical defences. An extensive review of the microbiology of the avian egg is explained by Board and Fuller (1994). Routes of transmission of microbes into the egg contents are transovarian or oviducal contamination (vertical transmission) or trans-shell (horizontal transmission) (Bruce and Drysdale, 1994; Howard et al., 2012; Musgrove, 2011; Raspoet et al., 2011). The main organism of concern associated with transovarian or oviducal transmission into table eggs is *Salmonella enterica* serovar *enteritidis*. Other serovars of

Salmonella, which are found on eggs, have not been conclusively shown to be transmitted vertically. In Australia, *Salmonella enteritidis* is not endemic in layer flocks, so horizontal transmission is the only route of infection of table eggs. However, in most parts of the world, including Europe and the United States, *S. enteritidis* is endemic in layer flocks, so both vertical and horizontal infection of eggs occur.

The anti-microbial properties of the avian egg are summarized in Board (2000) and Musgrove (2011). Lines of defence for the avian egg from external microbial penetration are the cuticle, ES with organic matrix, inner and outer shell membranes, egg albumen, vitelline membrane (VM) and yolk. Despite the fact that bacteria multiply relatively well in EY, the yolk does contain anti-microbial molecules.

Rehault-Godbert et al. (2011) divided egg anti-microbial molecules into categories, including where in the egg or oviduct these molecules are found: ES, EW, EY, VM, uterus (Ut) or eggshell membranes (SM). Molecules that degrade microbial components include hydrolases (lysozyme – ES, SM, EY, VM), an acyloxyacylhydrolase-like protein (EW), cathepsin L1 (ES) and lysozyme protective protein (Ut). Anti-microbial peptides include avian β-defensins – AvBD-11 (ES, EW, VM), AvBD-10 (ES), Gallin (EW), AvBD-9 (Ut) and histone proteins (14 in total found variously in ES, EW, EY). Lipopolysaccharide-binding and bactericidal/permeability-increasing proteins include ovocalyxin-36 (Ut, ES), TENP (EW, EY, VM), BPIL2 (EW, VM) and similar-to-BPI (EW). C-type lectin-like proteins include ovocleidin 17 (EW, VM, ES), tetranectin (EY), collagen XVIII (ES) and DEC-205 protein (ES). Molecules that decrease the bioavailability of iron and vitamins include iron chelators, including ovotransferrin (all compartments of the egg), vitellogen 1 and 2 (EY) and phosvitin; vitamin chelators include avidin (EW, VM, ES), similar to avidin (EY), riboflavin-binding protein (all egg compartments), retinol-binding protein 4 and vitamin-D-binding protein (EY and/or ES).

Molecules that inhibit the activity of microbial proteases include inhibitors of serine proteases such as kazal-type inhibitors ovoinhibitor and ovomucoid (EW), the serpins ovalbumin-related protein X and ovalbumin-related protein Y (EW), heparin factor II (all compartments except VM), antithrombin (EY), α-2 antiplasmin (ES?), other serpins found in EY, pancreatic trypsin inhibitors (EW, EY, ES), ovocalyxin 32 (Ut), ovomucin (EW, VM), inhibitors of cysteine proteases such as cystatin (all egg compartments) and similar to prekininogen (EY); metalloprotease inhibitors include tissue inhibitors of metalloprotease type 2 and 3 and broad spectrum inhibitors include ovostatin (all egg compartments). In addition, there are members of the immunoglobulin (Ig) superfamily that include immunoglobins and proteins with Ig-like domains and cytokines and other mediators of immune responses. Identification of anti-microbial molecules in eggs is on-going, but the examples listed provide an indication of the extensive anti-microbial properties of all compartments of the avian egg.

The presence of food-borne pathogens on the ES and, for *S. enteritidis*, inside the egg, has been confirmed by many studies (Humphrey et al., 1991; Martelli and Davies, 2012; Musgrove et al., 2004, 2005a,b,c). When eggs are stored, the anti-microbial properties of the egg change over time and are affected by storage conditions. *S. enteritidis* does not grow rapidly in eggs stored at room temperature until eggs have been stored for approximately 3 weeks and this appears to be associated with the changes in the VM (Humphrey and Whitehead, 1993). If eggs are to be stored at ambient temperature (25°C), there is evidence that an initial low temperature shock of 4°C for 4 hours increases the shelf-life of eggs (Theron et al., 2003). This study also found that storage and transport of eggs at a constant temperature of 8°C did not produce safer eggs when compared to storage and transport at 25°C.

For eggs infected by vertical transmission, preventative treatments include minimizing or eliminating the levels of *Salmonella* in the flock. Eggs that contain *Salmonella* when they are laid may be treated by pasteurization and refrigeration. For eggs that are laid with pathogenic bacteria on the outside of the shell, various treatments may be applied to reduce the microbial load and thus reduce the incidence of ingress of microbes into the egg contents. ES quality factors had relatively little effect on bacterial penetration under experimental conditions, although penetrated shells had lower cuticle deposition than those that were not penetrated in the study of de Reu et al. (2006). However, any ES defect has the potential to facilitate microbial entry into the egg contents.

4 Technologies for increasing the shelf-life of eggs

In addition to the treatments described in earlier chapters, a range of other technologies has been investigated for improving the safety and quality of eggs and therefore the shelf-life (Gould, 2000b; Morris et al., 2007). Some of these have potential for improving the shelf-life of table eggs.

4.1 Refrigeration and cryogenic cooling

The changes that occur in eggs during storage occur more slowly and to a lesser extent if eggs are refrigerated. Temperature and storage time and, to a lesser extent, the age of the flock are the most important factors influencing albumen quality (Williams, 1992). Refrigeration of eggs also inhibits the growth of bacteria on or in the eggs (Humphrey, 1994). However, it has been shown that eggs take a considerable time to cool once they are placed in the cool room and that the rate of cooling is influenced by the packaging of the eggs (Mellor et al., 1975a,b; Mellor and Gardner, 1975; Jones et al., 2002). Attempts have been made to cool eggs quickly initially by the use of cryogenic cooling (Curtis et al., 1995; Jones et al., 2002; Keener et al., 2000). Curtis et al. (1995) found that eggs cooled by cryogenic cooling with carbon dioxide gas had higher Haugh Unit after 30 days of storage than eggs cooled traditionally (transferred to a 10°C cool room and allowed to cool passively). Keener et al. (2000) used rapid cooling in either air or carbon dioxide followed by storage in either air or carbon dioxide. The best results were obtained by rapid cooling in carbon dioxide followed by storage in carbon dioxide with the shelf-life of eggs being extended to more than 14 weeks. Jones et al. (2002) used liquid nitrogen, gaseous nitrogen or gaseous carbon dioxide as the cooling medium and compared the results with traditional cooling of eggs. The cryogenically cooled eggs had higher Haugh Units and VM strength than the eggs cooled traditionally. This improvement was greatest for the eggs from older flocks that had poorer initial quality. Rapid gaseous cooling caused a slightly higher incidence of cracked eggs in the study of Jones et al. (2002), as had been reported previously by Fajardo et al. (1995), but the authors concluded that this could be reduced with improved design of the cooling chamber.

4.2 Pulsed visible light

Pulsed light involves brief but intense pulses of broad spectrum light (Barbosa-Canovas et al., 2000; Gould, 2000b; Morris et al., 2007). A company, PurePulse Technologies, Inc., in California, developed the technologies of pulsed light (Purebright®) and pulsed electric field (CoolPure™) for killing of microorganisms in food, including whole eggs and liquid egg (Dunn, 1996). The pulsed light technology uses short-duration (few hundred milliseconds) flashes of broad spectrum high-intensity (20 000 times the intensity of sun light) white light at a rate of 1–20 flashes per second and is suitable for use on whole eggs. The pulsed electric field uses multiple, short-duration, high-intensity electric field pulses but is suitable only for products that are 'pumpable', including liquid egg. Eggs were inoculated by immersion in a suspension of *S. enteritidis*, treated with PureBright® and the wash-in-bag recovery method used to evaluate the effectiveness of the treatment. Eggs that were flashed with eight flashes at 0.5 J/cm^2 per flash for a total of 4 J/cm^2 had up to an 8 log reduction in levels of *S. enteritidis*. In a second experiment, temperature was carefully controlled such that the inoculum was either 2°C or 27°C colder than the egg or 23°C hotter than the egg. Inoculum colder than the egg promotes aspiration of the inoculum into the pores of the ES, as has been shown when eggs are cooled rapidly (Fajardo et al., 1995). PureBright® still reduced the levels of *S. enteritidis* in the eggs, indicating that it is able to penetrate the shell pores at least to some extent.

4.3 Electrolysed water

As outlined in an earlier chapter, sterilizing agents may be added to water used to wash eggs. A novel extension of this process is the use of electrolysed water (Cao et al., 2009; Russell, 2003). Russell (2003) investigated the use of electrolysed oxidative (EO) water applied using electrostatic spraying to decontaminate hatching eggs, as an alternative to treatments such as formaldehyde or hydrogen peroxide. An electrostatic spray-charging system was found to be more effective in distributing the sanitizer over the surface of the eggs, compared with a conventional commercial fogging system. The EO water was prepared by electrolysis of a 20% saline solution with a final pH of 2.1 and an oxidation–reduction potential of 1150 mV. The electrolysis resulted in the production of low concentrations of anti-microbial substances such as chlorine ions, chlorine dioxide, ozone and hydrogen peroxide, which are, most likely, the active ingredients. Eggs were dipped in inoculation solutions of *Salmonella typhimurium*, *Staphylococcus aureus*, *Listeria monocytogenes* and *Escherichia coli*. ES and membranes were treated to neutralize the sanitizers and then tested for the presence of bacteria. The experiment was conducted four times. The EO treatment resulted in a minimum of 4 $\log_{10}$ reduction in *S. typhimurium*, 3 $\log_{10}$ reduction in *S. aureus*, 1 $\log_{10}$ reduction in *L. monocytogenes* and 2 $\log_{10}$ for *E. coli*. A later study by Cao et al. (2009), using table eggs, compared the bactericidal activity at 4, 29 and 45°C of slightly acidic electrolysed water (SAEW), sodium hypochlorite solution (NaClO) and acidic electrolysed water (AEW) in inactivating *S. enteritidis*. The available chlorine, pH and oxidation–reduction potential of each treatment solution were controlled or measured. A reduction of the levels of *S. enteritidis* on shell eggs by 6.5 $\log_{10}$ CFU/g was achieved by SAEW containing 15 mg/L available chlorine for 3 min, compared with a reduction of 0.9–1.2 $\log_{10}$ CFU/g for the control treatment. The result for SAEW compared with that for the NaClO solution and was better than for the AEW. SAEW has the advantages of being noncorrosive, more stable than AEW, inexpensive and less hazardous to workers.

4.4 Modified-atmosphere packaging

General aspects of packaging of eggs are covered in an earlier chapter. Modified-atmosphere packaging of food has been used to reduce microbial growth and extend shelf-life (Church, 1994; Farber, 1991). A concern with the use of this technology has been that it may inhibit spoilage organisms, the growth of which alerts consumers, while not inhibiting the growth of more serious pathogens. Eggs are reported to have been stored in an atmosphere of 20% carbon dioxide and 80% nitrogen (Farber, 1991).

4.5 Ultraviolet radiation

Ultraviolet light is well known for its ability to inactivate microorganisms (Bintis et al., 2000; Guerrero-Beltran and Barbosa-Canovas, 2004; Morris et al., 2007). Kuo et al. (1997) found that short-wave ultraviolet radiation (UVC) was effective in reducing the total aerobic and mould counts and the levels of *Salmonella typhimurium* on the surface of ES. However, these authors did not test the effectiveness of UVC on *S. enteritidis*. Ultraviolet radiation is also suitable for treatment of liquid foods, such as liquid egg (Guerrero-Beltran and Barbosa-Canovas, 2004).

4.6 Ionizing radiation

Food irradiation involves brief exposure of foods to a radiant energy source such as gamma rays, x-rays or electron beams (Anon., 2010b; Farkas, 1998; Thakur and Singh, 1994). When conducted correctly, irradiation is safe and has little effect on nutritional and organoleptic properties of food (Morehouse, 2002; Morris et al., 2007). Serrano et al. (1997) reported that a minimal dose of 0.5 kGy was sufficient to eliminate five different isolates of *S. enteritidis* from the surface of shell eggs. However, the bacteria were more resistant to irradiation when present inside the eggs, with a dose of 1.5 kGy necessary to significantly reduce counts by 4 $\log_{10}$ inside shell eggs or in liquid egg. Neal (1965) concluded that radiation treatment was suitable for pasteurizing liquid egg (even in frozen form). Morehouse (2002) reports that the Food and Drug Administration (FDA) recommends a maximum dose of 3 kGy for shell eggs for pathogen control.

4.7 Whole egg pasteurization

While pasteurization is more commonly conducted with liquid egg (Herald and Smith, 1989), it is also applicable to intact shell eggs (Hou et al., 1996; Schuman et al. 1997). While other methods such as egg washing, rapid chilling, irradiation and ultrasonic treatment inactivate bacteria on the outside of the egg, they do not destroy *Salmonella* inside the egg (Catalano and Knabel, 1994). The study by Vanlith et al. (1995) reported that irradiation at levels that can destroy bacteria inside the egg (6 kGy) also lowered the quality of the irradiated eggs. The functional properties of eggs such as coagulation, foaming and emulsification are very sensitive to heat. Therefore, any heat treatment aimed at destroying microbes must avoid the loss of these functional properties. In the study of Hou et al. (1996), *S. enteritidis* was injected into the yolk of eggs prior to pasteurization. Following the pasteurization treatment, the liquid whole eggs were then sampled for the presence of live bacteria. Two types of equipment were used for the pasteurization, a water bath at 57°C and a hot air oven at 55°C. Some eggs were treated first in the water bath and then in the hot air oven. The eggs were also evaluated for functional properties, Haugh Unit, pH, viscosity, colour and turbidity of egg albumen, foaming ability, foam stability, lysozyme activity, yolk index and colour and storage stability, following the pasteurization treatments. Pasteurization in the 57°C circulating water bath for 25 min reduced the levels of *S. enteritidis* by 3 log, while heating in the hot air oven for 180 min gave a 5 log reduction of *S. enteritidis*. A combination of the two (25 min in water bath then 60 min in hot air oven) resulted in a 7 log reduction in *S. enteritidis*. Following the pasteurization treatments, the functional properties of the eggs remained 'acceptable'.

4.8 Ultrasonic treatment

The use of ultrasound to destroy bacteria was first reported in 1929 (Harvey and Loomis, 1929). Since then, ultrasound has been used on its own or in combination with heat (and pressure in the case of liquid foods) to inactivate bacteria (Cabeza et al., 2004; Gould, 2000b; Morris et al., 2007; Wrigley and Llorca, 1992). Treatment of at least 20 000 vibrations per second is necessary to achieve a bactericidal effect (Morris et al., 2007). The use of ultrasonic wave treatment, combined with a temperature of 50°C, resulted in 1–3 log CFU reductions in *Salmonella typhimurium* in liquid whole egg (Wrigley and Llorca, 1992). Cabeza et al. (2004) attempted the more difficult task of applying thermoultrasonication to intact shell eggs. Ultrasonic treatment (24 kHz, 400 W, 60 μm) was applied in a cell disrupter inside a custom-designed jacketed vessel. Eggs were removed and dried in cold air flow, and the bacterial count of the shell was determined. The ultrasonic treatment combined with heat resulted in a 99.5% reduction in the bacterial load, compared with a 90% reduction for the heat treatment alone. The authors discuss the importance of maintaining a relatively low volume of the treatment media in order to optimize the reduction in bacterial load.

5 Coatings to increase the shelf-life of eggs

A range of substances has been used to coat the outside of shell eggs, with the aim of improving the shelf-life by maintaining internal quality during storage and reducing microbial ingress into the eggs. These coatings include proteins (Biladeau and Keener, 2009; Caner, 2005a,b; Cho et al., 2002; Herald et al., 1995; Rhim et al., 2004; Wardy et al., 2010; Wong et al., 1996; Xie et al., 2002), carbohydrates such as cellulose (Suppakul et al., 2010) and chitosan (Bhale et al., 2003; Caner and Cansiz, 2007, 2008; Jin et al., 2013; Kim et al., 2006, 2008b; No et al., 2005; Wardy et al., 2010; Xie et al., 2002), oils (Biladeau and Keener, 2009; Jirangrat et al., 2010; Kamel et al., 1980; Knight et al., 1972; Obanu and Mpieri, 1984; Sabrani and Payne, 1978; Waimaleongora-Ek et al., 2009; Wardy et al., 2010, 2013), emulsions of chitosan and oil (Torrico et al., 2010, 2011a,b, 2014; Wardy et al., 2013) and synthetic polymers (Meyer and Spencer, 1973).

5.1 Oils and waxes

Lipid films provide a good moisture barrier (Heath, 1977), but their permeability and mechanical characteristics are affected by their fatty acid composition (Wardy et al., 2010). Mineral oil is commonly applied to the blunt end of the ES following commercial washing of eggs to seal the pores, reducing carbon dioxide and water vapour loss and reducing the incidence of microbial contamination (Stadelman, 1995b). Oils tested include mineral oil (Goodwin et al., 1962; Homler and Stadelman, 1963; Kamel et al., 1980; Schwall et al., 1961; Stadelman and Wilson, 1958), polydimethylsiloxane fluids (Knight et al., 1972), linseed oil (Sabrani and Payne, 1978), groundnut oil, cottonseed oil and coconut oil (Obanu and Mpieri, 1984) and canola, corn, grape seed, olive, soybean and sunflower oils (Ryu et al., 2011) (see Table 1).

Waimaleongora-Ek et al. (2009) investigated the effect of mineral oil of different viscosities (7, 11, 14, 18, 22, 26 cP) on quality and shelf-life of table eggs. Eggs in all groups lost weight over the 5-week storage at 25°C, but the extent of the weight loss was greatly reduced by the mineral oil treatments, with the 26 cP viscosity oil resulting in a weight loss over 5 weeks that was approximately one-tenth of the control, uncoated eggs. The mineral oil coating improved the Haugh Unit, yolk index and albumen pH, compared with the control, with the 26 cP treatment again having the best result. A U.S. Grade AA was maintained until 2 weeks of storage for the 14, 18 and 26 cP oils and all treatments. However, with the 7 cP treatment, A-grade eggs were maintained until 3 weeks of storage and the control eggs were classified as B grade after only one week of storage. There was no effect of any of the mineral oil treatments on emulsion capacity of yolk, indicating that mineral oil of any viscosity is suitable for coating eggs. Biladeau and Keener (2009) coated eggs with mineral oil and paraffin wax, stored them at 7°C for 12 weeks and measured Haugh Units, pH of albumen and yolk, albumen CO_2 content, VM strength, egg water loss, shell strength, shell colour, foam volume of albumen, angel cake volume measurements and emulsion stability. The coated eggs had higher Haugh Units, CO_2 content and VM strength and lower albumen pH than the control eggs. Shell strength was higher for the wax-coated eggs than for the control and oil-coated eggs, which may have been due to the thickness of the wax coating. The authors concluded that coating of eggs extended the shelf-life of the eggs 'beyond 6 weeks'. The study by Jirangrat et al. (2010) used both refrigerated storage (4°C) for 15 weeks and storage at 25°C for 5 weeks of eggs coated with mineral oil. When eggs were stored at 25°C for 5 weeks, weight loss was only 0.85% for the coated eggs compared with 9.23% for the control eggs. The coated eggs also had higher Haugh Units (55.8 vs 20.13), yolk index (0.37 vs 0.21), egg grade (B vs C) and lower albumen pH (8.64 vs 9.42) compared with the control eggs. For eggs stored at 4°C, all measurements were improved at 5 weeks of storage, compared with storage at 25°C. The control eggs maintained AA grade to 5 weeks and A grade to 15 weeks, whereas the coated eggs maintained AA grade until 15 weeks of storage. Following 15 weeks of storage at 4°C, the coated eggs had lower weight loss (1.19 vs 12.44%) and albumen pH (7.96 vs 8.53), and higher Haugh Units (74.44 vs 62.90) and yolk index (0.43 vs 0.41). These results indicate that a combination of oil coating and refrigerated storage provides the longest shelf-life of table eggs. If refrigerated storage is not available, coating of eggs with oil provides a significant improvement in shelf-life, compared with uncoated eggs stored at room temperature.

Table 1 Effect of oil coating on egg weight loss and Haugh Unit in stored eggs

Treatment	Storage conditions	Weight loss	Haugh Unit	Reference
		% control	% control	
Silicone 50 cSt	19°C, 59d	9.1	213.2	Knight et al., 1972
Silicone 100 cSt	19°C, 59d	9.2	178.0	
Silicone 350 cSt	19°C, 59d	11.3	203.3	
Silicone 50 cSt	12.5°C, 59d	4.4	137.0	
Silicone 100 cSt	12.5°C, 59d	8.5	137.0	
Silicone 350 cSt	12.5°C, 59d	21.2	132.1	
Linseed oil	12°C, 60–70% RH, 24d	17.8	126	Sabrani and Payne, 1978
	28°C, 60–80% RH, 24d	10.6	202.2	
Mineral oil	25°C, 75–80% RH, 12d	31.6	140.1	Kamel et al., 1980
	37°C, 75–80% RH, 12d	7.8	158.1	
	42°C, 75–80% RH, 12d	4.9	298.3	
Groundnut oil	25–32°C, 40–85% RH, 36d	7.0	NA	Obanu and Mpieri, 1984
Cottonseed oil		7.0	NA	
Coconut oil		8.6	NA	
*Mineral oil	Room Temp, 4 wk	82.9	93.6	Wong et al., 1996
Mineral oil 7 cP	25°C, 5 wk	39.3	120.7	Waimaleongora-Ek et al., 2009
Mineral oil 11 cP		16.9	219.3	
Mineral oil 14 cP		17.3	245.2	
Mineral oil 18 cP		13.4	269.3	
Mineral oil 22 cP		11.9	264.5	
Mineral oil 26 cP		9.7	282.5	
Paraffin wax	7°C, 12 wk	0.71	106.7	Biladeau and Keener, 2009
Mineral oil		10.0	111.3	
Mineral oil	4°C, 15 wk	9.6	118.3	Jirangrat et al., 2010
	25°C, 5 wk	9.2	277.2	
Mineral oil	25°C, 5 wk	10.6	216.1	Wardy et al., 2010
Mineral oil	4°C, 20 wk	9.0	134.5	
Soybean oil	25°C, 5 wk	9.6	217.2	
Soybean oil	4°C, 20 wk	7.7	138.6	
Soybean oil	25°C, 7 wk	7.9	186.0	Wardy et al., 2013
Soybean oil	4°C, 20 wk	10.3	119.9	
Glycerol	25°C, 5 wk	96.4	123.4	Nongtaodum et al., 2013
Coconut oil		11.1	287.2	
Rice bran oil		9.3	291.4	
Soybean oil		9.3	297.7	
Palm oil		7.2	305.1	

*Commercial eggs; NA, not available.

Nongtaodum et al. (2013) investigated the use of alternatives to mineral oil for the coating of eggs, using coconut, palm, rice bran and soybean oils (SOs) as well as glycerol. These authors emphasized that refrigerated storage of eggs is not practised in many countries of the world and that oils other than mineral oil may be more readily available and less expensive in some countries. Eggs coated with one of the oils, or glycerol, were stored at 25°C for 5 weeks and weight loss, Haugh Unit, albumen pH and viscosity, gloss value and sensory properties were evaluated. All oil coatings resulted in lower weight loss, albumen pH, air cell height and higher Haugh Unit and gloss value than the control and glycerol-coated eggs. Albumen viscosity was higher than the control only for SO and palm oil. All eggs coated with oil maintained AA grade to 3 weeks and A grade to 5 weeks of storage and the SO and palm oil maintained AA grade to 4 weeks of storage. The authors concluded that the edible oils used in the study were all suitable for extending the shelf-life of eggs in unrefrigerated storage. This finding is similar to earlier studies (Ryu et al., 2011; Obanu and Mpieri, 1984), which reported that vegetable oils are a satisfactory alternative to mineral oil for the coating of table eggs.

5.2 Carbohydrate coatings

Films made from polysaccharides have limited moisture barrier properties but are very effective in controlling the diffusion of gases (Butler et al., 1996; Wardy et al., 2010). Table 2 summarizes the results of studies investigating the effect of carbohydrate-based coatings on egg internal quality. Of the carbohydrates that have been tested as coatings on ES, the most studied are chitosan-based coatings. These coatings have the advantage that they not only maintain the egg internal quality, as measured by albumen quality, but also have anti-bacterial properties (Wang, 1992). Chitin is 'the second most abundant biopolymer in the world after cellulose' (Friedman and Juneja, 2010) and resembles cellulose in chemical structure. Chitosan is a deacetylated product of chitin, which possesses anti-bacterial and antioxidant properties. A chitosan–lysozyme coating inhibited microbial growth on hard-boiled eggs (Kim et al., 2008a), extended the shelf-life of shell eggs (Kim et al., 2009; Su et al., 2007) and inhibited food spoilage organisms in egg-containing mayonnaise (Oh et al., 2001). Chitosan–glycerol coatings applied by brushing, dipping or spraying, with glycerol, propylene glycol or sorbitol as plasticizer, extended the shelf-life of eggs by 3 weeks in the study of Kim et al. (2008b).

Jin et al. (2013) used the biopolymers polylactic acid (PLA) and chitosan in solutions to coat eggs with the aim of reducing microbial contamination. Chitosan is regarded as an edible polymer, whereas PLA is a food contact polymer. The coating solutions contained the natural anti-microbials nisin, allyl isothiocyanate (AIT – 20, 60 mL) and lauric arginate ester (LAE – 0.1, 0.5, 1.0%) as well as organic (acetic, lactic and levulinic) acids. The PLA coating also contained PLA resin in methylene chloride. ES were inoculated with *S. enteritidis* by spot inoculation at 10 locations and then allowed to dry. The coatings were brushed onto the eggs, which were then placed into a biosafety hood at room temperature to dry. For storage testing, eggs were maintained at 7°C. The ES with membranes attached were tested for microbiological analysis. The coatings reduced weight loss during the 12 weeks of storage and both PLA and chitosan coatings reduced the levels of *Salmonella* to varying extents.

A chitosan coating (molecular weight 223 kDa – 2% w/v concentration in 1% v/v acetic acid) was less effective than mineral and vegetable oils in maintaining egg internal quality as measured by weight loss and Haugh Unit in the study of Wardy et al. (2010). However,

Table 2 Effect of carbohydrate coating on egg weight loss and Haugh Unit in stored eggs

Treatment	Storage conditions	Weight loss	Haugh Unit	Reference
		% control	% control	
Chitosan 2%	25°C, 5 wk			Bhale et al., 2003
High MW 1100 kDa		91.5	>371	
Med MW 746 kDa		89.0	>411	
Low MW 470 kDa		85.3	>370	
Chitosan[a]	25°C, 5 wk			No et al., 2005
30–0		82.0	127.2	
30–5		93.7	138.2	
30–15		84.5	128.3	
20–15		91.2	132.4	
10–15		89.7	102.9	
0–15		110.2	94.6	
0–0		100.5	86.7	
Chitosan	25°C, 4 wk	62.1	131.7	Caner, 2005a
Shellac		11.1	135.1	
Chitosan[b]	25°C, 5 wk			Kim et al., 2006
0% glycerol – AA		91.9	153.8	
0.5% glyc – AA		78.7	183.6	
1.0% glyc – AA		88.0	160.3	
1.5% glyc – AA		81.1	180.1	
2.0% glyc – AA		75.8	187.9	
2.0% glyc – LA		69.0	158.3	
Chitosan	25°C, 4 wk			Kim et al., 2007
α-CH 1110 kDa		87.3	126.9	
α-CH 1110 kDa		72.8	151.3	
α-CH 1110 kDa		80.2	177.9	
α-CH 1110 kDa		76.0	188.5	
β-chitosan		80.0	109.9	
CH-glycerol	25°C, 5 wk	48.6	156.3	Kim et al., 2008b
CH-propylene glycol		57.1	161.1	
CH-sorbitol		40.0	160.4	
Cellulose-based	26°C, 65% RH, 4 wk	48.5	139.4	Suppakul et al., 2010
Chitosan	25°C, 5 wk	102.3	144.5	Wardy et al., 2010
	4°C, 20 wk	93.0	108.0	
Chitosan	25°C, 5 wk	84.8	71.4	Wardy et al., 2013
	4°C, 20 wk	94.9	100.5	

CH, chitosan.
[a]Different deproteinization and demineralization times in preparation of chitosan.
[b]Chitosan in either acetic (AA) or lactic acid (LA) with different % glycerol.

the chitosan eggs had higher Haugh Units than the control for eggs stored at 25°C for 5 weeks and Haugh Units tended to be higher in the coated eggs stored at 4°C for 20 weeks. These findings differ from those of Bhale et al. (2003), who used three chitosans (1100 kDa, 746 kDa and 470 kDa) at two concentrations: 1% w/v (dissolved in 1% v/v acetic acid) and 2% w/v (dissolved in 2% v/v acetic acid). Weight loss was lower for the coated eggs and lowest for the low molecular weight chitosan (470 kDa) at both 1% and 2% concentrations. Haugh Units were significantly higher for the coated eggs, compared with the control and were higher for the 2% solutions than for the 1% solutions of chitosan.

Caner and Cansiz (2007) prepared solutions from chitosan (89.9% acetylation) with 1% of acetic, lactic or proprionic acid and glycerol added as a plasticizer. Eggs were immersed in the solutions and allowed to dry before being stored for 4 weeks at 24°C. All coated eggs had lower weight loss and higher Haugh Units and yolk index than the control eggs. Haugh Units were higher for lactic and proprionic acids than for acetic acid. The control eggs maintained A grade only to 1 week of storage, whereas all coated eggs remained at A grade until 3 weeks of storage, with lactic and proprionic acid groups being at A grade up to 4 weeks of storage. In a later study, Caner and Cansiz (2008) coated eggs with chitosan containing an organic acid (acetic, lactic or proprionic), but in this experiment, albumen pH, shell strength and albumen viscosity were also measured. Albumen pH was lower for all coated groups and lower for the lactic and proprionic acid groups than for the acetic acid group. Albumen viscosity was higher for all coated groups, with the viscosity in lactic and proprionic acid groups being higher than in the acetic acid group. Shell strength was higher for all coated groups, with the percentage in the lactic acid group being the highest.

Kim and co-workers (Kim et al., 2006, 2007, 2008b) conducted several studies to investigate the optimal chitosan solution for coating eggs. The first experiment of the 2006 study (Kim et al. 2006) investigated the effect of 2% w/v chitosan in 1% v/v acetic acid with glycerol (plasticizer) added at different levels: 0, 0.5, 1.0, 1.5 and 2.0% v/v. In a second experiment, chitosan solutions were prepared in 1% acetic acid and/or 2% lactic acid with 2% glycerol. Coatings were applied to eggs by using a sponge brush and eggs were then stored for 5 weeks at 25°C. Coated eggs had lower weight loss and higher Haugh Unit and yolk index. In the second experiment, the acetic acid group tended to have higher Haugh Unit and yolk index, although the differences were not statistically significant. The 2007 study (Kim et al., 2007) investigated two types of chitosan, α-chitosan (four different molecular weights 282, 440, 746 and 1110 kDa) and β-chitosan (MW 577 kDa) and three pH of solutions (4.5, 5.0, 5.5) on the shelf-life of eggs. The pH of the solution did not affect egg internal quality. However, weight loss was lower for all treatment groups, except 1110 kDa α-chitosan, compared with the control group. Haugh Unit was higher for the 282, 440 and 746 kDa α-chitosan but not for the 1110 kDa α-chitosan or the β-chitosan, compared with the control. The anti-bacterial activity of all coating treatments was significantly higher than for the control.

The 2008 study (Kim et al., 2008b) investigated the effect of plasticizer type and coating method on the shelf-life of eggs. Three different plasticizer types (glycerol, propylene glycol and sorbitol) were used in the first experiment and then sorbitol as a plasticizer, along with three different coating methods (brushing, dipping, spraying) was used in the second experiment. A 2% solution of chitosan (MW 282 kDa) was prepared in a 1% acetic acid solution with a 2% concentration of plasticizer. Eggs were stored for 5 weeks at 25°C. In the first experiment, all coated groups had significantly lower weight loss, albumen pH and percentage yolk and higher percentage albumen, Haugh Unit and yolk index.

Although there were no significant differences among the plasticizers, sorbitol was chosen as the plasticizer for the second experiment. In the comparison of coating methods, the highest Haugh Units were achieved when the coating was applied by brushing or dipping. Haugh Unit for the sprayed eggs was not significantly different from the control.

Suppakul et al. (2010) investigated cellulose-based coatings for enhancing the shelf-life of eggs stored at ambient conditions for 28 days. The carbohydrate biopolymers, methylcellulose (MC – 2.0 w/v) plus hydroxypropyl methycellulose (HPMC – 1.0% w/v), were used with varying amounts of polyethylene glycol-400 as a plasticizer and a stearic/palmitic acid blend as a lipid source to produce carbohydrate-based films for coating eggs by immersion. Coated eggs had lower weight loss and albumen pH and higher Haugh Unit than uncoated eggs. The coated eggs maintained AA grade to 14 days of storage and A grade to 28 days compared with the uncoated eggs that maintained AA grade to 5 days and A grade to 14 days. Xie et al. (2002) reported that coating of eggs with carboxymethyl cellulose, following washing with water, sodium carbonate or sodium hypochlorite, reduced dye penetration into the eggs, compared with the washed control eggs.

5.3 Oil–chitosan emulsions

The use of oil–chitosan emulsions has the benefit of combining the ability of oils to restrict movement of moisture and gases with the anti-microbial properties of chitosans (see Table 3). Oils are relatively slow to dry and also change the colour and glossiness of the ES. Torrico and co-workers conducted a series of experiments using mineral oil with chitosan (Torrico et al., 2010, 2011a,b, 2014), whereas the study of Wardy and co-workers (Wardy et al., 2013) used SO. Torrico et al. (2010) prepared three emulsions of food-grade mineral oil (MO – viscosity 26 mPa s) mixed with chitosan (CH – MW 223 kDa, 2% chitosan dissolved in 1% acetic acid) with MO:CH ratios of 72:25, 50:50 and 25:75. Eggs were coated using a sponge brush and dried overnight at 25°C. Seven treatment groups were prepared: an uncoated control group, 100% MO unwiped or wiped; MO:CH = 75:25, 50:50, 25:75 and 100% CH. Fifty eggs per treatment were stored at 25°C and 60–65% RH for 5 weeks. Ten eggs per treatment were removed every week for measurement of weight loss, Haugh Unit, yolk index and albumen pH. Sensory testing was conducted on day 0 and microbiological analysis of egg contents was conducted at day 0 and following 5 weeks of storage. Mineral oil was clearly the best coating for prevention of weight loss as the greater the proportion of MO in the coating, the lower the weight loss. Weight loss of the 100% CH treatment group did not differ from that of the uncoated control. The same pattern was observed for Haugh Unit, yolk index and albumen pH, which were improved by all coatings containing mineral oil but with the 100% CH treatment not different from the uncoated control. Consumers were unable to significantly distinguish between the coated and uncoated eggs, although only the 25:75 MO:CH treatment was judged not to be more glossy than uncoated eggs. No *Salmonella* was detected in the contents of eggs from any of the groups and total plate counts were either not detectable or very low, both before and after storage. The study concluded that the three emulsions of MO and CH were equally effective as MO alone in preserving egg internal quality and more effective than CH alone.

Torrico et al. (2011a) used MO:CH coatings at a ratio of 25:75 with one of four commercially available emulsifiers (Tandem® 552K, Tic® Pretested Ticaloid®210S, Tween, Eficacia – 1%), compared with a coating of MO alone. Tandem and Tween are oil-miscible emulsifiers, whereas Tic and Eficacia are water-miscible. Coatings were applied and eggs stored as described in Torrico et al. (2010), except that eggs were stored at either 4°C

Table 3 Effect of chitosan–oil emulsions on egg weight loss and Haugh Unit in stored eggs

Treatment	Storage conditions	Weight loss % control	Haugh Unit % control	Reference
MO (U)	25°C, 5 wk	9.7	241.3	Torrico et al., 2010
MO:CH 75:25		12.3	245.7	
MO:CH 50:50		12.6	223.7	
MO:CH 25:75		14.4	221.2	
CH		95.5	125.2	
MO	25°C, 5 wk	<11.6	>160.1	Torrico et al., 2011a
MO	4°C, 20 wk	12.0	111.7	
MO:CH 25:75 + Tandem	25°C, 5 wk	<12.9	>155.7	
MO:CH 25:75 + Tandem	4°C, 20 wk	16.7	114.9	
MO:CH 25:75 + Tic	25°C, 5 wk	<14.0	>163.5	
MO:CH 25:75 + Tic	4°C, 20 wk	13.3	108.9	
MO:CH 25:75 + Tween	25°C, 5 wk	<17.8	>157.8	
MO:CH 25:75 + Tween	4°C, 20 wk	16.7	116.9	
MO:CH 25:75 + Eficacia	25°C, 5 wk	<12.8	>179.6	
MO:CH 25:75 + Eficacia	4°C, 20 wk	13.0	115.8	
HHU MO	25°C, *3 wk	15.0	216.1	Torrico et al., 2011b
HHU MO:CH 25:75		12.4	215.1	
MHU MO		6.7	183.2	
MHU MO:CH 25:75		12.9	209.7	
LHU MO		8.1	158.7	
LHU MO:CH 25:75		11.4	119.8	
SO:CH 60:40	25°C, 7 wk	18.1	185.2	Wardy et al., 2013
	4°C, 20 wk	31.6	114.0	
SO:CH 50:50	25°C, 7 wk	21.8	182.5	
	4°C, 20 wk	34.7	112.4	
SO:CH 40:60	25°C, 7 wk	26.2	179.5	
	4°C, 20 wk	37.5	114.4	

SO:CH, soybean oil:chitosan emulsion; MO, mineral oil; HHU, high initial Haugh Unit; MHU, medium initial Haugh Unit; LHU, low initial Haugh Unit.
*Haugh Units not measured in weeks 4 and 5 because very low.

or 25°C. Once again, eggs were sampled weekly for 5 weeks and weight loss, Haugh Unit and yolk index measured. The emulsifiers were equally effective in maintaining egg internal quality at both storage temperatures and there was a synergistic effect between refrigeration and all types of coating.

The suggestion that initial albumen quality may influence the results obtained was tested by Torrico et al. (2011b). The same MO and MO:CH 25:75 coatings were used with eggs stored for 5 weeks at 25°C, but three groups of initial albumen quality (Haugh Unit H (high)=87.8, M (medium) = 75.6, L (low) = 70.9) were used. The authors concluded that all coatings preserved egg internal quality, with the H albumen group extending shelf-life by 4 weeks and the M and L groups by 3 weeks. The later study (Torrico et al., 2014) combined the factors studied in the earlier experiments by using MO:CH = 25:75 emulsions with four different emulsifiers, two storage temperatures 4°C and 25°C and two initial albumen qualities (Haugh Unit H = 87.8 and L = 70.9). Again, all coated eggs maintained egg internal quality for 5 weeks storage at 25°C and 20 weeks storage at 4°C, regardless of the initial albumen quality. There were no differences among emulsifiers at 4°C storage, but there were marginal differences at 20°C storage. Once again, a synergistic effect between coating and refrigeration was observed. Wardy et al. (2013) used SO, chitosan (CH) and emulsions of the two (SO:CH = 60:40, 50:50, 40:60) to coat eggs, which were then stored at either 4°C or 25°C for 5 weeks. Egg internal quality, as measured by weight loss, albumen pH and yolk index was improved by coating with the SO alone, which was generally better than the emulsions. As SO is cheaper than mineral oil, it can be considered a viable alternative.

5.4 Proteins

Table 4 summarizes studies on the effect of protein coatings on egg internal quality. Biladeau and Keener (2009) coated eggs with soybean protein isolate (SPI) and whey

Table 4 Effect of protein coating on egg weight loss and Haugh Unit in stored eggs

Treatment	Storage conditions	Weight loss % control	Haugh Unit % control	Reference
Wheat gluten film	Room Temp, 4 wk	NA	166	Herald et al., 1995
Egg albumen	Room Temp, 4 wk	71.2	113.3	Wong et al., 1996
SPI		58.6	157.7	
Wheat gluten		37.8	161.4	
Corn zein		27.9	162.3	
WPI	25°C, 4 wk	62.7	119.2	Caner, 2005a
WPI 6%	25°C, 4 wk	73.7	117.7	Caner, 2005b
WPI 12%		65.7	125.4	
WPI 18%		64.1	127.4	
SPI	7°C, 12 wk	86.0	109.6	Biladeau and Keener, 2009
WPI		81.4	111.4	
WPI	25°C, 5 wk	91.9	123.9	Wardy et al., 2010
	4°C, 20 wk	88.4	105.0	

WPI, whey protein isolate; SPI, soy protein isolate; NA, not available.

protein isolate (WPI), stored them at 7°C for 12 weeks and measured Haugh Units, pH of albumen and yolk, albumen CO_2 content, VM strength, egg water loss, shell strength, shell colour, foam volume of albumen, angel cake volume measurements and emulsion stability. The protein coatings resulted in higher Haugh Units and albumen CO_2 content, and lower albumen and yolk pH. VM strength was not significantly different from the control for the SPI treatment but was higher than the control for the WPI coating. Weight loss over the 12 weeks of eggs coated with WPI and SPI (2.7 and 2.9 g, respectively) was lower than the control eggs (3.4 g) but significantly higher than for eggs coated with oil (0.35 g) or wax (0.03 g). Coating of eggs with protein isolate, particularly WPI, improved the shelf-life of table eggs (as indicated by Haugh Units), a finding similar to that of the earlier study by Caner (2005b).

Herald et al. (1995) used degradable films from spray-dried and flash-dried wheat gluten solutions (at different pH, salt levels, temperatures and shearing times) for coating eggs and found that the coatings maintained Haugh Units above those of the uncoated eggs, with eggs being A grade quality after 30 days storage at room temperature. Xie et al. (2002) reported that coating eggs with SPI, WPI, carboxymethyl cellulose or wheat gluten enhanced the puncture strength of eggs and inhibited microbial contamination as indicated by post-wash dye penetration. Wong et al. (1996) coated eggs with solutions of egg albumen, soy protein isolate, wheat gluten or corn zein and compared the results with uncoated eggs. All the protein coatings tested reduced moisture loss from the eggs and increased the Haugh Unit, with the corn zein giving the best results, including an increase in shell strength. A whey protein coating was less effective than mineral and vegetable oils in maintaining egg internal quality as measured by weight loss and Haugh Unit (Wardy et al., 2010).

5.5 Synthetic polymers

Meyer and Spencer (1973) coated eggs with a range of synthetic polymers: acrylic resin, polyvinyl alcohol, polyvinylidene chloride, polyvinyl acetate, casein, Zein and Epolene wax. Eggs were immersed in the coating solution using a glass loop and allowed to dry at 24°C. The thickness and weight of the coating achieved by this method varied among the polymers. Egg internal quality following 1 or 7 days storage, as measured by Haugh Unit, was not significantly improved by the coatings, although polyvinyl alcohol, acrylic resin and Zein improved shell strength significantly. The polymer coatings reduced the peelability of hard cooked eggs. Cooking caused the polyvinyl alcohol eggs to become 'gummy' on the surface and the acrylic resin came off from the shell. No more recent studies appear to be available on the use of synthetic polymers as coatings for shell eggs.

6 Summary and future trends

Commercial application of some of the technologies reported in this chapter is relatively uncommon, apart from the routine use of mineral oil following egg washing. However, currently, a combination of whole shell egg pasteurization and coating with food-grade wax is being used by Safest Choice™ eggs (Anon., 2015e). According to the Safest Choice™ eggs web site, eggs are submerged in a water bath at a particular temperature and for a particular length of time (details not provided on the web site). The eggs are then 'sealed with an FDA-approved, food-grade wax'. Eggs are stored under refrigeration below 45°F. This 'precision-pasteurization process' is claimed to extend the shelf-life of the eggs and to guarantee microbiological safety. Although the details of the methodology used are not made public, egg processors seeking to use this technology could use publicly available scientific data (e.g. Hou et al., 1996) as a starting point for their own commercial development.

Egg shelf-life is determined by a combination of egg internal quality and egg microbiological quality. The factors affecting these two properties are not necessarily the same. Egg internal quality is measured by albumen height and Haugh Unit and the strength of the VM. These properties are best in fresh eggs from young hens but can be retained by refrigerated storage of eggs and by the use of various coatings to reduce the rate of loss of moisture and carbon dioxide from the egg after lay. Egg microbial safety is influenced by the microbiological status of the flock, collection of eggs and handling and processing of shell eggs. Correctly controlled washing of eggs can reduce the number of microbes on the outside of the egg. However, a treatment that can penetrate the egg such as whole egg pasteurization or ionizing radiation is necessary to destroy microbes present inside the egg. Coating of the outside of the egg with oil or wax, cellulose-based coatings, chitosan-based coatings, emulsions of oils and chitosans, and proteins of synthetic polymers can reduce the weight loss and reduction in albumen quality occurring during egg storage. Some of these coatings such as the chitosans have anti-microbial properties. Innovative technologies must guide commercial egg operations in various parts of the world so that egg shelf-life can be improved in the context of the market situation prevailing in those parts.

The experimental findings summarized in this chapter are available for use by large commercial companies to be refined and developed for commercial use. Additional experimental research is likely to occur to further define the optimal combination of technologies suitable for improving both egg internal quality and egg microbiological safety. The application of technology to produce products such as Safest Choice™ eggs is likely to be extended to other commercially available products.

7 Where to look for further information

American Egg Board web site: http://www.aeb.org.

Australian Egg Corporation Limited (AECL) web site: http://www.aecl.org; Food Safety Advice https://www.aecl.org/assets/Resources/Food-Safety-Advice-eggs.1.pdf; Code of Practice for Shell eggs https://www.aecl.org/assets/Uploads/Resources/Code-of-Practice-for-Shell-egg-production-grading-packing-and-distribution-2009.pdf.

European Food Safety Authority web site: http://www.efsa.europa.eu/; Cooling of table eggs http://www.efsa.europa.eu/sites/default/files/scientific_output/files/main_documents/biohaz_op_ej957_cooling_eggs_en%2C3.pdf.

Food Standards Agency, U.K. web site: https://www.food.gov.uk; egg production site: https://www.food.gov.uk/sites/default/files/multimedia/pdfs/eggenforcefeb09.pdf.

National Pasteurized Eggs, Inc. web site: http://www.safeeggs.com.

United States Department of Agriculture (USDA) Food Safety and Inspection Service website: http://www.fsis.usda.gov.

8 References

Akyurek, H. and Okur, A. A. (2009), 'Effect of storage time, temperature and hen age on egg quality in free-range layer hens', *J. Anim. Vet. Adv.*, 8(10), 1953–8.

Anon. (2010a), 'Code of practice for shell egg, production, grading, packing and distribution', Australian Egg Corporation Limited, https://www.aecl.org/assets/Uploads/Resources/Code-of-Practice-for-Shell-egg-production-grading-packing-and-distribution-2009.pdf (accessed 7 January 2016).

Anon. (2010b), 'Food irradiation', *Health Physics Society Fact Sheet*, https://hps.org/documents/Food_Irradiation_Fact_Sheet.pdf (accessed 29 October 2015).

Anon. (2015a), American Egg Board (AEB), http://www.aeb.org (accessed 29 October 2015).

Anon. (2015b), United States Department of Agriculture Food Safety and Inspection Service, http://www.fsis.usda.gov (accessed 29 October 2015).

Anon. (2015c), Food Standards Agency, U.K., https://www.food.gov.uk (accessed 30 November 2015).

Anon. (2015d), European Food Safety Authority, http://www.efsa.europa.eu (accessed 30 November 2015).

Anon. (2015e), National Pasteurized Eggs, Inc., http://www.safeeggs.com (accessed 30 November 2015).

Barbosa-Canovas, G. V., Schaffner, D. W., Pierson, M. D. and Zhang, Q. H. (2000), 'Pulsed light technology', *J. Food Sci. Suppl: Kinet. Microb. Inactivation for Alternative Food Process. Technol.*, 65, 82–85.

Bhale, S., No, H. K., Prinyawiwatkul, W., Farr, A. J., Nadarajah, K and Meyers, S. P. (2003), 'Chitosan coating improves shelf life of eggs', *J. Food Sci.* 68(7), 2378–83.

Biladeau, A. M. and Keener, K. M. (2009), 'The effects of edible coatings on chicken egg quality under refrigerated storage', *Poultry Sci.*, 88, 1266–74.

Bintis, T., Litopoulou-Tzanetaki, E. and Robinson, R. K. (2000), 'Existing and potential applications of ultraviolet light in the food industry – a critical review', *J. Sci. Food Agr.*, 80, 637–45.

Board, R. G. (2000), 'Eggs and egg products', In B. M. Lund, T. C. Baird-Parker and G. W. Gould (eds), *The Microbiological Safety and Quality of Food*, Vol II, Aspen Publishers, Inc., pp. 590–619

Board, R. G., Clay, C., Lock, J. and Dolman, J. (1994), 'The egg: a compartmentalized aseptically packaged food', In R. G. Board and R. Fuller (eds), *Microbiology of the Avian Egg*, Chapman and Hall, pp. 43–61.

Board, R. G. and Fuller, R. (1994), *Microbiology of the Avian Egg*. Chapman and Hall.

Bruce, J. and Drysdale, E. M. (1994), 'Trans-shell transmission', In R. G. Board and R. Fuller (eds), *Microbiology of the Avian Egg*, Chapman and Hall, pp. 63–91.

Butler, B. L., Vergano, P. J., Testin, R., Bunn, J. M. and Wiles, J. L. (1996), 'Mechanical and barrier properties of edible chitosan films as affected by composition and storage', *J. Food Sci.*, 61, 953–6.

Cabeza, M. C., Ordonez, J. A., Cambero, I., de la Hoz, L. and Garcia, M. L. (2004), 'Effect of thermoultrasonication on *Salmonella enterica* serovar Enteritidis in distilled water and intact shell eggs', *J. Food Protect.*, 67(9), 1886–91.

Caner, C. (2005a), 'The effect of edible eggshell coatings on egg quality and consumer perception', *J. Sci. Food Agr.*, 85, 1897–902.

Caner, C. (2005b), 'Whey protein isolate coating and concentration effects on egg shelf-life', *J. Sci. Food Agr.*, 85, 2143–8.

Caner, C. and Cansiz, O. (2007), 'Effectiveness of chitosan-based coating in improving shelf-life of eggs', *J. Sci. Food Agr.*, 87, 227–32.

Caner, C. and Cansiz, O. (2008), 'Chitosan coating minimizes eggshell breakage and improves egg quality', *J. Sci. Food Agr.*, 88, 56–61.

Cao, W., Zhu, Z. W., Shi, Z. X., Wang, C. Y. and Li, B. M. (2009), 'Efficiency of slightly acidic electrolyzed water for inactivation of *S. enteritidis* and its contaminated shell eggs', *Int. J. Food Microbiol.*, 130, 88–93.

Catalano, C. R. and Knabel, S. J. (1994), 'Destruction of *S. enteritidis* in high pH and rapid chilling during simulated commercial egg processing', *J. Food Protect.*, 57, 592–5.
Cho, J. M., Park, S. K., Lee, Y. S. and Rhee, C. O. (2002), 'Effects of soy protein isolate coating on egg breakage and quality of eggs during storage', *Food Sci. Biotechnol.*, 11, 392–6.
Church, N. (1994), 'Developments in modified-atmosphere packaging and related technologies', *Trends Food Sci. Tech.*, 51, 345–52.
Curtis, P. A., Anderson, K. E. and Jones, F. T. (1995), 'Cryogenic gas for rapid cooling of commercially processed shell eggs before packaging', *J. Food Protect.* 58, 389–94.
D'Ambrosio, C., Arena, S., Scaloni, A., Guerreri, L., Boschetti, E., Mendieta, M. E., Citterio, A. and Righetti, P. G. (2008), 'Exploring the chicken egg white proteome with combinatorial peptide ligand libraries', *J. Proteome Res.*, 7, 3461–74.
de Reu, K., Grijspeerdt, K., Messens, W., Heyndrickx, M., Uyttendaele, M., Debevere, J. and Herman, L. (2006), 'Eggshell factors influencing eggshell penetration and whole egg contamination by different bacteria, including *S. enteritidis*', *Int. J. Food Microbiol.*, 112, 253–60.
Dunn, J. (1996), 'Pulsed light and pulsed electric field for foods and eggs', *Poultry Sci.*, 75, 1133–6.
Fajardo, T. A., Anantheswaran, Puri, V. M. and Knabel, S. J. (1995), 'Penetration of *S. enteritidis* into eggs subjected to rapid cooling', *J. Food Protect.* 58(5), 473–7.
Farber, J. M. (1991), 'Microbiological aspects of modified-atmosphere packaging technology – a review', *J. Food Protect.*, 54(1), 58–70.
Farkas, J. (1998), 'Irradiation as a method for decontaminating food: a review', *Int. J. Food Microbiol.*, 44, 189–204.
Feeney, R. E., Ducay, E. D., Silva, R. B. and MacDonnell, L. R. (1952), 'Chemistry of shell egg deteroriation', *Poultry Sci.*, 31(4), 639–47.
Friedman, M and Juneja, V. K. (2010), 'Review of anti-microbial and antioxidative activities of chitosans in food', *J. Food Protect.*, 73(9), 1737–61.
Gautron, J., Rehault-Godberg, S., Nys, Y., Mann, K. and Righeti, P. G. (2011), 'Use of high-throughput technology to identify new egg components', In Y. Nys, M. Bain and F. V. Immerseel (eds), *Improving the Safety and Quality of Eggs and Egg Products: Egg Safety and Nutritional Quality*, Vol 1, Woodhead Publishing, pp. 133–49.
Goodwin, T. L., Wilson, M. L. and Stadelman, W. J. (1962), 'Effects of oiling time, storage position, and storage time on the condition of shell eggs', *Poultry Sci.*, 41, 840–4.
Gould, G. W. (2000a), 'Strategies for food preservation', In B. M. Lund, T. C. Baird-Parker and G. W. Gould (eds), *The Microbiological Safety and Quality of Food*, Vol I, Aspen Publishers, Inc., pp. 19–35.
Gould, G. W. (2000b), 'New and emerging physical methods of preservation', In B. M. Lund, T. C. Baird-Parker and G. W. Gould (eds), *The Microbiological Safety and Quality of Food*, Vol I, Aspen Publishers, Inc., pp. 277–93.
Guerrero-Beltran, J. A. and Barbosa-Canovas, G. V. (2004), 'Review: advantages and limitations on processing foods by UV light', *Food Sci. Technol. Int.*, 10(3), 137–47.
Harvey, E. and Loomis, A. (1929), 'The destruction of luminous bacteria by high frequency sound waves', *J. Bacteriol.*, 17, 373–9.
Heath, J. L. (1977), 'Chemical and related osmotic changes in egg albumen during storage', *Poultry Sci.*, 56, 822–8.
Herald, T. J., Gnanasambandam, R., McGuire, B. H. and Hachmeister, K. A. (1995), 'Degradable wheat gluten films: preparation, properties and applications', *J. Food Sci.*, 60, 1147–50, 1156.
Herald, T. J. and Smith, D. M. (1989), 'Functional properties and composition of liquid whole egg proteins as influenced by pasteurization and frozen storage', *Poultry Sci.*, 68, 1461–9.
Hidalgo, A., Lucisano, M., Comelli, E. M. and Pompei, C. (1996), 'Evolution of chemical and physical yolk characteristics during the storage of shell eggs', *J. Agr. Food Chem.*, 44, 1447–52.
Homler, B. F. and Stadelman, W. J. (1963), 'The effect of oiling before and after cleaning in maintaining the albumen condition', *Poultry Sci.*, 42, 190–4.

Hou, H., Singh, R. K., Muriana, P. M. and Stadelman, W. J. (1996), 'Pasteurization of intact shell eggs', *Food Microbiol.*, 13, 93–101.

Howard, Z. R., O'Bryan, C. A., Crandall, P. G. and Ricke, S. C. (2012), '*S. enteritidis* in shell eggs: current issues and prospects for control', *Food Res. Int.*, 45, 755–64.

Humphrey, T. J. (1994), 'Contamination of egg shell and contents with *S. enteritidis*: a review', *Int. J. Food Microbiol.*, 21, 31–40.

Humphrey, T. J. and Whitehead, A. (1993), 'Egg age and the growth of *S. enteritidis* PT4 in egg contents', *Epidemiol. Infect.*, 111, 209–19.

Humphrey, T. J., Whitehead, A., Gawler, A. H. L., Henley, A. and Rowe, B. (1991), 'Numbers of *S. enteritidis* in the contents of naturally contaminated hens' eggs', *Epidemiol. Infect.*, 106, 489–96.

Jin, T. Z., Gurtler, J. B. and Li, S.-Q. (2013), 'Development of anti-microbial coatings for improving the microbiological safety and quality of shell eggs', *J. Food Protect.*, 76(5), 779–85.

Jirangrat, W., Torrico, D. D., No, J., No, H. K. and Prinyawiwatkul, W. (2010), 'Effects of mineral oil coating on internal quality of chicken eggs under refrigerated storage', *Int. J. Food Sci. Tech.*, 45, 490–5.

Jones, D. R. and Musgrove, M. T. (2005), 'Effects of extended storage on egg quality factors', *Poultry Sci.*, 84, 1771–7.

Jones, D. R., Tharrington, J. B., Curtis, P. A., Anderson, K. E., Keener, K. M. and Jones, F. T. (2002), 'Effects of cryogenic cooling of shell eggs on egg quality', *Poultry Sci.*, 81, 727–33.

Kahraman-Dogan, H., Bayindirli, L. and Ozilgen, M. (1994), 'Quality control charts for storage of eggs', *J. Food Quality*, 17, 495–501.

Kamel, B., Bond, C. and Diab, M. (1980), 'Egg quality as affected by storage and handling methods', *J. Food Quality*, 3, 261–73.

Kato, A., Imoto, T. and Yagishita, K. (1975), 'The binding groups in ovomucin-lysozyme interaction', *Agr. Biol. Chem. Tokyo*, 39(2), 541–4.

Kato, A., Ogino, K., Kuraamoto, Y. and Kobayashi, K. (1979), 'Degradation of the O-glycosidically linked carbohydrate units of ovomucin during egg white thinning', *J. Food Sci.*, 44(5), 439–40.

Keener, K. M., LaCrosse, J. D., Curtis, P. A., Anderson, K. E and Farkas, B. E. (2000), 'The influence of rapid air cooling and carbon dioxide cooling and subsequent storage in air and carbon dioxide on shell egg quality', *Poultry Sci.*, 79, 1067–71.

Kim, K. Y., Daeschel, M. and Zhao, Y. (2008a), 'Edible coatings for enhancing microbial safety and extending shelf life of hard-boiled eggs', *J. Food Sci.*, 73, M227–35.

Kim, S. H., No, H. K. and Prinyawiwatkul, W. (2008b), 'Plasticizer types and coating methods affect quality and shelf life of eggs coated with chitosan', *J. Food Sci.*, 73, S111–17.

Kim, S. H., No, H. K., Kim, S. D. and Prinyawiwatkul, W. (2006), 'Effect of plasticizer concentration and solvent types on shelf-life of eggs coated with chitosan', *J. Food Sci.* 71(4), S349–53.

Kim, S. H., No, H. K. and Prinyawiwatkul, W. (2007), 'Effect of molecular weight, type of chitosan, and chitosan solution pH on the shelf-life and quality of coated eggs', *J. Food Sci.*, 72(1), S44–8.

Kim, S. H., Youn, D. K., No, H. K., Choi, S. W. and Prinyawiwatkul, W. (2009), 'Effects of chitosan coating and storage position on quality and shelf life of eggs', *Int. J. Food Sci. Tech.*, 44, 1351–9.

Knight, D. W., Bowrey, M. and Cooke, D. J. (1972), 'Preservation of internal egg quality using silicone fluids', *British Poultry Sci.*, 13, 587–93.

Kuo, T.-L., Carey, J. B. and Ricke, S. C. (1997), 'UV irradiation of shell eggs: effect on population of aerobes, molds, and inoculated Salmonella typhimurium', *J. Food Protect.*, 60(6), 639–43.

Li-Chan, E. and Nakai,S. (1989), 'Biochemical basis for the properties of egg white', *CRC Cr. Rev. Poultry Biol.*, 2, 21–58.

Mann, K. (2007), 'The chicken egg white proteome', *Proteomics*, 7, 3558–68.

Martelli, F. and Davies, R. H. (2012), '*Salmonella* serovars isolated from table eggs: an overview', *Food Res. Int.*, 45, 745–54.

Mellor, D. B. and Gardner, F. A. (1975), 'An evaluation of interior quality of shell eggs packed in the carton while still warm', *Poultry Sci.*, 54, 461–5.

Mellor, D. B., Gardner, F. A. and Campos, F. A. (1975a), 'Effect of type of packaging and storage temperature on interior quality of shell treated eggs', *Poultry Sci.*, 54, 7242–6.

Mellor, D. B., Gardner, F. A. and Miller, M. M. (1975b), 'Interior quality decline in eggs cartooned and sealed in plastic bags', *Poultry Sci.*, 54, 1998–2001.

Meyer, R. and Spencer, J. V. (1973), 'The effect of various coatings on shell strength and egg quality', *Poultry Sci.*, 52, 703–11.

Morehouse, K. M. (2002), 'Food irradiation – US regulatory considerations', *Radiat. Phys. Chem.*, 63, 281–4.

Morris, C., Brody, A. L. and Wicker, L. (2007), 'Non-thermal food processing/preservation technologies: a review with packaging implications', *Packag. Technol. Sci.*, 20, 275–86.

Mueller, W. J. (1958), 'Shell porosity of chicken eggs', *Poultry Sci.*, 37, 437–44.

Mueller, W. J. (1959), 'Factors affecting the quality loss in egg albumen during storage', *Poultry Sci.*, 38, 843–6.

Musgrove, M. T. (2011), 'Microbiology and safety of table eggs', In F. V. Immerseel, Y. Nys and M. Bain (eds), *Improving the Safety and Quality of Eggs and Egg Products*, Vol 2, Woodhead Publishing, pp. 3–33.

Musgrove, M. T., Jones, D. R., Northcutt, J. K., Cox, N. A. and Harrison, M. A. (2004), 'Identification of *Enterobacteriaceae* from washed and unwashed commercial shell eggs', *J. Food Protect.*, 67(11), 2613–16.

Musgrove, M. T., Jones, D. R., Northcutt, J. K., Cox, N. A. and Harrison, M. A. (2005a), 'Shell rinse and shell crush methods for the recovery of aerobic microorganisms and *Enterobacteriaceae* from shell eggs', *J. Food Protect.*, 68(10), 2144–8.

Musgrove, M. T., Jones, D. R., Northcutt, J. K., Harrison, M. A. and Cox, N. A. (2005b), 'Impact of commercial processing on the microbiology of shell eggs', *J. Food Protect.*, 68(11), 2367–75.

Musgrove, M. T., Jones, D. R., Northcutt, J. K., Harrison, M. A., Cox, N. A., Ingram, K. D. and Hinton, A. J. (2005c), 'Recovery of *Salmonella* from commercial shell eggs by shell rinse and shell crush methodologies', *Poultry Sci.*, 84, 1955–8.

Nakamura, R., Hirai, M. and Takemori, Y. (1980), 'Some differences noted between the properties of ovalbumin and s-ovalbumin in native state', *Agr. Biol. Chem. Tokyo*, 44(1), 149–53.

Nakamura, R. and Ishimaru, M. (1981), 'Changes in the shape and surface hydrophobicity of ovalbumin during its transformation to s-ovalbumin', *Agr. Biol. Chem. Tokyo*, 45(12), 2775–80.

Neal, W. T. L. (1965), 'Prospects for radiation preservation of foods in Britain', In *Radiation Preservation of Foods*, National Academy of Sciences, National Research Council, pp. 25–30.

No, H. K., Prinyawiwatkul, W. and Meyers, S. P. (2005), 'Comparison of shelf life of eggs coated with chitosans prepared under various deproteinization and demineralization times', *J. Food Sci.*, 70, S377–82.

Nongtaodum, S., Jangchud, A., Jangchud, K., Dhamvithee, P., No, H. K. and Prinyawiwatkul, W. (2013), 'Oil coating affects internal quality and sensory acceptance of selected attributes of raw eggs during storage', *J. Food Sci.*, 78(2), S329–35.

Nys, Y. and Guyot, N. (2012), *How to Control Albumen Quality: State of the Art, Recent Developments and Putative Innovation*, Unpublished Report for the Poultry Cooperative Research Centre, p. 74.

Obanu, Z. A. and Mpieri, A. A. (1984), 'Efficiency of dietary vegetable oils in preserving the quality of shell eggs under ambient tropical conditions', *J. Sci. Food Agr.*, 35, 1311–17.

Oh, H. I., Kim, Y. J., Chang, E. J. and Kim, J. Y. (2001), 'Anti-microbial characteristics of chitosans against food spoilage microorganisms in liquid media and mayonnaise', *Biosci. Biotech. Biochem.*, 65, 2378–83.

OzFoodNet Working Group (2012), 'Monitoring the incidence and causes of diseases potentially transmitted by food in Australia: Annual Report of the OzFoodNet network 2010', *Commun. Dis. Intell.*, 36(3), 213–41.

Raspoet, R., Gantois, I., Devloo, R., Pasmans, F., Haesebrouck, F., Ducatelle, R. and Van Immerseel, F. (2011), 'Internal contamination of eggs by *S. enteritidis*', In F. V. Immerseel, Y. Nys and M. Bain (eds), *Improving the Safety and Quality of Eggs and Egg Products*, Vol 2, Woodhead Publishing, pp. 46–61.

Rehault-Godbert, R., Herve-Grepinet, V., Gautron, J., Cabau, C and Nys, Y. (2011), 'Molecules involved in chemical defence of the chicken egg', In Y. Nys, M. Bain and F. V. Immerseel (eds), *Improving the Safety and Quality of Eggs and Egg Products*, Vol 1, Woodhead Publishing, pp. 183–208.

Reinke, W. C., Spencer, J. V. and Tryhew, L. J. (1973), 'The effect of storage upon the chemical, physical and functional properties of chicken eggs', *Poultry Sci.*, 52, 692–702.

Rhim, J. W., Weller, C. L. and Gennadios, A. (2004), 'Effects of soy protein coating on shell strength and quality of shell eggs', *Food Sci. Biotechnol.*, 13, 455–9.

Roberts, J. R. (2006), 'Egg and eggshell quality guidelines for the Australian egg industry', In J. S. Sim and H. H. Sunwoo (eds), *The Amazing Egg*, University of Alberta, pp. 499–508.

Robinson, D. S. (1972), 'Egg white glycoproteins and the physical properties of egg white', In B. M. Freeman and P. E. Lake (eds), *Egg Formation and Production*, British Poultry Science, Edinburgh, pp. 65–86.

Robinson, D. S. and Monsey, J. B. (1972), 'Changes in the composition of ovomucin during liquefaction of thick egg white', 23(1), 29–38.

Russell, S. M. (2003), 'The Effect of electrolyzed oxidative water applied using electrostatic spraying on pathogenic and indicator bacteria on the surface of eggs', *Poult. Sci.*, 82, 158–62.

Ryu, K. N., No, H. K. and Prinyawiwatkul, W. (2011), 'Internal quality and shelf life of eggs coated with oils from different sources', *J. Food Sci.*, 76, S235–9.

Sabrani, M. and Payne, G. G. (1978), 'Effect of oiling on internal quality of eggs stored at 28 and 12°C', *British Poult. Sci.*, 19, 567–71.

Schuman, J. D., Sheldon, B. W., Vandepopuliere, J. M. and Ball, Jr., H. R. (1997), 'Immersion heat treatments for inactivation of *S. enteritidis* with intact eggs', *J. App. Microbiol.*, 83, 438–44.

Schwall, D. V., Gardner, F. A. and Parnell, E. D. (1961), 'Effects of oil treating on shell egg quality during short term refrigerated storage', *Poult. Sci.*, 40, 583–8.

Scott, T. A. and Silversides, F. G. (2000), 'The effect of storage and strain of hen on egg quality', *Poult. Sci.*, 79, 1725–9.

Serrano, L. E., Murano, E. A., Shenoy, K. and Olson, D. G. (1997), 'D values of *S. enteritidis* isolates and quality attributes of shell eggs and liquid whole eggs treated with irradiation', *Poult. Sci.*, 76, 202–6.

Silversides, F. G. and Scott, T. A. (2001), 'Effect of storage and layer age on quality of eggs from two lines of hens', *Poult. Sci.*, 80, 1240–5.

Smith, M. B. and Back, J. F. (1968), 'Studies on ovalbumin. IV. Trypsin digestion and the cysteine peptides of ovalbumin and S-ovalbumin', *Aust. J. Biol. Sci.*, 21(3), 549–58.

Stadelman, W. J. (1995a), 'Quality identification of shell eggs', In W. J. Stadelman and O. J. Cotterill (eds), *Egg Science and Technology*, AVI Publishing, Westport, CT, pp. 39–66.

Stadelman, W. J. (1995b), 'The preservation of quality in shell eggs', In W. J. Stadelman and O. J. Cotterill (eds), *Egg Science and Technology*, AVI Publishing, Westport, CT, pp. 67–79.

Stadelman, W. J. and Wilson, M. L. (1958), 'An oil dispensing aerosol for quality preservation of shell eggs', *Poult. Sci.*, 37, 731–3.

Su, H. K., Hong, K. N. and Prinyawiwatkul, W. (2007), 'Effect of molecular weight, type of chitosan, and chitosan solution pH on the shelf life and quality of coated eggs', *J. Food Sci.*, 72, S44–8.

Suppakul, P., Jutakorn, K. and Bangchokedee, Y. (2010), 'Efficacy of cellulose-based coating on enhancing the shelf life of fresh eggs', *J. Food Eng.*, 98, 207–13.

Thakur, B. R. and Singh, R. K. (1994), 'Food irradiation – chemistry and applications', *Food Rev. Int.*, 10(4), 437–73.

Theron, H., Venter, P. and Lues, J. F. R. (2003), 'Bacterial growth in chicken eggs in various storage environments', *Food Res. Int.*, 36, 969–75.

Torrico, D. D., Jirangrat, W., No, H. K., Prinyawiwatkul, W., Ge, B. and Ingram, D. (2010), 'A novel emulsion coating and its effects on internal quality and shelf life of eggs during room temperature storage', *Food Sci. Tech.*, 45, 2241–9.

Torrico, D. D., No, H. K., Prinyawiwatkul, W., Janes, M., Corredor, J. A. H. and Osorio, L. F. (2011a), 'Mineral oil-chitosan emulsion coatings affect quality and shelf-life of coated eggs during refrigerated and room temperature storage', *J. Food Sci.*, 26(4), S262–8.

Torrico, D. D., No, H. K., Sriwattana, S., Ingram, D. and Prinyawiwatkul, W. (2011b), 'Effects of initial albumen quality and mineral oil-chitosan emulsion coating on internal quality and shelf-life of eggs during room temperature storage', *Int. J. Food Sci. Tech.*, 46, 1783–92.

Torrico, D. D., Wardy, W., Carabante, K. M., Pujols, K. D., Xu, Z., No, H. K. and Prinyawiwatkul, W. (2014), 'Quality of eggs coated with oil-chitosan emulsion: combined effects of emulsifier types, initial albumen quality, and storage', *LWT-Food Sci. Technol.*, 57, 35–41.

Vanlith, L. A., Putirulan, J. T. and Mulder, R. W. A. (1995), 'Pasteurization of table eggs to eliminate salmonellai', *Archiv. Fur Geflugelkunde*, 59, 157–60.

Waimaleongora-Ek, P., Garcia, K. M., No, H. K., Prinyawiwatkul, W. and Ingram, D. R. (2009), 'Selected quality and shelf life of eggs coated with mineral oil with different viscosities', *J. Food Sci.*, 74(9), S423–9.

Wang, G. H. (1992), 'Inhibition and inactivation of five species of foodborne pathogens by chitosan', *J. Food Protect.*, 55, 916–19.

Wardy, W., Torrico, D. D., Corredor, J. A. H., No, H. K., Zhang, X., Xu, Z. and Prinyawiwarkul, W. (2013), 'Soybean oil-chitosan emulsion affects internal quality and shelf-life of eggs stored at 25 and 4°C', *Int. J. Food Sci. Tech.*, 48, 1148–56.

Wardy, W., Torrico, D. D., No, H. K., Prinyawiwatkul, W. and Saalia, F. K. (2010), 'Edible coating affects physico-functional properties and shelf life of chicken eggs during refrigerated and room temperature storage', *Int. J. Food Sci. Tech.*, 45, 2659–68.

Williams, K. C. (1992), 'Some factors affecting albumen quality with particular reference to Haugh Unit score', *World's Poult. Sci. J.* 48, 5–16.

Wong, Y. C., Herald, T. J. and Hachmeister, K. A. (1996), 'Evaluation of mechanical and barrier properties of protein coatings on shell eggs', *Poultry Sci.*, 75, 417–22.

Wrigley, D. M. and Llorca, N. G. (1992), 'Decrease of Salmonella typhimurium in skim milk and egg by heat and ultrasonic wave treatment', *J. Food Protect.*, 55(9), 678–80.

Xie, L., Hettiarachchy, N. S., Ju, Z. Y., Meullenet, J., Wang, H., Slavik, M. F. and Janes, M. E. (2002), 'Edible film coating to minimize eggshell breakage and reduce post-wash bacterial contamination measured by dye penetration in eggs', *J. Food Sci.*, 67, 280–4.

Chapter 14

The nutritional role of eggs

Tia M. Rains and Mitch Kanter, Egg Nutrition Center, USA

1 Introduction

Eggs have been a staple in human diet for thousands of years. Once maligned for their cholesterol content, decades of nutrition research have advanced the understanding of the complex relationships between dietary cholesterol, blood cholesterol and cardiovascular disease (CVD). As such, most public health and government organizations around the world have eliminated cholesterol limits from dietary recommendations (Fernandez and Calle 2010).

In recent years, research on eggs has expanded beyond CVD and blood cholesterol studies to the positive attributes of whole eggs and egg constituents on health and disease. It is not surprising that eggs would favourably affect health, given that they are a source of high-quality protein and other important nutrients, including choline, vitamin D and various antioxidants. Such nutrients have been previously shown to confer benefits against a multitude of diseases. Moreover, newly discovered non-nutrient compounds in eggs may contribute to optimal health, furthering the value of eggs in the prevention of chronic disease.

A variety of eggs from different species of birds are commercially available in different parts of the world. However, the majority of eggs consumed today are chicken eggs. As such, the overwhelming majority of scientific inquiry into the nutritional effects of eggs has utilized those from the chicken. This chapter will focus solely on the nutritional attributes of chicken eggs as they relate to human health.

2 The nutrient composition of chicken eggs

Eggs are one of the most affordable, accessible and nutrient-dense foods available around the world. The proteins, lipids, carbohydrates, vitamins, minerals and other functional

http://dx.doi.org/10.19103/AS.2016.0012.16

Table 1 Macronutrient distribution in raw chicken egg (per 50 g large egg)

	Whole egg	Egg white	Egg yolk
Weight (%)	100	66	34
Water (g)	38.1	28.9	8.9
Energy			
–kJ	301.2	71.1	230.1
–kcal	72	17	55
Protein (g)	6.3	3.6	2.7
Lipid (g)	4.8	0.1	4.5
Saturated fatty acids	1.6	0	1.6
Monounsaturated fatty acids	1.8	0	2.0
Polyunsaturated fatty acids	1.0	0	0.7
Cholesterol (mg)	186	0	184
Sugars (g)	0.4	0.2	0.1

US Department of Agriculture, Agricultural Research Service, Nutrient Data Laboratory. USDA National Nutrient Database for Standard Reference, Release 28. Version Current: September 2015. Internet: http://www.ars.usda.gov/ba/bhnrc/ndl.

nutrients in an egg are sufficient to not only support the transition from fertilized cell to newborn chick, but to also provide unique nutrient contributions to human diet.

The whole egg is composed of 76.8% water, 12.6% protein and 9.5% lipid, with the remainder consisting of vitamins, minerals and a small amount of carbohydrates (Table 1). The exact nutrient composition of an egg can vary slightly subject to a number of factors, including age, breed or strain of the hen, as well as the composition of the hen's diet, and the season of the year. Although most variations are relatively minor, the lipid-soluble components in particular are sensitive to dietary alterations. This relationship has been exploited in the production of eggs with enhanced nutritional attributes such as increased long-chain omega-3 fatty acids, vitamin D and vitamin E.

2.1 Protein

The most quintessential nutritional feature of eggs is the protein fraction. Egg protein, like that from other animal foods, is considered a 'complete' or 'high-quality' protein in that it contains all nine essential amino acids (Table 2) as well as many non-essential amino acids in optimal proportions for the human body. Several methods are used to determine the 'quality' of a protein source, including the Chemical Score, protein efficiency ratio, biological value and Protein Digestibility Corrected Amino Acid Score (PDCAAS). Recently, the Food and Agriculture Organization recommended the Digestible Indispensable Amino Acid Score (DIAAS) rating system as the preferred method for measuring protein quality (Rutherford et al., 2015). Unlike the PDCAAS method, the DIAAS method does not truncate values at 100, more accurately reflecting the differences in the ability of proteins to supply amino acids to the body. By any method, eggs are considered one of the

Table 2 Amino acid content of a large egg

Amino acids	Grams per large egg	Grams per 100 g whole egg
Alanine	0.37	0.74
Arginine	0.41	0.82
Aspartic acid	0.66	1.33
Cystine	0.14	0.27
Glutamic acid	0.84	1.67
Glycine	0.22	0.43
Histidine	0.15	0.31
Isoleucine	0.34	0.67
Leucine	0.54	1.09
Lysine	0.46	0.91
Methionine	0.19	0.38
Phenylalanine	0.34	0.68
Proline	0.26	0.51
Serine	0.49	0.97
Threonine	0.28	0.56
Tryptophan	0.08	0.17
Tyrosine	0.25	0.50
Valine	0.43	0.86

US Department of Agriculture, Agricultural Research Service, Nutrient Data Laboratory. USDA National Nutrient Database for Standard Reference, Release 28. Version Current: September 2015. Internet: http://www.ars.usda.gov/ba/bhnrc/ndl.

highest-quality protein sources available in the human diet, and in the case of biological value, eggs are the standard of comparison for measuring protein quality of other protein sources (Rutherford et al., 2015).

Eggs are especially rich in the essential amino acid leucine (600 mg in one large egg), which has been demonstrated to act as a 'signal' for protein synthesis within muscle cells (Layman, 2009). In fact, leucine appears to be the essential factor necessary to generate a post-meal anabolic response in skeletal muscle (Phillips et al., 2016). Accumulating evidence suggests that meals with at least 25–30 g of high-quality protein are required to provide adequate leucine to maximize the anabolic response in skeletal muscle (Phillips et al., 2016). This is particularly important during the recovery phase following exercise and for older adults to offset the age-related decline in muscle mass (Pasiakos et al., 2014).

Sarcopenia, the decline of muscle mass and function, can lead to weakness, frailty and higher risk of falls and injury in older adults (Fielding et al., 2011). Protein undernutrition is a major contributor to the development of sarcopenia (Berger and Doherty, 2010; Fielding et al., 2011; von Haehling et al., 2010; Paddon-Jones et al., 2008). A diet that includes 25–30 g of protein per meal is recommended to maximize muscle protein synthesis/ prevent muscle loss in the elderly (Paddon-Jones, 2015). Although plant-based proteins provide essential amino acids and can be combined to ensure that no essential amino acids are limiting, animal-based proteins have a higher proportion of not only leucine, but all essential amino acids in the optimal ratios (Rutherford, 2015). However, regular consumption of high-quality protein can be challenging for older adults, as evidenced by the fact that protein intakes decline with age (Paddon-Jones and Rasmussen, 2009). Eggs, by virtue of their texture, affordability and versatility, may be a viable solution to the deficit between protein intakes and recommendations. Research has also shown that dietary protein is important for bone health, another important consideration during senescence (Heaney and Layman, 2008).

2.2 Lipids

Within the yolk, the lipid fraction is composed of triacylglycerides (65%), phospholipids (31%) and cholesterol (4%). Of the total phospholipid amount, phosphatidylcholine (lecithin) is the largest fraction, accounting for 26%. Although the exact fatty acid composition of egg yolk lipid is determined by the source of fat of the hen's diet, the majority of the fatty acids present are unsaturated (Table 3). One egg yolk also contains ~185 mg of cholesterol, a value that has decreased over time. There is speculation as to why eggs now have less cholesterol than they did in the past. Some researchers suggest that the diet of the hen has been modified over time, bringing about a change in egg cholesterol levels. Others point to changes in breeding practices, different rates of lay among modern hens and methodological differences in the determination of egg cholesterol.

Table 3 Egg yolk lipid profile per large egg

Lipids	Amount
Fatty acids, total saturated	1.56 g
8:0–14:0	0.02 g
16:0	1.12 g
18:0	0.41 g
20:0–24:0	0.01 g
Fatty acids, total monounsaturated	1.83 g
16:1	0.10 g
18:1	1.71 g
20:1	0.01 g
Fatty acids, total polyunsaturated	0.96 g
18:2	0.78 g
18:3	0.02 g
20:4	0.09 g
20:5 n-3	0.00 g

(Continued)

Table 3 (*Continued*)

Lipids	Amount
22:5 n-3	0.01 g
22:6 n-3	0.03 g
Cholesterol	186 mg
Carotene, beta	0 µg
Carotene, alpha	0 µg
Cryptoxanthin, beta	4.0 µg
Lutein + zeaxanthin	252 µg

US Department of Agriculture, Agricultural Research Service, Nutrient Data Laboratory. USDA National Nutrient Database for Standard Reference, Release 28. Version Current: September 2015. Internet: http://www.ars.usda.gov/ba/bhnrc/ndl.

2.3 Vitamins and minerals

Eggs contain all essential vitamins for human health, except vitamin C. This is because vitamin C is not needed in the developing chick and adult hens can endogenously synthesize vitamin C. The yolk contains all of the fat-soluble vitamins as well as the majority of the water-soluble vitamins with the exception of riboflavin and niacin (Fig. 1). The vitamin profile of eggs has some unique qualities relative to other protein-containing foods. Eggs are one of the few natural sources of vitamin D, providing 41 IU per large egg. Eggs are one of the richest dietary sources of choline (147 mg per large egg), an essential nutrient that plays an important role in foetal and infant brain development, affecting the areas of the brain responsible for memory and life-long learning ability (Zeisel, 2006; Zeisel and da Costa, 2009). Women with diets low in choline have a four times greater risk of having infants with neural tube defects than women with adequate intakes of choline (Shaw et al., 2004, 2009).

Egg yolk contains two important yellow-orange pigments, lutein and zeaxanthin, that act as antioxidants once consumed. A large egg has 252 µg of lutein and zeaxanthin; however, like other fat-soluble compounds, the content of these xanthophylls is dependent upon the hen's diets. Egg yolk lutein levels can be increased five- to tenfold through modification of the feed by yellow corn, alfalfa meal, corn gluten meal, dried algae meal or marigold petal meal (Leeson and Caston, 2004; Skřivan, 2015).

Eggs contain a relatively small amount of lutein and zeaxanthin compared to plant sources such as corn, squash and spinach. However, studies have shown that egg yolk xanthophylls have a higher bioavailability compared to plant sources due to the presence of fat within the yolk that facilitates absorption of fat-soluble compounds (Chung et al., 2004; Goodrow et al., 2006). This increased bioavailability results in significant increases in plasma levels of lutein and zeaxanthin, as well as increased macular pigment densities with egg feeding in humans, and has been associated with a decreased risk of cataracts and macular degeneration (Abdel-Aal et al., 2013; Burke et al., 2005; Giordano et al., 2012).

Eggs contain small amounts of several minerals essential to human health, most of which reside within the yolk. Of particular importance is iron, which is often under consumed in

Figure 1 Nutrient content of the egg white and yolk fractions.

women and children. In addition, eggs contain calcium, phosphorus, sodium, potassium, magnesium, zinc, copper and manganese in small amounts.

3 The relationship between egg consumption and human health and disease

3.1 Cardiovascular disease

Over the past 100 years, the perceived healthfulness of whole egg consumption has shifted several times. Once a common staple in the home, research in the 1960s and 1970s suggested that eggs were a major contributor to hypercholesterolaemia and as such, coronary heart disease (CHD). In 1968, the American Heart Association established a recommendation of no more than 300 mg/day of dietary cholesterol for individuals with hypercholesterolaemia, a level that at the time was justified primarily by studies in animals and observational evidence in humans that did not adjust for other dietary components (Committee on Nutrition AHA: Diet and Heart Disease 1968) The 300 mg/day limit was extended to healthy adults by several public health organizations around the world shortly thereafter (Fernandez and Calle, 2010). This prompted a steady decline in egg consumption in the decades following, as well as an increase in scientific inquiry to better define the impact of dietary cholesterol on health and disease, independent of other dietary components often co-consumed with cholesterol that influence CHD risk (e.g. saturated fat) (Kanter et al., 2012).

Over the past 40 years, hundreds of peer-reviewed papers, including prospective cohort studies, cross-sectional analyses and randomized clinical trials, have evaluated the independent relationships between dietary cholesterol, CHD and blood lipids. The overwhelming body of evidence demonstrates a lack of association between egg intake and CHD risk (Eckel et al., 2013; Shin et al., 2013; Rong et al., 2013; Li et al., 2013). Additionally, there is an acknowledgement among experts that factors such as smoking, hypertension, diabetes, elevated blood lipids, obesity, inactivity, age and genetics have the strongest influence on CHD risk and mortality (Eckel et al., 2013).

Elevated serum lipoprotein concentrations are an accepted biomarker for the development of CHD (Lloyd-Jones, 2010). There have been numerous studies evaluating the lipoprotein lipid response to dietary cholesterol, frequently provided in research studies as egg yolks, given that they are one of the few foods that contain cholesterol but little saturated fat (Kanter et al., 2012). It is generally accepted that the dataset is inconsistent (Eckel et al., 2013; Lloyd-Jones et al., 2010). Based on several older meta-analyses, it has been estimated that every 100 mg/day increment in dietary cholesterol would produce an average increase of 2.2–2.5 mg/dL in total cholesterol, 1.9–2.1 mg/dL in low-density lipoprotein cholesterol (LDL-C) and 0.30–0.44 mg/dL in high-density lipoprotein cholesterol (HDL-C), though it should be noted that the majority of studies upon which these numbers are based evaluated dietary cholesterol levels several fold higher than typical current intakes (Clarke et al., 1997; Howell et al., 1997; Weggemans et al., 2001).

Inconsistency in the cholesterol literature is presumed to be heavily influenced, among other reasons, by variability in cholesterol absorption between individuals (Griffin and Lichtenstein, 2013). It is estimated that 15–25% of the population experiences an increase in plasma lipoprotein lipid concentrations following dietary cholesterol intakes, a subgroup termed hyper-responders (Griffin and Lichtenstein, 2013). Genetic factors such as the presence of certain alleles in the apolipoprotein E gene may influence this response (Virtanen, 2016). Among those individuals, both LDL-C (2.84 mg/dL increase for an increase of 100 mg/day of dietary cholesterol) and, to a smaller degree, HDL-C increase in response to dietary cholesterol, maintaining the LDL-C:HDL-C ratio, a biomarker previously shown to reflect CVD risk (Fernandez and Webb, 2008; Greene et al., 2005). The plasma lipid response to dietary intake of cholesterol is minimal in those characterized as hypo-responders (Fernandez and Webb, 2008; Greene et al., 2005).

On the basis of lipoprotein cholesterol concentrations alone, it would appear that hyper-responders represent a vulnerable group that may be negatively impacted by increased cholesterol intakes. However, recent evidence from the Kuopio Ischaemic Heart Disease Risk Factor Study did not show increased CHD risk in those individuals with the apolipoprotein E4 allele associated with hyper-responders. These results may be explained, at least in part, by a growing appreciation for differences in cardiovascular risk related to the number of atherogenic lipoprotein particles and size of such particles versus cholesterol concentrations measured in a standard lipid panel (Davidson et al., 2011; Diffenderfer and Schaefer, 2014; Lamarche et al., 1999). A greater number of atherogenic particles (those carrying apolipoprotein B, particularly LDLs) are associated with increased CHD event rates in a linear manner (Davidson et al., 2011). Similarly, small, dense LDL particles have several properties that increase their ability to traverse the artery wall, disproportionately increasing their atherogenicity and resulting in three times the risk for developing CHD in those individuals with greater concentrations of small, dense LDL versus larger LDL particles (Barona and Fernandez, 2012).

Numerous studies have evaluated the effects of egg intake on lipoprotein size and number in the general population (Blesso et al., 2013; Greene et al., 2005; Harman et al., 2008; Ischerwood et al., 2010; Mutungi et al., 2008; Vislockey et al., 2009) and in those characterized as hyper-responders (Herron et al., 2002, 2004). Collectively, results are consistent in demonstrating either no change in LDL particle number or apolipoprotein B concentrations (Vislockey et al., 2009), an increase or no change in LDL particle size (Blesso et al., 2013; Greene et al., 2005; Harman et al., 2008; Herron et al., 2002; Mutungi et al., 2008) or both (Herron et al., 2004; Isherwood et al., 2010) following consumption of the equivalent of two to four eggs per day (an incremental 370–740 mg/day). Further, there is also evidence that egg intake favourably affects HDL-C concentrations and cholesterol efflux (Andersen et al., 2013). Such results may explain the absence of an association between egg consumption and CHD and stroke incidence and mortality.

Recently, relationships between egg intake and other biomarkers of CHD beyond serum lipid levels have been explored. For example, Katz and colleagues (2005) compared the effects of two eggs per day to oatmeal for six weeks in 49 healthy adults in a crossover design. There were no changes from baseline or between conditions in flow-mediated dilation (FMD) in the brachial artery using a high-frequency ultrasound. The same researchers conducted a similar study in 40 hyperlipidemic adults, and results showed no differences in FMD following an acute intake of three eggs or chronic (four weeks) feeding of two eggs per day (Njike et al., 2010). Recent observational research studies found no association between egg intake and markers of CHD, including calcified atherosclerotic plaque in the coronary arteries and the burden associated with coronary atherosclerosis (Goldberg et al., 2014; Robbins et al., 2014). A similar study failed to show any associations between egg or cholesterol intakes and increased common carotid intima-medial thickness, a predictor of CHD risk (Virtanen et al., 2015).

3.2 Type 2 diabetes

Commensurate with the global obesity epidemic, the rates of type 2 diabetes (T2D) and related metabolic disorders are increasing worldwide (Hu, 2011). A number of factors increase the risk for the development of such disorders, including diet and lifestyle factors as well as age, race and family history (Hu, 2011). T2D is a leading risk factor for CVD, the top cause of death in the United States and other industrialized nations.

In 2009, Djousse and colleagues identified a positive association between intakes of dietary cholesterol/eggs and incident T2D. Since then, several other prospective cohort analyses have replicated these results (Djousse et al., 2010; Ericson, 2013). Other studies have shown that high egg intakes increase the risk for the development of CHD in those with diagnosed T2D (Hu et al., 1999; Quershi, 2007). Although no mechanism has been identified, it has been postulated that dietary cholesterol, protein or other nutrients in eggs may play a role (Shin et al., 2013).

Despite these studies condemning egg intake, there are other studies using similar methodology that have shown either no association or an inverse association between egg intake and development of T2D or CHD in those with T2D (Lajous et al., 2015; Tran, 2014). For example, in a prospective study of over 2000 middle-aged men in Finland that were followed for nearly 20 years, those in the highest versus the lowest quartile of egg intake (>45 g/day vs. <14 g/day, respectively) had a 38% lower risk of incident T2D after adjustment for potential confounders (Virtanen et al., 2015). Similarly, a prospective analysis of over 4500 African Americans with T2D concluded that there was no association between egg intake and incidence of T2D (Djousse et al., 2015). A recent meta-analysis of 12 prospective cohort studies showed no association between egg intake and risk of T2D among studies conducted outside the United States, but an elevated risk with consumption of >3 eggs/week among studies conducted on U.S. populations. This suggests that foods often consumed with eggs in the United States (e.g. processed meats) or other diet or lifestyle behaviours characteristic of frequent egg consumers may be driving the positive association observed in the United States (Djousse et al., 2016).

Randomized controlled intervention studies in which daily egg intake is evaluated in the context of a prescribed dietary approach have failed to show modifications in blood lipid or glucose parameters in individuals with diagnosed T2D, pre-diabetes or related metabolic conditions. In fact, in several of these studies, favourable shifts in various biomarkers were demonstrated, which challenge results from the prospective cohort studies showing a positive association between disease risk and egg consumption (Andersen et al., 2013; Blesso et al., 2013; Hu et al., 1999; Pearce et al., 2011; Radzeviciene and Ostrauskas, 2012; Shi et al., 2011). For example, in a clinical trial among middle-aged Mexican men and women with T2D, those who consumed one egg for breakfast compared to 40 g of oatmeal made with lactose-free milk every day for five weeks showed no differences in plasma glucose, insulin, glycosylated haemoglobin or other indicators of glucose metabolism and cardiometabolic syndrome (Ballesteros et al., 2015). Another clinical trial found that, when consumed as part of an energy-restricted diet, two eggs per day improved a number of key health metrics for people with T2D, such as glycaemic control, lipid profiles and blood pressure (Pearce et al., 2011). In addition, individuals with T2D who consumed two eggs per day had increased plasma folate and lutein, nutrients important in the management of T2D. In another clinical trial, patients with T2D who consumed two eggs per day, six days a week for three months had no difference in glycaemic control or blood lipids compared with patients who consumed more than two eggs per week (Fuller et al., 2015). In this study, people in the group consuming more eggs reported less hunger and a greater feeling of fullness.

It is well known that associations identified in observational studies may not translate into cause–effect relationships (Maki et al., 2014). In some instances, methodological limitations in observational studies may skew results. Further, public health organizations and others continue to include eggs as part of diets for individuals at risk for the development of T2D or with existing T2D (Campbell and Rains, 2015; Fox et al., 2015). It is clear that additional testing is necessary to better define the relationship, if any, between egg intake and T2D.

3.3 Eggs, weight loss and satiety

Overweight and obesity continue to be of pandemic proportions (Ng et al., 2014). Obesity is a multi-factorial and complex health issue that involves both genetics and behaviour (Ng et al., 2014). Behavioural factors include dietary intake, physical activity levels and use of medications, along with societal cues. One dietary strategy that has shown promise for weight loss and weight loss maintenance is consumption of higher intakes of protein (Layman, 2004; Leidy et al., 2015).

There are several hypotheses for why dietary protein may influence body weight regulation. One of the best studied is the impact of protein-rich meals on appetite (Li et al., 2016). Both acute and long-term studies have reported increased feelings of fullness and reduced energy intake at subsequent meals following high intakes of protein relative to meals high in carbohydrate or fat (Li et al., 2016). In fact, compared to carbohydrate or fat, dietary protein is the most satiating macronutrient particularly when consumed at a breakfast meal (Leidy et al., 2015). It is generally accepted that dietary protein influences gastrointestinal hormones that regulate appetite (Leidy et al., 2013). Specially, meals high in protein have been shown to reduce the hunger-stimulating hormone ghrelin and increase the satiety-stimulating hormone, peptide YY (PYY) (Leidy et al., 2013).

Several feeding studies have directly assessed the effects of breakfast meals containing a high amount of egg protein from whole eggs in children, adolescents and adults (Baum et al., 2015; Kral et al., 2016; Leidy et al., 2013; Rains et al., 2015; Ratliff et al., 2010; Vander Wal et al., 2005). Vander Wal et al. (2005) conducted one of the earliest acute studies that evaluated the appetitive effects of an egg-based breakfast (18 g of total protein; 17 g of total fat) compared to an isocaloric bagel-based breakfast (13 g of total protein; 11 g of total fat) in overweight and obese adults. Not only were hunger scores, as measured by visual analog scales, reduced in the period following breakfast, but *ad libitum* energy intake at the subsequent lunch was 1100 kJ less following the egg-based meal. This difference persisted over the 24-h period following consumption of the test breakfast. Similar results in adults have been reported by others (Rains et al., 2015; Ratliff et al., 2010).

In recent years, investigations have focussed on the satiety effects of eggs in children, particularly given the rampant increase in childhood obesity worldwide (WHO, 2016). For example, a study in adolescent girls showed that an egg-containing breakfast not only reduced hunger, but also led to fewer food cravings and consumption of fewer high-fat snacks later in the evening (Leidy et al., 2013). Kral and colleagues (2016) have recently compared the appetitive effects of an egg-based breakfast to ready-to-eat breakfast cereal and oatmeal in normal and overweight children. Energy intakes at the subsequent lunch were reduced by 293 kJ following the egg breakfast relative to both other conditions. Other studies have similarly reported the satiating benefits of egg-based breakfasts in children (Baum, 2015), although not all have corroborated these results (Liu et al., 2015).

Dietary protein may also influence body weight regulation by additional mechanisms beyond satiety (Westerterp-Plantenga et al., 2012). Studies have shown that energy-restricted diets with higher intakes of protein preserve lean mass during weight loss, which is the major determinant of resting and 24-h energy expenditure (Astrup et al., 2015; Li et al., 2016). Lean mass also plays an important role in protein and glucose metabolism, promoting an overall metabolic health (Wycherley et al., 2012).

There is also evidence that dietary protein increases the thermic effect of food (Astrup et al., 2015). Many reviews have been published on the benefits of higher-protein weight loss diets (see Astrup et al., 2015 or Wycherley et al., 2012), with the consistent conclusions being that higher-protein (25–35% energy) diets promote a modest, but significant benefit on weight and fat mass loss.

Eggs are a natural fit for higher-protein diets and are often included in controlled weight loss trials as a source of protein. However, there is limited evidence on the specific effects of eggs on weight loss. In a study in overweight adults, calorie-restricted diets that included daily egg breakfasts for eight weeks resulted in 65% more weight loss than a bagel-based breakfast matched for total energy and energy density (Vander Wal et al., 2008). Those in the egg breakfast group consumed an average of 330 fewer calories per day and reported feeling more energetic than those who ate a bagel for breakfast. Future studies are warranted to better understand if eggs provide any unique benefit relative to other sources of protein.

4 Future trends: emerging health benefits

Over the past several years, there has been a burgeoning interest into the presence and function of bioactive components within eggs. Such bioactive components have been discovered within the protein fractions of the white and yolk, as well as within the shell membrane. For example, protein hydrolysates derived from the enzymatic treatment of egg white have been shown in animal studies to have anti-hypertensive properties by reducing angiotensin-converting enzyme (ACE) activity (Miguel, 2005). Different enzymes have been shown to produce different bioactive fractions depending on the site of enzymatic activity within the egg protein (Nimalaratne and Bandara, 2015). Other peptides with anti-hypertensive activity have been derived from enzymatic hydrolysis of egg yolk (Yoshii, 2001). Feeding studies in spontaneously hypertensive rats have shown that oral administration of such yolk peptides promotes reductions in systolic and diastolic blood pressures (Yoshii, 2001).

Other peptide products of protein hydrolysis from the white and yolk protein fractions have shown antioxidant activity (Chen et al., 2012; Yousr and Howell, 2015; Nimalaratne and Bandara, 2015), as well as antimicrobial, anti-adhesive, immunomodulatory and antithrombotic activities, and increased mineral absorption (Samaraweera et al., 2011). Egg shell membranes, a by-product of the egg industry, have also been shown to possess biological activity, including immunomodulatory and antimicrobial activity (Makkar et al., 2015).

Subfractions of the lipid component in the yolk may impart physiological effects. For example, a yolk oil fraction has been shown to reduce the inflammatory process in a cellular model of wound healing (Bereiter-Hahn, 2014). Indeed, animals treated with yolk oil showed improved recovery in response to a burn versus a control group (Rastegar et al., 2011). These discoveries hold great potential for human applications as well as in commercial animal agriculture, but additional work is needed.

5 Summary

Eggs are an affordable, nutrient-dense source of high-quality protein and 13 other essential nutrients. They are widely available and an established component in the dietary patterns of cultures around the world. In some countries, eggs were ill-advised for decades due to concerns about dietary cholesterol negatively impacting cardiovascular health. However, more recent evidence has shown that consistent egg intake does not contribute to the risk of heart disease or stroke. In fact, public health organizations in many parts of the world encourage egg intake because of their many nutritional attributes, many of which may be protective against various chronic diseases.

6 Where to look for further information

There are a number of recent review papers that provide a more in depth analysis of the scientific literature on the health benefits of eggs, including:

- Smith, A. and Gray, J. (2016), Considering the benefits of egg consumption for older people at risk of sarcopenia. *Br. J. Community Nurs.*, 21(6), 305–9. doi:10.12968/bjcn.2016.21.6.305.
- Fuller, N. R., Sainsbury, A., Caterson, I. D. and Markovic, T. P. (2015), Egg consumption and human cardio-metabolic health in people with and without diabetes. *Nutrients*, 7(9), 7399–420. doi:10.3390/nu7095344.

- Andersen, C. J. (2015), Bioactive egg components and inflammation. *Nutrients*, 7(9), 7889–913. doi:10.3390/nu7095372.
- Miranda, J. M., Anton, X., Redondo-Valbuena, C., Roca-Saavedra, P., Rodriguez, J. A., Lamas, A., Franco, C. M. and Cepeda, A. (2015), Egg and egg-derived foods: effects on human health and use as functional foods. *Nutrients*, 7(1), 706–29. doi:10.3390/nu7010706.
- Nimalaratne, C. and Wu, J. (2015), Hen egg as an antioxidant food commodity: a review. *Nutrients*, 7(10), 8274–93. doi:10.3390/nu7105394.

Several books have also been previously published on the health aspects of eggs:

- Watson, R. R. and De Meester, F. (eds) (2015), *Handbook of Eggs in Human Function*, United Kingdom, Wageningen Academic Publishers.
- Watson, R. R. (ed.) (2002), *Eggs and Health Promotion*, Ames, Iowa: Iowa State Press.

7 References

Abdel-Aal, S. M., Akhtar, H., Zaheer, K. and Ali, R. (2013), Dietary sources of lutein and zeaxanthin carotenoids and their role in eye health. *Nutrients*, 5(4), 1169–85.

Andersen, C. J., Blesso, C. N., Lee, J., Barona, J., Shah, D., Thomas, M. J. and Fernandez, M. L. (2013), Egg consumption modulates HDL lipid composition and increases the cholesterol-accepting capacity of serum in metabolic syndrome. *Lipids*, 48, 557–67.

Astrup, A., Raben, A. and Geiker, N. (2015), The role of higher protein diets in weight control and obesity-related comorbidities. *Int. J. Obes.*, 39, 721–6.

Ballesteros, M. N., Valenzuela, F., Robles, A. E., Artalejo, E., Aguilar, D., Andersen, C. J., Valdez, H. and Fernandez, M. L. (2015), One egg per day improves inflammation when compared to an oatmeal-based breakfast without increasing other Cardiometabolic risk factors in diabetic patients. *Nutrients*, 7(5), 3449–63.

Barona, J. and Fernandez, M. L. (2012), Dietary cholesterol affects plasma lipid levels, the intravascular processing of lipoproteins and reverse cholesterol transport without increasing the risk for heart disease. *Nutrients*, 4, 1015–25.

Baum, J. I., Gray, M. and Binns, A. (2015), Breakfasts higher in protein increase postprandial energy expenditure, increase fat oxidation, and reduce hunger in overweight children from 8 to 12 years of age. *J. Nutr.*, 145(10), 2229–35.

Bereiter-Hahn, J., Bernd, A., Beschmann, H., Eberle, I., Kippenberger, S., Rossberg, M., Strecker, V. and Zölle, N. (2014), Cellular responses to egg-oil (Charismon©). *Acta Medica (Hradec Králové)*, 57(2), 41–8.

Berger, M. J. and Doherty, T. J. (2010), Sarcopenia: prevalence, mechanisms, and functional consequences. *Interdiscip. Top Gerontol.*, 37, 94–114.

Blesso, C. N., Andersen, C. J., Barona, J., Volek, J. S. and Fernandez, M. L. (2013), Whole egg consumption improves lipoprotein profiles and insulin sensitivity to a greater extent than yolk-free egg substitute in individuals with metabolic syndrome. *Metabolism*, 62, 400–10.

Burke, J. D., Curran-Celentano, J. and Wenzel, A. J. (2005), Diet and serum carotenoid concentrations affect macular pigment optical density in adults 45 years and older. *J. Nutr.*, 135, 1208–14.

Campbell, A. P. and Rains, T. M. (2015), Dietary protein is important in the practical management of prediabetes and type 2 diabetes. *J. Nutr.*, 145(1), 164S–9S.

Chen, C., Chi, Y. J., Zhao, M. Y. and Lv, L. (2012), Purification and identification of antioxidant peptides from egg white protein hydrolysate. *Amino Acids*, 43(1), 457–66.

Chung, H. Y., Rasmussen, H. M. and Johnson, E. J. (2004), Lutein bioavailability is higher from lutein-enriched eggs than from supplements and spinach in men. *J. Nutr.*, 134, 1887–93.

Clarke, R., Frost, C., Collins, R., Appleby, P. and Peto, R. (1997), Dietary lipids and blood cholesterol: quantitative meta-analysis of metabolic ward studies. *BMJ*, 314, 112.

Committee on Nutrition, American Heart Association. (1968), *Diet and Heart Disease. American Heart Association*, Dallas, TX.

Davidson, M. H., Ballantyne, C. M., Jacobson, T. A., Bittner, V. A., Braun, L. T., Brown, A. S., Brown, W. V., Cromwell, W. C., Goldberg, R. B., McKenney, J. M., et al. (2011), Clinical utility of inflammatory markers and advanced lipoprotein testing: advice from an expert panel of lipid specialists. *J. Clin. Lipidol.*, 5, 338–67.

Diffenderfer, M. R. and Schaefer, E. J. (2014), The composition and metabolism of large and small LDL. *Curr. Opin. Lipidol.*, 25, 221–6.

Djoussé, L., Gaziano, J. M., Buring, J. E. and Lee, I. M. (2009), Egg consumption and risk of type 2 diabetes in men and women. *Diab. Care*, 32(2), 295–300.

Djoussé, L., Petrone, A. B., Hickson, D. A., Talegawkar, S. A., Dubbert, P. M., Taylor, H. and Tucker, K. L. (2015), Egg consumption and risk of type 2 diabetes among African Americans: the Jackson heart study. *Clin. Nutr.*, 35(3), 679–84.

Djoussé, L., Khawaja, O. A. and Gaziano, J. M. (2016), Egg consumption and risk of type 2 diabetes: a meta-analysis of prospective studies. *Am. J. Clin. Nutr.*, 103(2), 474–80.

Eckel, R. H., Jakicic, J. M., Ard, J. D., de Jesus, J. M., Houston Miller, N., Hubbard, V. S., et al. (2014), 2013 AHA/ACC guideline on lifestyle management to reduce cardiovascular risk: a report of the American College of Cardiology/American Heart Association Task Force on Practice Guidelines. *J. Am. Coll. Cardiol.*, 63, 2960–84.

Ericson, U., Sonestedt, E., Gullberg, B., Hellstrand, S., Hindy, G., Wirfält, E., Orho-Melander, M. (2013), High intakes of protein and processed meat associate with increased incidence of type 2 diabetes. *Br. J. Nutr.*, 109(6), 1143–53.

Fernandez, M. L. and Webb, D. (2008), The LDL to HDL cholesterol ratio as a valuable tool to evaluate coronary heart disease risk. *J. Am. Coll. Nutr.*, 27, 1–5.

Fernandez, M. L. and Calle, M. (2010), Revisiting dietary cholesterol recommendations: does the evidence support a limit of 300 mg/d? *Curr. Atheroscler. Rep.*, 12, 377–83.

Fielding, R. A., Vellas, B., Evans, W. J., Bhasin, S., Morley, J. E., Newman, A. B, Abellan van Kan, G., Andrieu, S., Bauer, J., Breuille, D., et al. (2011), Sarcopenia: an undiagnosed condition in older adults. Current consensus definition: prevalence, etiology, and consequences. International working group on sarcopenia. *J. Am. Med. Dir. Assoc.*, 12, 249–56.

Fox, C. S., Golden, S. H., Anderson, C., Bray, G. A., Burke, L. E., de Boer, I. H., Deedwania, P., Eckel, R. H., Ershow, A. G., Fradkin, J., Inzucchi, S. E., Kosiborod, M., Nelson, R. G., Patel, M. J., Pignone, M., Quinn, L., Schauer, P. R., Selvin, E., Vafiadis, D. K.; American Heart Association Diabetes Committee of the Council on Lifestyle and Cardiometabolic Health; Council on Clinical Cardiology, Council on Cardiovascular and Stroke Nursing, Council on Cardiovascular Surgery and Anesthesia, Council on Quality of Care and Outcomes Research; American Diabetes Association. (2015), Update on prevention of Cardiovascular disease in adults with type 2 diabetes mellitus in light of recent evidence: a scientific statement from the American heart association and the American diabetes association. *Diab. Care*, 38, 1777–803.

Fuller, N. R., Caterson, I. D., Sainsbury, A., Denyer, G., Fong, M., Gerofi, J., Baqleh, K., Williams, K. H., Lau, N. S. and Markovic, T. P. (2015), The effect of a high-egg diet on cardiovascular risk factors in people with type 2 diabetes: the Diabetes and Egg (DIABEGG) study-a 3-mo randomized controlled trial. *Am. J. Clin. Nutr.*, 101, 705–13.

Giordano, P., Scicchitano, P., Locorotondo, M., Mandurino, C., Ricci, G., Carbonara, S., Gesualdo, M., Zito, A., Dachille, A., Caputo, P., et al. (2012), Carotenoids and cardiovascular risk. *Curr. Pharm. Des.*,18, 5577–89.

Goldberg, S., Gardener, H., Tiozzo, E., Ying Kuen, C., Elkind, M. S., Sacco, R. L. and Rundek, T. (2014), Egg consumption and carotid atherosclerosis in the Northern Manhattan study. *Atherosclerosis*, 235, 273–80.

Goodrow, E. F., Wilson, T. A., Houde, S. C., Vishwanathan, R., Scollin, P. A., Handelman, G. and Nicolosi, R. J. (2006), Consumption of one egg per day increases serum lutein and zeaxanthin concentrations in older adults without altering serum lipid and lipoprotein cholesterol concentrations. *J. Nutr.*, 136, 2519–24.

Greene, C. M., Zern, T. L., Wood, R. J., Shrestha, S., Aggarwal, D., Sharman, M. J., Volek, J. S. and Fernandez, M. L. (2005), Maintenance of the LDL cholesterol: HDL cholesterol ratio in an elderly population given a dietary cholesterol challenge. *J. Nutr.*, 135, 2793–8.

Griffin, J. D. and Lichtenstein, A. H. (2013), Dietary cholesterol and plasma lipoprotein profiles: randomized-controlled trials. *Curr. Nutr. Rep.*, 2, 274–82.

Harman, N. L., Leeds, A. R. and Griffin, B. A. (2008), Increased dietary cholesterol does not increase plasma low density lipoprotein when accompanied by an energy-restricted diet and weight loss. *Eur. J. Nutr.*, 47, 287–93.

Heaney, R. P. and Layman, D. K. (2008), Amount and type of protein influences bone health. *Am. J. Clin. Nutr.*, 87, 1567S–70S.

Herron, K. L., Vega-Lopez, S., Conde, K., Ramjiganesh, T., Roy, S., Shachter, N. S. and Fernandez, M. L. (2002), Pre-menopausal women, classified as hypo- or hyperresponders, do not alter their LDL/HDL ratio following a high dietary cholesterol challenge. *J. Am. Coll. Nutr.*, 21, 250–8.

Herron, K. L., Lofgren, I. E., Sharman, M., Volek, J. S. and Fernandez, M. L. (2004), High intake of cholesterol results in less atherogenic low-density lipoprotein particles in men and women independent of response classification. *Metabolism*, 53, 823–30.

Howell, W. H., McNamara, D. J., Tosca, M. A., Smith, B. T. and Gaines, J. A. (1997), Plasma lipid and lipoprotein responses to dietary fat and cholesterol: a meta-analysis. *Am. J. Clin. Nutr.*, 65, 1747–64.

Hu, F. B., Stampfer, M. J., Rimm, E. B., Manson, J. E., Ascherio, A., Colditz, G. A., Rosner, B. A., Spiegelman, D., Speizer, F. E., Sacks, F. M., Hennekens, C. H. and Willett, W. C. (1999), A prospective study of egg consumption and risk of cardiovascular disease in men and women. *JAMA*, 281, 1387–94.

Hu, F. B. (2011), Globalization of diabetes: the role of diet, lifestyle, and genes. *Diab. Care*, 34(6), 1249–57.

Isherwood, C., Wong, M., Jones, W. S., Davies, I. G. and Griffin, B. A. (2010), Lack of effect of cold water prawns on plasma cholesterol and lipoproteins in normo-lipidaemic men. *Cell Mol. Biol. (Noisy-le-grand)*, 56, 52–8.

Kanter, M. M., Kris-Etherton, P. M., Fernandez, M. L., Vickers, K. C. and Katz, D. L. (2012), Exploring the factors that affect blood cholesterol and heart disease risk: is dietary cholesterol as bad for you as history leads us to believe? *Adv. Nutr. Int. Rev. J.*, 3, 711–17.

Katz, D. L., Evans, M. A., Nawaz, H., Njike, V. Y., Chan, W., Comerford, B. P. and Hoxley, M. L. (2005), Egg consumption and endothelial function: a randomized controlled crossover trial. *Int. J. Cardiol.*, 99(1), 65–70.

Kral, T. V., Bannon, A. L., Chittams, J. and Moore, R. H. (2016), Comparison of the satiating properties of egg- versus cereal grain-based breakfasts for appetite and energy intake control in children. *Eat. Behav.*, 20, 14–20.

Lajous, M., Bijon, A., Fagherazzi, G., Balkau, B., Boutron-Ruault, M. C. and Clavel-Chapelon, F. (2015), Egg and cholesterol intake and incident type 2 diabetes among French women. *Br. J. Nutr.*, 114(10), 1667–73.

Lamarche, B., Lemieux, I. and Despres, J. P. (1999), The small, dense LDL phenotype and the risk of coronary heart disease: epidemiology, patho-physiology and therapeutic aspects. *Diab. Metab.*, 25, 199–211.

Layman, D. K. (2004), Protein quantity and quality at levels above RDA improves adult weight loss. *J. Am. Coll. Nutr.*, 6, 631S–6S.

Layman, D. K. (2009), Dietary Guidelines should reflect new understandings about adult protein needs. *Nutr. Metab.*, 6, 12.

Leeson. S. and Caston, L. (2004), Enrichment of eggs with lutein. *Poult. Sci.*, 83(10), 1709–12.

Leidy, H. J. and Racki, E. M. (2010), The addition of a protein-rich breakfast and its effect on acute appetite control and food intake in 'breakfast skipping' adolescents. *Int. J. Obes.*, 34, 1125–33.

Leidy, H. J., Ortinau, L. C., Douglas, S. M. and Hoertel, H. A. (2013), Beneficial effects of a higher-protein breakfast on the appetitive, hormonal, and neural signals controlling energy intake regulation in overweight/obese, 'breakfast-skipping', late-adolescent girls. *Am. J. Clin. Nutr.*, 97, 677–88.

Leidy, H. J., Clifton, P. M., Astrup, A., Wycherley, T. P., Westerterp-Plantenga, M. S., Luscombe-Marsh, N. D., Woods, S. C. and Mattes, R. D. (2015), The role of protein in weight loss and maintenance. *Am. J. Clin. Nutr.* [Epub ahead of print]

Li, Y., Zhou, C., Zhou, X. and Li, L. (2013), Egg consumption and risk of cardiovascular diseases and diabetes: a meta-analysis. *Atherosclerosis*, 229, 524–30.

Li, J., Armstrong, C. L. and Campbell, W. W. (2016), Effects of dietary protein source and quantity during weight loss on appetite, energy expenditure, and Cardio-Metabolic responses. *Nutrients*, 8, 2.

Liu, A. G., Puyau, R. S., Han, H., Johnson, W. D., Greenway, F. L. and Dhurandhar, N. V. (2015), The effect of an egg breakfast on satiety in children and adolescents: a randomized crossover trial. *J. Am. Coll. Nutr.*, 34(3), 185–90.

Lloyd-Jones, D. M., Hong, Y., Labarthe, D., Mozaffarian, D., Appel, L. J., Van Horn, L., Greenlund, K., Daniels, S., Nichol, G., Tomaselli, G. F., Arnett, D. K., Fonarow, G. C., Ho, P. M., Lauer, M. S., Masoudi, F. A., Robertson, R. M., Roger, V., Schwamm, L. H., Sorlie, P., Yancy, C. W., Rosamond, W. D.; American Heart Association Strategic Planning Task Force and Statistics Committee. (2010), Defining and setting national goals for cardiovascular health promotion and disease reduction: the American Heart Association's strategic Impact Goal through 2020 and beyond. *Circulation*, 121, 586–613.

Maki, K. C., Slavin, J. L., Rains, T. M. and Kris-Etherton, P. M. (2014), Limitations of observational evidence: implications for evidence-based dietary recommendations. *Adv. Nutr.*, 5, 7–15.

Makkar, S., Liyanage, R., Kannan, L., Packialakshmi, B., Lay, J. O. Jr and Rath, N. C. (2015), Chicken egg shell membrane associated proteins and peptides. *J. Agric. Food Chem.*, 63(44), 9888–98.

Miguel, M., López-Fandiño, R., Ramos, M. and Aleixandre, A. (2005), Short-term effect of egg-white hydrolysate products on the arterial blood pressure of hypertensive rats. *Br. J. Nutr.*, 94(5), 731–7.

Mutungi, G., Ratliff, J., Puglisi, M., Torres-Gonzalez, M., Vaishnav, U., Leite, J. O., Quann, E., Volek, J. S. and Fernandez, M. L. (2008), Dietary cholesterol from eggs increases plasma HDL cholesterol in overweight men consuming a Carbohydrate-restricted diet. *J. Nutr.*, 138, 272–6.

Ng, M., Fleming, T., Robinson, M., Thomson, B., Graetz, N., Margono, C., et al. (2014), Global, regional, and national prevalence of overweight and obesity in children and adults during 1980-2013: a systematic analysis for the Global Burden of Disease Study 2013. Erratum in: *Lancet*. 30 August 2014, 384(9945): 766–81.

Nimalaratne, C., Bandara, N. and Wu, J. (2015), Purification and characterization of antioxidant peptides from enzymatically hydrolyzed chicken egg white. *Food Chem.*, 188, 467–72.

Njike, V., Faridi, Z., Dutta, S., Gonzalez-Simon, A. L. and Katz, D. L. (2010), Daily egg consumption in hyperlipidemic adults—effects on endothelial function and cardiovascular risk. *Nutr. J.*, 9, 28.

Paddon-Jones, D., Short, K. R., Campbell, W. W., Volpi, E. and Wolfe, R. R. (2008), Role of dietary protein in the sarcopenia of aging. *Am. J. Clin. Nutr.*, 87, 1562S–6S.

Paddon-Jones, D. and Rasmussen, B. B. (2009), Dietary protein recommendations and the prevention of sarcopenia. *Curr. Opin. Clin. Nutr. Metab. Care.*, 12, 86–90.

Paddon-Jones, D., Campbell, W. W., Jacques, P. F., Kritchevsky, S. B., Moore, L. L., Rodriguez, N. R. and van Loon, L. J. (2015), Protein and healthy aging. *Am. J. Clin. Nutr.* [Epub ahead of print]

Pasiakos, S. M., Lieberman, H. R. and McLellan, T. M. (2014), Effects of protein supplements on muscle damage, soreness and recovery of muscle function and physical performance: a systematic review. *Sports Med.*, 44(5), 655–70.

Pearce, K. L., Clifton, P. M. and Noakes, M. (2011), Egg consumption as part of an energy-restricted high-protein diet improves blood lipid and blood glucose profiles in individuals with type 2 diabetes. *Br. J. Nutr.*, 30, 584–92.

Phillips, S. M., Chavalier, S. and Leidy, H. J. (2016), Protein 'requirements' beyond the RDA: implications for optimizing health. *Applied Physiology, Nutrition, and Metabolism.*, 41(5), 565–72 [E-pub ahead of print].

Qureshi, A. I., Suri, F. K., Ahmed, S., Nasar, A., Divani, A. A. and Kirmani, J. F. (2007), Regular egg consumption does not increase the risk of stroke and cardiovascular diseases. *Med. Sci. Monit.*, 13(1), CR1–8.

Radzeviciene, L. and Ostrauskas, R. (2012), Egg consumption and the risk of type 2 diabetes mellitus: a case-control study. *Public Health Nutr.*, 15, 1437–41.

Rains, T. M., Leidy, H. J., Sanoshy, K. D., Lawless, A. L and Maki, K. C. (2015), A randomized, controlled, crossover trial to assess the acute appetitive and metabolic effects of sausage and egg-based convenience breakfast meals in overweight premenopausal women. *Nutr. J.*, 14, 17.

Ratliff, J., Leite, J. O., de Ogburn, R., Puglisi, M. J., VanHeest, J. and Fernandez, M. L. (2010), Consuming eggs for breakfast influences plasma glucose and ghrelin, while reducing energy intake during the next 24 hours in adult men. *Nutr. Res.*, 30, 96–103.

Rasmussen, B. B., Tipton, K. D., Miller, S. L., Wolf, S. E. and Wolfe, R. R. (2000), An oral essential amino acid-carbohydrate supplement enhances muscle protein anabolism after resistance exercise. *J. Appl. Physiol.*, 88, 386–92.

Rastegar, F., Azarpira, N., Amiri, M. and Azarpira, A. (2011), The effect of egg yolk oil in the healing of third degree burn wound in rats. *Iran. Red Crescent Med. J.*, 13(10), 739–43.

Robbins, J. M., Petrone, A. B., Ellison, R. C., Hunt, S. C., Carr, J. J., Heiss, G., Arnett, D. K., Gaziano, J. M. and Djoussé, L. (2014), Association of egg consumption and calcified atherosclerotic plaque in the coronary arteries: the NHLBI family heart study. *ESPEN J.*, 9(3), e131–5.

Rong, Y., Chen, L., Zhu, T., Song, Y., Yu, M., Shan, Z., Sands, A., Hu, F. B. and Liu, L. (2013), Egg consumption and risk of coronary heart disease and stroke: dose-response meta-analysis of prospective cohort studies. *Brit. Med. J.*, 346, e8539.

Rutherfurd, S. M., Fanning, A. C., Miller, B. J. and Moughan, P. J. (2015), Protein digestibility-corrected amino acid scores and digestible indispensable amino acid scores differentially describe protein quality in growing male rats. *J. Nutr.*, 145(2), 372–9.

Samaraweera, H., Zhang, W. G., Lee, E. J. and Ahn, D. U. (2011), Egg yolk phosvitin and functional phosphopeptides—review. *J. Food Sci.*, 76(7), R143–50.

Shaw, G. M., Carmichael, S. L., Yang, W., Selvin, S. and Schaffer, D. M. (2004), Periconceptional dietary intake of choline and betaine and neural tube defects in offspring. *Am. J. Epidemiol.*, 160, 102–9.

Shaw, G. M., Finnell, R. H., Blom, H. J., Carmichael, S. L., Vollset, S. E., Yang, W. and Ueland, P. M. (2009), Choline and risk of neural tube defects in a folate-fortified population. *Epidemiology*, 20, 714–19.

Shi, Z., Yuan, B., Zhang, C., Zhou, M. and Holmboe-Ottesen, G. (2011), Egg consumption and the risk of diabetes in adults, Jiangsu, China. *Nutrition*, 27, 194–8.

Shin, J. Y., Xun, P., Nakamura, Y. and He, K. (2013), Egg consumption in relation to risk of cardiovascular disease and diabetes: a systematic review and meta-analysis. *Am. J. Clin. Nutr.*, 98(1), 146–59.

Skřivan, M., Englmaierová, M., Skřivanová, E. and Bubancová, I. (2015), Increase in lutein and zeaxanthin content in the eggs of hens fed marigold flower extract M. *Czech J. Anim. Sci.*, 60, 89–96.

Tran, N. L., Barraj, L. M., Heilman, J. M. and Scrafford, C. G. (2014), Egg consumption and cardiovascular disease among diabetic individuals: a systematic review of the literature. *Diab. Metab. Syndr. Obes.*, 7, 121–37.

Vander Wal, J. S., Marth, J. M., Khosla, P., Jen, K. L. and Dhurandhar, N. V. (2005), Short-term effect of eggs on satiety in overweight and obese subjects. *J. Am. Coll. Nutr.*, 24(6), 510–5.

Vander Wal, J. S., Gupta, A., Khosla, P., et al. (2008), Egg breakfast enhances weight loss. *Int. J. Obes.* (Lond.), 32, 1545–51.

Virtanen, J. K., Mursu, J., Tuomainen, T. P., Virtanen, H. E. and Voutilainen, S. (2015), Egg consumption and risk of incident type 2 diabetes in men: the Kuopio Ischaemic heart disease risk factor study. *Am. J. Clin. Nutr.*, 101(5), 1088–96.

Vislocky, L. M., Pikosky, M. A., Rubin, K. H., Vega-Lopez, S., Gaine, P. C., Martin, W. F., Zern, T. L., Lofgren, I. E., Fernandez, M. L. and Rodriguez, N. R. (2009), Habitual consumption of eggs does not alter the beneficial effects of endurance training on plasma lipids and lipoprotein metabolism in untrained men and women. *J. Nutr. Biochem.*, 20, 26–34.

von Haehling, S., Morley, J. E. and Anker, S. D. (2010), An overview of sarcopenia: facts and numbers on prevalence and clinical impact. *J. Cachexia Sarcopenia Muscle.*, 1, 129–33.

Weggemans, R. M., Zock, P. L. and Katan, M. B. (2001), Dietary cholesterol from eggs increases the ratio of total cholesterol to high-density lipoprotein cholesterol in humans: a meta-analysis. *Am. J. Clin. Nutr.*, 73, 885–91.

Westerterp-Plantenga, M. S., Lemmens, S. G. and Westerterp, K. R. (2012), Dietary protein – its role in satiety, energetics, weight loss and health. *Br. J. Nutr.*, 108(Suppl. 2), S105–12.

World Health Organization. (2016), Commission on ending childhood obesity. Available at: http://www.who.int/end-childhood-obesity/en/ (accessed 12 April 2016).

Wycherley, T. P., Moran, L. J., Clifton, P. M., Noakes, M. and Brinkworth, G. D. (2012), Effects of energy-restricted high-protein, low-fat compared with standard-protein, low-fat diets: a meta-analysis of randomized controlled trials. *Am. J. Clin. Nutr.*, 96(6), 1281–98.

Yoshii, H., Tachi, N., Ohba, R., Sakamura, O., Takeyama, H. and Itani, T. (2001), Antihypertensive effect of ACE inhibitory oligopeptides from chicken egg yolks. *Comp. Biochem. Physiol. C Toxicol. Pharmacol.*, 128(1), 27–33.

Yousr, M. and Howell, N. (2015), Antioxidant and ACE inhibitory bioactive peptides purified from egg yolk proteins. *Int. J. Mol. Sci.*, 16(12), 29161–78.

Zeisel, S. H. (2006), The fetal origins of memory: the role of dietary choline in optimal brain development. *J. Pediat.*, 149, S131–6.

Zeisel, S. H. and da Costa, K. A. (2009), Choline: an essential nutrient for public health. *Nutr. Rev.*, 67, 615–23.

Chapter 15

Nutraceutical benefits of eggs

Hoon H. Sunwoo and Naiyana Gujral, University of Alberta, Canada

1 Introduction

In recent years, there has been growing consumer interest in mainlining good health and preventing disease. This has led to an interest in 'natural' products which may help to prevent disease. These natural products, with health benefits over and above their existing nutritional value, are known as nutraceuticals. They play crucial roles in preventive health care of the ageing population with lifestyle-related diseases. The global nutraceutical market has grown significantly in the past two decades, influenced by health specialists and media, both as means of reducing negative health effects of existing health conditions as well as risks of future diseases.

The hen's egg is considered to be nature's perfect food and has been recognized as an excellent source of essential nutrients: protein, lipids, vitamins and minerals. When a fertilized egg is provided with essential nutrients and optimal energy in the form of heat during embryogenesis, it readily turns into a potential new life. Thus, all substances essential for the creation of the new chick are present in the egg, confirming the excellence of eggs regarding their biological values. Eggs, both intact and enriched with bioactive components, may be used to purify nutraceutical compounds due to high content and functionality of active elements, proteins, phospholipids and fatty acids, beneficial for human health.

Numerous studies have shown biological activities of substances contained in the egg, such as anti-bacterial, anti-viral, anti-fungal, anti-cancer and immune-modulating

http://dx.doi.org/10.19103/AS.2016.0012.17

properties (Schade, 2007; Kołaczkowska et al., 2010). Egg material may also be used for the production of new bioactive substances used in the prophylaxis of various diseases, such as cardiovascular diseases, cancers, bone diseases (osteoporosis) and so on (Schade, 2007; López-Fandiño, 2007). Eggshells, containing calcium and many microelements, may be used for calcium supplements, aimed at prophylaxis of osteoporosis (Rovenský et al., 2003). Various food proteins function as precursors of biopeptides. Among those, the hen egg proteins are known as the richest sources of bioactive peptides (Pellegrini et al., 2000). Among the protein egg components, there are both active forms, such as lysozyme, phosvitin, serine and cysteine proteinases inhibitors, immunoglobulins, as well as those that exhibit activities after proteolytic degradation (Mine and Kovacs-Nolan, 2006; Mine, 2007). Egg yolk lipids, most prominently lecithin with prophylactic effects, are characterized by anti-oxidative (Sugino et al., 1997) and lipid-lowering properties (Rampone, 1973).

The use of hens and their eggs as live bioreactors, for the production of nutraceutical and biomedical preparations, is constantly gaining in popularity. Some existing chemical composition and biological features of eggs may be further modified by fortification of desired components using several techniques (Surai and Sparks, 2000; Gładkowski et al., 2011). In the biological and economic aspect, eggs are a perfect and relatively inexpensive reservoir of biocomponents (Schade, 2007; Shapira, 2010) for the development of biologically active nutraceutical compounds.

An egg constitutes three main components: eggshell (10%), egg white (60%) and yolk (30%). An egg contains 75% water, 12% proteins, 12% lipids and the remaining 1% is carbohydrates and minerals (Kovacs-Nolan and Mine, 2005). Proteins in the egg are present in both egg white and yolk, whereas lipids are mainly concentrated in the yolk. Eighty-eight percentage of the egg white constitutes water and the remaining egg white consists of protein (11%), ash and trace amounts of lipids (1%). The yolk consists of approximately 50% water, 16% protein, 33% lipids and 1% carbohydrate. Egg components with nutraceutical benefits are described below.

2 The eggshell and its membrane

2.1 The eggshell

Eggshell is composed of 95% minerals, with calcium being the major element, which constitutes about 98%. Eggshell powder is a natural source of calcium and microelements (magnesium, fluorine, strontium, copper, manganese and zinc), which have a positive effect on bone density (Eaton-Evans, 1994; Meunier, 2001). One whole medium-sized eggshell yields about 750–800 mg of elemental calcium. Among the microelements, strontium is present in a significant amount and has an anabolic effect on bone health (Marie et al., 1985). A bioavailability study in piglets showed that calcium from eggshell was similar to, or better than, calcium carbonate (purified food grade) (Schaafsma and Beelen, 1999). A clinical study of eggshell powder in postmenopausal women with osteoporosis showed reduced pain and bone resorption, as well as increased mobility and bone density (Schaafsma, 2000). Pavlikova and coworkers (1995) showed that eggshell powder also had *in vitro* stimulatory effect on chondrocyte differentiation and cartilage growth. Carbohydrates in eggshell are composed of anionic polysaccharides, glycosaminoglycans, consisting of hyaluronic acid (48%), and galactosminoglycan (52%) (Nakano et al., 2001), which have the positive effect in strengthening of skin and cartilage. Both experimental and clinical studies have shown that eggshell powder has positive effects on bone and cartilage health and is suitable for the prevention and treatment of osteoporosis.

2.2 The eggshell membrane

Eggshell membrane is comprised of two protein fibre membranes between the egg albumen and the eggshell. The fibres contain predominantly Type I and Type V collagen fibres encapsulated in proteoglycans and macromolecules (Wong et al., 1984). Other components in the eggshell membranes are glycosaminoglycans, such as dermatan sulphate and hyaluronic acid (Long et al., 2005), chondroitin sulphate (Baker and Balch, 1962), sialic acid (Nakano et al., 2003), desmosine and isodesmosine (Starcher and King, 1980). The eggshell membrane contains several bacteriolytic lysozyme and *N*-acetyl glucosaminidase and membrane proteins are beneficial in treating injuries (Hincke et al., 2000; Ahlborn et al., 2006). The membrane-derived peptides have been reported to have a stimulatory effect on skin fibroblasts *in vitro* (Suguro et al., 2000). A randomized, multi-centred, double-blind, placebo-controlled clinical trial proved that dietary supplementation with 500 mg per day egg shell membrane significantly reduced both joint pain and stiffness compared to placebo after a minimum of 10 days continued use (Ruff et al., 2009). Both *in vitro* and *in vivo* studies support the potential use of eggshell membrane in therapeutic and nutraceutical applications for maintaining healthy articular cartilage and synovium.

3 Egg white proteins

Egg white proteins are recognized for their functional importance. The proteins in egg white contain 60% ovalbumin, 12% ovotransferrin, 11% ovomucoid, 3.5% lysozyme, 3.5% ovomucin, 0.05% avidin, 0.05% cystatin, 0.5% ovomacroglobulin, 0.8% ovoflavoprotein, 1.0% ovoglycoprotein and 1.5% ovoinhibitor (Kovacs-Nolan and Mine, 2005). The proteins with bioactivity, discussed below, may be extracted from egg white and used alone or in combination as nutraceutical components.

3.1 Ovalbumin

Ovalbumin is the main protein (classed as a phosphoglycoprotein) in the egg white, constituting approximately 60% of total protein content. Its proteolytic degraded peptide has substantial bioactivity. Enzymatic digest of ovalbumin was found to be strongly active against *B. subtilus* and to a lesser extent against *E. coli*, *B. bronchiseptica*, *P. aeruginosa*, *S. marcescens*, and *C. albicans* (Pellegrini et al., 2004). Ovokinin, a pepsin digest of ovalbumin, has been shown to lower the systolic blood pressure of hypertensive rats. On consumption of whole eggs, ovokinin will be released from ovalbumin in the stomach, emulsified with egg yolk and absorbed (Davalos et al., 2004; Miguel and Aleixandre, 2006), which may also contribute to the control of blood pressure by improving endothelial dysfunction. Ovalbumin-modified derivative has been found to induce the release of tumour necrosis factor (TNF) (Fan et al., 2003), increase phagocytic activity of macrophages and enhance immune responses for cancer immunotherapy (Vidovic et al., 2002; He et al., 2003). Nakamura and Kato (2000) reported increased lipid affinity when ovalbumin was glycosylated which resulted in the enhancement of the radical-scavenging ability.

3.2 Ovotransferrin

Ovotransferrin is a monomeric glycoprotein sized 78 kDa of the transferrin family with a specific binding site for iron. The most recognized function of ovotransferrin is its antibacterial activity, related to its ability to bind iron (Fe^{3+}), hindering bacterial growth (Alderton et al., 1946; Bullen et al., 1978). Ovotransferrin demonstrated *in vivo* antibacterial activity in newborn guinea pigs orally infected with *E. coli* 0111 B4 (Valenti et al., 1983). It also confirmed the iron deprivation effect on *Pseudomonas* spp., *E. coli*, *Streptococcus* spp. (Valenti et al., 1983) and *Salmonella* spp. (Superti et al., 2007).

Ovotransferrin is a superoxide dismutase-like protein possessing superoxide radical scavenging activity (Ibrahim et al., 2006, 2007). Self-cleaved ovotransferrin also expressed high antioxidative capacity against stress-induced DNA damage in human leucocytes (Kim et al., 2012). Recently, enzymatically hydrolyzed ovotransferrin peptides were found to attenuate TNF-α-induced inflammatory responses and oxidative stress in vascular endothelial cells (Majumder et al., 2013). It has been reported that ovotransferrin thiol-linked auto-cleavage after reduction possessed substantial anticancer effects towards colon and breast cancer by inducing apoptosis (Ibrahim and Kiyono, 2009). The ovotransferrin also showed antihypertensive function by displaying both angiotensin-converting enzyme inhibitory and vasodilatory activities (Wu and Acero-Lopez, 2012). Ovotransferrin-derived conjugates are, therefore, desirable as potential nutraceutical ingredients with metal-chelating, antibacterial, antioxidant, antihypertensive and anti-inflammatory properties.

3.3 Ovomucin

Ovomucin is a highly glycosylated protein with a molecular mass of approximately 165 kDa, consisting of peptide-rich α-subunit and a carbohydrate-rich β-subunit. Ovomucin possesses virus haemagglutination inhibition activity (Tsuge et al., 1996), anti-tumour activity (Oguro et al., 2001) and cholesterol uptake inhibition activity (Nagaoka et al., 2002). Ovomucin and ovomucin-derived peptides have also demonstrated antiviral activity against human influenza virus *in vitro* (Tsuge et al., 1996). The enzymatic digested ovomucin glycopeptides (OGP) were reported to possess binding sites for *E. coli* and, therefore, are protective against *E. coli* O157:H7 infection (Kobayashi et al., 2004). Ovomucin peptides have also shown macrophage-stimulating activity *in vitro* (Tanizaki et al., 1997), and *in vivo* anti-tumour activity in mice (Watanabe et al., 1998), suggesting the ability to inhibit tumour growth (Oguro et al., 2001).

3.4 Ovomucoid

Ovomucoid is a 28 kDa highly glycosylated protein. Ovomucoid is a trypsin inhibitor and considered the main food allergen in eggs. However, the elimination of allergenicity and enhancement of bioactivity is achieved when ovomucoid is hydrolysed into smaller peptides. It has been reported that both ovomucoid and its peptides can control

Streptomyces erythraeus growth (Nagata and Yoshida, 1984). Ovomucoid peptides have been demonstrated to enhance cellular immune system by induction of cytokines interleukin (IL)-4, IL-6, IL-10, IL-13 and interferon (IFN)-gamma secretion (Holen et al., 2001) as well as exhibiting ACE-inhibitory, antioxidant and iron-chelating activities (Abeyrathne et al., 2015).

3.5 Avidin

Avidin is a glycoprotein containing four identical subunits with a combined mass of 67 kDa. Avidin possesses specific binding ability to biotin (a water-soluble vitamin), and inhibits the growth of biotin-requiring bacteria and yeasts (Banks et al., 1986). Avidin has been reported to bind to both gram-negative and gram-positive bacteria, including *E. coli* K-12, *K. pneumoniae*, *S. marcescens*, *P. aeruginosa*, *S. aureus* and *S. epidermis* (Korpela et al., 1984), thus inhibiting their growth.

3.6 Lysozyme

Lysozyme is a 14.3 kDa strong basic polypeptide consisting of 129 amino acid residues, and is known for its bacteriostatic and bacteriocidal activities against major food-borne pathogens, *L. monocytogens* and *C. botulinum* (Radziejewska et al., 2008). Lysosome-modified derivative represented increased antiviral, antimicrobial and anti-inflammatory effects by exposing its functional sites. The food safety organizations in many countries allow the use of lysozyme in preserved foods (Mine et al., 2004). Lysozyme can also be used as an anticancer agent due to its immune-modulating and tumour-cells-suppressing capabilities (Kovacs-Nolan and Mine, 2005).

3.7 Cystatin

Cystatin in egg white belongs to the type 2 cystatin, having about 115 amino acids and two disulphide bonds, with no carbohydrates. Secreted cystatin has a molecular weight of 13 kDa. It has been reported to possess antibacterial activity, preventing the growth of group A *Streptococcus* (Björck et al., 1989), *S. typhimurium* (Nakai, 2000) and the periodontitis-causing *P. gingivalis* (Travis et al., 1997). Cystatin can also induce the synthesis of cytokines, which results in upregulated release of nitric oxide in mouse peritoneal macrophages (Verdot et al., 1996), and inhibits tumour invasion in Ras-transformed breast epithelial cells (Premzl et al., 2001). This evidence suggests potential roles for chicken cystatin in antimicrobial and cancer therapies.

4 Egg yolk proteins

The proteins in egg yolk are composed of 36% lipovitellins (high-density lipoprotein, HDL), 38% livetins, 8% phosvitin and 17% low-density lipoproteins (LDL). They are primarily present in the form of lipoproteins, which can be separated into granule and plasma fractions (Sugino et al., 1997). The granule fraction contains lipovitellins (HDL), phosvitin and LDL (Burley and Cook, 1961). The plasma fraction contains LDL and livetins (lipid-free globular proteins) (Li-Chan et al., 1995). LDL is the major protein that accounts for 65% of the total yolk proteins. The HDL exists as a complex with phosphoprotein known as 'phosvitin'.

4.1 Lipoprotein

Lipovitellin (LDL) is synthesized in the liver of laying hen and consists of protein (80%) and lipid (20%), including phospholipids (60% of the lipid, primarily lecithin), triacylglycerols (40%), and small amounts of cholesterol, sphingomyelin and other lipids. LDL apoproteins consist of α- and β-lipovitellins, which have distinct amino acid composition as well as bound phosphorous and carbohydrate. The proportion of α- and β-lipovitellins in yolk granules appears to be genetically based. Another lipoprotein fraction is comprised of lipovitellinin (HDL) present in dimers of five main apoproteins.

Mine and Kovac-Nolan (2006) have documented *in vitro* antimicrobial activity of egg yolk lipoproteins against *Streptococcus mutans*. HDL and their peptides were reported to inhibit the colonization of *S. typhimurium*, *C. jejuni* and *E. coli* O157:H7 due to the presence of anti-adhesive factors. Digestion of vitellenin, the apoprotein of HDL, with pronase gives two glycopeptides A and B. Glycopeptide A has a high content of sialic acid (*N*-acetylneuraminic), expressing blood protein half-life regulation, toxin neutralization, regulation of cellular adhesion and inhibition of cell cytolysis (Abdou et al., 2013). Egg yolk lipoproteins, therefore, function as lipid-mediated antimicrobial and antioxidative agents (Yamamoto and Omori, 1994).

4.2 Livetins

Livetins are lipid-free water-soluble globular proteins that account for 30% of the plasma proteins, consisting of α-livetin (serum albumin), β-livetin (α2-glycoprotein) and γ-livetins (immunoglobulin Y, IgY) (Sugino et al., 1997). Among the three livetin fractions, γ-livetins are of great interest and are extensively studied due to their distinct functional properties.

Egg yolk IgY antibody is accumulated at a concentration of ~25 mg/mL from serum antibody (IgG) transported from the mother hen as acquired passive immunity. The specificity of IgY is based on mother hen exposure to foreign antigens. Mimicking the natural process of antibody production, the hen can be immunized with certain immunogenic antigens in order to produce antibody specific to those particular antigens in the eggs. Chicken egg yolk IgY is readily obtained in large quantities, as well as a cost-effective, convenient and hygienic source of antibody as compared to mammalian antibodies. Studies have shown that IgY antibody has antimicrobial activity owing to the ability to neutralize several food-borne pathogens such as rotavirus (Sarker et al., 2001), *E. coli* O157:H7 (Sunwoo et al., 2002), *Salmonella enteritis* (Lee et al., 2002) and *Clostridium perfringens* (Song et al., 2009).

Our research team has recently developed IgY specific to gliadin as a therapeutic alternative to the only effective gluten-free diet treatment in celiac disease (Gujral et al., 2012a). The anti-gliadin IgY is investigated for its potential use as oral passive antibody therapy to neutralize the toxic gliadin peptides before they enter the small intestine, and so to preempt the celiac disease cascade of autoimmune inflammatory events. The formulation produced proved to be highly stable under simulated gastrointestinal conditions. The IgY polyclonal antibody produced against wheat gliadin was used as a universal antibody to neutralize all CD causing toxic peptides in grains (wheat, rye, barley and possibly oat), due to the peptide cross-reactivity (Gujral et al., 2012b). Anti-gliadin IgY inhibited the production of gliadin-induced inflammatory markers (TNF-α and IL-1β) and gliadin absorption in human carcinoma cell culture (an *ex vivo model* for celiac disease) (Gujral et al., 2015). Our preclinical *in vivo* mouse and anti-gliadin IgY formulation feeding study with spiked gliadin suggested that anti-gliadin prevented >99% absorption of gliadin (Gujral et al., 2012a). Our recent safety clinical trial proved that anti-gliadin IgY is safe and may improve CD-related outcome measures in patients on a gluten-free diet. A larger study to address efficacy will commence towards the end of 2016.

Both experimental and clinical studies have indicated that orally administered IgY has effective immune-modulating properties including anti-toxin (Lee et al., 2015), antimicrobial (Wang et al., 2014; Sui et al., 2011) and anti-allergen effects (Gujral et al., 2015). In particular, IgY improves the health and function of the gastrointestinal tract and greatly reduces systemic and intestinal inflammation (Vega et al., 2011; Sarker et al., 2007). Therefore, the aforementioned unique biological attributes make IgY an effective antimicrobial and immunotherapeutic nutraceutical agent.

4.3 Phosvitin

Phosvitin is a phosphoglycoprotein with α- and β-phosvitin subunits containing approximately 2–9% phosphorus, respectively. Phosvitin contains 54% of serine residues; therefore, it is one of the most phosphorylated proteins occurring in nature. Due to high phosphoric acid bound to serine residues, phosvitin is highly water soluble, and resistant to proteolysis (Juneja and Kim, 1997). It is the presence of highly negatively charged phosphate groups in serine residues that makes phosvitin one of the strongest naturally occurring metal-binding biomolecule for bivalent cations such as calcium, iron and magnesium (Grizzuti and Perlmann, 1973; Taborsky, 1963). Phosvitin contains over 100 phosphoserine residues (Castellani et al., 2004) and has the ability to bind approximately 60 iron molecules (Taborsky, 1963; Albright et al., 1984). The ability of phosvitin in removal of an essential factor such as iron has proved to inhibit growth of food-borne pathogens (Sattar Khan et al., 2000) and iron-induced oxidation. Therefore, phosvitin shows several functionalities such as antioxidant and antimicrobial effects.

Phosvitin can capture free Fe^{2+} and Fe^{3+} to inhibit the production of reactive oxygen species. PV can also efficiently inhibit Fe^{2+}- and Cu^{2+}-mediated phospholipid oxidations by chelating metals. It showed an inhibitory effect on phospholipid oxidation in the presence of Fe^{2+} and Cu^{2+} (Lu and Baker, 1986). Also, PV inhibited lipid peroxidation induced by UV irradiation in a dose-dependent manner in the presence of excess iron (Ishikawa et al., 2004). These results suggest that PV might be useful to prevent other iron-mediated oxidative stress-related diseases such as colorectal cancer (Ishikawa et al., 2009; Katayama et al., 2006).

Because iron is an essential element for all living organisms, PV is considered an antimicrobial agent owing to its high iron-chelating capacity (Taborsky, 1963). Most pathogenic gram-negative bacteria have lipopolysaccharide (LPS) present in the outer membrane acting as a permeability barrier against external agents. Phosvitin acts as a chelator in the biological system, affecting the outer membrane of bacteria and inhibiting bacterial growth. Phosvitin at a concentration of 0.1% has been reported to eliminate *E. coli* at 50°C for 20 min (Sattar Khan et al., 2000). Wang et al. (2011) demonstrated that PV inhibited the growth of *E. coli*, *A. hydrophilia* and *S. aureus*.

Despite being a biofunctional protein, intact phosvitin is considered nutritionally undesirable due to the formation of insoluble iron complex upon iron chelation, hindering iron absorption in the digestive tract (Guerin-Dubiard et al., 2007; Sato et al., 1987). The limitations including poor dissolution and stability of phosvitin, therefore, restrict its applications (Lee et al., 2014). Phosvitin phosphopeptides produced by enzymatic hydrolysis have demonstrated enhanced solubility, metal chelation, antioxidant and antimicrobial properties. Jiang and Mine (2001) reported that tryptic digested fragments of these peptides sized 1–3 kDa showed a higher ability in solubilizing calcium than commercial casein phosphopeptides. Feng and Mine (2006) reported that phosvitin phosphopeptide facilitated iron uptake in a $CaCO_2$ cell monolayer model leading to increased iron absorption.

Phosvitin phosphopeptides can also be exploited as suppressors of inflammatory response, oxidative stress and lipid peroxidation in living cells. Katayama et al. (2006) found that phosvitin phosphopeptides resulted in significantly reduced production of oxidative stress indicators, such as IL-8, a proinflammatory mediator, compared with the control and intact phosvitin. Phosvitin phosphopeptides showed stronger antioxidant and radical-scavenging activities than intact phosvitin (Xu et al., 2007).

5 Egg lipids

Lipids are the main components of the egg yolk solids. Chicken eggs contain approximately 66% triglycerides, 28% phospholipids and 5% cholesterol of total lipids by weight (Kovacs-Nolan and Mine, 2005). Factors including hen age, genotype and diet can affect the composition of yolk lipids.

5.1 Fatty acids

Fatty acids in egg yolk are found at different chain lengths, degrees of saturation and configurations. There are three classes of fatty acids including saturated fatty acids (SFA), monounsaturated fatty acids (MUFA) and polyunsaturated fatty acids (PUFA). The degree of unsaturation of fatty acids is of interest because of its positive effects on health. The content of MUFA and PUFA is dependent on the age and type of fatty acid composition in the hen diet. PUFA contents were significantly higher in eggs from young hens than from older ones, whereas MUFA contents were significantly higher in eggs from older hen (93-week-old). The unsaturated to saturated fatty acid ratio in eggs produced from hens at 27 and 39 weeks of age was lower than that produced at 51 weeks. Enhancement of the hen's feed with MUFA (oleic acid) has been reported to double the content of oleic acid as compared to regular eggs (Cherian and Sim, 1993). In contrast, a significant decrease in egg oleic acid was determined when diets contained PUFA (conjugated linoleic acids, CLA) (Aydin et al., 2001; Cherian et al., 2002).

The omega-6 and omega-3 fatty acids are the two families of PUFA in eggs. The content of long-chain omega-6 PUFA (LCPUFA) (>20-carbon) varies from 1 to 2%, and is reflected by the type of laying hen diet (Cherian and Sim, 1991). The content of omega-3 fatty acids in eggs is made of α-linoleic acids (<1%) and docosahexaenoic acids (DHA, 1–3%). Enhancement with flax, fish oil and marine algae in the laying hen diet was reported to significantly increase α-linoleic acid and DHA content in eggs (Cherian and Sim, 1991). Egg products enriched with PUFA need to be enriched with antioxidants to prevent the risk of damage from autoxidation. Grune et al. (2001) suggested that vitamin E supplementation of laying hen feed had a preventive effect against increased cytotoxic aldehydic lipid peroxidation in omega-3 PUFA-enriched eggs. As well, dietary supplementation of hen feed with vitamin E resulted in a decrease of PUFA, SFA and total lipids in fresh yolk lipids, whereas MUFA remained unchange. Furthermore, dietary E supplement slowed down the process of oxidation of egg yolk fatty acid during storage. Szymczyk and Pisulewski (2005) demonstrated that dietary vitamin E increased the rate of egg laying and production and may alleviate fatty acid composition of CLA-enriched eggs. Omega-3-rich eggs fortified with antioxidants (selenium and vitamin E) proved effective in preventing oxidation of omega-3 fatty acids under harsh conditions (Ren et al., 2013).

Oleic acid (C18:1) is the major MUFA in chicken eggs, reported to have both cholesterol and triacylglycerol lowering activities with no effect on HDL cholesterol in hyperlipidaemia patients (Mattson and Grundy, 1985). Omega-3 can be enriched in eggs by using a fortified hen diet. Omega-3 has beneficial effects on brain and visual function in humans (Maki et al., 2003). Other health benefits of omega-3 fatty acids include reduction of blood pressure, triglyceride, inflammatory markers, platelet aggregation, vasoconstriction, tumour growth and risk of sudden cardiac death (Juturu, 2008).

5.2 Phospholipids

The major components of egg yolk phospholipids are lecithin (81% phosphatidylcholine and 12% phosphatidylethanolamine), sphingomyelin, lysophosphatidylcholine and lysophosphatidylethanolamine. The fatty acid compositions of egg yolk lecithin are oleic (35%), palmitic (30%), linolenic (18%), stearic (15%) and arachidonic (5%) acids. Kivini et al. (2004) showed that supplementing the hen's diet with essential fatty acid from vegetable or fish oil had a significant effect on phosphatidylcholine composition. Phosphatidylcholine contains ω-3 fatty acids as the non-polar component, and choline as the polar component. Choline has been reported as a crucial nutrient for brain development, liver function and cancer prevention (Gutierrez et al., 1997). Yolk phospholipids contain more than three-fold higher phosphatidylcholine than soy phospholipids. It has been demonstrated that consumption of yolk phospholipids tends to ameliorate Alzheimer disease-related symptoms (Juneja, 1997). Clinical studies have shown that lecithin has antioxidation (Sugino et al., 1997) and cholesterol absorption inhibiting activities (Rampone, 1973).

6 Summary

Egg components are unique and offer several biofunctional properties. With efficient separation and purification, they have high potential to be used in the nutraceutical industries. The egg white and yolk proteins possess antimicrobial, anticancer, immunomodulating, ACE-inhibitory and antioxidant properties. Lipids present solely in the egg yolk also possess functional properties, such as cholesterol and blood pressure lowering, brain and proper liver functioning, antioxidation and anti-tumour effects. Several egg yolk properties such as antibody specificity and omega-3 content may be modified by hen artificial antigen exposure and feed fortification, respectively. Some proteins in eggs need further processing to enhance their functional properties, for example egg yolk phosvitin and egg white proteins when hydrolyzed have added metal-chelating, antimicrobial or antioxidant properties.

The egg components with their potential nutraceutical application listed in Table 1 may be extracted from regular or designed eggs. These components may be further purified and used as valuable nutraceutical compounds. These components may be utilized as a

Table 1 Potential nutraceutical applications for egg components

Potential nutraceutical application	Egg component
Anti-inflammatory	Ovotransferrin Ovomucoid Lysozymeγ-livetins (IgY) Omega-3 fatty acid
Anti-toxin/allergen	γ-livetins (IgY)
Antihypertensive	Ovalbumin Ovotransferrin Ovomucoid Omega-3 fatty acid
Antimicrobial	Ovalbumin Ovotransferrin Ovomucin Ovomucoid Avidin Lysozyme Cystatin Lipovitellinin (HDL) γ-livetins (IgY) Phosvitin
Antioxidant	Ovotransferrin Ovomucoid Lipovitellinin (HDL) Phosvitin Phospholipid
Blood cholesterol and lipid-lowering	Ovomucin Oleic acid Omega-3 fatty acid Phospholipid
Bone strengthening	Egg shell Phosvitin (peptide conjugated to calcium)
Brain proper functioning	Phospholipid Omega-3 fatty acid
Cardioprotective	Omega-3 fatty acid
Cartilage strengthening	Egg shell Egg shell membrane
Immune modulatory	Ovalbumin Ovomucin Ovomucoid Lysozyme Cystatin γ-livetins (IgY) Phosvitin Omega-3 fatty acid
Increase metal ion absorption	Ovotransferrin Ovomucoid Phosvitin (peptide conjugated to metal ion)
Visual acuity enhancing	Omega-3 fatty acid

sole or combined to obtain their targeted functional properties in health and diseased persons.

7 Future trends in research

Further research needs to identify new and existing biological compounds and functions of the hen egg. These include the development of new methods to improve purity and yield of the value-added biologically active compounds with a particular role in therapy and prevention of chronic and infectious diseases. Recently, special attention has been paid to the development of integrated technological processes, such as membrane techniques, industrial chromatography and extractions.

In order to identify egg protein components, studies on the elaboration of the egg biochemistry (proteomics) and molecular biology (such as cDNA libraries, chicken genome bioinformatic analysis, transcriptomics) are warranted. Studies on the development of separation methods of egg components have been developed since the early 1900s, but the scale-up of preparation methods for commercial applications is still under way. The main aim is to obtain simple and scalable sequential non-toxic chemical separation methods. The current focus has been given primarily to the innovative environmentally friendly supercritical fluids extraction system. This method is widely used in a global setting in the egg industry, especially in egg lipid extraction of cholesterol and phospholipids. The supercritical system allows for precise fractionation of lipid compounds and facilitates further processing of valuable post-extraction proteins.

Enzymatic hydrolysis is one of the most widely used technologies in obtaining bioactive peptides from egg proteins. The peptides derived from proteolytic enzyme hydrolysis have various beneficial health effects. High hydrostatic pressure processing has experienced a tremendous growth when used in combination with protease treatments for the production of bioactive peptide hydrolysates. The innovative high hydrostatic pressure enzymatic hydrolysis for further modification of the extracted protein components of egg white and yolk on an industrial scale requires further study.

To augment nutraceutical benefits of egg components, more research on identification of novel substances with biofunction, efficient methods for large-scale isolation and purification of biologically active constituents of eggs are required. The development of sequential separation processing of egg components remains crucial to minimize waste, maximize yield and purity, leading to an overall reduced cost of egg value-added components. The new bioactive substances from eggs, with various pro-health qualities, obtained from an efficient large-scale processing may find their way from bench top to over-the-counter nutraceutical applications.

8 Where to look for further information

Please refer to the listed references for further information. Authors, NG (email: gujral@ualberta.ca) and HS (hsunwoo@ualberta.ca), can provide further information upon readers' request.

9 References

Abdou, A. M., Kim, M. and Sato, K. (2013), 'Functional proteins and peptides of hen's egg origin'. In Hernandez-Ledesma, B. and Hsieh, C. C. (Eds), *Bioactive Food Peptides in Health and Disease*, pp. 115–44. CCBY z.o license.

Abeyrathne, E. D., Lee, H. Y., Jo, C., Suh, J. W. and Ahn, D. U. (2015), 'Enzymatic hydrolysis of ovomucoid and the functional properties of its hydrolysates', *Poult. Sci.*, 94(9), 2280–7.

Ahlborn, G. J., Clare, D. A. and Sheldon, B. W. (2006), 'Identification of eggshell membrane proteins and purification of ovotransferrin and β-NAGase from hen egg white', *Prot. J.*, 25, 71.

Albright, K. J., Gordon, D. T. and Cotterill, O. J. (1984), 'Release of iron from phosvitin by heat and food additives', *J. Food Sci.*, 49, 78–81.

Alderton, G., Ward, W. H. and Fevold, H. L. (1946), 'Identification of the bacteria-inhibiting iron-binding protein of egg white as conalbumin', *Arch. Biochem.*, 11, 9–13.

Aydin, R., Pariza, M. W. and Cook, M. E. (2001), 'Olive oil prevents the adverse effects of dietary conjugated linoleic acid on chick hatchability and egg quality', *J. Nutr.*, 131(3), 800–6.

Baker, J. R. and Balch, D. A. (1962), 'A study of the organic material of hen's eggshell', *Biochem. J.*, 82(2), 352–61.

Banks, J. G., Board, R. G. and Sparks, N. H. C. (1986), 'Natural antimicrobial systems and their potential in food preservation of the future', *Biotechnol. Appl. Biochem.*, 8, 103–47.

Björck, L., Akesson, P., Bohus, M., Trojnar, J., Abrahamson, M., Olafsson, I. and Grubb, A. (1989), 'Bacterial growth blocked by a synthetic peptide based on the structure of a human proteinase inhibitor', *Nature*, 337(6205), 385–6.

Bullen, J. J., Rogers, H. J. and Griffiths, E. (1978), 'Role of iron in bacterial infection', *Curr. Top. Microbiol. Immunol.*, 80, 1–35.

Burley, R. W. and Cook, W. H. (1961), 'Isolation and composition of avian egg yolk granules and their constituents D- and E-lipovitellins', *Can. J. Biochem. Physiol.*, 39, 1295–307.

Castellani, O., Guerin-Dubiard, C., David-Driand, E. and Anton, M. (2004), 'Influence of physiochemical conditions and technological treatments on the iron binding capacity of eggyolk phosvitin', *Food Chem.*, 85, 569–77.

Cherian, G. and Sim, J. S. (1991), 'Effect of feeding full fat flax and canola seeds to laying hens on the fatty acid composition of eggs, embryos and newly hatched chicks', *Poult. Sci.*, 70 (4), 917–22.

Cherian, G. and Sim, J. S. (1993), 'Omega-3 fatty acid and cholesterol content of newly hatched chicks from alphalinolenic acid enriched eggs', *Lipids*, 27, 706–10.

Cherian, G., Goeger, M. P. and Ahn, D. U. (2002), 'Dietary conjugated linoleic acid with fish oil alters yolk n-3 and trans fatty acid content and volatile compounds in raw, cooked, and irradiated eggs', *Poult. Sci.*, 81, 1571–7.

Davalos, A., Miguel, M., Bartolome, B. and Lo´pez-Fandiño, R. (2004), 'Antioxidant activity of peptides derived from egg white proteins by enzymatic hydrolysis', *J. Food Prot.*, 67, 1939–44.

Eaton-Evans, J. (1994), 'Osteoporosis and the role of diet', *Br. J. Biomed. Sci.*, 51(4), 358–70.

Fan, X., Subramaniam, R., Weiss, M. F. and Monnier, V. M. (2003), 'Methylglyoxal-bovine serum albumin stimulates tumor necrosis factor alpha secretion in RAW 264.7 cells through activation of mitogen-activating protein kinase, nuclear factor kappaB and intracellular reactive oxygen species formation', *Arch. Biochem. Biophys.*, 409, 274–86.

Feng, F. and Mine, Y. (2006), 'Phosvitin phosphopeptides increase iron uptake in a Caco-2 cell monolayer model', *Int. J. Food Sci. Tech.*, 41, 455–8.

Gładkowski, W., Kiełbowicz, G., Chojnacka, A., Gil, M., Trziszka, T., Dobrzanski, Z. and Wawrzeńczyk, C. (2011), 'Fatty acid composition of egg yolk phospholipid fractions following feed supplementation of Lohmann Brown hens with humic-fat preparations', *Food Chem.* 126, 1013–18.

Grizzuti, N. and Perlman, G. E. M. (1973), 'Binding of magnesium and calcium ions to the phosphoglycoprotein phosvitin', *Biochemistry*, 12, 4399–403.

Grune, T., Kramer, K., Hoppe, P. P. and Siems, W. (2001), 'Enrichment of eggs with n-3 polyunsaturated fatty acids: Effects of vitamin E supplementation', *Lipids*, 36, 833–8.

Guerin-Dubiard, C., Castellani, O. and Anton, M. (2007), 'Egg compounds with antioxidant and mineral binding properties'. In Huopalahti, R., Lopez-Fandino, R., Anton, M. and Schard, R. (Eds), *Bioactive Egg Compounds*, pp. 223–8. Springer, Berlin, New York.

Gujral, N., Löbenberg, R., Suresh, M. and Sunwoo, H. (2012a), 'In-vitro and in-vivo binding activity of chicken egg yolk immunoglobulin Y (IgY) against gliadin in food matrix', *J. Agric. Food Chem.*, 60(12), 3166–72.

Gujral, N., Suh, J. W. and Sunwoo, H. H. (2015), 'Effect of anti-gliadin IgY antibody on epithelial intestinal integrity and inflammatory response induced by gliadin', *BMC Immunol*, 16, 41.

Gujral, N., Suresh, M. R. and Sunwoo, H. H. (2012b), 'Quantitative double antibody sandwich ELISA for the determination of gliadin', *J. Immunoassay Immunochem.*, 33, 339–51.

Gutierrez, M. A., Takahashi, H. and Juneja, L. R. (1997), 'Nutritive evaluation of hen eggs'. In Yamamoto, T., Juneja, L. R., Hatta, H. and Kim, M. (Eds), *Hen Eggs: Their Basic and Applied Science*, pp. 25–71. CRC Press, New York.

He, X., Tsang, T. C., Luo, P., Zhang, T. and Harris, D. T. (2003), 'Enhanced tumor immunogenicity through coupling cytokine expression with antigen presentation', *Cancer Gene Ther.*, 10, 669–77.

Hincke, M. T., Gautron, J., Panheleux, M., Garcia-Ruiz, J., McKee, M. D. and Nys, Y. (2000), 'Identification and localization of lysozyme as a component of eggshell membranes and eggshell matrix', *Matrix Biol.*, 19, 443–53.

Holen, E., Bolann, B. and Elsayed, S. (2001), 'Novel B and T cell epitopes of chicken ovomucoid (Gal d 1) induce T cell secretion of IL-6, IL-13, and IFN-gamma', *Clin. Exp. Allergy*, 31, 952–64.

Ibrahim, H. R., Haraguchi, T. and Aoki, T. (2006), 'Ovotransferrin is a redox-dependent autoprocessing protein incorporating four consensus self-cleaving motifs flanking the two kringles', *Biochim. Biophys. Acta*, 1760, 347–55.

Ibrahim, H. R., Hoq, M. I. and Aoki, T. (2007), 'Ovotransferrin possesses SOD-like superoxide anion scavenging activity that is promoted by copper and manganese binding', *Int. J. Biol. Macromol.*, 41, 631–40.

Ibrahim, H. R. and Kiyono, T. (2009), 'Novel anticancer activity of the autocleaved ovotransferrin against human colon and breast cancer cells', *J. Agric. Food Chem.*, 57, 11383–90.

Ishikawa, S., Yano, Y., Arihara, K. and Itoh, M. (2004), 'Egg yolk phosvitin inhibits hydroxyl radical formation from the fenton reaction', *Biosci. Biotechnol. Biochem.*, 68, 1324–31.

Ishikawa, S. I., Asano, T., Takenoshita, S., Nozawa, Y., Arihara, K. and Itoh, M. (2009), 'Egg yolk proteins suppress azoxymethane-induced aberrant crypt focci formation and cell proliferation in the colon of rats', *Nutr. Res.*, 29, 64–9.

Jiang, B. and Mine, Y. (2001), 'Phosphopeptides derived from hen egg yolk phosvitin: Effect of molecular size on the calcium-binding properties', *Biosci. Biotechnol. Biochem.*, 65(5), 1187–90.

Juneja, L. R. and Kim, M. (1997), 'Egg yolk proteins'. In Yamamoto, T., Juneja, L. R., Hatta, H. and Kim, M. (Eds), *Hen Eggs: Their Basic and Applied Science*, pp. 57–71. CRC Press, New York.

Juneja, L. R. (1997), 'Egg yolk yolk lipids'. In Yamamoto, T., Juneja, L., Hatta, H. and Kim, M. (Eds), *Hen Eggs: Their Basic and Applied Science*, pp. 73–98. CRC Press, New York.

Juturu, V. (2008). 'Omega-3 fatty acids and the cardiometabolic syndrome', *J. Cardiometab. Syndr.*, 3(4), 244–53.

Katayama, S., Xu, X. M., Fan, M. Z. and Mine, Y. (2006), 'Antioxidative stress activity of oligophosphopeptides derived from hen egg yolk phosvitin in Caco-2 cells', *J. Agric. Food Chem.*, 54(3), 773–8.

Kim, J., Moon, S. H., Ahn, D. U., Paik, H. D. and Park, E. (2012), 'Antioxidant effects of ovotransferrin and its hydrolysates', *Poult. Sci.*, 91, 2747–54.

Kivini, H., Järvenpää, E. P., Aro, H., Huopalahti, R. and Ryhä-nen E. L. (2004), 'Qualitative and quantitative liquid chro-matographic analysis methods for the determination of the effects of feed supplements on hen egg yolk phospholipids', *J. Agri. Food Chem.*, 52, 4289–95.

Kobayashi, K., Hattori, M., Hara-Kudo, Y., Okubo, T., Yamamoto, S., Takita, T. and Sugita-Konishi, Y. (2004), 'Glycopeptide derived from hen egg ovomucin has the ability to bind enterohemorrhagic Escherichia coli O157:H7', *J. Agric. Food Chem.*, 52, 5740–6.

Kołaczkowska, A., Sokołowska, A., Miecznikowska, H., Rolka, K. and Polanowski, A. (2010), 'The antifungal properties of chicken egg cystatin against *Candida* yeast isolates showing different levels of azole resistance', *Mycoses*, 53, 314–20.

Korpela, J., Salonen, E. M., Kuusela, P., Sarvas, M. and Vaheri, A. (1984), 'Binding of avidin to bacteria and to the outer membrane porin of Escherichia coli', *FEMS Microbiol. Lett.*, 22, 3–10.

Kovacs-Nolan, P. and Mine, Y. (2005), 'Advances in the value of eggs and egg components for human health', *J. Agric. Food Chem.*, 53(22), 8421–31.

Lee, C. H., Lee, Y. C., Liang, M. H., Leu, S. J., Lin, L. T., Chiang, J. R. and Yang, Y. Y. (2015), 'Antibodies against venom of the snake Deinagkistrodon acutus', *Appl. Environ. Microbiol.*, 82(1), 71–80.

Lee, E. N., Sunwoo, H. H., Menninen, K. and Sim, J. S. (2002), 'In vitro studies of chicken egg yolk antibody (IgY) against Salmonella enteritidis and Salmonella typhimurium', *Poult. Sci.*, 81, 632–41.

Lee, H. Y., Abeyrathne, E. D., Choi, I., Suh, J. W. and Ahn, D. U. (2014), 'Sequential separation of immunoglobulin Y and phosvitin from chicken egg yolk without using organic solvents', *Poult. Sci.*, 93(10), 2668–77.

Li-Chan, E. C. Y., Powrie, W. D. and Nakai, S. (1995), 'The chemistry of eggs and egg products'. In Stadelman, W. J. and Cotterill, O. J. (Eds), *Egg Science and Technology*, 4th edition, pp. 105–75. Food Products Press, New York.

Long, F. D., Adams, R. G., DeVore, D. P. and Inventors (2005), 'Preparation of hyaluronic acid from eggshell membrane', US Patent No. 6,946,551.

López-Fandiño, R. (2007), 'Use of egg compounds for human nutrition'. In Huopalahti, R., López-Fandiño, R., Anton, M. and Schade, R. (Eds), *Bioactive Egg Compounds*, pp. 117–90. Springer-Verlag, Berlin Heidelberg, Germany.

Lu, C. L. and Baker, R. C. (1986), 'Characteristics of egg yolk phosvitin as an antioxidant for inhibiting metal-catalyzed phospholipid oxidations', *Poult. Sci.*, 65(11), 2065–70.

Majumder, K., Chakrabarti, S., Davidge, S. T. and Wu, J. (2013), 'Structure and activity study of egg protein ovotransferrin derived peptides (IRW and IQW) on endothelial inflammatory response and oxidative stress', *J. Agric. Food Chem.*, 61, 2120–9.

Maki, K. C., Van Elswyk, M. E., Mccarthy, D., Seeley, M. A., Veith, P. E., Hess, S. P., Ingram, K. A., Halvorson, J. J., Calaguas, E. M. and Davidson, M. H. (2003), 'Lipid responses in mildly hypertriglyceridemic men and women to consumption of docosahexaenoic acid-enriched eggs', *Int. J. Vitam. Nutr. Res.*, 73, 357–68.

Marie, P. J., Garba, M. T., Hott, M. and Miravet, L. (1985), 'Effect of low doses of stable strontium on bone metabolism in rats', *Miner. Electrolyte Metab.*, 11(1), 5–13.

Mattson, F. H. and Grundy, S. M. (1985), 'Comparison of effects of dietary saturated, monounsaturated, and polyunsaturated fatty acids on plasma lipids and lipoproteins in man', *J. Lipid Res.*, 26(2), 194–202.

Meunier, P. J. (2001), 'Anabolic agents for treating postmenopausal osteoporosis', *Joint Bone Spine*, 68(6), 576–81.

Miguel, M. and Aleixandre, A. (2006), 'Antihypertensive peptides derived from egg proteins', *J. Nutr.*, 136, 1457–60.

Mine, Y. and Kovacs-Nolan, J. (2006), 'New insights in biologically active proteins and peptides derived from hen egg', *World Poult. Sci. J.*, 62, 87.

Mine, Y., Ma, F. and Lauriau, S. (2004), 'Antimicrobial peptides released by enzymatic hydrolysis of hen egg white lysozyme', *J. Agric. Food Chem.*, 52, 1088–94.

Mine, Y. (2007), 'Egg proteins and peptides in human health – chemistry, bioactivity and production', *Curr. Pharm. Des.*, 13, 875–84.

Nagaoka, S., Masaoka, M., Zhang, Q., Hasegawa, M. and Watanabe, K. (2002), 'Egg ovomucin attenuates hypercholesterolemia in rats and inhibits cholesterol absorption in Caco-2 cells', *Lipids*, 37 (3), 267–72.

Nagata, K. and Yoshida, N. (1984), 'Interaction between Trypsin-like enzyme from Streptomyces erythraeus and chicken ovomucoid', *J. Biochem.*, 96, 1041–9.

Nakai, S. (2000), 'Molecular modification of egg proteins for functional improvement'. In Sim, J. S., Nakai, S. and Guenter, W. (Eds), *Egg Nutrition and Biotechnology*, pp. 205–17. CAB Int., Oxon, UK.

Nakamura, S. and Kato, A. (2000), 'Multi-functional biopolymer prepared by covalent attachment of galactomannan to egg-white proteins through naturally occurring Maillard reaction', *Nahrung*, 44(3), 201–6.

Nakano, T., Ikawa, N. I. and Ozimak, L. (2001), 'Extraction of glycosaminoglycans from chicken eggshell', *Poult. Sci.*, 80, 681–4.

Nakano, T., Ikawa, N. I. and Ozimek, L. (2003), 'Chemical composition of chicken eggshell and shell membranes', *Poult. Sci.*, 82, 510–14.

Oguro, T., Ohaki, Y., Asano, G., Ebina, T. and Watanabe, K. (2001), 'Ultrastructural and immunohistochemical characterization on the effect of ovomucin in tumor angiogenesis', *Jpn. J. Clin. Electron. Microsc.*, 33(2), 89–99.

Pavlikova, D., Rovensky, J., Marek, J., Schreiberova, O. and Rovenske, E. (1995), 'The effect of Biomin H on in vitro chicken embryonic cartilage growth', *Rheumatolgoica*, 4, 185–92.

Pellegrini, A., Hülsmeier, A. J., Hunziker, P. and Thomas, U. (2004), 'Proteolytic fragments of ovalbumin display antimicrobial activity', *Biochim. Biophys. Acta*, 1672(2), 76–85.

Pellegrini, A., Thomas, U., Wild, P., Schraner, E. and Von Fellenberg, R. (2000), 'Effect of lysozyme or modified lysozyme fragments on DNA and RNA synthesis and membrane permeability of *Echerichia coli*', *Microbiol. Res.*, 155, 69–77.

Premzl, A., Puizdar, V., Zavasnik-Bergant, V., Kopitar-Jerala, N., Lah, T. T., Katunuma, N., Sloane, B. F., Turk, V. and Kos, J. (2001), 'Invasion of ras-transformed breast epithelial cells depends on the proteolytic activity of cysteine and aspartic proteinases', *Biol. Chem.*, 382(5), 853–7.

Radziejewska, R. C., Les´nierowski, G. and Kijowski, J. (2008), 'Properties and application of egg white lysozyme and its modified preparations – A review', *Polish, J. Food Nutr. Sci.*, 58, 5–10.

Rampone, A. J. (1973), 'The effect of lecithin on intestinal cholesterol uptake by rat intestine in vitro', *J. Physiol.*, 229(2), 505–14.

Ren, Y., Perez, T. I., Zuidhof, M. J., Renema, R. A. and Wu, J. (2013), 'Oxidative stability of omega-3 polyunsaturated fatty acids enriched eggs', *J. Agric. Food Chem.*, 27, 61(47), 11595–602.

Rovenský, J., Stancíková, M., Masaryk, P., Svík, K. and Istok, R. (2003), 'Eggshell calcium in the prevention and treatment of osteoporosis', *Int. J. Clin. Pharmacol. Res.*, 23(2–3), 83–92.

Ruff, K. J., Winkler, A., Jackson, R. W., DeVore, D. P. and Ritz, B. W. (2009), 'Eggshell membrane in the treatment of pain and stiffness from osteoarthritis of the knee: A randomized, multicenter, double-blind, placebo-controlled clinical study', *Clin. Rheumatol.*, 28(8), 907–14.

Sarker, S. A., Casswall, T. H., Juneja, L. R., Hoq, E., Hossain, I., Fuchs, G. J. and Hammarström, L. (2001), 'Randomized,placebo-controlled,clinicaltrialofhyperimmunizedchickeneggyolkimmunoglobulin in children with rotavirus diarrhea', *J. Pediatr. Gastroenterol. Nutr.*, 32, 19–25.

Sarker, S. A., Pant, N., Juneja, L. R. and Hammarström, L. (2007), 'Successful treatment of rotavirus-induced diarrhoea in suckling mice with egg yolk immunoglobulin', *J. Health Popul. Nutr.*, 25(4), 465–8.

Sato, R., Noguchi, T. and Naito, H. (1987), 'The effect of feeding demineralized egg-yolk protein on the solubility of intra-intestinal iron', *Nutr. Reports Intern.*, 36(3), 593–602.

Sattar Khan, M. A., Nakamura, S., Ogawa, M., Akita, E., Azakami, H. and Kato, A. (2000), 'Bactericidal action of egg yolk phosvitin against Escherichia coliunder thermal stress', *J. Agric. Food Chem.*, 48(5), 1503–6.

Schaafsma, A. and Beelen, G. M. (1999), 'Eggshell powder, a comparable or better source of calcium than purified calcium carbonate: Piglet studies', *J. Sci. Food Agric.*, 79(12), 1596–600.

Schaafsma, A. (2000), *Chicken Eggshell Powder in Nutritional Intervention of Age Related Bone Loss.* Maastricht University, Netherlands

Schade, R. (2007), 'Use of eggs for human/animal health and biotechnology'. In Huopalahti, R., López-Fandiño, R., Anton, M. and Schade, R. (Eds), *Bioactive Egg Compounds*, pp. 191–290. Springer-Verlag, Berlin Heidelberg, Germany.

Shapira, N. (2010), 'Every egg may have a targeted purpose: Toward a differential approach to egg according to composition and functional effect', *World Poult. Sci. J.*, 66, 271–84.

Song, M. S., Kim, C. J., Cho, W. I. and Sunwoo, H. H. (2009), 'Growth inhibition of Clostridium perfringens vegetative cells and spores using chicken immunoglobulin Y', *J. Food Safety*, 29, 511–20.

Starcher, B. C. and King, G. S. (1980), 'The presence of desmosine and isodesmosine in eggshell membrane protein', *Connect. Tissue Res.*, 8, 53–5.

Sugino, H., Nitoda, T. and Juneja, L. R. (1997), 'General chemical composition of hen eggs'. In Yamamoto, T., Juneja, L. R., Hatta, H. and Kim, M. (Eds), *Hen Eggs, Their Basic and Applied Science*, pp. 13–24. CRC Press Inc., New York.

Suguro, N., Horiike, S., Masuda, Y., Kunou, M. and Kokubu, T. (2000), 'Bioavailability and commercial use of eggshell calcium, membrane proteins and yolk lecithin products'. In Sim, S., Nakai, N. and Guenter, W. (Eds), *Egg Nutrition and Biotechnology*, pp. 219–32. CABI, New York.

Sui, J., Cao, L. and Lin, H. (2011), 'Antibacterial activity of egg yolk antibody (IgY) against Listeria monocytogenes and preliminary evaluation of its potential for food preservation', *J. Sci. Food Agric.*, 91(11), 1946–50.

Sunwoo, H. H., Lee, E. N., Menninen, K., Suresh, M. R. and Sim, J. S. (2002), 'Growth inhibitory effect of chicken egg yolk antibody (IgY) on Escherichia coli O157:H7', *J. Food Sci.*, 67, 1486–94.

Superti, F., Ammendolia, M. G., Berlutti, F. and Valenti, P. (2007), 'Ovotransferrin'. In Huopalahti, R. and Lopez-Fandino, R. (Eds), *Bioactive Egg Compounds*, pp. 43–8. Springer-Verlag, Berlin, Germany.

Surai, P. F. and Sparks, H. N.C. (2000), 'Designer eggs: From improvement of egg composition to functional food', *Trends Food Sci. Tech.*, 12, 7–13.

Szymczyk, B. and Pisulewski, P. M. (2005), 'Effects of dietary conjugated linoleic acid isomers and vitamin E on fatty acid composition and cholesterol content of hen egg yolks', *J. Anim. Feed Sci.*, 14, 109–23.

Taborsky, G. (1963), 'Interaction between phosvitin and iron and its effect on a rearrangement of phosvitin structure', *Biochemistry*, 2, 266–71.

Tanizaki, H., Tanaka, H., Iwata, H. and Kato, A. (1997), 'Activation of macrophages by sulphated glycopeptides in ovomucin, yolk membrane, and chalazae in chicken eggs', *Biosci. Biotechnol. Biochem.*, 61, 1883–9.

Travis, J., Pike, R., Imamura, T. and Potempa, J. (1997), 'Porphyromonas gingivalis proteinases as virulence factors in the development of periodontitis', *J. Periodontal Res.*, 32(1 Pt 2), 120–5.

Tsuge, Y., Shimoyamada, M. and Watanabe, K. (1996), 'Binding of egg white proteins to viruses', *Biosci. Biotechnol. Biochem.*, 60, 1503–4.

Valenti, P., Antonini, G., von Hunolstein, C., Visca, P., Orsi, N. and Antonini, E. (1983), 'Studies of the antimicrobial activity of ovotransferrin', *Int. J. Tissue React.*, 5, 97–105.

Vega, C., Bok, M., Chacana, P., Saif, L., Fernandez, F. and Parreño, V. (2011), 'Egg yolk IgY: Protection against rotavirus induced diarrhea and modulatory effect on the systemic and mucosal antibody responses in newborn calves', *Vet. Immunol. Immunopathol.*, 142(3–4), 156–69.

Verdot, L., Lalmanach, G., Vercruysse, V., Hartmann, S., Lucius, R., Hoebeke, J., Gauthier, F. and Vray, B. (1996), 'Cystatins up-regulate nitric oxide release from interferon-gamma-activated mouse peritoneal macrophages', *J. Biol. Chem.*, 271(45), 28077–81.

Vidovic, D., Graddis, T., Chen, F., Slagle, P., Diegel, M., Stepan, L. and Laus, R. (2002), 'Antitumor vaccination with HER-2-derived recombinant antigens', *Int. J. Cancer*, 102, 660–4.

Wang, B., Yang, J., Cao, S., Wang, H., Pan, X., Zhu, J., Zhou, Y., Gao, L., Li, W. and Li, M. (2014), 'Preparation of specific anti-Helicobacter pylori yolk antibodies and their antibacterial effects', *Int. J. Clin. Exp. Pathol.*, 7(10), 6430–7.

Wang, S., Wang, Y., Ma, J., Ding, Y. and Zhang, S. (2011), 'Phosvitin plays a critical role in the immunity of zebrafish embryos via acting as a pattern recognition receptor and an antimicrobial effector', *J. Biol. Chem.*, 286(25), 22653–64.

Watanabe, K., Tsuge, Y., Shimoyamada, M., Ogama, N. and Ebina, T. (1998), 'Antitumor effects of Pronase-treated fragments, glycopeptides, from ovomucin in hen egg white in a double grafted tumor system', *J. Agric. Food Chem.*, 46, 3033–8.

Wong, M., Hendrix, M. J.C., von der Mark, K., Little, C. and Stern, R. (1984), 'Collagen in the Egg Shell Membranes of the Hen', *Dev. Biol.*, 104, 28–36.

Wu, J. and Acero-Lopez, A. (2012), 'Ovotransferrin: Structure, bioactivities and preparation', *Food Res. Int.*, 46, 480–7.

Xu, X., Katayama, S. and Mine, Y. (2007), 'Antioxidant activity of tryptic digests of hen egg yolk phosvitin', *J. Sci. Food Agric.*, 87(14), 2604–8.

Yamamoto, Y. and Omori, M. (1994), 'Antioxidative activity of egg yolk lipoproteins', *Biosci. Biotech. Biochem.*, 58 (9), 1711–13.

Chapter 16

Enhancing the nutritional profile of eggs

Erin M. Goldberg and Neijat Mohamed, University of Manitoba, Canada and James D. House, University of Manitoba and the Canadian Centre for Agri-Food Research in Health and Medicine, Canada

1 Introduction

The hen's egg represents one of the most economical single-food sources of complete protein and is an important component of the human diet. From the perspective of current egg usage for human consumption, the edible components of the egg can be considered either in their entirety (whole egg) or its constituent parts (vitellus/yolk and albumen/white). The chicken egg yolk is particularly nutrient dense, being rich in fatty acids (FAs) (Cherian and Sim 1991), fat-soluble compounds, such as vitamins A, D and E (Kovacs-Nolan et al. 2005), and carotenoids (Anton 2007). Furthermore, egg yolk contains proteins that act as carriers for important water-soluble nutrients, including folate (House et al. 2002). While egg yolk contains balanced amounts of the most essential nutrients for a developing embryo, it can still be enhanced, primarily through the manipulation of the hen's diet, to contain bio-active compounds that are known to have diverse biological functions beyond basic nutrition in humans. As such, this chapter will focus on providing an overview towards the opportunities and challenges in the enhancement of hen eggs

http://dx.doi.org/10.19103/AS.2016.0012.18

with bio-actives, in particular omega-3 polyunsaturated FAs (n-3 PUFA), but also vitamins, minerals and carotenoids based on different dietary modifications.

2 Egg nutrient composition and role in the human diet

An egg contains all the essential nutrients, including lipids, amino acids and vitamins, required for a new life (Farrell 1998). Hen eggs consist of a protective porous shell (0.2–0.4 mm thick), albumen (white) and vitellus (yolk) contained within various thin membranes. The yolk is surrounded by the egg white, which is an aqueous, gel-like liquid that is mostly composed of water and protein. The yolk, which is suspended in the centre of the egg by the chalazae, is the part of an egg that feeds the developing embryo, and contains most of the nutrients. When the hen's diet is altered, it is the yolk's composition that is generally affected, not the albumen.

The average egg contains 6 g of lipid in the form of (a) triglycerides (TG) (66%), esters derived from glycerol and three FAs; (b) phospholipids (PL) (30%), major components of all cell membranes as they form PL bilayers, most of which contain a diglyceride, a phosphate group and a simple organic molecule; and (c) free cholesterol (5%), a waxy steroid of fat manufactured in the liver or intestines. Other minor lipid components consist of cholesterol esters and free FAs (Rose 1997).

Similar to humans, the chicken's liver is the primary site for FA synthesis (Leveille et al. 1975). Yolk lipid composition is the result of a combination of *de novo* lipogenesis and the transfer of lipid components from the diet. Upon absorption in the intestinal lumen, hydrolysed products of lipid digestion such as long-chain (LC) FAs and monoacylglyerols must be re-esterified within the endoplasmic reticulum of enterocytes prior to transport (Yannakopoulos 2007). The resultant TG are packaged with cholesterol, PL and proteins to form lipoproteins called portomicrons, instead of chylomicrons found in humans, because they are transferred to the hepatic portal circulation. Short-chain FAs and free glycerol are transported in this manner as well. The lipids reform and are included in particles of very low-density lipoprotein (VLDL), passing to the Golgi apparatus where they acquire PL and further glycosylation. Finally, completed particles of VLDL concentrate in secretory vesicles and are then discharged into the blood. The VLDL are then carried by the blood to the ovarian follicles where they are taken up via receptor-mediated endocytosis through the oolemma.

3 Omega-3 PUFA: biological factors influencing the n-3 composition of eggs

As depicted in Fig. 1, the essential FA precursors in the n-6 (linoleic acid (LA)) and n-3 (alpha-linolenic acid (ALA)) pathway are converted into FAs of longer chain length and a higher degree of unsaturation within the liver by alternating desaturation and chain elongation reactions (Sprecher 1981). After consumption, ALA can be converted into eicosapentaenoic acid (EPA), docosapentaenoic acid (DPA) and docosahexaenoic acid (DHA) by the hen's liver, and then secreted into the yolk in smaller amounts (Beynen 2004). The key desaturase enzymes, delta-5 (D5DE) and D6DE are encoded by the genes

n-6 Series		n-3 Series
C18:2n6 LA	*D6DE (FADS2)*	C18:3n3 ALA
C20:3n6 GLA	*Elongase (ELOVL5)*	C18:4n3 SDA
C18:3n6 DGLA	*D5DE (FADS1)*	C20:4n3 ETA
C20:4n6 AA	*Elongase (ELOVL2)*	C20:5n3 EPA
C22:4n6 DTA	*Elongase (ELOVL2)*	C22:5n3 DPA
C24:4n6 TTA	*D6DE (FADS2)*	C24:5n3 TPA
C24:5n6 TPA	*β-oxidation*	C24:6n3 THA
C22:5n6 DPA		C22:6n3 DHA

Figure 1 Essential fatty acid conversion in humans (Simopoulos and De Meester, 2009).

FA desaturase 1 and 2 (*FADS1* and *FADS2*, respectively) (de Schrijver and Privett 1982; Cho et al. 1999; Schaeffer et al. 2006). The genes elongation of very LC FA 2 (*ELOVL2*) and elongation of very LC FA 5 (*ELOVL5*) are responsible for the elongation reactions (Gregory et al. 2013). Desaturation adds a double bond by the removal of hydrogen, which is catalysed by delta-desaturase whereas elongation reactions add two carbon atoms. These pathways are well established and observed across kingdoms (Nakamura and Nara 2004).

More specifically, ALA is first desaturated to stearidonic acid (SDA) by D6DE. This is followed by elongation to eicosatetraenoic acid (ETA) by an elongase, which is then desaturated to form EPA. In chickens, EPA is elongated and desaturated and the further elongation of DPA to tetracopentaenoic acid involves the combined effect of *ELOVL5* plus *ELOVL2*, enabling a better efficiency of conversion than any other species (Gregory et al. 2013). Furthermore, C24:5n-3 undergoes a desaturation to tetracohexaenoic acid (C24:6n-3), which generates DHA requiring an additional chain-shortening step (i.e. β-oxidation) in the peroxisome that involves multifunctional enzyme delta-4-desaturase as described by Sprecher (2000). In addition, a retroconversion of DHA into EPA takes place (Sprecher et al. 1995; Willumsen et al. 1996; Sprecher 2000). Utilizing the same enzymes as in the n-3 series, a similar desaturation–elongation cycle occurs in the conversion of LA to arachidonic acid (AA) (Cherian and Sim 1995).

The enrichment of the egg yolk with LCPUFA can occur through two main pathways; the direct deposit of these FAs from the diet (i.e. pre-formed DHA) or as a final synthesis from its primary precursor (i.e. ALA) through the synthesis of the intermediary products, a result of the *de novo* synthesis (Leskanich and Noble 1997; Baucells et al. 2000; Yalçyn et al. 2007). ALA supplementation results in a non-proportional increase in yolk LCPUFA (Grobas et al. 2001; Gakhar et al. 2012). The accumulation of EPA in the egg yolk is largely unaffected by levels of dietary n-3 FAs, particularly of plant origin, with yolk levels shifting within a narrow range (refer to Table 2). This is because the majority of the EPA is converted to the LCPUFA such as DPA and DHA (Aymond and Van Elswyk 1995; Cachaldora et al. 2008a). The alternative route to impart n-3 PUFA into eggs is through sources of pre-formed DHA, including fish oil or dried marine algae, which provide significant linear enrichment of LCPUFA (Gonzalez-Esquerra and Leeson 2001; Alvarez et al. 2004; Šefer et al. 2011). Regardless of the dietary substrate, n-3 PUFA from vegetable sources are retained achieving a steady state of accumulation in yolk fat just as efficiently as from marine sources (Garcia-Rebollar et al. 2008); however, saturation of LCPUFA in the egg yolk is eventually achieved, as defined by a plateau in enrichment. For example, flaxseed used at 10% or 20% in poultry diets can increase the ALA content of egg yolk fat from 0.4% in the conventional egg to 4.6% and 8.9% in the enriched egg, respectively (Flax Council of Canada 2014). However, no increase in n-3 LCPUFA was reported between eggs from hens fed 8% compared to 16% flaxseed (Cherian and Sim 1992). Because, the LCPUFA are predominantly incorporated into the PL, in particular the phosphatidylethanolamine (PE) fraction, distinct changes in the pattern of incorporation of the LCPUFA are evident in these lipid fractions. In a previous study, Schreiner et al. (2004) found that, when seal blubber oil was fed to hens in excess of 1.25% of the diet, n-3 LCPUFA in the sn-2 position of phosphatidylcholine (PC) and PE plateaued. Feeding beyond 1.25% also caused an increase of n-3 LCPUFA in the TG, albeit a much smaller increase compared to the alteration of these FAs in the PL. The authors concluded that, when more unsaturated fat is fed to hens beyond possible enrichment threshold, the hen has the capacity to saturate the fat because the unsaturation of yolk lipids must be within certain limits for biological and reproductive reasons.

Conversion of ALA into LCPUFA in the yolk is not highly effective, and largely caused by the low activity of the desaturase enzymes (Fraeye et al. 2012). This conversion could be inhibited by several other factors, including heredity, age and diet of the laying hen (Scheideler et al. 1997; Fredriksson et al. 2006).

LA, and particularly ALA, absorption in the hen and subsequent deposition in the yolk has been found to increase as hens age (Scheideler et al. 1997). In fact, the yolk n-6:n-3 ratio fluctuates in the process of the laying period despite no change in diet. It has been hypothesized that older hens, having larger livers, produce larger yolks, which results in an increased ability to absorb lipids and for a more effective conversion of ALA into LCPUFA (Fredriksson et al. 2006). However, contradictory evidence has shown that younger hens produced yolks with a greater level of LCPUFA (Nielsen 1998). The authors suggested that, because these hens produce eggs with smaller yolks and also have lower laying frequency, the yolk contains compensatory amounts of LCPUFA. Both theories have yet to be fully elucidated.

Because of competition between n-6 and n-3 PUFA for D5DE and D6DE, an excess of LA in the diet can potentially inhibit the extent of ALA conversion to EPA and DHA (Cachaldora et al. 2008b). Therefore, the activities of these enzymes can be altered by the diet (Brenner 1989). In diets high in n-3, most of D5DE is used for the n-3 pathway and thus little is available to convert DGLA into AA while, in a low n-3 diet, most of D5DE is available for the conversion of DGLA to AA (Emken et al. 1992). As an example, flaxseed's inherently low n-6:n-3 PUFA ratio of 0.3 ensures a lowering of the ratio from 7.5 in conventional egg yolk to approximately 2–3 in n-3 enriched egg yolk for various flaxseed diets (Botsoglou et al. 1998). One study revealed that single-nucleotide polymorphisms (SNPs) of the FADS1 (position 391 (C->A) and 468 (C->T)) in Vietnamese and European hen breeds were significantly associated with myristic acid (MYRIST), palmitic acid (PALM), palmitoleic acid (PALMO) and LA while FADS2 SNPs were associated with AA, DHA and the n-6:n-3 ratio (Khang 2006). This may also explain why MYRIST increases desaturase enzyme activity significantly (Jan et al. 2004). Also, D6DE preferably receives C18 substrates for the catalytic mechanism and also allows preferential desaturation of MYRIST (Shappell et al. 2001). Another cause for decreased ALA conversion is an excess of EPA and DHA in the diet. It is evident that a plateau is reached in the yolk whereby PUFA content in the sn-2 position of the PL fraction cannot exceed a certain amount, and after this point the LCPUFA are increasingly stored in TG (Schreiner et al. 2004). Addition of dietary antioxidants also seems to modulate the elongation – desaturation pathway favourably, by significantly reducing yolk AA content and increasing n-3 LCPUFA compared to hens fed flaxseed with no antioxidants. One study found that adding 100 IU vitamin E or 50 mg/kg Butylated hydroxytoluene (BHT) to diets containing 10% flaxseed, increased total n-3 PUFA from 4.93 mg/g yolk to 6.30 and 7.97 mg/g yolk, respectively (Hayat et al. 2009).

4 Omega-3 PUFA: enhancing the n-3 profile of eggs

4.1 Strategies to enhance the n-3 profile of eggs

Cruickshank (1934) was the first to report that the yolk FA composition could be modified through dietary manipulation. Decades later, the first designer egg rich in both n-3 PUFA and antioxidants was developed (Qi and Sim 1998). In Canada, a minimum of at least 300 mg n-3 PUFA in 1 egg is required for an n-3 content claim, and according to the Canadian Egg Marketing Agency, n-3-enriched eggs currently account for 15% of the Canadian egg market (Flax Council of Canada 2014). Table 1 shows the composition of n-3-enriched eggs sold in Canada. An enriched egg can provide upwards of 500 mg total n-3 PUFA, which may contain up to 290 mg of combined EPA and DHA. This could provide adults with half of the necessary recommendation (500 mg EPA and DHA per day) to reduce the risk of developing Cardiovascular disease (CVD) (Harris et al. 2008).

Currently, canola oil, fish oil and flaxseed are the only available sources for the Canadian egg industry to incorporate n-3 PUFA in the hen's diet. As depicted in Table 2, ALA can be significantly increased in vegetable diets with limited changes in DHA, whereas fish oil diets increase DHA levels. It is also clear that DHA plateaus occur quickly, particularly in vegetable-based diets. Flaxseed can be ground in order to increase lipid absorption and deposition of FA into egg yolk slightly (Scheideler and Froning 1996); however, increased oxidation of the feed must be considered. Incorporation of additional antioxidants such as vitamin E (alpha-tocopherol) and/or microencapsulation of sensitive lipids may offer more protection from oxidation (Kim et al. 2000). Since both ALA and LCPUFA confer health benefits, diets combining both vegetable and marine sources of n-3 PUFA are ideal in order to yield equal deposition of desirable PUFA.

Incorporation of n-3 PUFA in hen diets into the egg has direct implications not only for the egg industry with respect to the developing progeny, which can improve chick growth, health and development (Cherian and Sim 1991) but also for the development of the speciality egg market, aimed at boosting human intakes of n-3 FAs. In addition, n-3 enhancement of egg yolk lipids, in particular the PL and TG fractions, has a potential use in the food industry. For example, egg PL have been utilized in infant formulas (Makrides et al. 2002), while egg TG molecules (based on their FA composition) have ultimate use in

Table 1 Comparison of the nutrient content of enriched and conventional eggs[1]

Nutrient	n-3 egg	Conventional egg[2]
Energy (calories)	70	70
Protein (g)	6.0	6.0
Total fat (g)	5.0	5.0
Saturated fat (g)	1.5	1.5
Monounsaturated fat (g)	2.0	2.0
Total n-6 fatty acids (g)	0.8	0.6
Linoleic acid (mg)	640[5]	540
Arachidonic acid (mg)	30[5]	80
Total n-3 fatty acids (g)	0.4[3]	0.1
Alpha-linolenic acid (mg)	174[5]	31
Eicosapentaenoic acid (mg)	17[5]	1
Docosapentaenoic acid (mg)	10[5]	4
Docosahexaenoic acid (mg)	108[5]	34[4]
Cholesterol (mg)	195	195

[1]Abbreviations: g = grams, mg = milligrams.
[2]Nutrient content for one large hard-boiled chicken egg (Canadian Nutrient File 2014).
[3]On average, a commercially available n-3-enriched egg contains about 75 mg DHA for an egg from a flax-fed hen, and 125 mg DHA from a hen fed both flaxseed and fish oil (Burnbrae Farms 2014).
[4]Taber et al. 1998.
[5]Gillingham et al. 2005.

food applications (Osborn and Akoh 2002). In a study involving the use of hemp products in laying hen diets (Neijat et al. in press), the total PL as well as the TG of the yolk lipids fraction were found to be enhanced with n-3 PUFA and reduced n-6/n-3 ratio.

Although plant ingredients rich in n-3 FA, ALA (e.g. flaxseeds, canola), have been used to enrich eggs with n-3 FA, it is primarily the ALA and to some extent the DHA that is enriched, but not EPA; unlike the inclusion of pre-formed LCPUFA such as marine oils (Van Elswyk et al. 1995) and algae (Herber and Van Elswyk 1996) which have been shown to provide a significant enrichment of eggs with LCPUFA including EPA (Šefer et al. 2011). Even then, the latter sources preferentially deposit higher levels of DHA than EPA (Gonzalez-Esquerra and Leeson 2001; Cachaldora et al. 2006). Generally, the cause for the low incorporation of EPA in egg yolk is still unclear; however, endogenous synthesis of n-3 LCPUFA from precursor ALA may relate to the close association between the levels of LC metabolites and ALA in the liver PL fraction (Tu et al. 2010; Neijat et al. unpublished), possibly controlling homeostatic levels of LCPUFA metabolites in the body. Hence, the overall, conversion efficiency from ALA to DHA may relate to the total amount of n-3 PUFA in the diet, as well as on their relative proportions (Cachaldora et al. 2008b).

Table 2 Alpha-linolenic acid and docosahexaenoic acid content of egg yolk from hens fed various dietary treatments

	% of total fatty acids			
Dietary treatment	ALA	EPA	DHA	References
Vegetable oil				
4% sunflower oil	0.19	0.11	0.48	Baucells et al. 2000; Farrell 1998[ac]
4% rapeseed oil	0.73	0.08	1.01	Baucells et al. 2000[a]
1% flaxseed oil	1.39	0.15	1.44	Rizzi et al. 2009[a]
4% flaxseed oil	4.87	0.25	1.56	Baucells et al. 2000[a]
6% flaxseed oil	6.49	NR	1.45	Antruejo et al. 2011[a]
6% chia seed oil	10.98	NR	1.15	Antruejo et al. 2011[a]
Oilseed				
25% rapeseed	1.49	NR	1.24	Antruejo et al. 2011[a]
5% flaxseed	2.01	NR	1.83	Scheideler and Froning 1996[a]
10% flaxseed	4.87	0.20	1.70	Ferrier et al. 1995; Scheideler and Froning 1996[ac]
15% flaxseed	7.07	NR	1.78	Antruejo et al. 2011; Scheideler and Froning 1996[ac]
20% flaxseed	10.70	0.20	1.80	Ferrier et al. 1995[a]
7% chia seed	2.76	NR	5.48	Ayerza and Coates 1999[a]
14% chia seed	5.17	NR	3.97	Ayerza and Coates 1999[a]
21% chia seed	9.04	NR	3.08	Ayerza and Coates 1999[a]
25% chia seed	10.48	NR	0.93	Antruejo et al. 2011[a]
28% chia seed	16.52	NR	1.24	Ayerza and Coates 1999[a]
Fish oil				
0.5% fish oil	0.28	0.11	2.07	Van Elswyk et al. 1995[b]
1.5% fish oil	0.38	0.25	3.09	Van Elswyk et al. 1995[b]; Scheideler and Froning 1996[a]
2% fish oil (microencapsulated)	1.49	0.27	2.20	Lawlor et al. 2010[b]
3% fish oil	0.36	0.43	3.73	Van Elswyk, Dawson, Sams 1995[b]
4% fish oil	0.44	0.92	3.18	Baucells et al. 2000[a]
4% fish oil (microencapsulated)	1.84	0.60	2.25	Lawlor et al. 2010[b]
5% fish oil	0.36	1.00	5.27	Farrell 1998[a]
6% fish oil (microencapsulated)	1.94	1.11	4.49	Lawlor et al. 2010[b]
Combination				
3% fish oil + 1% rapeseed oil	0.51	0.64	2.84	Baucells et al. 2000[a]
3% fish oil + 1% sunflower oil	0.32	0.58	2.72	Baucells et al. 2000[a]
3% fish oil + 1% flaxseed oil	2.26	0.58	3.80	Farrell 1998[a]
3% fish oil + 1% flaxseed oil	1.81	0.63	2.83	Baucells et al. 2000[a]
2% fish oil + 2% flax	2.37	0.66	2.62	Baucells et al. 2000[a]
2% fish oil + 1% flaxseed oil + 1% canola oil	2.32	0.45	3.38	Farrell 1998[a]
2% fish oil + 2% rapeseed oil	0.58	0.43	2.48	Baucells et al. 2000[a]
2% fish oil + 2% sunflower oil	0.31	0.39	2.30	Baucells et al. 2000[a]
1% fish oil + 3% rapeseed oil	0.66	0.18	1.76	Baucells et al. 2000[a]
1% fish oil + 3% sunflower oil	0.26	0.17	1.61	Baucells et al. 2000[a]
1% fish oil + 3% flaxseed oil	3.71	0.33	1.99	Baucells et al. 2000[a]

NR = value not reported.
[a]Unchanged reported values (% of fatty acids).
[b]Calculated % of total fatty acids ((fatty acid/total fatty acids)*100).
[c]Average of reported values.

4.2 Plant sources of n-3

Table 3 displays the n-3 FA content of some common food sources that can be incorporated into hen diets to increase the n-3 PUFA content. n-3 PUFA can be produced by various plants, including canola, soya bean, walnuts and wheatgrass (Harris et al. 2009). Flax, a crop grown abundantly on the Canadian prairies which contains upwards of 41% lipid in the whole seed, is one of the richest sources of n-3 PUFA, with 18% ALA present in the whole seed, and 56% in the seed oil (Scheideler and Froning 1996). The 25–35% oil that is contained within whole hempseed consists of more than 90% unsaturated FAs, with LA at around 57% and ALA at 17–19% (Oomah et al. 2002). Less common sources of n-3 PUFA include camelina (*Camelina sativa* Crantz), perilla, kiwi and chia (*Salvia hispanica* L.), at 35, 58, 62 and 64% ALA in the seed oil, respectively (Zubr 1997). Pearl millet is rich in oil compared to common cereals, having a typical fat content above 5%, with ALA comprising 4% of the total FAs (Rooney 1978; Collins et al. 1997). Although the chloroplast membranes of green leafy plants are very rich in n-3s, the amount as a whole food is modest (Bezard et al. 1994). Smaller amounts can also be found in legumes and cauliflower.

Marine microalgae are the original source of n-3 PUFA in the diet of fish, obtained via fermentation, referred to as heterotrophic microalgae (Barclay et al. 1994; Laca et al. 2009), are rich in DHA (Herber and Van Elswyk 1996). Others, the autotrophic microalgae, derived from organic compounds (carbon dioxide and sunlight), may also contain EPA depending on the microalgae species (Fraeye et al. 2012).

4.3 Marine sources of n-3

LCPUFA are not found in plants, but are ubiquitous in mammalian tissue, achieving highest concentrations in the membranes of neural and other excitable tissue (Brenna

Table 3 n-3 fatty acid content of various plant and fish oils as potential sources of enrichment in the hen diet

	% ALA	% EPA	% DPA	% DHA
Plant oils				
Flax	56	0	0	0
Hemp	18	0	0	0
Walnut	12	0	0	0
Canola	9	0	0	0
Soya bean	7	0	0	0
Corn	1	0	0	0
Fish oils				
Salmon	1	13	3	18[a]
Sardine	1	10	2	11[a]
Cod liver	1	7	1	11[a]
Menhaden	1	13	5	9[a]
Herring	1	6	1	4[a]

[a] Canadian Nutrient File 2014; USDA Nutrient Database 2014.

and Diau 2007). The richest animal source of n-3 PUFA, mainly in the form of EPA and DHA, includes fatty fish such as salmon, tuna, mackerel, herring, trout and sardines (Simopoulos 1991). Crustaceans and molluscs also contain a small amount. Between fish species there are large variations in n-3 PUFA content, with some containing substantially higher levels of EPA compared to DHA and vice versa. For example, pink salmon canned in water (100 g) may contain 1665 mg LCPUFA, while the same amount of canned tuna contains only 229 mg (Meyer 2011). Other minor sources include dairy, eggs, chicken, pork and beef. However, the amount of n-3 PUFA in these foods is substantially lower than in fish, at times negligible, and highly variable depending on the diet of the animal.

Despite a high content of n-3 LCPUFA, a limitation of using fish oils as a feed ingredient in poultry diets is not only cost but also safety. Many potentially toxic pollutants are fat soluble, and feeding fish oil to layers may result in increased elevations of such marine pollutants in subsequent egg yolk (Domingo 2014). Furthermore, fish consumption advisories meant to protect human health do not extend to fish products fed to farmed animals such as laying hens (Dórea 2006). In Canada and the United States, approximately 20% of stocks are overfished which leads to disrupted ecosystems (National Research Council 2013).

5 Omega-3 PUFA: sensory attributes of enriched eggs

Sensory analysis is a multidisciplinary science that uses human panellists and their senses of sight, smell, taste, touch and hearing to measure the sensory characteristics and acceptability of food products (Martens 1999). Sensory evaluation of any food is essential because there is no instrument that can entirely replace the human response. It can be applicable in many aspects of the food industry, including product development, product improvement, quality control, storage studies and process improvement. It is important that functional foods maintain similar sensory characteristics compared to regular foods in order to remain palatable to consumers. Otherwise, the likelihood of consumption is drastically decreased. This is most problematic when functional foods are developed for the sole purpose of reducing the risk of disease development and progression.

Since the eggshell is porous, allowing for gaseous exchange in and out of the egg, migration of volatile compounds from packaging materials or the surrounding environment through the shell can result in altered egg aroma and flavour (Matiella and Hsieh 1991). However, altering the hen's diet can also have a profound effect on egg sensory quality and a potential downside of egg enrichment is the possibility of a sensory profile that is different, or more concerning, worsened, compared to its conventional counterpart. It was once believed that strongly flavoured feeds resulted in strongly flavoured eggs; however, it is now understood that this is a much more complex issue. There could also be diet–gene interactions that impact sensory characteristics. For instance, it has been demonstrated that not all breeds or chickens within a breed produce off-flavoured eggs, that even feeds without fishmeal can produce similar negative sensory effects, and that the level of ingredient inclusion into the diet has an effect on the resulting eggs.

Typically, in the case of enriched eggs, the greatest sensory concern lies in the potential for off-aromas and off-flavours, usually described as 'fishy' due to n-3 PUFA incorporation. Other bio-active enrichment, such as vitamin D, tends to not significantly impact sensory or functional properties, therefore issues in consumer acceptance is unlikely (Mattila et al. 2003).

Feeding fish oil to hens has often resulted in fishy taint in eggs. It was once believed that processing these oils using a method known as deodourization to remove ketones and benzene-containing compounds would solve this problem. However, results from Gonzalez-Esquerra and Leeson (2000) showed that deodourized fish oil in the layer diet was no different than regular fish oil in terms of aftertaste or off-flavours in the eggs. After this was demonstrated, this hypothesis was dismissed and the majority of subsequent research has focused more on oxidation of feed components and specific dietary composition.

An increase in the content of n-3 PUFA in the egg may increase its susceptibility to lipid oxidation, which some researchers believe to result in off-flavour and off-aroma (Van Elswyk et al. 1995; Cherian et al. 1996). Fishiness is thought to be more of a problem during egg processing (Woods and Fearon 2009).

To attempt to inhibit feed oxidation, which ultimately may lead to impaired sensory quality through the deposition of oxidized lipids into the egg (Galobart et al. 2001; Hayat et al. 2010b), several methods have been employed with varying degrees of success. Most commonly used has been the supplementation of antioxidants into an enriched diet, including mixed tocopherols, which has resulted in a significant improvement in oxidative stability of eggs, compared with those without tocopherols (Cherian, Wolfe, Sim 1996).

In fact, Cherian et al. (1996) and Galobart et al. (2001) found that the inclusion of tocopherols (200 mg/kg) resulted in a significant reduction in thiobarbituric acid reactive substances (TBARS) in enriched eggs. Unfortunately, in other studies, there was no significant improvement in lipid stability and/or sensorial characteristics resulting from antioxidant supplementation (Leeson, Caston, MacLaurin 1998; Hayat et al. 2010a,b). In fact, contrary results have showed that lower levels of vitamin E may be better at preventing these off-flavours (i.e. 10 vs. 100 IU vitamin E) (Leeson et al. 1998). Also, several authors have demonstrated that vitamin E over 120 ppm in conventional diets can act as a pro-oxidant in egg yolk, resulting in an elevation in TBARS (Chen et al. 1998; Gebert et al. 1998; Franchini et al. 2002). Large doses of vitamin E should be used sparingly in general, but research suggests that this tactic may be helpful when antioxidants are paired with sufficient amounts of unstable lipids, like n-3 PUFA. This effect should be investigated with other antioxidants, as this pro-oxidant effect is likely to occur. Furthermore, protecting n-3 oils through microencapsulation may prove to be an efficient method of preserving FA and protecting them from oxidation, which would result in decreased sensory quality (Lawlor et al. 2010). Regardless of the level of antioxidant inclusion in the diet, off-aromas and flavours still occur most likely due to the egg's FA composition with increased n-3 levels.

Contrary to those who believe that off-aromas and flavours in enriched eggs are the direct results of lipid oxidation, other researchers have proposed that the culprit is more likely to be a combination of lipid and non-lipid substances in the hen diet. These components can affect the concentration of volatile compounds and the presence of trimethylamine (TMA) in addition to lipid oxidation products within the egg (Jiang et al. 1992; Van Elswyk et al. 1992, 1995; Scheideler et al. 1997; Leskanich and Noble 1997; Parpinello et al. 2006).

The laying hen diet that is chosen to yield an enriched egg is ultimately dependent on the desired FA profile of the egg yolk. If the more healthful LCPUFA such as DHA are desired, then fish or algal ingredients are typically used in industry to achieve this outcome. In this approach, the FAs are highly sensitive to oxidation due to the increased number of double bonds; however, smaller quantities are typically needed in the diet (Aymond and Van Elswyk 1995). If oxidation does occur, these FAs and their oxidized products are not only undesirable but also known to cause adverse biological effects in humans (Aymond and Van Elswyk 1995).

An immediate concern with supplementing a hen's diet with fish oil or fish meal is the negative effects on organoleptic quality as eggs procured from these hens have been found to possess a pronounced fishy aroma and flavour, depending on the dietary concentration and the particular type of fish used (Surai and Sparks 2001). Aro et al. (2011) found that the use of 5% fish oil in hen diets yielded eggs with significantly lower sensory scores (higher off-aroma and flavour) but were still considered generally acceptable compared to the control eggs. Ceylan et al. (2011) had similar findings, but these authors used a lower level of fish oil inclusion at 3%. Hammershøj (1995) found that taste and general sensory impression were of lower grades for eggs procured from hens consuming diets with 1.5 and 3% fish oil compared with animal fat diets. Differences in source of fish oil or fishmeal were also observed, likely in part due to differences in their FA profile. For example, the addition of sardine oil did not influence egg flavour (Vondell and Putnam 1945), but menhaden fish oil did (Holdas and May 1966). Diets containing British Columbia herring meal had higher flavour quality than the use of hake meal and Canadian Atlantic herring meal (Koehler and Bearse 1975). Peruvian anchovy oil at a low inclusion level of 1.5% produced off-flavoured eggs (Koehler and Bearse 1975). Menhaden fish oil has been investigated extensively for its use in layer diets and has been found to be generally acceptable at levels less than 1.5% and 8% for oil and meal, respectively (Van Elswyk et al. 1992; Marshall et al. 1994; Van Elswyk et al. 1995; Scheideler et al. 1997; Van Elswyk 1997; Gonzalez-Esquerra and Leeson 2000). Other types of fish, such as Cantabrian blue fish oil at 1.5% have also been found acceptable (Laca et al. 2009). In another study, researchers found that only 1 out of 8 trained panellists were able to distinguish between eggs from hens fed 1.25% seal blubber oil or conventional eggs (Schreiner et al. 2004). As the level of inclusion rose, so did the panellists' ability to pick out the treatment eggs. However, because a plateau in n-3 deposition was observed beyond 1.25%, it would not be worthwhile to feed greater amounts, and in turn affect sensory quality.

On the other hand, protecting these delicate oils from oxidation has been successful, as demonstrated by Lawlor et al. (2010) who showed that as much as 6% microencapsulated fish oil (MFO) in the layer diet was acceptable when eggs were scrambled. However, when these same eggs were hard-boiled, off-flavour and sulphur flavour were increased significantly compared to the 4% MFO group, which was not different from the control group. Therefore, commonly used cooking practices should be taken into consideration when designing diets to ensure that deleterious flavours or aromas do not occur.

Despite the fact that greater increases in egg DHA (3.5%) and EPA (0.52%) have been achieved using lower amounts of fish oil than with greater numbers of oleaginous seeds (Hargis and Van Elswyk 1993), there are a number of advantages in using vegetarian feed. The idea of using flaxseed for n-3 egg production began when sensory studies reported strong 'fishy flavours' from fish oil-based n-3-enriched eggs (Leskanich and Noble 1997). Flaxseed has been a popular choice as a vegetable source of n-3 PUFA for animals but there have been reports that eggs from hens offered flaxseed also have a fishy aroma or flavour, similar to that found in eggs from hens offered fish products. In such terrestrial diets, the plant oils contain a substantial amount of ALA, which deposit into the egg yolk, along with a small proportion of longer chain n-3 PUFA such as DHA, which the hen liver converts to this form. The benefit of such diets is that they are comparatively inexpensive and less prone to oxidation than the former method. Many would argue, however, that important long-chain n-3 PUFA such as DHA that would be deposited in this case would not be as abundant in eggs from fish-fed hens. However, if the threshold for fish oil in hen diets is to be respected to avoid sensory impairment (i.e. 1.5% for fish oil), yolk DHA deposition is likely to not exceed 138 mg/egg (Herber and Van Elswyk 1996), which can already be achieved using vegetarian ingredients (i.e. flaxseed). In fact, eggs were acceptable to consumers with as much as 5% FO in the layer diet (Cloughley et al. 1997). However, another study found that the taste of Madeira cakes made with eggs from hens fed 2% FO were worse, as compared to those cakes which included eggs from hens fed palm butter, grapeseed oil or lard (control) at the same level of inclusion (Tallarico et al. 2002). In a study which incorporated 10% flaxseed in the layer diet, it was determined that the resultant eggs were acceptable to consumer panellists; however, a trained panel was able to detect differences in flavour, aroma, off-flavour and overall difference (Hayat et al. 2010a). In another study using 10% *ground* flaxseed, sensory acceptability was also maintained (Mridula et al. 2012). Levels of flaxseed greater than 10%, however, have showed lower acceptability along with 'fishy' and off-flavours compared to conventional eggs (Caston et al. 1994; Scheideler et al. 1997; Leeson et al. 1998). Moreover, Scheideler et al. (1997) found that appearance and flavour scores were slightly lower for 10–15% ground flaxseed compared to the same level in the form of whole seed. This could be attributed to either enhanced absorption of FA in the hen, or oxidation occurring in the feed before it is consumed.

Overall, the conservative approach to prevent fishy taint is to include no more than 5% flaxseed, or 1.5% fish oil in layer diets (Scheideler et al. 1997; Surai and Sparks 2001; Woods and Fearon 2009). However, the sensory outcome is highly dependent on the level of inclusion, type (oil/seed), processing (whole/ground) and mixture with other ALA ingredients. For example, Ayerza and Coates (2002) found that combining flaxseed with chia seed in the hens' diet, but keeping the flaxseed content below 5% of the diet, successfully increased yolk ALA without any adverse effect on egg flavour or aroma (Woods and Fearon 2009).

However, other promising ingredients rich in desirable FA may be used as an alternative to fish and flax. For example, marine algae have been found to be a promising ingredient to enrich DHA in eggs, without compromising the sensory profile (Herber-McNeill and Van Elswyk 1998). In fact, the latter authors found that the inclusion of 4.8% of a marine micro algal product was as acceptable as control eggs in terms of egg flavour. Along with 2% sardine oil, 10% of each of the following marine algae were included in the laying ration; *Macrocystis pyrifera, Sargassum sinicola* and *Enteromorpha* spp. (Carrillo et al. 2008). The authors found no flavour differences, but warned that using algae inclusion levels above 10% should be avoided as this could affect the acceptance of the feed, lower the egg production or cause diarrhoea due to the high mineral content in the seaweed.

The inclusion of hempseed and hempseed oil at maximum inclusion levels (20 and 12%, respectively) was found to not significantly impact the sensory profile of eggs (Goldberg et al. 2012). Since the n-3 PUFA content in hemp products is much lower than that of flax, these results are not exceptional. Another study involving the use of chia in hen diets also resulted in similar sensory characteristics as conventional eggs and, even at a maximum inclusion level of 30%, it did not change the aroma or flavour profile of cooked eggs. This is especially important because chia seed contains more n-3 PUFA than flaxseed. Moreover, chia-fed hens produced eggs with a lower score and hence a milder flavour compared to the control (Ayerza and Coates 1999). This finding is similar to results from Rokka et al. (2002), who found that feeding 5% camelina seed oil resulted in an improved sensory profile compared to a control diet. More interesting is that the sensory profile was improved compared to a flaxseed oil diet, suggesting it may be a good alternative to flax, at least from a sensory point of view (Rokka et al. 2002). Dietary levels up to 10% of extruded camelina meal were deemed acceptable; however, two panellists noted some fishiness (Kakani et al. 2012).

6 The vitamin profile of eggs

Vitamins are a complex group of organic compounds that are generally present in small quantities in foodstuffs. Vitamins A, D, E and K are lipid-soluble and consequently deposited in the yolk, whereas vitamins B_1 (thiamine) and B_{12} (cobalamin) are water-soluble and situated in the albumen (McDowell 2000; Cheeke 2005). Vitamins B_2 (riboflavin) and B_9 (folate) are hydrosoluble and distributed among both yolk and albumen, although folate mainly deposits in the yolk fraction (Anton et al. 2006).

6.1 Fat-soluble vitamins

Since the potential for increasing vitamin K levels in eggs has not been adequately addressed, this section will focus on vitamins D, E and A as well as carotenoids. Two main strategies exist for enriching eggs with vitamin D2 (ergocalciferol) and/or vitamin D3 (cholecalciferol): (1) dietary enrichment of vitamin D and (2) exposing hens to ultraviolet B radiation. A combination of these two approaches is likely ideal for achieving high levels in egg yolk. In one study, the average vitamin D concentration in egg yolk during weeks 3 to 40 and the deposition rate of cholecalciferol during the first 3 weeks were both linearly increased with the dietary cholecalciferol level when the feed contained no more than 24700 IU/kg cholecalciferol (Yao et al. 2013). In general, the transfer of vitamin D into egg yolk has been shown to be more effective using vitamin D_3 compared to vitamin D_2 (Mattila et al. 2004). In another study, which compared a dietary strategy for enrichment and sunlight exposure, hens receiving exposure to UVB radiation for 3 hours daily for 4 weeks laid eggs with more vitamin D_3 in the yolk (17.5 $\pm$ 7.2 μg/100 g dry matter (DM) compared to eggs from hens fed diets with 3000 IU vitamin D_3/kg (5.2 $\pm$ 2.4 μg/100 g DM, $P < 0.01$) (Schutkowski et al. 2013). Vitamin D_3 content in yolk of vitamin D_3-supplemented hens could be further increased by UVB radiation (32.4 $\pm$ 10.9 μg/100 g DM). A recent study had similar findings when the authors found that vitamin D_3 content of egg yolk was three- to fourfold higher in hens that were exposed to sunlight compared to hens receiving no sunlight ($P < 0.001$) (Kühn et al. 2014). In both of these studies, the 25-hydroxyvitamin D (25[OH]D_3) content of egg yolk was also influenced by sunlight exposure, although the effect of sunlight was less pronounced than the vitamin D_3 content ($P < 0.05$) (Schutkowski et al. 2013; Kühn et al. 2014). The consumption of one vitamin D-enriched egg (approximately 60 g) could provide up to 2.8 μg of vitamin D, which is approximately 3 times more than the intake from typical eggs and corresponds to about 50% of the recommended 5 μg/d allowance (Food and Nutrition Board 1999; Mattila et al. 1999, 2004).

There are fewer studies investigating vitamin A deposition in egg. Yolk retinol concentration was enhanced from 24.6 IU/g for eggs from the control group, to 33.6 and 37.7 IU/g of yolk when hens were fed 15000 and 30000 IU retinyl acetate/kg of diet (Mori et al. 2003). In another study, the effects of tomato powder supplementation (5 or 10 g/kg) were investigated. Not surprisingly, since tomato powder is a rich source of carotenoids, vitamin A precursors, authors found that concentrations of yolk lycopene, β-carotene, lutein as well as vitamin A increased for both diets (Akdemir et al. 2012).

The intensity of yolk colour is due to the presence of yellow and red oxycarotenoids, which are xanthophyll pigments found in some plants (Karunajeewa et al. 1984; Palmer and Kempster 1919). There are two main classes of naturally occurring carotenoids: (1) carotenes (β-carotene, α-carotene), which are hydrocarbons, are either linear or cyclized at one or both ends of the molecule, and (2) xanthophylls, the oxygenated derivatives of carotenes (Sajilata et al. 2008). Egg yolks range in colour from pale yellow to deep orange depending on the concentration of these natural pigments. For example, a typical Canadian wheat/barley-based diet will result in a pale yellow yolk, while a European corn/lucerne/alfalfa-based diet yields a darker yellow yolk. In addition to such ingredients, pigments to impart richer yolk colours can be naturally obtained from many sources, including marigold, tomato, carrots, red pepper, paprika, annatto seed, microalgae and *Chlorella vulgaris* (Fletcher and Halloran 1983; Nys 2000; Moeini et al. 2013; Hammershøj et al. 2010; Gouveia et al. 1996) or synthesized artificially, although this is seen as negative practice in industry in some countries.

Although macronutrients remain the same regardless of yolk colour, there may be small changes in some of the micronutrients. Hydroxyl compounds like xanthophylls are normally absorbed from the feed in the following order: lutein, luteinmono- and diester, 3′-oxolutein and zeaxanthin (Singh et al. 2012). These carotenoids not only provide colour, but can also exhibit antioxidant function contributing to eye health (Carpentier et al. 2009). Lutein in particular has demonstrated biological activities that have attracted great attention in relation to human health. In particular, high lutein intakes have been associated with a lower risk of developing CVD, several types of cancer, cataracts and age-related maculopathy (Granado, Olmedilla, Blanco 2003). Both lutein and zeaxanthin have been found to be helpful in the prevention of age-related macular degeneration, a leading cause of blindness in the elderly, and have been associated with lower risk of cataract extraction (Moeller, Jacques, Blumberg 2000).

Some carotenoids, like beta-carotene, have nutritional value as a precursor to vitamin A, which is essential for healthy growth and reproduction (Chytil and Ong 1976; Zile and Cullum 1983). However, dark coloured egg yolks only indicate the presence of total carotenoids, not necessarily the presence of pro-vitamin A compounds alone. In fact, beta-carotene content in the yolk reaches saturation at 1%, regardless of the beta-carotene content of the hen's feed (Herber-McNeill and Van Elswyk 1998).

Carotenoid consumption through egg yolk confers the additional benefit of being a highly bioavailable source of antioxidants, since these lipid-soluble compounds are taken up by diffusion of micelles and transported to the liver in blood (Handelman et al. 1999). In fact, results from an intervention study showed that, after 10 days of ingesting 6 mg lutein per day, the absolute serum lutein concentration was significantly higher from egg yolk (528.5 ± 56.3 nmol/L) compared to lutein supplements (288.1 ± 31.5 nmol/L) or spinach (326.1 ± 43.5 nmol/L) (Chung et al. 2004).

Naturally occurring vitamin E exists in various forms, and alpha-tocopherol is the form of vitamin E that is preferentially absorbed and accumulated in humans (Traber 2006). Vitamin E is commonly supplemented in animal diets for the purpose of increasing the oxidative stability of unsaturated FAs and delaying the development of off-aromas or flavours. The effectiveness of increasing the vitamin E content of eggs through greater fortification of hens' diets has been demonstrated in numerous studies (Jiang et al. 1994; Mori et al. 2003; Shahriar et al. 2008). Meluzzi et al. (2000) showed that yolk alpha-tocopherol content increased linearly as fortification of the hens' diet increased from 0 to 200 IU of vitamin E per kilogram of feed. An egg with a vitamin E content of 300 mg per gram would provide roughly 50–100% of the recommended daily allowances published by various nations for this essential nutrient.

6.2 Water-soluble vitamins

With respect to egg modification, fewer studies have focused on the water-soluble vitamins. Using a protocol that examined the enrichment of hen feed with levels 3 to 10 times those considered to be the industry standard, Leeson and Caston (2003) identified the potential for substantive enhancement of eggs with vitamin B_{12}. Other vitamins were less responsive, when compared to the 'regular egg'; however, this may have reflected the relatively generous fortification of water-soluble vitamins present in the control feed, thus blunting further enhancement. In order to gain further insight into the regulatory mechanisms controlling egg nutrient content and the potential for enhanced eggs to make meaningful contributions to the human diet, we undertook a series of experiments using folic acid as a model nutrient. House et al. (2002) showed that a 10 mg/kg inclusion of folic acid increased folate incorporation, predominantly as the biologically available 5-methyltetrahydrofolate form, into egg yolk (41.0 μg/egg) over that of an un-supplemented diet (17.5 μg/egg), and that egg folate content reached saturation. Subsequent studies (Hebert et al. 2005; Dickson et al. 2010) provided evidence that this enrichment was stable over the production cycle of modern laying hens. Furthermore, the observed response profile of egg folate enrichment mimicked that observed for plasma folate, providing strong evidence that factors regulating plasma folate concentrations were indirectly regulating egg folate content (Hebert et al. 2005). Further research (Tactacan et al. 2010, 2011, 2012) provided evidence that mechanisms related to intestinal folate uptake and hepatic metabolism were key control points regulating egg folate deposition.

One folate-enriched egg (60 g) can provide enough folate to account for 10–15% of the recommended daily allowance of 400 μg for adults or 13–16% of the estimated average requirements of 320 μg/day (Institute of Medicine 1998). Recent work in other laboratories has provided similar data (Sugiyama et al. 2012). Of particular importance, the hen converts crystalline folic acid, present in the feed, into the biologically available 5-methyltetrahydrofolate form. The folate in egg yolk was shown to have a greater relative bioavailability than crystalline folic acid, when explored in a rodent bioassay (House et al. 2003). Furthermore, through fractionation techniques, Naderi et al. (2014) have shown that the plasma fraction of egg yolk is devoid of folate, whereas the granule fraction is concentrated by a threefold factor in comparison with non-enriched egg yolk. This latter work supports the development of an enhancement paradigm strategy that couples research into efficient egg enhancement strategies with modern fractionation technologies and processes. Such strategies may lead to egg-derived streams of bio-active and functional fractions for ultimate usage in food, natural health product, or other applications.

7 Minerals and trace elements in eggs

Among minerals and trace elements, egg enrichment with selenium (Se), iodine, manganese, chromium, zinc, copper, cobalt and boron has been investigated. Se can be deposited in the albumen, but the bulk of deposition occurs in the yolk fraction (Jiakui and Xiaolong 2004). The differences between supplying hens with organic (Se-malt, Se-enriched yeast, DL-selenomethionine [SM], 2-hydroxy-4-methylselenobutanoic acid [HMSeBA]) or inorganic forms (sodium selenite) of Se have been investigated extensively, indicating similar metabolic routes of Se in either form (Jiakui and Xiaolong 2004), with higher concentrations in egg from organic dietary sources (Payne, Lavergne and Southern 2005; Utterback et al. 2005; Pan et al. 2007; Pavlović et al. 2009; Čobanová et al. 2011). Among organic sources of Se, egg concentrations were greatest in hens fed HMSeBA (+28.78%; $P < 0.01$), followed by Se-enriched yeast (Jlali et al. 2013). In another study, it was observed that hens fed DL-SM (an organic source of Se) accumulated more Se in egg albumen, than hens fed diets with Se-yeast ($P < 0.05$) (Jing et al. 2015). Whether HMSeBA or SM will be more efficiently deposited into eggs has yet to be evaluated. Dietary Se-enriched probiotics also increased ($P < 0.05$) egg Se content (Pan et al. 2011). There may also be differences in which fraction more Se is deposited based on whether it is in the organic or inorganic form. For instance, eggs from hens fed selenite showed significantly higher Se deposition in yolk than in albumen ($P < 0.01$). On the other hand, the feed supplementation with Se doses 0.4 or 0.9 mg/kg DM from Se-enriched yeast resulted in significantly higher Se concentration in albumen than in yolk (both $P < 0.001$) (Čobanová et al. 2011). Overall, organic Se sources in layer diets at levels that provide 0.3–0.5 mg/kg Se in the feed offer adequate levels in the egg (Fisinin et al. 2009) and feeding up to 5.1 µg/g of Se will not affect egg production or the welfare of the laying hen and is a practical way of producing Se-enriched eggs for consumers (Bennett and Cheng 2010).

Numerous studies have demonstrated that iodine content can be increased in eggs through dietary supplementation (Garber et al. 1993; Kaufmann et al. 1998; Rottger et al. 2008; Durmus et al. 2010). Hen diets were supplemented with graded concentrations of iodine in the form of calcium iodate, up to a maximum level of 24 mg/kg iodine. The iodine concentrations in both yolk and albumen increased with increased iodine supplementation and the authors concluded that 3–6 mg/kg iodine supplied in the layer diet could enrich eggs with iodine without having any adverse effect on performance or egg traits (Yalçin et al. 2004). In a very recent study (Słupczyńska et al. 2014), young hens were fed diets enriched with potassium iodide (KI) or potassium iodate (KIO_3) up to 5 mg of supplemented iodine/kg of feed. Enhanced iodine supply from KI significantly increased accumulation in eggs ($P = 0.000$) after 3 and 5 months of the experiment from 260 and 310 to 1011 and 1256 µg/kg of wet egg content, respectively. The concentration of iodine in the egg yolk from hens fed 1 or 2 mg iodine (iodine-enriched yeast) per kg feed was about 80 and 90% higher, respectively, compared to yolk from hens fed 1 mg of iodine/kg of feed in the form of $Ca(IO_3)_2.H_2O$ (Opalinski et al. 2012).

Manganese, zinc, copper and boron have also been included in layer diets, with varying impact on egg quality and indices of performance, but many of these studies do not include results for egg mineral deposition (Gheisari et al. 2011; Ma et al. 2014; Olgun et al. 2013). Mabe et al. (2003) showed that addition of zinc, manganese and copper to a basal corn–soya bean meal diet increased all of the mineral concentrations in yolk and improved eggshell quality.

The chromium content in eggs was five times higher for chromium-supplemented hens than for control ones; however, egg production and weight were negatively affected by heavy metals at high dosages (Meluzzi et al. 1996), indicating that, although it is possible to accumulate chromium in eggs to supply it as a chromium source for humans, more studies should be performed in order to produce an economical chromium-enriched egg. One study that supplemented chromium (up to 600 μg/kg chromium) in the form of chromium propionate did not affect egg chromium deposition ($P > 0.05$) (Ma et al. 2014). A different study, which supplemented two marine macroalgae: *Enteromorpha prolifera* and *Cladophora* sp., enriched with organic microelements [Cu(II), Zn(II), Co(II), Mn(II), Cr(III)] to the diet of laying hens resulted in higher microelement transfer to eggs compared to the control group, in which microelements were supplemented in the inorganic form (Michalak et al. 2011).

Other studies have identified an association between the dietary iron concentration of laying hens and the iron content of eggs. Park et al. (2004) found that the iron content of eggs could be increased by 5–18% by the addition of organic or inorganic iron. However, Revell et al. (2009) found the supplementation of an organic form of iron did not significantly increase iron concentration in egg yolk. Supplementation of the basal diet with iron from 93 to 120 mg/kg increased iron concentration in egg yolk and white by 6.3 and 2.2%, respectively.

Overall, it has been shown that egg mineral content can easily be manipulated through the diet; however, organic or inorganic forms offer varying degrees of deposition. The extent of egg deposition of other minerals and trace elements warrants further investigation.

8 Dietary combinations

Many studies have analysed synergistic effects between multiple vitamins and minerals. For instance, the combination of iron with zinc and copper increased iron concentration in the yolk and white by 36.7 and 34.9%, respectively (Skrivan et al. 2005). As the dietary concentration of DL-alpha-tocopheryl acetate rose, the efficiency of uptake of alpha-tocopherol from feed to yolk decreased linearly ($P < 0.001$). In the same study, in a different experiment, a significantly lower egg yolk alpha-tocopherol concentration was found in hens fed diets including the highest level of vitamin A supplementation (40 000 IU vitamin A/kg diet) ($P < 0.001$), and no interaction of dietary DL-alpha-tocopheryl acetate and vitamin A concentrations was observed on egg yolk alpha-tocopherol concentration (Grobas et al. 2002). The influence of different forms of dietary Se on alpha-tocopherol and Se contents of egg yolk was investigated. Treatments compared the effects of inorganic Se supplementation (selenite) to those of organic (Se-enriched yeast, Se-enriched alga *Chlorella*) supplements. Dietary Se supplementation increased the alpha-tocopherol content of yolks from 297 mg/kg DM in treatment without supplementation to 311 mg/kg when Se was supplemented as selenite, and to 370–375 mg/kg when organic supplements were used (Skrivan et al. 2008). In some cases, FA enrichment can be improved by including antioxidants in the diet, likely by protecting against oxidation. For example, in a recent study it was found that dietary Se and vitamin E supplementation acted synergistically in increasing n-3 FA levels in the egg (Sujatha and Narahari 2011).

9 Conclusion

Hen eggs are an excellent functional food that can be used to deposit desirable bio-active compounds, mainly in the yolk. The enrichment of n-3 PUFAs, vitamins and minerals into eggs has the ability to capture the market by enhancing human health and appealing to a segment of consumers who are willing to pay for these desirable modifications. In general, optimal enrichment strategies are influenced by a number of factors, including (1) the transfer efficiency of bio-actives from the diet into egg; (2) the potential impact on hen health and performance; (3) the potential impact on the quality and sensory characteristics of egg; and (4) cost and feasibility of the feed constituents.

10 Where to look for further information

For further information, the reader is referred to key books, including *Bioactive Egg Compounds* by M. Anton. Furthermore, peer-reviewed publications and conference proceedings from national and international poultry science societies, including Poultry Science Association, offer important outlets for the presentation of egg-related research.

11 References

Akdemir, F., Orhan, C., Sahin, N., Sahin, K. and Hayirli A. 2012. Tomato powder in laying hen diets: effects on concentrations of yolk carotenoids and lipid peroxidation. *Br. Poult. Sci.* 53(5): 675–80.

Alvarez, C., Cachaldora, P., Méndez, J., García-Rebollar, P. and De Blas, J. 2004. Effects of dietary conjugated linoleic acid and fish oil supplementation on performance and egg quality in laying hens. *Br. Poult. Sci.* 45(4): 524–9.

Anton, M. 2007. *Bioactive Egg Compounds*. Berlin, Heidelberg: Springer Berlin Heidelberg.

Anton, M., Nau, F. and Nys, Y. 2006. Bioactive egg components and their potential uses. *World's Poult. Sci. J.* 62(3): 429–38.

Antruejo, A., Azcona, J. O., Garcia, P. T., Gallinger, C., Rosmini, M., Ayerza, R., Coates, W. and Perez, C. 2011. n-3 enriched egg production: The effect of α-linolenic ω-3 fatty acid sources on laying hen performance and yolk lipid content and fatty acid composition. *Br. Poult. Sci.* 52(6): 750–60.

Aro, H., Rokka, T., Valaja, J., Hiidenhovi, J., Huopalahti, R. and Ryhänen, E. 2011. Functional and sensory properties of hen eggs with modified fatty acid compositions. *Food & Function* 2(11): 671–7.

Ayerza, R. and Coates, W. 1999. An ω-3 fatty acid enriched chia diet: Influence on egg fatty acid composition, cholesterol and oil content. *Can. J. Anim. Sci.* 79(1): 53–8.

Ayerza, R. and Coates, W. 2002. Dietary levels of chia: Influence on hen weight, egg production and sensory quality, for two strains of hens. *Br. Poult. Sci.* 43(2): 283–90.

Aymond, W. and Van Elswyk, M. 1995. Yolk thiobarbituric acid reactive substances and n-3 fatty acids in response to whole and ground flaxseed. *Poult. Sci.* 74(8): 1388–94.

Barclay, W., Meager, K. and Abril, J. 1994. Heterotrophic production of long chain n-3 fatty acids utilizing algae and algae-like microorganisms. *J. Appl. Phycol.* 6(2): 123–9.

Baucells, M., Crespo, N., Barroeta, A., López-Ferrer, S. and Grashorn, M. 2000. Incorporation of different polyunsaturated fatty acids into eggs. *Poult. Sci.* 79(1): 51–9.

Belitz, H., Grosch, W. and Schieberle, P. 2009. Eggs. In H. Belitz, W. Grosch and P. Schieberle (Eds), 4th revised edition, *Food Chemistry*, 546–62. Springer: Berlin.

Bennett, D. C. and Cheng, K. M. 2010. Selenium enrichment of table eggs. *Poult. Sci.* 89(10): 2166–72.

Beynen, A. 2004. Fatty acid composition of eggs produced by hens fed diets containing groundnut, soya bean or linseed. *NJAS-Wagen J. Life Sci.* 52(1): 3–10.

Bezard, J., Blond, J., Bernard, A. and Clouet, P. 1994. The metabolism and availability of essential fatty acids in animal and human tissues. *Reprod. Nutr. Dev.* 34(6): 539–68.

Botsoglou, N. A., Yannakopoulos, A. L., Fletouris, D. J., Tserveni-Goussi, A. S. and Psomas, I. E. 1998. Yolk fatty acid composition and cholesterol content in response to level and form of dietary flaxseed. *J. Agric. Food Chem.* 46(11): 4652–6.

Brenna, J. T. and Diau, G. 2007. The influence of dietary docosahexaenoic acid and arachidonic acid on central nervous system polyunsaturated fatty acid composition. *PLEFA* 77(5): 247–50.

Brenner, R. 1989. Factors influencing fatty acid chain elongation and desaturation. In A. J. Vergroesen and M. Crawford (Eds), *The Role of Fats in Human Nutrition*, 45–79. Academic Press: London.

Cachaldora, P., Garcia-Rebollar, P., Alvarez, C., De Blas, J. and Mendez, J. 2008a. Effect of type and level of basal fat and level of fish oil supplementation on yolk fat composition and n-3 fatty acids deposition efficiency in laying hens. *Anim. Feed Sci. Technol.* 141(1): 104–14.

Cachaldora, P., García-Rebollar, P., Alvarez, C., De Blas, J. C. and Méndez, J. 2006. Effect of type and level of fish oil supplementation on yolk fat composition and n-3 fatty acids retention efficiency in laying hens. *Br. Poult. Sci.* 47(1): 43–9.

Cachaldora, P., García-Rebollar, P., Alvarez, C., Méndez, J. and De Blas, J. C. 2008b. Double enrichment of chicken eggs with conjugated linoleic acid and n-3 fatty acids through dietary fat supplementation. *Anim. Feed Sci. Technol.* 144(3–4): 315–26.

Canadian Nutrient File [Internet]. Available from http://webprod3.hc-sc.gc.ca/cnf-fce/index-eng.jsp. (Accessed 4 July 2014).

Carpentier, S., Knaus, M. and Suh, M. 2009. Associations between Lutein, Zeaxanthin, and Age-Related Macular Degeneration: An Overview. *Crit. Rev. Food Sci. Nutr.* 49: 313–26.

Carrillo, S., López, E., Casas, M. M., Avila, E., Castillo, R. M., Carranco, M. E., Calvo, C. and Pérez-Gil, F. 2008. Potential use of seaweeds in the laying hen ration to improve the quality of n-3 fatty acid enriched eggs. *J. Appl. Phycol.* 20(5): 721–8.

Caston, L., Squires, E. and Leeson, S. 1994. Hen performance, egg quality, and the sensory evaluation of eggs from SCWL hens fed dietary flax. *Can. J. Anim. Sci.* 74(2): 347–53.

Ceylan, N., Ciftci, I., Mizrak, C., Kahraman, Z. and Efil, H. 2011. Influence of different dietary oil sources on performance and fatty acid profile of egg yolk in laying hens. *J. Anim. Feed Sci.* 20(1): 71–83.

Cheeke, P. R. 2005. *Applied Animal Nutrition*. 3rd ed. Upper Saddle River, NJ: Pearson Prentice Hall.

Chen, J., Latshaw, J., Lee, H. and Min, D. 1998. α-Tocopherol content and oxidative stability of egg yolk as related to dietary α-Tocopherol. *J. Food Sci.* 63(5): 919–22.

Cherian, G. and Sim, J. S. 1991. Effect of feeding full fat flax and canola seeds to laying hens on the fatty acid composition of eggs, embryos, and newly hatched chicks. *Poult. Sci.* 70(4): 917–22.

Cherian, G. and Sim, J. S. 1992. n-3 fatty acid and cholesterol content of newly hatched chicks from α-linolenic acid enriched eggs. *Lipids* 27(9): 706–10.

Cherian, G. and Sim, J. S. 1995. Dietary. alpha.-linolenic acid alters the fatty acid composition of lipid classes in swine tissues. *J. Agric. Food Chem.* 43(11): 2911–6.

Cherian, G., Wolfe, F. and Sim, J. 1996. Feeding dietary oils with tocopherols: Effects on internal qualities of eggs during storage. *J. Food Sci.* 61(1): 15–8.

Cho, H. P., Nakamura, M. T. and Clarke, S. D. 1999. Cloning, expression, and nutritional regulation of the mammalian delta-6 desaturase. *J. Biol. Chem.* 274(1): 471–7.

Chung, H. Y., Rasmussen, H. M. and Johnson, E. J. 2004. Lutein bioavailability is higher from lutein-enriched eggs than from supplements and spinach in men. *J. Nutr.* 134(8): 1887–93.

Chytil, F. and Ong, D. E. 1976. Mediation of retinoic acid-induced growth and anti-tumour activity. *Nature* 260: 49–51.

Cloughley, J., Noble, R., Speake, B. and Sparks, N. 1997. Manipulation of docosahexaenoic (22: 6n-3) acid in chicken's egg. *PLEFA* 57: 222.

Čobanová, K., Petrovič, V., Mellen, M., Arpášova, H., Grešáková, L. and Faix, Š. 2011. Effects of dietary form of selenium on its distribution in eggs. *Biol. Trace Elem. Res.* 144(1–3): 736–46.

Collins, V., Cantor, A., Pescatore, A., Straw, M. and Ford, M. 1997. Pearl millet in layer diets enhances egg yolk n-3 fatty acids. *Poult. Sci.* 76(2): 326–30.

Cruickshank, E. 1934. Studies in fat metabolism in the fowl: The composition of the egg fat and depot fat of the fowl as affected by the ingestion of large amounts of different fats. *Biochem. J.* 28(3): 965–77.

de Schrijver, R. and Privett, O. S. 1982. Effects of dietary long-chain fatty acids on the rat biosynthesis of unsaturated fatty acids in the rat. *J. Nutr.* 112(4): 619–26.

Dickson, T. M., Tactacan, G., Hebert, K., Guenter, W. and House, J. D. 2010. Optimization of folate deposition in eggs through dietary supplementation of folic acid over the entire production cycle of Hyline W36, Hyline W98 and CV20 laying hens. *J. Appl. Poult. Res.* 19: 80–91.

Domingo, J. L. 2014. Nutrients and chemical pollutants in fish and shellfish. balancing health benefits and risks of regular fish consumption. *Crit. Rev. Food Sci. Nutr.* (just-accepted).

Dórea, J. G. 2006. Fish meal in animal feed and human exposure to persistent bioaccumulative and toxic substances. *Food Protect* 69(11): 2777–85.

Durmus, I., Goger, H., Demirtas, S. E. and Yurtogullari, S. 2010. Comparison of rapid and slow feathering egg layers with respect to egg production and hatchability parameters. *Asian J. Anim. Vet. Adv.* 5: 66–71.

Emken, E., Adlof, R., Rohwedder, W. and Gulley, R. 1992. Comparison of linolenic and linoleic acid metabolism in man: Influence of dietary linoleic acid. In A. Sinclair and R. Gibson (Eds), *Essential Fatty Acids and Eicosanoids*, 23–5. AOCS Press: Champaign.

Farrell, D. J. 1998. Enrichment of hen eggs with n-3 long-chain fatty acids and evaluation of enriched eggs in humans. *Am. J. Clin. Nutr.* 68(3): 538–44.

Ferrier, L. K., Caston, L. J., Leeson, S., Squires, J., Weaver, B. J. and Holub, B. J. 1995. Alpha-linolenic acid- and docosahexaenoic acid-enriched eggs from hens fed flaxseed: Influence on blood lipids and platelet phospholipid fatty acids in humans. *Am. J. Clin. Nutr.* 62(1): 81–6.

Fisinin, V. I., Papazyan, T. T. and Surai, P. F. 2009. Producing selenium-enriched eggs and meat to improve the selenium status of the general population. *Crit. Rev. Biotechnol.* 29(1): 18–28.

Flax Council of Canada: Flax in egg production [Internet]. Available from http://www.flaxcouncil.ca/english/index.jsp?p=food5&=food (Accessed 30 August 2014).

Fletcher, D. and Halloran, H. 1983. Egg yolk pigmenting properties of a marigold extract and paprika oleoresin in a practical type diet. *Poult. Sci.* 62(7): 1205–10.

Food and Nutrition Board. 1999. Dietary reference intakes for calcium, phosphorus, magnesium, vitamin D, and fluoride. Standing Committee on the Scientific Evaluation of Dietary Reference Intakes, Institute of Medicine. National Academy Press, Washington, DC.

Fraeye, I., Bruneel, C., Lemahieu, C., Buyse, J., Muylaert, K. and Foubert, I. 2012. Dietary enrichment of eggs with n-3 fatty acids: A review. *Food Res. Int.* 48(2): 961–9.

Franchini, A., Sirri, F., Tallarico, N., Minelli, G., Iaffaldano, N. and Meluzzi, A. 2002. Oxidative stability and sensory and functional properties of eggs from laying hens fed supranutritional doses of vitamins E and C. *Poult. Sci.* 81(11): 1744–50.

Fredriksson, S., Elwinger, K. and Pickova, J. 2006. Fatty acid and carotenoid composition of egg yolk as an effect of microalgae addition to feed formula for laying hens. *Food Chem.* 99(3): 530–7.

Gakhar, N., Goldberg, E., Jing, M., Gibson, R. and House, J. D. 2012. Effect of feeding hemp seed and hemp seed oil on laying hen performance and egg yolk fatty acid content: Evidence of their safety and efficacy for laying hen diets. *Poult. Sci.* 91(3): 701–11.

Galobart, J., Barroeta, A. C., Baucells, M. D. and Guardiola, F. 2001. Lipid oxidation in fresh and spray-dried eggs enriched with n3 and n6 polyunsaturated fatty acids during storage as affected by dietary vitamin E and canthaxanthin supplementation. *Poult. Sci.* 80(3): 327–37.

Garber, D. W., Henkin, Y., Osterlund, L. C., Woolley, T. W. and Segrest, J. P. 1993. Thyroid function and other clinical chemistry parameters in subjects eating iodine-enriched eggs. *Food Chem. Toxicol.* 31: 247–51.

Garcia-Rebollar, P., Cachaldora, P., Alvarez, C., De Blas, C. and Mendez, J. 2008. Effect of the combined supplementation of diets with increasing levels of fish and linseed oils on yolk fat composition and sensorial quality of eggs in laying hens. *Anim. Feed Sci. Technol.* 140(3): 337–48.

Gebert, S., Messikommer, R., Pfirter, H., Bee, G. and Wenk, C. 1998. Dietary fats and vitamin E in diets for laying hens. *Agrarforschung Schweiz* 5(7): 325–8.

Gheisari, A. A., Sanei, A., Samie, A., Gheisari, M. M. and Toghyani, M. 2011. Effect of diets supplemented with different levels of manganese, zinc, and copper from their organic or inorganic sources on egg production and quality characteristics in laying hens. *Biol. Trace Elem. Res.* 142(3): 557–71.

Gillingham, L., Caston, L., Leeson, S., Hourtovenko, K. and Holub, B. J. 2005. The effects of consuming docosahexaenoic acid (DHA)-enriched eggs on serum lipids and fatty acid compositions in statin-treated hypercholesterolemic male patients. *Food Res. Int.* 38(10): 1117–23.

Goldberg, E. M., Gakhar, N., Ryland, D., Aliani, M., Gibson, R. A. and House, J. D. 2012. Fatty acid profile and sensory characteristics of table eggs from laying hens fed hempseed and hempseed oil. *J. Food Sci.* 77(4): S153–60.

Gonzalez-Esquerra, R. and Leeson, S. 2000. Effect of feeding hens regular or deodourized menhaden oil on production parameters, yolk fatty acid profile, and sensory quality of eggs. *Poult. Sci.* 79(11): 1597–602.

Gonzalez-Esquerra, R. and Leeson, S. 2001. Alternatives for enrichment of eggs and chicken meat with n-3 fatty acids. *Can. J. An. Sci.* 81(3): 295–305.

Gouveia, L., Veloso, V., Reis, A., Fernandes, H., Novais, J. and Empis, J. 1996. Chlorella vulgaris used to colour egg yolk. *J. Sci. Food Agric.* 70(2): 167–72

Granado, F., Olmedilla, B. and Blanco, I. 2003. Nutritional and clinical relevance of lutein in human health. *Br. J. Nutr.* 90(03): 487–502.

Gregory, M. K., Geier, M. S., Gibson, R. A. and James, M. J. 2013. Functional characterization of the chicken fatty acid elongases. *J. Nutr.* 143(1): 12–6.

Grobas, S., Mendez, J., Lazaro, R., De Blas, C. and Mateo, G. 2001. Influence of source and percentage of fat added to diet on performance and fatty acid composition of egg yolks of two strains of laying hens. *Poult. Sci.* 80(8): 1171–9.

Grobas, S., Méndez, J., Lopez, B. C., De, B. C. and Mateos, G. G. 2002. Effect of vitamin E and A supplementation on egg yolk alpha-tocopherol concentration. *Poult. Sci.* 81(3): 376–81.

Hammershøj, M. 1995. Effects of dietary fish oil with natural content of carotenoids on fatty acid composition, n-3 fatty acid content, yolk colour and egg quality of hen eggs. *Archiv Für Geflügelkunde* 59(3): 189–97.

Hammershøj, M., Kidmose, U. and Steenfeldt, S. 2010. Deposition of carotenoids in egg yolk by short-term supplement of coloured carrot (Daucus carota) varieties as forage material for egg-laying hens. *J. Sci. Food Agric.* 90(7): 1163–71.

Handelman, G. J., Nightingale, Z. D., Lichtenstein, A. H., Schaefer, E. J. and Blumberg, J. B. 1999. Lutein and zeaxanthin concentrations in plasma after dietary supplementation with egg yolk. *Am. J. Clin. Nutr.* 70(2): 247–51.

Hargis, P. S. and Van Elswyk, M. E. 1993. Manipulating the fatty acid composition of poultry meat and eggs for the health conscious consumer. *Worlds Poult. Sci. J.* 49(03): 251–64.

Harris, W. S., Miller, M., Tighe, A. P., Davidson, M. H. and Schaefer, E. J. 2008. n-3 fatty acids and coronary heart disease risk: Clinical and mechanistic perspectives. *Atherosclerosis* 197(1): 12–24.

Harris, W. S., Mozaffarian, D., Rimm, E., Kris-Etherton, P., Rudel, L. L., Appel, L. J., Engler, M. M., Engler, M.B. and Sacks, F. 2009. n-6 fatty acids and risk for cardiovascular disease: A science advisory from the american heart association nutrition subcommittee of the council on nutrition, physical activity, and metabolism; council on cardiovascular nursing; and council on epidemiology and prevention. *Circulation* 119(6): 902–7.

Hayat, Z., Cherian, G., Pasha, T., Khattak, F. and Jabbar, M. 2009. Effect of feeding flax and two types of antioxidants on egg production, egg quality, and lipid composition of eggs. *J. Appl. Poult. Res.* 18(3): 541–51.

Hayat, Z., Cherian, G., Pasha, T., Khattak, F. and Jabbar, M. 2010a. Sensory evaluation and consumer acceptance of eggs from hens fed flax seed and 2 different antioxidants. *Poult. Sci.* 89(10): 2293–8.

Hayat, Z., Cherian, G., Pasha, T. N., Khattak, F. M. and Jabbar, M. A. 2010b. Oxidative stability and lipid components of eggs from flax-fed hens: Effect of dietary antioxidants and storage. *Poult. Sci.* 89(6): 1285–92.

Hebert, K., House, J. D. and Guenter, W. 2005. Effect of dietary folic acid supplementation on egg folate content and the performance and folate status of two strains of laying hens. *Poult. Sci.* 84(10): 1533–8.

Herber, S. and Van Elswyk, M. 1996. Dietary marine algae promotes efficient deposition of n-3 fatty acids for the production of enriched shell eggs. *Poult. Sci.* 75(12): 1501–7.

Herber-McNeill, S. and Van Elswyk, M. 1998. Dietary marine algae maintains egg consumer acceptability while enhancing yolk colour. *Poult. Sci.* 77(3): 493–6.

Holdas, A. and May, K. 1966. Fish oil and fishy flavour of eggs and carcasses of hens. *Poult. Sci.* 45(6): 1405–7.

House, J. D., Braun, K., Balance, D. M., O'Connor, C. P. and Guenter, W. 2002. The enrichment of eggs with folic acid through supplementation of the laying hen diet. *Poult. Sci.* 81: 1332–7.

House, J. D., O'Connor, C. P. and Guenter, W. 2003. Plasma homocysteine and glycine are sensitive indices of folate status in a rodent model of folate depletion and repletion. *J. Agric. Food Chem.* 51(15): 4461–7.

Institute of Medicine, F.a.N.B.N.A.P. 1998. Dietary Reference Intakes for Thiamin, Riboflavin, Niacin, Vitamin B6, Folate, Vitamin B12, Pantothenic Acid, Biotin, and Choline. National Academy Press, Washington, DC.

Jan, S., Guillou, H., D'Andréa, S., Daval, S., Bouriel, M., Rioux, V. and Legrand, P. 2004. Myristic acid increases D6-desaturase activity in cultured rat hepatocytes. *Reprod. Nutr. Dev.* 44: 131–40.

Jiakui, L. and Xiaolong, W. 2004. Effect of dietary organic versus inorganic selenium in laying hens on the productivity, selenium distribution in egg and selenium content in blood, liver and kidney. *J. Trace Elem. Med. Biol.* 18(1): 65–8.

Jiang, Y. H., McGeachin, R. B. and Bailey, C. A. 1994. Alpha-tocopherol, beta-carotene, and retinol enrichment of chicken eggs. *Poult. Sci.* 73: 1137.

Jiang, Z., Ahn, D., Ladner, L. and Sim, J. 1992. Influence of feeding full-fat flax and sunflower seeds on internal and sensory qualities of eggs. *Poult Sci* 71(2): 378–82.

Jing, C. L., Dong, X. F., Wang, Z. M., Liu, S. and Tong, J. M. 2015. Comparative study of DL-selenomethionine vs sodium selenite and seleno-yeast on antioxidant activity and selenium status in laying hens. *Poult. Sci.* 94(5): 965–75.

Jlali, M., Briens, M., Rouffineau, F., Mercerand, F., Geraert, P. A. and Mercier, Y. 2013. Effect of 2-hydroxy-4-methylselenobutanoic acid as a dietary selenium supplement to improve the selenium concentration of table eggs. *J. Anim. Sci.* 91(4): 1745–52.

Kakani, R., Fowler, J., Haq A-, Murphy, E. J., Rosenberger, T. A., Berhow, M. and Bailey, C. A. 2012. Camelina meal increases egg n-3 fatty acid content without altering quality or production in laying hens. *Lipids* 47(5): 519–26.

Karunajeewa, H., Hughes, R. J., McDonald, M. W. and Shenstone, F. S. 1984. A review of factors influencing pigmentation of egg yolks. *World's Poult. Sci. J.* 40: 52–65.

Kaufmann, S., Wolfram, G., Delange, F. and Rambeck, W. A. 1998. Iodine supplementation of laying hen feed: a supplementary measure to eliminate iodine deficiency in humans? *Z Ernahrungswiss* 37: 288–93.

Khang, N. T. K. 2006. Genetic factors affecting the n-3 and n-6 fatty acid variation in egg yolk. Dissertation. 1–108.

Kim, S. J., Park, G. B., Kang, C. B., Park, S. D., Jung, M. Y., Kim, J. O. and Ha, Y. L. 2000. Improvement of oxidative stability of conjugated linoleic acid (CLA) by microencapsulation in cyclodextrins. *J. Agric. Food Chem.* 48(9): 3922–9.

Koehler, H. H. and Bearse, G. E. 1975. Egg flavour quality as affected by fish meals or fish oils in laying rations. *Poult. Sci.* 54(3): 881–9.

Kovacs-Nolan, J., Phillips, M. and Mine, Y. 2005. Advances in the value of eggs and egg components for human health. *J. Agric. Food Chem.* 53(22): 8421–31.

Kühn, J., Schutkowski, A., Kluge, H., Hirche, F. and Stangl, G. I. 2014. Free-range farming: a natural alternative to produce vitamin D-enriched eggs. *Nutrition* 30(4): 481–4.

Laca, A., Paredes, B. and Díaz, M. 2009. Quality characteristics of n-3 polyunsaturated fatty acid-enriched eggs. *J. Anim Feed Sci.* 18(1): 101–12.

Lawlor, J. B., Gaudette, N., Dickson, T. and House, J. D. 2010. Fatty acid profile and sensory characteristics of table eggs from laying hens fed diets containing microencapsulated fish oil. *Anim. Feed Sci. Technol.* 156(3–4): 97–103.

Leeson, S., Caston, L. and MacLaurin, T. 1998. Organoleptic evaluation of eggs produced by laying hens fed diets containing graded levels of flaxseed and vitamin E. *Poult. Sci.* 77(9): 1436–40.

Leeson, S. and Caston, L. J. 2003. Vitamin enrichment of eggs. *J. Appl. Poult. Res.* 12(1): 24–6.

Leskanich, C. and Noble, R. 1997. Manipulation of the n-3 polyunsaturated fatty acid composition of avian eggs and meat. *Worlds Poult. Sci. J.* 53(02): 155–83.

Leveille, G. A., Romsos, D. R., Yeh, Y. and O'Hea, E. K. 1975. Lipid biosynthesis in the chick. A consideration of site of synthesis, influence of diet and possible regulatory mechanisms. *Poult. Sci.* 54(4): 1075–93.

Ma, W., Gu, Y., Lu, J., Yuan, L. and Zhao, R. 2014. Effects of chromium propionate on egg production, egg quality, plasma biochemical parameters, and eggchromium deposition in late-phase laying hens. *Biol. Trace Elem. Res.* 157(2): 113–9.

Mabe, I., Rapp, C., Bain, M. M. and Nys, Y. 2003. Supplementation of a corn-soybean meal diet with manganese, copper, and zinc from organic or inorganic sources improves eggshell quality in aged laying hens. *Poult. Sci.* 82: 1903–13.

Makrides, M., Hawkes, J. S., Neumann, M. A. and Gibson, R. A. 2002. Nutritional effect of including egg yolk in the weaning diet of breast-fed and formula-fed infants : a randomized controlled trial. *Am. J. Clin. Nutr.* 75(6): 1084–92.

Marshall, A., Sams, A. and Van Elswyk, M. 1994. Oxidative stability and sensory quality of stored eggs from hens fed 1.5% menhaden oil. *J. Food Sci.* 59(3): 561–3.

Martens, M. 1999. A philosophy for sensory science. *Food Qual. Pref.* 10(4): 233–44.

Matiella, J. E. and Hsieh, T. 1991. Volatile compounds in scrambled eggs. *J. Food Sci.* 56(2): 387–90.

Mattila, P., Lehikoinen, K., Kiiskinen, T. and Piironen, V. 1999. Cholecalciferol and 25-hydroxycholecalciferol content of chicken egg yolk as affected by the cholecalciferol content of feed. *J. Agric. Food Chem.* 47: 4089–92.

Mattila, P., Rokka, T., Könkö, K., Valaja, J., Rossow, L. and Ryhänen, E. L. 2003. Effect of cholecalciferol-enriched hen feed on egg quality. *Agric. Food Chem.* 51(1): 283–7.

Mattila, P., Valaja, J., Rossow, L., Venäläinen, E. and Tupasela, T. 2004. Effect of vitamin D2- and D3-enriched diets on egg vitamin D content, production, and bird condition during an entire production period. *Poult. Sci.* 83(3): 433–40.

McDowell, L. R. 2000. Vitamins in animal and human nutrition. 2nd ed. Iowa State University Press, Ames, IA.

Meluzzi, A., Simoncini, F., Sirri, F., Vandi, L. and Giordani, G. 1996. Feeding hens diets supplemented with heavy metals (chromium, nickel and lead). *Arch. Geflügelk.* 60: 119–25.

Meluzzi, A., Sirri, F., Manfreda, G., Tallarico, N. and Franchini, A. 2000. Effects of dietary vitamin E on the quality of table eggs enriched with n-3 long-chain fatty acids. *Poult. Sci.* 79: 539–45.

Mennicken, L., Ponsuksili, S., Tholen, E., Thi Kim Khang, N., Steier, K., Petersen, J., Schellander, K. and Wimmers, K. 2005. Divergent selection for n3: Omega6 polyunsaturated fatty acid ratio in quail eggs. *Archiv Fur Tierzucht* 48(5): 527.

Meyer, B. 2011. Are we consuming enough long chain n-3 polyunsaturated fatty acids for optimal health? *PLEFA* 85(5): 275–80.

Michalak, I., Chojnacka, K., Dobrzański, Z., Górecki, H., Zielińska, A., Korczyński, M. and Opaliński, S. 2011. Effect of macroalgae enriched with microelements on egg quality parameters and mineral content of eggs, eggshell, blood, feathers and droppings. *J. Anim. Physiol. Anim. Nutr. (Berl).* 95(3): 374–87.

Moeini, M. M., Ghazi, S. H., Sadeghi, S. and Malekizadeh, M. 2013. The effect of red pepper (Capsicum annuum) and marigold flower (Tageteserectus) powder on egg production, egg yolk color and some blood metabolites of laying hens. *Iranian Journal of Applied Animal Science* 3(2): 301–5.
Moeller, S. M., Jacques, P. F. and Blumberg, J. B. 2000. The potential role of dietary xanthophylls in cataract and age-related macular degeneration. *J. Am. Coll. Nutr.* 19(Suppl. 5): 522S–7S.
Mori, A. V., Mendonca, Jr. C. X., Almeida, C. R. M. and Pita, M. C. G. 2003. Supplementing hen diets with vitamins A and E affects egg yolk retinol and αCtocopherol levels. *J. Appl. Poult. Res.* 12: 106–14.
Mridula, D., Kaur, D., Nagra, S., Barnwal, P., Gurumayum, S. and Singh, K. 2012. Effect of dietary flaxseed supplementation on egg production and quality in laying hens. *Indian. J. Poult. Sci.* 47(1): 40–7.
Naderi, N., House, J. D. and Pouliot, Y. 2014. Scaling-up a process for the preparation of folate-enriched protein extracts from hen egg yolks. *J. Food Engineer.* 141: 85–92.
Nakamura, M. T. and Nara, T. Y. 2004. Structure, function, and dietary regulation of Δ6, Δ5, and Δ9 desaturases. *Nutr* 24.
National Research Council. 2013. Evaluating the effectiveness of fish stock rebuilding plans in the United States. Washington, DC: National Academies Press.
Neijat, M., Suh, M., Neufeld, J. and House, J. 2016. Hempseed products fed to hens effectively increased n-3 polyunsaturated fatty acids in total lipids, triacylglycerol and phospholipid of egg yolk. *Lipids*: 1–14 (In press).
Nielsen, H. 1998. Hen age and fatty acid composition of egg yolk lipid. *Br. Poult. Sci.* 39(1): 53–6.
Nys, Y. 2000. Dietary carotenoids and egg yolk colouration-a review. *Archiv Für Geflügelkunde* 64(2): 45–54.
Olgun, O., Yazgan, O. and Cufadar, Y. 2013. Effect of supplementation of different boron and copper levels to layer diets on performance, egg yolk and plasma cholesterol. *J. Trace Elem. Med. Biol.* 27(2): 132–6.
Oomah, B. D., Busson, M., Godfrey, D. V. and Drover, J. C. 2002. Characteristics of hemp (Cannabis sativa L.) seed oil. *Food Chem.* 76(1): 33–43.
Opalinski, S., Dolinska, B., Korczynski, M., Chojnacka, K., Dobrzanski, Z. and Ryszka, F. 2012. Effect of iodine-enriched yeast supplementation of diet on performance of laying hens, egg traits, and egg iodine content. *Poult. Sci.* 91(7): 1627–32.
Osborn, H. T. and Akoh, C. C. 2002. Structured lipids-novel fats with medical, nutraceutical, and food applications. *Compr. Rev. Food Sci. Food Saf.* 1(3): 110–20.
Palmer, L. S. and Kempster, H. L. 1919. The influence of specific feeds and certain pigments on the colour of the egg yolk and body fat of fowls. *J. Biol. Chem.* 39: 331–7.
Pan, C., Huang, K., Zhao, Y., Qin, S., Chen, F. and Hu, Q. 2007. Effect of selenium source and level in hen's diet on tissue selenium deposition and egg selenium concentrations. *J. Agric. Food Chem.* 55(3): 1027–32.
Pan, C., Zhao, Y., Liao, S. F., Chen, F., Qin, S., Wu, X., Zhou, H. and Huang, K. 2011. Effect of selenium-enriched probiotics on laying performance, egg quality, egg selenium content, and eggglutathione peroxidase activity. *J. Agric. Food Chem.* 59(21): 11424–31.
Park, S. W., Namkung, H., Ahn, H. J. and Paik, I. K. 2004. Production of iron enriched eggs of laying hens. Asian-Australas. *J. Anim. Sci.* 17: 1725–8.
Parpinello, G. P., Meluzzi, A., Sirri, F., Tallarico, N. and Versari, A. 2006. Sensory evaluation of egg products and eggs laid from hens fed diets with different fatty acid composition and supplemented with antioxidants. *Food Res. Int.* 39(1): 47–52.
Pavlović, Z., Miletić, I., Jokić, Z. and Sobajić, S. 2009. The effect of dietary selenium source and level on hen production and egg selenium concentration. *Biol. Trace Elem. Res.* 131(3): 263–70.
Payne, R. L., Lavergne, T. K. and Southern, L. L. 2005. Effect of inorganic versus organic selenium on hen production and egg selenium concentration. *Poult. Sci.* 84(2): 232–7.
Qi, G. and Sim, J. 1998. Natural tocopherol enrichment and its effect in n-3 fatty acid modified chicken eggs. *J. Agric. Food Chem.* 46(5): 1920–6.

Revell, D. K., Zarrinkalam, M. R. and Hughes, R. J. 2009. Iron content of eggs from hens given diets containing organic forms of iron, serine and methyl group donors, or phytoestogens. *Br. Poult. Sci.* 50: 536–42.
Rhodes, D. N. and Lea, C. H. 1957. On the composition of hen's phospholipids. *Biochem. J.* 65: 526.
Rizzi, L., Bochicchio, D., Bargellini, A., Parazza, P. and Simioli, M. 2009. Effects of dietary microalgae, other lipid sources, inorganic selenium and iodine on yolk n-3 fatty acid composition, selenium content and quality of eggs in laying hens. *J. Sci. Food Agric.* 89(10): 1775–81.
Rokka, T., Alén, K., Valaja, J. and Ryhänen, E-. 2002. The effect of a camelina sativa enriched diet on the composition and sensory quality of hen eggs. *Food Res. Int.* 35(2–3): 253–6.
Rooney, L. 1978. Sorghum and pearl millet lipids. *Cereal Chem.* 55(5): 584–90.
Rose, S. 1997. *Principles of Poultry Science.* Cab International, Oxford University Press.
Rottger, A. S., Halle, I., Danicke, S., Wagner, H., Reeves, G. and Flachowsky, G. 2008. Long term effect of varying nutrient iodine on the performance of laying hen and the carry over into egg. *Proceedings of the 13th World Poultry Congress*, 29 June 2008, Brisbane, Queensland, Australia.
Sajilata, M., Singhal, R. and Kamat, M. 2008. The carotenoid pigment zeaxanthin—a review. *Compr. Rev. Food Sci. and Food Safety* 7(1): 29–49.
Schaeffer, L., Gohlke, H., Muller, M., Heid, I. M., Palmer, L. J., Kompauer, I., Demmelmair, H., Illig, T., Koletzko, B. and Heinrich, J. 2006. Common genetic variants of the FADS1 FADS2 gene cluster and their reconstructed haplotypes are associated with the fatty acid composition in phospholipids. *Hum Mol. Genet.* 15(11): 1745–56.
Scheideler, S. and Froning, G. 1996. The combined influence of dietary flaxseed variety, level, form, and storage conditions on egg production and composition among vitamin E-supplemented hens. *Poult. Sci.* 75(10): 1221–6.
Scheideler, S., Froning, G. and Cuppett, S. 1997. Studies of consumer acceptance of high n-3 fatty acid-enriched eggs. *J. Appl. Poult. Res.* 6(2): 137–46.
Schreiner, M., Hulan, H., Razzazi-Fazeli, E., Böhm, J. and Iben, C. 2004. Feeding laying hens seal blubber oil: Effects on egg yolk incorporation, stereospecific distribution of n-3 fatty acids, and sensory aspects. *Poult. Sci.* 83(3): 462–73.
Schutkowski, A., Krämer, J., Kluge, H., Hirche, F., Krombholz, A., Theumer, T. and Stangl, G. I. 2013. UVB exposure of farm animals: study on a food-based strategy to bridge the gap between current vitamin Dintakes and dietary targets. *PLoS ONE* 8(7).
Šefer, D., Andonov, A., Šobajić, S., Marković, R., Radulović, S., Jakić-Dimić, D. and Petrujkić, B. 2011. Effects of feeding laying hens diets supplemented with n 3 fatty acids on the egg fatty acid profile. *Biotech. in An Husb* 27(3): 679–86.
Shahriar, H. A., Adl, K. N., Nezhad, Y. E., Nobar, R. S. D. and Ahmadzadeh, A. 2008. Effect of dietary fat type and different levels of vitamin E on broiler breeder performance and vitamin E levels of egg. *Asian J. Anim. Vet. Adv.* 3: 147–54.
Shappell, S. B., Zhang, J., Page, R., Olson, S. J., Keeney, D. S. and Brash, A. R. 2001. 15-Lipoxygenase-2 expression in benign and neoplastic sebaceous glands and other cutaneous adnexa. *Journal of Investigative Dermatology* 117(1): 36–43.
Simopoulos, A. P. 1991. n-3 fatty acids in health and disease and in growth and development. *The American Journal of Clinical Nutrition* 54: 438–63.
Simopoulos, A. P. and De Meester, F. (Eds). 2009. A balanced n-6/n-3 fatty acid ratio, cholesterol and coronary heart disease. *World Rev Nutr Diet.* Basel: Karger.
Singh, V., Pathak, V. and Akhilesh, K. V. 2012. Modified or enriched eggs: A smart approach in egg industry: A review. *Am. J. Food Tech.* 7(5): 266–77.
Skrivan, M., Marounek, M., Dlouhá, G. and Sevcíková, S. 2008. Dietary selenium increases vitamin E contents of egg yolk and chicken meat. *Br. Poult. Sci.* 49(4): 482–6.
Skrivan, M., Skrivanova, V. and Marounek, M. 2005. Effects of dietary zinc, iron, and copper in layer feed on distribution of these elements in eggs, liver, excreta, soil, and herbage. *Poult. Sci.* 84: 1570–5.

Słupczyńska, M., Jamroz, D., Orda, J. and Wiliczkiewicz, A. 2014. Effect of various sources and levels of iodine, as well as the kind of diet, on the performance of young laying hens, iodine accumulation in eggs, egg characteristics, and morphotic and biochemical indices in blood. *Poult Sci.* 93(10): 2536–47.

Sprecher, H. 1981. Biochemistry of essential fatty acids. *Prog. Lipid Res.* 20: 13–22.

Sprecher, H. 2000. Metabolism of highly unsaturated <i> n-3 and <i> n-6 fatty acids. *Biochimica Et Biophysica Acta (BBA)-Molecular and Cell Biology of Lipids* 1486(2): 219–31.

Sprecher, H., Luthria, D. L., Mohammed, B. S. and Baykousheva, S. P. 1995. Reevaluation of the pathways for the biosynthesis of polyunsaturated fatty acids. *J. Lipid Res.* 36(12): 2471–7.

Sugiyama, A., Awaji, H., Horie, K., Kim, M. and Nakata, R. 2012. The Beneficial Effect of Folate-Enriched Egg on the Folate and Homocysteine Levels in Rats Fed a Folate-and Choline-Deficient Diet. *J. Food Sci.* 77(12): H268–72.

Sujatha, T. and Narahari, D. 2011. Effect of designer diets on egg yolk composition of 'White Leghorn' hens. *J. Food Sci. Technol.* 48(4): 494–7.

Surai, P. F. and Sparks, N. H. C. 2001. Designer eggs: from improvement of egg composition to functional food. *Trends Food Sci. and Tech.* 12: 7–16.

Taber, L., Chiu, C. H. and Whelan, J. 1998. Assessment of the arachidonic acid content in foods commonly consumed in the American diet. *Lipids* 33(12): 1151–7.

Tactacan, G. B., Jing, M., Thiessen, S., O'Connor, D. L., Rodriguez-Lecompte, J. C., Guenter, W. and House, J. D. 2010. Characterization of folate-dependent enzymes and indices of folate status in laying hens supplemented with dietary folic acid or 5-methyltetrahydrofolate. *Poult. Sci.* 89: 688–96.

Tactacan, G. B., Rodriguez-Lecompte, J. C., O, K. and House, J. D. 2011. Functional characterization of folic acid transport in the intestine of the laying hen. *Poult. Sci.* 90: 83–90.

Tactacan, G. B., Rodriguez-Lecompte, J. C., O, K. and House, J. D. 2012. The regulation of folate absorption in the intestine of the laying hen with increased dietary folic acid supplementation. *Poult. Sci.* 91: 121–8.

Tallarico, N., Sirri, F., Meluzzi, A., Pittia, P., Parpinello, G. P. and Franchini, A. 2002. Effect of dietary vegetable lipids on functional and sensory properties of chicken eggs. *Italian J. Food Sci.* 14(2): 159–66.

Traber, M. G. 2006. Vitamin E. In M. E. Shils, M. Shike, A. C. Ross, B. Caballero and R. Cousins (Eds), *Modern Nutrition in Health and Disease,* 396–411. 10th ed. Baltimore, MD: Lippincott Williams & Wilkins.

Tu, W. C., Cook-Johnson, R. J., James, M. J., Mühlhäusler, B. S. and Gibson, R. A. 2010. n-3 long chain fatty acid synthesis is regulated more by substrate levels than gene expression. *Prostaglandins Leukot. Essent. Fatty Acids.* 83(2): 61–8.

USDA Nutrient Database [Internet]. Available from http://ndb.nal.usda.gov/ndb/ (Accessed 4 July 2014).

Utterback, P. L., Parsons, C. M., Yoon, I. and Butler, J. 2005. Effect of supplementing selenium yeast in diets of laying hens on egg selenium content. *Poult. Sci.* 84(12): 1900–1.

Van Elswyk, M. E. 1997. Comparison of n–3 fatty acid sources in laying hen rations for improvement of whole egg nutritional quality: A review. *Br. J. Nutr.* 78(01): S61–9.

Van Elswyk, M. E., Dawson, P. and Sams, A. 1995. Dietary menhaden oil influences sensory characteristics and headspace volatiles of shell eggs. *J. Food Sci.* 60(1): 85–9.

Van Elswyk, M. E., Sams, A. and Hargis, P. 1992. Composition, functionality, and sensory evaluation of eggs from hens fed dietary menhaden oil. *J. Food Sci.* 57(2): 342–4.

Vondell, J. H. and Putnam, J. N. 1945. The effect of certain feedstuffs on the flavour, albumen condition and yolk colour of eggs. *Poult. Sci.* 24(3): 284–6.

Willumsen, N., Vaagenes, H., Lie, Ø., Rustan, A. C. and Berge, R. K. 1996. Eicosapentaenoic acid, but not docosahexaenoic acid, increases mitochondrial fatty acid oxidation and upregulates 2, 4-dienoyl-CoA reductase gene expression in rats. *Lipids* 31(6): 579–92.

Woods, V. B. and Fearon, A. M. 2009. Dietary sources of unsaturated fatty acids for animals and their transfer into meat, milk and eggs: A review. *Livestock Sci.* 126(1–3): 1–20.
Yalçin, S., Kahraman, Z., Yalçin, S., Yalçin, S. S. and Dedeoğlu, H. E. 2004. Effects of supplementary iodine on the performance and egg traits of laying hens. *Br. Poult. Sci.* 45(4): 499–503.
Yalçyn, H., Kemal Ünal, M. and Basmacyoolu, H. 2007. The fatty acid and cholesterol composition of enriched egg yolk lipids obtained by modifying hens' diets with fish oil and flaxseed. *Grasas Aceites* 58(4): 372–8.
Yannakopoulos, A. 2007. Egg enrichment in n-3 fatty acids. In *Bioactive Egg Compounds*, Springer, 159.
Yao, L., Wang, T., Persia, M., Horst, R. L. and Higgins, M. 2013. Effects of vitamin D(3)-enriched diet on egg yolk vitamin D(3) content and yolk quality. *J. Food Sci.* 78(2): C178–83.
Zile, M. H. and Cullum, M. E. 1983. The function of vitamin A: Current concepts. *Proc. Soc. Exp. Biol. Med.* 172(2): 139–52.
Zubr, J. 1997. Oil-seed crop: camelina sativa. *Industrial Crops and Products* 6(2): 113–9.

Chapter 17

Molecular breeding techniques to improve egg quality

Anna Wolc, Iowa State University, and Hy-Line International, USA; and Janet E. Fulton, Hy-line International, USA

1 Introduction

Since the release of the chicken genome sequence in 2004 (Hillier et al., 2004), the tool box available for genetic improvement of the multiple traits that define egg quality has improved tremendously. That initial sequence has been revised and improved several times with Gallus_gallus-4.0 now available in multiple publicly accessible websites. An improved Gallus_gallus-5.0 version is soon to be released. Genome sequence comparisons between mammalian and avian species have allowed for the identification and subsequent annotation of genes, and initiated the understanding of the functions of the genes and the interdependent relationships between the thousands of genes that comprise multicellular organisms. The recent sequencing of multiple breeds of chickens, both research and commercially utilized lines, revealed the presence of millions of genetic variants and resulted in the development of a commercially available chip that can simultaneously query 600 000 single DNA base sequence variants (Kranis et al., 2013). Numerous methodologies are now available that allow for detection of gene expression in specific tissues or relative gene expression between different treatments or individuals (Rapaport et al., 2013). To aid in the analysis of the large amounts of genetic information that is generated, various computer programmes have been developed that allow for the integration and eventual application of this information into a breeding programme (Wolc et al., 2016).

http://dx.doi.org/10.19103/AS.2016.0012.19

Molecular breeding methodologies for improvement of egg quality traits can be broadly divided into two different, but interconnected approaches. The whole-genome approach involves an extensive examination of the entire genome, encompassing genome-wide association studies (GWAS) and genomic selection. Targeted (single gene) genetics encompasses a more focused view of the variation at individual genes and the specific effects of these variants on the traits of interest. These two methods feed into each other; the whole-genome approach identifies genomic regions that influence the traits of interest, which then allows focus to be placed on the smaller subset of genes within those regions. These two approaches can be subsequently combined through gene expression and pathway analyses to develop an integrated understanding of the genetic control of the multiple traits that influence egg quality in commercial production.

There are two aspects of egg-quality traits, external and internal quality. External egg quality involves the size and shape of the egg, as well as the structure, thickness and strength of the shell (Bain, 2005) but also shell colour (both intensity and its surface distribution), as the world market for eggs has been clearly differentiated into white and brown eggs. External quality is critical for maintaining egg integrity during transportation and for maintaining food safety by providing protection from bacterial infestation. Internal egg quality involves quality of the albumen, measured in the Haugh units (HU) (Haugh, 1937), the relative size and composition of the various internal components and the integrity of the vitelline membrane. Internal quality has an impact on the egg's suitability for processing, as well as on consumer appeal. For both internal and external quality, the absence of defects is also important (e.g. fishy taint and blood and meat inclusions). The quality aspects that are economically important have been under genetic improvement for decades, but more recently the trait measurements have become more accurate and increasingly sophisticated. In addition, with the development of molecular tools, marker-assisted selection or genomic selection can be used to more effectively improve egg quality. Molecular technology tools such as gene editing have also been tested to produce chickens that lay eggs with unique properties such as ovalbumin knockouts (Park et al., 2014).

In this chapter, we will review how egg quality can be improved through selection and how this process can be enhanced using molecular information. Molecular methods will be separated into two major approaches: whole-genome methods including GWAS and genomic selection, and detailed studies of specific candidate genes as identified from either genome-wide studies or gene function and pathways studies, though information from these two methods is very much integrated.

2 Whole-genome methods to improve egg quality

Selection for egg quality was traditionally based on combining pedigree information and multiple individual performance records collected at various ages. These methods assumed that egg quality is influenced by many genes, each with a very small effect (infinitesimal model) and environmental factors. With the recent availability of molecular information, much insight has been gained into the genetic determination of these traits.

Numerous GWAS have identified regions of the genome associated with variation in the different egg quality traits. These regions are called quantitative trait loci (QTL), as it is presumed that they contain genes that influence the quantitative traits. Several review articles have summarized QTL studies in the chicken that include multiple egg quality traits (Abasht et al., 2006; Vilkki, 2007; Sazanov et al., 2010). However, the field is rapidly changing and new studies are reported every year. Published animal QTL studies are summarized and available in a publicly accessible database (www.animalgenome.org). The information presented in Table 1 is current as of 8 December 2015. This database includes 41 studies published between 2002 and 2015 and summarizes 458 genomic locations associated with egg quality. The greatest number of QTL was found on chromosomes 2, 4 and Z. Chromosome 4 is the fourth largest in the chicken genome, containing 9% of the genetic material, yet it contains 14% of the reported QTL. This chromosome is particularly interesting because of reports from multiple populations and breeds that it contains QTL influencing both egg weight and the relative size of various egg components.

Earlier QTL studies utilized microsatellite markers with sparse coverage, generally between 200 and 400 markers. This made it difficult to narrow the region containing the QTL, resulting in large regions containing putative QTL. With the recent development of the publicly available 600 000 single nucleotide polymorphisms (SNPs) chip (Kranis et al., 2013), genomic locations influencing traits of interest can be located more precisely. More accurate identification of narrower genome regions allows for more focused candidate gene investigations. The usefulness of QTL information for breeding programmes depends on the size of the QTL effect, its frequency within specific population and the distance from the identified marker to the causative mutation (Dekkers and van der Werf, 2007; Lahav et al., 2006). For traits determined by many genes with small effects, individual markers provide limited support to selection decisions. In contrast, genomic selection methods that consider the whole genome simultaneously are more applicable. One of the aspects that must be considered is that the same marker may have a different effect in different genetic backgrounds (different breeds or lines) and under different environmental conditions (Romé et al., 2015). Thus, validation within the target population is necessary before implementing marker-assisted selection.

2.1 Egg production

Saleable egg production is the single most important trait in selection of layers and also a basis for any egg-quality measurements. It is a function of oviposition rate and egg quality. Despite its low-to-moderate heritability (Nurgiartiningsih et al., 2005; Wolc et al., 2012), egg production has shown remarkable response to selection over the past two decades. Analysis of the trend in egg production in commercial crosses shows an improvement of 4.9 and 3.6 eggs per hen per generation for white shell and brown shell layers, respectively (Anderson and Havenstein, 2012). Egg production shows unfavourable genetic correlations with some quality traits, such as egg weight and shell strength (Wolc et al., 2012).

GWAS for egg production (Atzmon et al., 2008; Wardecka et al., 2002; Goraga, et al., 2012; Goto et al., 2011; Hansen et al., 2005; Liu et al., 2011; Sasaki et al., 2004; Schreiweis et al., 2006; Tuiskula-Haavisto et al., 2002, 2004; Wolc et al., 2014; Yuan et al., 2015; Zhang et al., 2015), defined either as egg number or production rate, have identified regions associated with this trait on almost every chromosome of the chicken (Table 1), with each region explaining a relatively small proportion of genetic variance. This supports the hypothesis of the complex, polygenic nature of this trait. Yuan et al. (2015) identified over 300 genes within 1 Mb from SNPs significantly associated with egg production and

Table 1 Summary of QTL associated with egg quality included in www.animalgenome.org database

	chr1	chr2	chr3	chr4	chr5	chr6	chr7	chr8	chr9	chr10	chr11	chr12	chr13	chr14
Albumen height	8	5	2	2	1	1	5		10	2			1	
Albumen %		1					1	4						
Albumen weight		1	1	6			1	3						
Haugh units	2	11		2			4							1
Total albumen quality	10	18	3	10	1	1	11	7	10	2			1	1
Egg shell colour	1	7	1	6		2		1	17			5		
Egg shell lightness					2	1					1			
Egg shell percentage		2		1	1		1							3
Egg shell redness	2			1	3	1					1			
Egg shell yellowness	2	1		1	3	1			1		2			
Total shell colour	5	10	1	9	9	5	1	1	18		4	5		3
Shell stiffness	2	4	1	2		2			1	1				2
Shell strength	1	4	2	3	2	2	1	1		1	1	2		1
Shell thickness	4	6		2	2		1	6	2		1			3
Shell weight	1	2	2	2	1	1			1		3	1		
Specific gravity	1	1	1	1										
Total shell strength	9	17	6	10	5	5	2	7	4	2	5	3		6
Yolk colour	2	2	2	3				3	1					
Yolk height				1	1		1	3						
Yolk index		1						7						
Yolk weight	5	1		5	3	1		3	3		2			
Total yolk quality	7	4	2	9	4	1	1	16	4		2			
Egg shell shape		1		2				2						
Long length of egg				1			1		1			1		
Short length of egg				1					1	1		1		
Total egg shape		1		4			1	2	2	1		2		
Egg weight	7	9	5	22	5	2	3	5	8	3			1	2
Blood and meat spots	1	1		1					2					
Egg aftertaste								1						
Total egg quality	39	60	17	65	24	14	19	39	48	8	11	10	2	12
Egg production	13	15	17	7	9		3	1	7	1	3	1	6	2
Chr size (Mb)*	195	149	110	90.2	59.6	35	36.2	28.8	23.4	19.9	19.4	19.9	17.8	15.2

*Chromosome size in Mb according to http://www.ncbi.nlm.nih.gov/genome/111.

chr15	chr16	chr17	chr18	chr19	chr20	chr21	chr22	chr23	chr24	chr25	chr26	chr27	chr28	chrZ	Total
1								1						3	42
												1			7
															12
					1						1		1	2	25
1					1			1			1	1	1	5	86
					2	1					1		1		45
															4
						3									11
															8
															11
					2	4					1		1		79
														4	19
				1							1			6	29
		2				1			1			1		5	37
		1							1					4	20
															4
		3		1		1			2		1	1		19	109
												1			14
1														2	9
		1													9
1							1				1			2	28
2		1					1				1	1		4	60
															5
														1	5
														2	6
															16
1		2			1	1						1		2	80
														22	27
															1
4	0	6	0	1	4	6	1	1	2	0	4	4	2	52	458
		4		1	1		2	1	2			1		3	100
12.7	0.54	10.5	11.2	9.98	14.3	6.8	4.08	5.72	6.32	2.19	5.33	5.21	4.74	82.4	

based on their known function suggested general transcription factor IIA (*GTF2A1*), claspin (*CLSPN*), phenylalanyl-tRNA synthetase (*FARSB*), *KIAA1549* and calmodulin 1 (*CALM1*) to be potential candidate genes. A GWAS by Zhang et al. (2015) using Jinghai Yellow Chicken identified fibulin 5 (*FBLN5*) gene in proximity of SNPs associated with egg number.

3 Improving external egg quality

3.1 Egg size

Egg weight is a trait for which neither extreme is desired. Small eggs have a low market value and neither small nor very large eggs are suitable for automatic packing equipment. Also, extreme egg sizes result in poor hatchability in both pure lines and crossbred hens during the multiplication process (grandparent and parent stocks) to produce commercial layer hybrids. Furthermore, very large eggs can have decreased shell quality and may cause welfare problems by inducing oviduct prolapse in the hens. Negative correlations between large egg size and other traits have been reported, including total egg numbers and chick hatchability (Nurgiartiningsih et al., 2005; Abiola et al., 2008). Egg weight has moderate to high heritability (Nurgiartiningsih et al., 2005, Wolc et al., 2011b), which allows for high accuracies of estimated breeding values, even with traditional pedigree-based analysis. Further complications with breeding for optimal egg weight are that egg weight increases with age of birds (Shalev and Pasternak, 1993) and measurements at different ages are highly correlated. This makes it difficult to increase early egg weight without excessive weight increases at older ages. A study looking at repeated measurements of egg weight across ages showed that the genomic regions with the largest effects show the same direction of effects at different ages. The presence of a polygenic component was suggested in the same study (Yi et al., 2015). One approach to optimize egg weight profile is to use methods that exploit information on the weight trajectory such as random regression models (Arango et al., 2009). Direct selection could be applied to egg weight uniformity, but this trait has lower heritability than egg weight (Wolc et al., 2011a, 2012). The use of genomic information further improves accuracy of predicting breeding values for egg weight (Wolc et al., 2011b) and also allows prediction of heterosis for egg weight in different line crosses (Amuzu-Aweh et al., 2015).

Egg weight is a trait with the largest number of entries in the chicken QTL database among egg quality traits – 80 reported QTL (as of 8 December 2015). Although many QTL mapping studies are specific to the experimental populations in which they were discovered, the QTL for egg weight on chromosome 4 has been independently identified in multiple studies (Goraga et al., 2012; Tuiskula-Haavisto et al., 2002; Sasaki et al., 2004; Schreiweis et al., 2006; Wolc et al., 2012; Yi et al., 2015). The exact causative mutation is not yet confirmed, but several positional candidates have been suggested, including non-SMC condensin I complex, subunit G (*NCAPG*) (Yi et al., 2015) and cholecystokinin A receptor (*CCKAR*) (Dunn et al., 2013).

3.2 Egg shape

The oval shape of eggs has been long known to contribute to the resistance of eggs to breaking, with egg shape accounting for 15–35% of variability in crushing strength

(Richards and Swanson, 1965). In a summary of several hatchability studies, Narushin and Romanov (2002) concluded that eggs with normal shape hatch better than either elongated or rounded eggs. Within the normal range, rounder eggs tend to have lower hatchability than elongated eggs, with a genetic correlation between shape index and hatchability of 0.19–0.32 (Cavero et al., 2011). In a comparison between historic versus modern strains, Anderson et al. (2004) showed that modern strains produce eggs that are more round. This change in shape may be an indirect effect of selection for shell strength, which has been shown to be genetically correlated to shape index and has been under selection for multiple generations (Blanco et al., 2014). GWAS studies for egg shape index, egg width and egg length identified significant associations with chromosomes 4, 7, 8, 10, 12 and Z (Sasaki et al., 2004; Hansen et al., 2005; Goto et al., 2014a).

3.3 Shell quality

Eggshell is a complex bioceramic, with the structure consisting of crystal minerals and specific proteins, all of which must work in concert to produce a correctly crystallized and mineralized shell within the hen oviduct every 24 hours. The eggshell protects the integrity of the egg from external forces and, together with the cuticle, provides a barrier against microbes. It is also an important structure that allows gas exchange, and provides the minerals to form the bone structures of the developing embryo. Shell quality can be measured in multiple ways, including shell specific gravity, weight, thickness, breaking strength and dynamic stiffness (Bain, 2005; De Ketelaere et al., 2004). The heritability of shell strength was estimated to be moderate to high (Dunn et al., 2005; Blanco et al., 2014). Most QTL reported for shell quality also impacted one or more of the following five categories of shell traits: shell weight (Wardecka et al., 2002; Liu et al., 2011), shell thickness (Wright et al., 2006; Goto et al., 2014a; Rosochacki et al., 2013), shell percentage (Fan et al., 2013), shell strength (Sasaki et al., 2004; Abasht et al., 2009; Tuiskula-Haavisto, 2011, Wolc et al., 2014) and shell stiffness (Tuiskula-Haavisto et al., 2011). There are overlaps of QTL locations between studies of shell quality, with chromosomes Z, 2, 4 and 8 most frequently found to be associated with shell-quality traits. Based on GWAS, three positional candidate genes inositol 1,4,5-trisphosphate receptor 2 (*ITPR2*), phosphatidylinositol-4-phosphate 3-kinase 2 gamma (*PIK3C2G*) and *NCAPG* were suggested to explain variation in shell quality by Sun et al. (2015b). QTL regions identified by Wolc et al. (2014) supported the involvement of secreted phosphoprotein 1 (*SPP1*), ovalbumin (*SERPIN*) and parathyroid hormone (*PTH*) in the regulation of shell strength.

3.4 Shell colour

The colour of the eggshell does not affect the nutritional value of eggs, but there are definite regional market preferences for consistently white or brown shell eggs. Uniform dark brown colour desired for brown eggs can be achieved by selection, thanks to a moderate to high heritability of the trait (Arango et al., 2006). Factors affecting shell colour in brown-egg layers were recently reviewed by Samiullah et al. (2015). The most common objective measures of shell colour are based on the combination of three aspects: lightness, spectra in the red-green scale and yellow-blue scale, which may be reported as separate traits (Sasaki et al., 2004; Uemoto et al., 2009; Goto et al., 2014a) or combined into an index representing a single colour value (Wardecka et al., 2002; Schreiweis et al., 2006; Wolc et al., 2014; Wragg et al., 2012). In some markets, in addition to colour intensity, incidence of speckles in brown eggs is considered an important egg quality trait. Sometimes, though, eggs showing speckles are considered as not suited for retail sale and are diverted to the further process industry at a significantly lower value. For individual colour components, consistent QTL have been reported on chromosomes 1, 5, 6 and 11. Shell colour QTL mapping studies have been performed using variation between local African breeds (Wragg et al., 2012), designed crosses (Schreiweis et al., 2006; Wardecka et al., 2002) or within commercial populations (Fulton et al., 2012; Wolc et al., 2014). Chromosomes 2, 4, 6, 12 and 20 were most commonly reported to carry shell colour QTL.

4 Improving internal egg quality

4.1 Albumen quality

Albumen quality determines foaming and whipping properties and thus the usefulness of an egg for industrial processing (Silversides and Budgell, 2004). Albumen quality is known to deteriorate with both age of the hen and storage time of the egg (Monira et al., 2003) and also with increasing storage temperature. Both pedigree and marker-based heritabilities were estimated to be around 0.3 (Wolc et al., 2014). The presence of heritable variation in this trait was confirmed by significant changes in albumen height obtained following divergent selection experiments (Poggenpoel, 1982). The most commonly used measurement is HU, which is a function of albumen height and egg weight (Haugh, 1937). Multiple QTL regions influencing albumen quality have been reported and include chromosomes 1, 2, 4, 7, 8, 9 and Z (Abasht et al., 2009; Honkatukia et al., 2013; Wolc et al., 2014; Hansen et al., 2005; Liu et al., 2011). On chromosome 1, a QTL with parent-of-origin specific expression was found by Tuiskula-Haavisto et al. (2004). Causative mutations were not identified, but two positional candidates have been suggested by Wolc et al. (2014), including polypeptide N-acetylgalactosaminyltransferase 1 (*GALNT1*), a gene responsible for phosphorylation of proteins that is located in a QTL region on chromosome 2 (also identified as a signal for signature of selection (Li et al., 2012), and ovotransferrin gene (*TF*) located on chromosome 9 that encodes one of the most abundant egg white proteins (Mann, 2007).

4.2 Yolk quality

Yolk is the egg component with a high nutritional value, containing vitamins and micronutrients (Vieira, 2007). In the food-processing industry, yolk solids are positively correlated with mayonnaise force (Jones and Musgrove, 2006). Yolk weight heritability is estimated to be moderate and correlated with egg weight (Wolc et al., 2014; Zhang et al., 2005). Wolc et al. (2014) found the same major QTL to affect yolk weight at different ages, which is consistent with the high genetic correlation between different measurements of this trait (Wolc et al., 2015). On the other hand, Sun et al. (2015a) found that significant loci at different ages did not overlap despite very high genetic correlations between yolk weight measured at different ages. Thus, whether the QTLs are consistent or age-dependent may be population specific. QTL mapping studies of yolk quality have evaluated yolk colour (Goto et al., 2014b) and size (Sun et al., 2015b; Goto et al., 2014b; Wolc et al., 2014; Liu et al., 2011; Wardecka et al., 2002). Most of the reports pointed to chromosomes 1, 2, 4, 5, 8, 9 and Z. The largest QTL influence on yolk weight is on chromosome 4, but this may be a reflection of QTL at the same location for egg weight.

Multiple positional candidates, including DOT1-like histone H3K79 methyltransferase (*DOT1L*), stonin 2 (*STON2*), galactosylceramidase (*GALC*), replication protein A2 (*RPA2*), breast cancer 2 (*BRCA2*), zygote arrest 1 (*ZAR1*), echinoderm microtubule associated protein like 4 (*EML4*), potassium channel, voltage-gated modifier G3 (*KCNG3*), elongation factor RNA polymerase II (*ELL*), ENSGALG00000028314, *RGS3F* (Sun et al., 2015a) and apovitellenin 1 (*APOV1*), insulin-like growth factor 1 (*IGF1*), progesterone receptor (*PGR*) and activin receptor (*ACVR1*) (Wolc et al., 2014), were identified near the estimated QTL positions.

4.3 Blood and meat spots

Blood spots are inclusions attached to the yolk surface or the thick albumen and are produced by micro-haemorrhages during ovulation (Solomon, 1997). Meat spots are mainly due to tissue detritus released from the oviduct epithelium as the egg migrates downwards. Like shell colour, blood spots do not change egg quality but may affect the perception of the product, and in some cultures, eggs that contain inclusions are not acceptable for consumption. The frequency of blood spots in modern layers is generally low, but there is large variability between breeds and strains (Honkatukia et al., 2011). Only recently, inline egg graders can automatically detect blood inclusions during egg grading of white eggs. Accurate detection requires breaking the egg and inspecting the inside. However, historically, incidence of both types of internal inclusions (blood and meat) have been much higher in brown-egg lines and varieties than in White Leghorns, which make all the white-egg varieties (Jesus Arango, personal communication). Heritability estimates are low, but successful selection experiments have been performed (Noda et al., 2007). Honkatukia et al. (2011) and Uemoto et al. (2009) reported associations with frequency of blood and meat spots on chromosomes 1, 2, 4, 9 and Z. Romé et al. (2015) reported QTL on chromosomes 3, 8, 12, 20, 24 and Z, three of which showed significant interaction with diet. Variations found in tight junction protein ZO-2 (TJP2) and microRNA gga-mir-1556 (Honkatukia et al., 2011) were identified to predispose egg layers to blood and meat spot defects.

5 Genomic selection

Genomic selection relies on using whole-genome information to predict breeding values of animals rather than trying to pin down some specific genomic regions. Medium (~40–60K) to high-density (600K) SNP chips combined with imputation from low-density panels can be used for cost-efficient implementation of genomic selection (Wang et al., 2013). The main advantages confirmed in selection experiments include increase in the accuracy of estimated breeding values, especially for animals without their own phenotype (i.e. sex limited traits, within family selection for roosters or traits expressed late in life). These advantages translate into greater response to selection (Misztal et al., 2013; Heidaritabar et al., 2014; Wolc et al., 2015), the possibility of restructure breeding programmes (Hsu et al., 2015) and reduction of generation intervals (Wolc et al., 2015). Much of the genomic

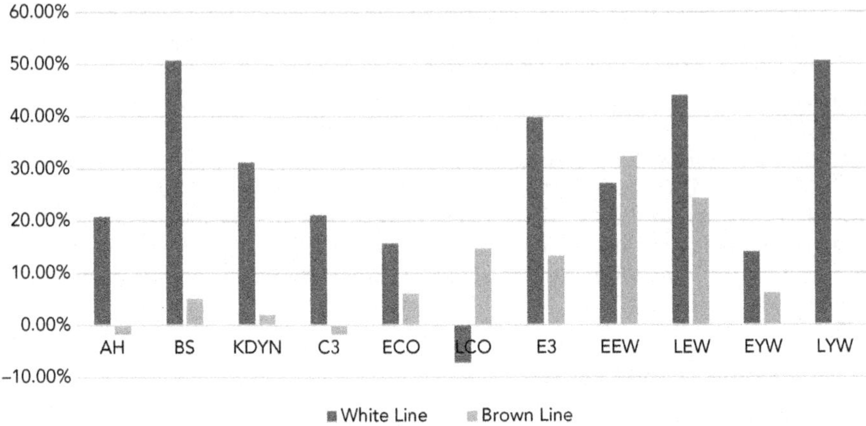

Figure 1 Relative change in accuracy of breeding values by including genomic information.

AH, albumen height; BS, breaking strength; KDYN, dynamic stiffness; C3, ECO, LCO, shell colour of first three eggs, at early and late age; E3, EEW, LEW, egg weight of first three eggs, at early and late age; EYW, LYW, yolk weight at early and late age.

selection research is done within breeding companies, with the results kept proprietary. Only one published study (Wolc et al., 2011b) has compared the accuracies of traditional pedigree based versus genomic breeding values for egg quality. The observed increase in accuracy was reflected in higher responses to selection in the genomic selected line compared to the pedigree-selected line in a multigenerational selection experiment. This improvement was seen for multiple traits, including egg weight, yolk weight, albumen height and shell colour (Wolc et al., 2015). Specific changes in the accuracy of genomic breeding values relative to the accuracy of pedigree-based breeding values for egg quality traits in one white and one brown line are shown in Fig. 1 (unpublished data). For almost all traits, genomic breeding values were more accurate than pedigree-based breeding values, with a more clear advantage in the white line, in which more genotypes were available. To benefit from these improved accuracies, genomic selection has already been implemented by major poultry breeding companies (see review by Wolc et al., 2016).

6 Targeting candidate genes for particular traits

The chicken genome sequence provides the basic information needed for the detailed examination of individual genes. Based on the function of genes determined from multiple species (comparative genomics), genes with potential to influence the traits of interest can be identified. These candidate genes are then investigated in more detail. The QTL studies previously mentioned identify regions of the genome in which DNA-level variation exists that can influence the traits studied, thus narrowing the area on which to focus efforts at understanding genetic influence on traits.

Variations in the DNA sequence can be identified from re-sequence information using unique lines or from variation reported within public databases (e.g. http://www.ncbi.nlm.

nih.gov/snp). Several types of variations are found in DNA. The most common genetic variants studied are single nucleotide polymorphisms (SNPs) due to recent development of numerous rapid and inexpensive detection methods. These SNPs can occur within genes and may either change the coded protein, with a subsequent effect on the function of the gene, or be a 'silent mutation' with no discernible effect on the protein sequence. Variation near genes or within the non-coding segments of the gene may affect regulation or expression of the gene, which can result in small changes in the amount of protein produced. These small changes can subsequently contribute to variation in the trait.

Recently, 15 million SNPs identified from commercially utilized lines (both meat and egg type stocks) and experimental lines have been reported (Gheyas et al., 2015). These SNPs were analysed to determine their location relative to known genes and their potential effect on both function and gene expression. This resource is likely to be rapidly utilized for the study of candidate genes that impact the multiple traits of interest for egg production.

Studies have been conducted to identify associations between variants within or near candidate genes and specific traits by using local or indigenous breeds of chickens or interbreed crosses. Frequently, the number of birds evaluated is low and the subset of birds with favourable variation is even lower. The identified associations may be breed specific, with the phenotypic effects being due to a combination of factors, including other genetic variations unique to the breed. Thus, caution needs to be applied when interpreting these results to apply globally to other chicken lines and breeds.

Early studies investigated the specific effects of only one or two genetic variants within or nearby to a gene, with many of these variants not predicted to change the encoded protein. Analysis of multiple SNPs within a gene allows for the identification of gene haplotypes (specific combinations of SNP variants that occur together on a chromosome) enabling a more thorough examination of the genetic variability found within a line. This multiple SNP analysis has become possible as the methodologies for detection of DNA variation have evolved to allow more rapid and inexpensive SNP detection.

In the following sections, examples of single gene variation and their reported effects on various egg-quality and production-related traits are summarized.

6.1 Egg production

Broody behaviour of chickens results in cessation of egg production and is related to the release of the hormone prolactin (Wilkanowska et al., 2014). The prolactin gene (*PRL*) and the prolactin receptor gene (*PRLR*) are potential candidate genes for control of broodiness (Dunn et al., 1998). Variation within both *PRL* and *PRLR* has been reported and investigated for potential effects on egg production traits. By using a local Blue-shell chicken breed, an association was found between variation within the promoter region of *PRL* and broodiness (Jiang et al., 2005). An association between other *PRL* promoter region variants and egg production was also reported in a local breed intercross (Cui et al., 2006). Neither of these studies report information on egg-quality traits. Associations of variation within the *PRL* promoter and HUs, albumen proportion and yolk index were reported in White Leghorn hens (Bhattacharya et al., 2011a,b). The *PRLR* gene is found on chromosome Z and is within QTL regions that are associated with shell strength (Tuiskula-Haavisto et al., 2011) blood and meat spots (Honkatukia et al., 2011) and egg weight (Wolc et al., 2012; Tuiskula-Haavisto., 2002).

Dopamine and the dopamine receptors are known to influence prolactin levels and thus can also play a role in broody behaviour (Youngren et al., 2002). The dopamine D2 receptor gene (*DRD2*) was examined for genetic variants in six indigenous chicken breeds. Multiple SNP variants were identified and significant association with broody traits was found for some SNP (Xu et al., 2010).

6.2 External egg quality

Through in-depth proteomics studies of eggshells and their associated membranes, multiple proteins and their corresponding genes involved with eggshell production have been identified (Marie et al., 2015). Each of these genes is a potential candidate for further investigation of gene variation that can influence the multiple egg-quality traits of interest. The examples given below show the depth of variation found in different populations for many of these genes. While primarily showing influences on various eggshell quality traits, many of these gene variants were also shown to affect other internal egg quality traits.

Variation in eggshell organic matrix genes and additional genes involved with the maintenance and function of the shell gland were studied in Rhode Island Red hens (Dunn et al., 2009). Multiple significant associations were found between variation in these selected genes, including ovocleidin-116 (*MEPE*) with shell thickness and stiffness, ovocalyxin-32 (*OCX32*) with mammillary layer thickness, osteopontin (*SPP1*) with fracture toughness and ovalbumin with breaking strength and shell thickness. Subsequent testing showed associations between variation in these genes and eggshell crystal size and/or orientation (Dunn et al., 2011).

In hybrid chickens produced by crossing White Leghorn and Rhode Island Red breeds, variants in *OCX32* were found to be associated with shell colour, egg production and meat spots (Uemoto et al., 2009). The gene was examined in more detail in eight commercially utilized egg layer pure lines, including in White Leghorn, Rhode Island Red and White Plymouth Rock breeds (Fulton et al., 2012). Multiple variants were found in this gene, resulting in 19 different *OCX32* protein variants across all lines. Many of these protein variants were line or breed specific. Significant associations were found between *OCX32* gene variation and several egg-production and egg-quality traits, including puncture score, egg weight, albumen height, shell colour and yolk weight. These associations between *OCX32* gene variation and trait were not seen for every line, confirming that caution must be applied in inferring similar relationships across multiple lines.

There are multiple ion transport proteins within the avian uterus with variable levels of expression depending upon the stage of shell formation (Jonchere et al., 2012). Variation in several genes from the sodium channel gene family (*SCNN1*) was found to be associated with multiple shell-quality traits, including eggshell strength, thickness and weight in White Leghorn hens (Fan et al., 2013). The sodium channel, non-voltage-gated 1 alpha gene *SCNN1a* is located on chromosome 1 within a QTL region associated with albumen height (Wolc et al., 2012). An analysis of the variation within 15 ion transport genes was conducted with Rhode Island Red hens (Duan et al., 2015). Eggshell traits measured included shell strength, shell weight and shell thickness. Significant associations were found between SNP variation in eight of these genes and eggshell properties.

Other genes found to contain variants that affect eggshell traits include *PTH*, which is involved with calcium homeostasis, ornithine decarboxylase (*ODC*), which is a key enzyme for polyamine synthesis, growth hormone receptor (*GHR*) and insulin-like growth factor 1 (*IGF-1*) (Jiang et al., 2010; Parsanejad et al., 2004; Li et al., 2010).

6.3 Internal egg quality

The elimination of the defect causing fishy taint in eggs is a perfect example of the application of molecular genetics techniques to improve internal egg quality. Fishy taint odour is due to the accumulation of trimethylamine (TMA) in the yolk. It occurred in brown eggshell-laying breeds when fed with rapeseed meal or high levels of choline. Defects in the *FMO3* (flavin-containing mono-oxygenase 3) gene are known to cause fishy body odour in humans (Dolphin et al., 1997) and fishy taint in cow's milk (Lundén et al., 2002). Sequence information for the *FM03* gene of fishy taint-affected birds revealed genetic variation that affected the functional enzymatic pocket, rendering the altered FM03 protein incapable of converting the malodorous TMA into the odourless form (Honkatukia et al., 2005). Birds with this genetic defect deposited the TMA in their egg yolk, resulting in the fishy odour. Genetic tests were developed to detect this variant and the fish taint defect has subsequently been eliminated from major commercially utilized brown-egg shell varieties.

Low-density lipoprotein receptor proteins (LRPs) play central roles in cholesterol metabolism, transport and cellular signalling (Stolt and Bock, 2006) and thus are considered strong candidate genes for influencing egg yolk quality. The restricted ovulator (R/O) defect was discovered within a commercial egg production line in hens descended from a single sire. These hens failed to lay eggs and had excessive serum levels of cholesterol and triglycerides. Genetic analysis determined that the defect was due to a single change in a gene subsequently called *LR8* (Elkin et al., 2012). This single genetic change results in lack of LR8 protein in the oocyte's plasma membrane, with subsequent failure of uptake of both very low-density lipoprotein and vitellogenin into the growing yolk follicle (Stifani et al., 1990). The ovarian follicles of R/O hens fail to mature or develop a normal follicular hierarchy (Elkin et al., 2012). Variation in the low-density lipoprotein receptor genes (*LRP2* and *LRP8*) was investigated in a dwarf line of layers. No relationship was found between SNP within either gene and yolk weight or yolk ratio, though an effect of *LRP8* was seen for yolk colour (Yao et al., 2010; Zhang et al., 2011).

Cathepsin D (*CTSD*) is an aspartic protease important for the processing of egg yolk within the follicle (Retzek et al., 1992). Associations between two SNP in this gene and yolk weight and height were reported in the Shandong breed of chickens (Sheng et al., 2013).

7 Gene expression and regulation

The thousands of genes that constitute the chicken genome do not work independently. They work together within genetic regulatory networks of interacting genes that influence gene expression. Proteomics is the study of large numbers of proteins, all of which may be expressed at specific times or in specific tissues. By combining information from genetic networks and from proteomic studies of specific tissues of interest, understanding can be gained on the phase when genes are turned on or off, in which tissues this occurs, in addition to providing insights into how this can affect various traits of interest for egg production.

Regulation of expression of specific genes within the oviduct has been shown to be associated with mechanical strain due to the presence of the egg. Genes regulated in this way include *SPPN*, *Na-K-ATPase* and *glypican-4* (Lavelin et al., 1998, 2001, 2002). Specific oviduct localization of genes and their expressed proteins has been studied for *SPPN* (Hincke et al., 2008) and collagen X (Wang et al., 2002).

There are several reviews of gene processes related to oviduct and egg protein and all of these point to genes of interest and exemplify the application of molecular biology tools to aid in understanding the multiple complex processes of eggshell development. The proteins of the ovalbumin gene family are important proteins found in egg yolk, egg white, vitelline membrane and egg shell. The gene structures, regulation and expression and protein function of ovalbumin genes were summarized by Da Silva et al. (2015). The mechanisms and regulation of calcium transport that are critical in the rapid transportation required for the production of the egg shell in the intestines and eggshell gland of birds were reviewed by Bar (2009).

Proteomics studies have focused on determining which genes are expressed in specific tissues and how these genes and their resultant proteins are regulated and interact. Several studies have examined gene expression in the oviduct (Brionne et al., 2014; Marie et al., 2015; Jeong et al., 2012; Jonchere et al., 2010, 2012; Liu et al., 2013).

8 Summary and future trends

The molecular tools described in this chapter – genome-wide analysis, genomic selection, candidate gene analysis, gene expression studies and pathway analysis, complement each other in achieving a better understanding of the genetic determination of egg-quality traits. Identification of causal mutations can be used to eliminate egg defects from populations by selective breeding, as was successfully achieved in some populations for the fishy taint mutation.

One of the challenges with breeding to improve egg-quality is that trait expression is limited to females. Males carry the genetic variation, but the impact of this variation cannot be measured directly in the males. Thus, selection of males with desirable genetic variation must be done indirectly through performance records of sisters, mothers and other female relatives included in the pedigree used in analysis for prediction of breeding values. If genetic variation can be identified in males at an early age, genetic selection can be applied directly to the males, thereby increasing the accuracy of male selection. Selection based on genetic variation within males can be attained by using the recently developed tools of molecular genetics.

The recently developed genetic manipulation tool Clustered Regularly Interspaced Short Palindromic Repeat (CRISPR) (Cong et al., 2013) allows rapid and highly specific editing of genes. The resulting animals are not transgenic, as they do not contain recombinant DNA. The genome is altered to produce a very specific and desired change in a gene to improve a particular trait. The introduced changes mimic the naturally occurring variations that are found in other breeds. This efficient gene editing technique has been successfully used to alter a viral receptor gene resulting in pigs that are genetically resistant to infection with the Porcine Reproductive and Respiratory Syndrome (PRRS) virus (Whitworth et al., 2015). Certain breeds of cattle are genetically hornless; the underlying gene variant causing this was introduced into horned dairy cattle, resulting in the production of hornless dairy cattle (Tan et al., 2013). This was achievable because the underlying genetic basis for these traits was well known. Most traits of importance for egg quality are likely to be influenced by variants with small, cumulative effects. In a recent simulation study, manipulation of multiple sites by gene editing in combination with genomic selection showed an improved response to selection compared with genomic selection alone (Jenko et al., 2015). For chickens, resistance to disease (such as avian influenza or Newcastle disease), enhanced phosphate utilization and egg quality are all traits of interest for commercial egg production. The genes involved with these traits and the desirable genetic variation must first be identified and their effects on the interrelated traits influencing egg production must be understood before these types of technologies can be utilized.

Use of gene editing in an efficient and responsible manner will require deep understanding of the biology of the bird, the physiology of egg production, the interrelationships of the approximately 20 000 genes and their resulting proteins, and the functioning of the whole organism as a connected system. Genomic selection is an intermediate tool that improves accuracy of selection without the need to identify causative mutations or to carry out technical interventions to the genome.

9 Where to look for further information

The information contained in this chapter represents a review of available literature on the topic of molecular breeding and egg quality traits in the chicken. The review of genes does not represent an exhaustive review of the literature, but rather examples of works published with focus on potential genes that influence egg quality traits. This material within this chapter was completed in December 2015. There is much interest in the improvement of egg quality traits, and the application of molecular methods to enhance selection progress for these traits. The public accessibility of the chicken genome sequence, the development of tools for analysis of the genome, and continual investigations into the function of genes within organisms ensures that research will continue in these newly developing arenas utilizing DNA based information. Many open access journals are publishing research in this area of study and many non-open access journals are accessible through searches of PubMed Central (http://www.ncbi.nlm.nih.gov/pmc/) which is a free archive of literature from multiple biomedical and life sciences publications collated by the US National Institute of Health. Database including published QTL and tools for comparative genomic analysis are available at www.animalgenome.org.

10 References

Abasht, B., Dekkers, J. C. M. and Lamont, S. J. (2006), 'Review of Quantitative Trait Loci identified in the chicken', *Poult. Sci.*, 85, 2079–96.

Abasht, B., Sandford, E., Arango, J., Settar, P., Fulton, J. E., O'Sullivan, N. P., Hassen, A., Habier, D., Fernando, R. L., Dekkers, J. C. M. and Lamont S. J. (2009), 'Extent and consistency of linkage disequilibrium and identification of DNA markers for production and egg quality traits in commercial layer chicken populations', *BMC Genom.*, 10(2), S2.

Abiola, S. S., Meshioye, O. O., Oyerinde, B. O. and Bamgbose, M. A. (2008), 'Effect of egg size on hatchability of broiler chicks', *Arch. Zootec.*, 57, 83–6.

Amuzu-Aweh, E. N., Bovenhuis, H., de Koning, D.-J. and Bijma, P. (2015), 'Predicting heterosis for egg production traits in crossbred offspring of individual White Leghorn sires using genome-wide SNP data', *Genet. Sel. Evol.*, 47, 27.

Anderson, K. E. and Havenstein, G. B. (2012), 'History of the North Carolina Layer Tests detailed description of housing and husbandry changes made from 1958 through 2009', North Carolina Cooperative Extension Service https://poultry.ces.ncsu.edu/wp-content/uploads/2012/04/NCLPMT-1958-2009-Historical-Summary.pdf?fwd=no (accessed 29 December 2015).

Anderson, K. E., Tharrington, J. B., Curtis, P. A. and Jones F. T. (2004), 'Shell characteristics of eggs from historic strains of single comb white leghorn chickens and relationship of egg shape to shell strength', *Int. J. Poult. Sci.*, 3, 17–19.

Arango, J., Settar, P., Arthur, J. A. and O'Sullivan, N. P. (2006), 'Relationship between shell color and incidence of speckles in brown egg lines', 12th European Poultry Conference, Verona, Italy, 10–14 September 2006. http://www.cabi.org/Uploads/animal-science/worlds-poultry-science-association/WPSA-italy-2006/10442.pdf (accessed 29 December 2015).

Arango, J., Turkmut, L., Settar, P. and O'Sullivan N. P. (2009), 'A random regression approach to model egg weight trajectory in two layer populations', *Proc. 6th European Poultry Genetics Symposium*, 30 September to 2 October 2009, Bedlewo near Poznan, Poland. http://www.up.poznan.pl/epgs2009/session2/selection/Arango.pdf (accessed 29 December 2015).

Atzmon, G., Blum, S., Feldman, M., Cahaner, A., Lavi, U. and Hillel, J. (2008), 'QTLs detected in a multigenerational resource chicken population', *J. Hered.*, 99(5), 528–38.

Bain, M. M. (2005), 'Recent advances in the assessment of eggshell quality and their future application', *World's Poult. Sci. J.*, 61, 268–77.

Bar, A. (2009), 'Calcium transport in strongly calcifying laying birds: mechanisms and regulation', *Comp. Biochem. Physiol. A Mol. Integr. Physiol.*, 152(4), 447–69. doi: 10.1016/j.cbpa.2008.11.020.

Bhattacharya, T. K., Chatterjee, R. N., Sharma, R. P., Rajkumar, U., Niranjan, M. and Reddy, B. L. (2011a), 'Association of polymorphism in the prolactin promoter and egg quality traits in laying hens', *Br. Poult. Sci.*, 52(5), 551–7.

Bhattacharya, T. K., Chatterjee, R. N., Sharma, R. P., Niranjan, M. and Rajkumar, U. (2011b), 'Associations between novel polymorphisms at the 5'-UTR region of the prolactin gene and egg production and quality in chickens', *Theriogenology*, 75(4), 655–61.

Blanco, A. E., Icken, W., Ould-Ali, D., Cavero, D. and Schmutz, M. (2014), 'Genetic parameters of egg quality traits on different pedigree layers with special focus on dynamic stiffness', *Poult. Sci.*, 93(10), 2457–63.

Brionne, A., Nys, Y., Hennequet-Antier, C. and Gautron, J. (2014), 'Hen uterine gene expression profiling during eggshell formation reveals putative proteins involved in the supply of minerals or in the shell mineralization process', *BMC Genom.*, 15, 220.

Cavero, D., Schmutz, M., Icken, W. and Preisinger, R. (2011), 'Improving hatchability in white egg layer strains through breeding', *Lohmann Info.* 46, 44–54.

Cong, L., Ran, A. F., Cox, D., Lin, S., Barretto, R., Habib, N., Hsu, P. D., Wu, X., Jiang, W., Marraffini, L. A. and Zhang, F. (2013), 'Multiplex Genome Engineering Using CRISPR/Cas Systems', *Science*, 339(6121), 819–23.

Cui, J. X., Du, H. L., Liang, Y., Deng, X. M., Li, N. and Zhang, X. Q. (2006), 'Association of polymorphisms in the promoter region of chicken prolactin with egg production', *Poult. Sci.*, 85, 26–31.

Da Silva, M., Beauclercq, S., Harichaux, G., Labas, V., Guyot, N., Gautron, J., Nys, Y. and Rehault-Godbert, S. (2015), 'The family secrets of avian egg-specific Ovalbumin and its related proteins Y and X', *Biol Reprod.*, 93(3), 71.

De Ketelaere, B., Bamelis, F., Kemps, B., Decuypere, E. and De Baerdemaeker, J. (2004), 'Non-destructive measurements of the egg quality', *World's Poult. Sci. J.*, 60, 289–302.

Dekkers, J. C. M. and van der Werf, J. (2007), 'Strategies, limitations and opportunities for marker-assisted selection in livestock', in E. P. Guimares, J. Ruane, B. D. Scherf, A. Sonnino and J. D. Dargie (eds), *Marker-Assisted Selection – Current Status and Future Perspectives in Crops, Livestock, Forestry and Fish*, pp. 167–228. Food and Agriculture Organization of the United Nations, Rome.

Dolphin, C. T., Janmohamed, A., Smith, R. L., Shephard, E. A. and Phillips, I. R. (1997), 'Missense mutation in flavin-containing mono-oxygenase 3 gene, FMO3, underlies fish-odour syndrome', *Nat. Genet.*, 17, 491–4.

Duan, Z., Chen, S., Sun, C., Shi, F., Wu, G., Liu, A., Xu, G. and Yang, N. (2015), 'Polymorphism in ion transport genes are associated with eggshell mechanical property', *PLoS ONE* 10(6), e0130160. doi:10.1371/jounalpone.0130160.

Dunn, I. C., McEwan, G., Okhubo, T., Sharp, P. J., Paton, I. R. and Burt, D. W. (1998), 'Genetic mapping of the chicken prolactin receptor gene: a candidate gene for the control of broodiness', *Br. Poultry Sci.*, 39, S23–S24.

Dunn, I. C., Bain, M., Edmond, A., Wilson, P. W., Joseph, N., Solomon, S., De Ketelaere, B., De Baerdemaeker, J., Schmutz, M., Preisinger, R. and Waddington, D. (2005), 'Heritability and genetic correlation of measurements derived from acoustic resonance frequency analysis; a novel method of determining eggshell quality in domestic hens', *Br. Poult. Sci.*, 46(3), 280–6.

Dunn, I. C., Joseph, N. T., Bain, M., Edmond, A., Wilson, P. W., Milona, P., Nys, Y., Gautron, J., Schmutz, M., Preisinger, R. and Waddington, D. (2009), 'Polymorphisms in eggshell organic matrix genes are associated with eggshell quality measurements in pedigree Rhode Island Red hens', *Anim. Genet.*, 40(1), 110–14.

Dunn, I. C., Rodriguez-Navarro, A. B., Mcdade, K., Schmutz, M., Preisinger, R., Waddington, D., Wilson, P. W. and Bain, M. M. (2011), 'Genetic variation in eggshell crystal size and orientation is large and these traits are correlated with shell thickness and are associated with eggshell matrix protein markers', *Anim. Genet.*, 43, 410–18.

Dunn, I. C., Meddle, S. L., Wilson, P. W., Wardle, C. A., Law, A. S., Bishop, V. R., et al. (2013), 'Decreased expression of the satiety signal receptor CCKAR is responsible for increased growth and body weight during the domestication of chickens', *Am. J. Physiol. Endocrinol. Metab.*, 304(9), E909–21.

Elkin, R. G., Bauer, R. and Schneider, W. J. (2012), 'The restricted ovulator chicken strain: an oviparous vertebrate model of reproductive dysfunction caused by a gene defect affecting an oocyte-specific receptor', *Anim. Reprod. Sci.*, 136, 1–13.

Fan, Y. F., Hou, Z. C., Yi, G. Q., Xu, G. Y. and Yang, N. (2013), 'The sodium channel gene family is specifically expressed in hen uterus and associated with eggshell quality traits', *BMC Genet.*, 14, 90.

Fulton, J. E., Soller, M., Lund, A. R., Arango, J. and Lipkin, E. (2012), 'Variation in the ovocalyxin-32 gene in commercial egg-laying chickens and its relationship with egg production and egg quality traits', *Anim. Genet.*, 43(S1), 102–13.

Gheyas, A. A., Boschiero, C., Eory, L., Ralph, H., Kuo, R., Woolliams, J. A. and Burt, D. W. (2015), 'Functional classification of 15 million SNPs detected from diverse chicken populations', *DNA Res.*, 22, 205–17.

Goraga, Z. S., Nassar, M. K. and Brockmann, G. A. (2012), 'Quantitative trait loci segregating in crosses between New Hampshire and White Leghorn chicken lines: I. egg production traits', *Anim. Genet.*, 43(2), 183–9.

Goto, T., Ishikawa, A., Onitsuka, S., Goto, N., Fujikawa, Y., Umino, T., Nishibori, M. and Tsudzuki, M. (2011), 'Mapping quantitative trait loci for egg production traits in an F2 intercross of Oh-Shamo and White Leghorn chickens', *Anim. Genet.*, 42(6), 634–41.

Goto, T., Ishikawa, A., Yoshida, M., Goto, N., Umino, T., Nishibori, M. and Tsudzuki, M. (2014a), 'Quantitative trait loci mapping for external egg traits in F2 chickens', *J. Poult. Sci.*, 51(2), 118–29.

Goto, T., Ishikawa, A., Goto, N., Nishibori, M., Umino, T. and Tsudzuki M. (2014b), 'Mapping of main-effect and epistatic quantitative trait loci for internal egg traits in an F2 resource population of chickens', *J. Poult. Sci.*, 51(4), 375–86.

Hansen, C., Yi, N., Zhang, Y. M., Xu, S., Gavora, J. and Cheng, H. H. (2005), 'Identification of QTL for production traits in chickens', *Anim. Biotechnol.*, 16(1), 67–79.

Haugh, R. R. (1937), 'The Haugh Unit for measuring egg quality', *US Egg Poult. Mag.*, 43, 552–5.

Heidaritabar, M., Vereijken A., Muir W. M., Meuwissen T., Cheng H., Megens H. J., Groenen, M. A. M. and Bastiaansen, J. W. M. (2014), 'Systematic differences in the response of genetic variation to pedigree and genome-based selection methods', *Heredity*, 113, 503–13.

Hillier, L. W., Miller, W., Birney, E., Warren, W., Hardison, R. C., Ponting, C. P., Bork, P., Burt, D. W., Groenen, M. A. M., Delany, M. E., Dodgson, J. B., Chinwalla, A. T., Cliften, P. F., Clifton, S. W., Delehaunty, K. D., Fronick, C., Fulton, R. S., Graves, T. A., Kremitzki, C., Layman, D., Magrini, V., McPherson, J. D., Miner, T. L., Minx, P., Nash, W. E., Nhan, M. N., Nelson, J. O., Oddy, L. G., Pohl, C. S., Randall-Maher, J., International Chicken Genome Sequencing Consortium, et al. (2004), 'Sequence and comparative analysis of the chicken genome provide unique perspectives on vertebrate evolution', *Nature*, 432, 695–716.

Hincke, M. T., Chien Y.-C., Gerstenfeld, L. C. and McKee, M. D. (2008), 'Colloidal-gold immunocytochemical localization of osteopontin in avian eggshell gland and eggshell', *J. Histochem. Cytochem.*, 56, 467–76.

Honkatukia, M., Reese, K., Preisinger, R., Tuiskula-Haavisto, M., Weigend, S., Roito, J., Mäki-Tanila, A. and Vilkki, J. (2005), 'Fishy taint in chicken eggs is associated with a substitution within a conserved motif of the FMO3 gene', *Genomics*, 86(2), 225–32.

Honkatukia, M., Tuiskula-Haavisto, M., Ahola, V., Uimari, P., Schmutz, M., Preisinger, R., Cavero, D., Vennerstrom, P., Arango, J. and O'Sullivan N. (2011), 'Mapping of QTL affecting incidence of blood and meat inclusions in egg layers', *BMC Genet.*,12(1), 55.

Honkatukia, M., Tuiskula-Haavisto, M., Arango, J., Tabell, J., Schmutz, M., Preisinger, R. and Vilkki, J. (2013), 'QTL mapping of egg albumen quality in egg layers', *Genet. Sel. Evol.*, 45, 31.

Hsu, W. L., Fernando, R. L., Dekkers, J. C. M. Arango, J., Settar, P., Fulton, J. E., O'Sullivan, N. P. and Wolc, A. (2015), 'A simulation study on the effect of nested vs factorial mating on response to pedigree and genomic selection', *Proc. Midwest Meeting of ADSA and ASAS*, Des Moines, IA. Poster 242.

Jenko, J., Gorjanc,G., Cleveland, M. A. Varshney, R. K., Whitelaw, C. B. A., Woolliams J. A. and Hickey, J. M. (2015), 'Potential of promotion of alleles by genome editing to improve quantitative traits in livestock breeding programs', *Genet. Sel. Evol.*, 47, 55.

Jeong, W., Lim, W., Jim, J., Ahn, S. E., Lee, H. C., Jeong, J.-W., Han, J. Y., Song, G. and Baser, F. W. (2012), 'Cell-specific and temporal aspects of gene expression in the chicken oviduct at different stages of the laying cycle', *Biol. Reprod.*, 86, 172.

Jiang, R. S., Xu, G. Y., Zhang, X. Q. and Tang, N. (2005), 'Association of polymorphisms for prolactin and prolactin receptor genes with broody traits in chickens', *Poult. Sci.*, 84, 839–45.

Jiang, R. S., Xie, Z., Chen, X. Y. and Geng, Z. Y. (2010), 'A single nucleotide polymorphism in the parathyroid hormone gene and effects on eggshell quality in chickens', *Poult. Sci.*, 89(10), 2101–5.

Jonchere,V. Rehault-Godbert, S., Hennequet-Antier, C., Cabau, C., Sibut, V., Cogburn, L. A., Nys, Y. and Gauron, J. (2010), 'Gene expression profiling to identify eggshell proteins involved in physical defense of the chicken egg', *BMC Genomics*, 11, 57.

Jonchere, V., Brionne, A., Gautron, J. and Nys, Y. (2012), 'Identification of uterine ion transporters for mineralization precursors of the avian eggshell', *BMC Physiol.*, 12, 10.

Jones, D. and Musgrove, M. (2006), 'Correlation of egg physical quality measurements and functional determinations', *12th European Poultry Conference*, Verona, Italy, 10–14 September 2006. http://www.cabi.org/cabdirect/FullTextPDF/2009/20093210368.pdf (accessed 29 December 2015).

Kranis, A., Gheyas, A. A., Boschiero, C., Turner, F., Yu, L., Smith, S., Talbot, R., Pirani, A., Brew, F., Kaiser, P., Hocking, P. M., Fife, M., Salmon, N., Fulton, J., Strom, T. M., Haberer, G., Weigend, S., Preisinger, R., Gholami, M., Qanbari, S., Simianer, H., Watson, K. A., Woolliams, J. A. and Burt, D. W. (2013), 'Development of a high density 600K SNP genotyping array for chicken', *BMC Genom.*, 14, 59.

Lahav, T., Atzmon, G., Blum, S., Ben-Ari, G., Weigend, S., Cahaner, A., Lavi, U. and Hillel, J. (2006), 'Marker-assisted selection based on a multi-trait economic index in chicken: Experimental results and simulation', *Anim. Genet.*, 37(5), 482–8.

Lavelin, I., Yarden, N., Ben-Bassat, S., Bar, A. and Pines, M. (1998), 'Regulation of osteopontin gene expression during egg shell formation in the laying hen by mechanical strain', *Matrix Biol.*, 17, 615–23.

Lavelin, I., Meiri, N., Genina, O., Alexiev, R. and Pines, M. (2001), 'NA$^+$-K$^+$-ATPase gene expression in the avian eggshell gland: distinct regulation in different cell types', *Am. J. Physiol.*, 281(4), 1169–76.

Lavelin, I., Meiri, N., Einat, M., Genina, O. and Pines, M. (2002), 'Mechanical strain regulation of the Chicken glypican-4 gene expression in the avian eggshell gland', *Am. J. Physiol.*, 283(4), 853–61.

Li, D. F., Liu, W. B., Liu, J. F., Yi, G. Q., Lian, L., Qu, L. J., Li, J. Y., Xu, G. Y. and Yang, N. (2012), 'Whole-genome scan for signatures of recent selection reveals loci associated with important traits in White Leghorn chickens', *Poult. Sci.*, 91, 1804–12.

Li, H., Zhu, W., Chen, K., Song, W., Shu, J. and Han, W. (2010), 'Effects of the polymorphisms of GHR gene and IGF-1 gene on egg quality in Wenchang chicken', *Res. J. Poult. Sci.*, 3, 19–22.

Liu, W., Li, D., Liu, J., Chen, S., Qu, L., Zheng, J., Xu, G. and Yang, N. (2011), 'A genome-wide SNP scan reveals novel loci for egg production and quality traits in White Leghorn and brown-egg dwarf layers', *PloS ONE*, 6(12), e28600.

Liu, Y., Qiu, N. and Ma, M. (2013), 'Comparative proteomic analysis of hen egg white proteins during early phase of embryonic development by combinatorial peptide ligand library and matrix-assisted laser desorption ionization-time of flight', *Poult. Sci.*, 92, 1897–904.

Lundén, A., Marklund, S., Gustafsson, V. and Andersson, L. (2002), 'A nonsense mutation in the FMO3 gene underlies fishy off-flavor in cow's milk', *Genome Res.*, 12, 1885–8.

Mann, K. (2007), 'The chicken egg white proteome', *Proteomics*, 7, 3558–68.

Marie, P., Labas, V., Brionne, A., Harichaux, G., Hennequet-Antier, C., Rodriguez-Navarro, A. B., Nys, Y. and Gautron, J. (2015), 'Quantitative proteomics provides new insights into chicken eggshell matrix protein functions during the primary events of mineralization and the active calcification phase', *J. Proteomics*, 126, 140–54.

Misztal, I., Aggrey, S. E. and Muir, W. M. (2013), 'Experiences with a single-step genome evaluation', *Poult. Sci.*, 92, 2530–4.

Monira, K. N., Salahuddin, M. and Miah, G. (2003), 'Effect of breed and holding period on egg quality characteristics of chicken', *Int. J. Poult. Sci.*, 2, 261–3.

Narushin, V. G. and Romanov M. N. (2002), 'Egg physical characteristics and hatchability', *World Poult. Sci. J.*, 58, 297–303.

Noda, K., Nakamura, A., Kino, K. and Kato, Y. (2007), 'Genetic improvement by selection for decrease of the meat spots in Nagoya breed', *Jpn. J. Poult. Sci.*, 44, 17–22.

Nurgiartiningsih, V. M. A., Mielenz, N., Preisinger, R., Schmutz, M. and Schüler, L. (2005), 'Heritabilities and genetic correlations for monthly egg production and egg weight of White Leghorn hens estimated based on hen-housed and survivor production', *Arch. Geflügelk.*, 69(3), 98–102.

Park, T. S., Lee, H. J., Kim, K. H., Kim, J. S. and Han, J. Y. (2014), 'Targeted gene knockout in chickens mediated by TALENs', *P. Natl. Acad. Sci. U.S.A.*, 111(35), 12716–21.

Parsanejad, R., Praslickova, D., Zadworny, D. and Kuhnlein, U. (2004), 'Ornithine decarboxylase: haplotype structure and trait associations in White Leghorn chickens', *Poult. Sci.*, 83(9), 1518–23.

Poggenpoel, D. G. (1982), 'Two-way selection for egg albumen quality', *S. Afr. J. Anim. Sci.*, 12, 389–91.

Rapaport, F., Khanin, R., Liang, Y., Pirun, M., Krek, A., Zumbo, P., Mason, C. E., Socci, N. D. and Betel, D. (2013), 'Comprehensive evaluation of differential gene expression analysis methods for RNA-seq data', *Genome Biol.*, 14, 95.

Retzek, H., Steyrer, E., Sanders, E. J., Nimpf, J. and Schneider, W. H. (1992), 'Molecular cloning and functional characterization of chicken cathepsin D, a key enzyme for yolk formation', *DNA Cell Biol.*, 11, 661–72.

Richards, J. F. and Swanson, M. H. (1965), 'The relationship of egg shape to shell strength', *Poult. Sci.*, 44(6), 1555–8.

Romé, H., Varenne, A., Hérault, F., Chapuis, H., Alleno, C., Dehais, P., Vignal, A., Burlot, T. and Le Roy, P. (2015), 'GWAS analyses reveal QTL in egg layers that differ in response to diet differences', *Genet. Sel. Evol.*, 47, 83.

Rosochacki, S. J., Olszewski, R., Wardecka, B., Jaszczak, K., Zieba, G., Juszczuk-Kubiak, E. and Poloszynowicz, J. (2013), 'Mapping of QTLs Controlling Egg Quality on Chromosomes 6-8,Z and Three Linkage Groups in Chickens', *Iran. J. Appl. Anim. Sci.*, 3(2), 369–78.

Samiullah, S. J., Roberts, R. and Chousalkar, K. (2015), 'Eggshell color in brown-egg laying hens -a review', *Poult. Sci.*, 94(10), 2566–75.

Sasaki, O., Odawara, S., Takahashi, H., Nirasawa, K., Oyamada, Y., Yamamoto, R., Ishii, K., Nagamine, Y., Takeda, H., Kobayashi, E. and Furukawa, T. (2004), 'Genetic mapping of quantitative trait loci affecting body weight, egg character and egg production in F2 intercross chickens', *Anim. Genet.*, 35(3), 188–94.

Sazanov, A., Sazanova, A., Barkova, O. and Jaszczak, K. (2010), 'QTL in chicken: a look back and forward - a review', *Anim. Sci. Pap. Rep.*, 28(4), 307–14.

Schreiweis, M. A., Hester, P. Y., Settar, P. and Moody, D. E. (2006), 'Identification of quantitative trait loci associated with egg quality, egg production, and body weight in an F2 resource population of chickens', *Anim. Genet.*, 37(2), 106–12.

Shalev, B. A. and Pasternak, H. (1993), 'Increment of egg weight with hen age in various commercial avian species', *Br. Poult. Sci.*, 34(5), 915–24.

Sheng, Q., Cao, D., Zhou, Y., Lei, Q., Han, H., Li, F., Lu, Y. and Wang, C. (2013), 'Detection of SNPs in the cathepsin D gene and their association with yolk traits in chickens', *PloS ONE*, 8(2), e56656.

Silversides, F. G. and Budgell, K. (2004), 'The relationships among measures of egg albumen height, pH, and whipping volume', *Poult. Sci.*, 83, 1619–23.

Solomon, S. E. (1997), 'Egg and egg shell quality', Munson Publishing. Veterinary Press. Iowa State University, Ames, p. 149.

Stifani, S., Barber, D. W., Nimpf, J. and Schneider, W. J. (1990), 'A single chicken oocyte plasma membrane protein mediates uptake of very low density lipoprotein and vitellogenin', *P. Natl. Acad. Sci.*, 87, 1566–959.

Stolt, P. C. and Bock, H. H. (2006), 'Modulation of lipoprotein receptor functions by intracellular adaptor proteins', *Cell. Signal.*, 18, 1560–71.

Sun, C., Lu, J., Yi, G., Yuan, J., Duan, Z., Qu, L., Xu, G., Wang, K. and Yang, N. (2015a), 'Promising loci and genes for yolk and ovary weight in chickens revealed by a Genome-Wide Association Study', *PloS ONE*, 10(9), e0137145.

Sun, C., Qu, L., Yi, G., Yuan, J., Duan, Z., Shen, M., Qu, L., Xu, G., Wang, K. and Yang, N. (2015b), 'Genome-wide association study revealed a promising region and candidate genes for eggshell quality in an F2 resource population', *BMC Genomics*, 16, 565.

Tan, W., Carlson, D. F., Lancto, C. A., Garbe, J. R., Webster, D. A., Hackett, P. B. and Fahrenkrug, S. C. (2013), 'Efficient nonmeiotic allele introgression in livestock using custom endonucleases', *P. Nat. Acad. Sci.*, 110, 16526–31.

Tuiskula-Haavisto, M., Honkatukia, M., Vilkki, J., de Koning, D. J., Schulman, N. F. and Maki-Tanila A. (2002), 'Mapping of quantitative trait loci affecting quality and production traits in egg layers', *Poult. Sci.*, 81(7), 919–27.

Tuiskula-Haavisto, M., de Koning, D. J., Honkatukia, M., Schulman, N. F., Maki-Tanila, A. and Vilkki, J. (2004), 'Quantitative trait loci with parent-of-origin effects in chicken', *Genet. Res.*, 84(1), 57–66.

Tuiskula-Haavisto, M., Honkatukia, M., Preisinger, R., Schmutz, M., de Koning, D. J., Wei, W. H. and Vilkki, J. (2011), 'Quantitative trait loci affecting eggshell traits in an F(2) population', *Anim. Genet.*, 42(3), 293–9.

Uemoto, Y., Suzuki, C., Sato, S., Sato, S., Ohtake, T., Sasaki, O., Takahashi, H. and Kobayashi, E. (2009), 'Polymorphism of the ovocalyxin-32 gene and its association with egg production traits in the chicken', *Poult. Sci.*, 88(12), 2512–17.

Vieira, S. L. (2007), 'Chicken embryo utilization of egg micronutrients', *Rev. Bras. Cien. Aví.*, 9(1), 1–8.

Vilkki, J. (2007), 'Application of genomic approaches in egg quality research', XVIII Europ. Symp. on the Quality of Poultry Meat and XII Europ. Symp. on the Quality of Eggs and Egg products, 2–5 September 2007, Prague, Czech Republic. http://www.cabi.org/Uploads/animal-science/worlds-poultry-science-association/WPSA-czech-republic-2007/4_Vilkki%20Johanna.pdf.

Wang, C., Habier, D., Peiris, B. L., Wolc, A., Kranis, A., Watson, K. A., Avendano, S., Garrick, D. J., Fernando, R. L., Lamont, S. J. and Dekkers, J. C. M. (2013), 'Accuracy of genomic prediction using an evenly spaced, low-density single nucleotide polymorphism panel in broiler chickens', *Poult. Sci.*, 92, 1712–23.

Wang, X., Ford, B. C., Praul, C. A. and Leach, R. M. (2002), 'Collagen X expression in oviduct tissue during the different stages of the egg laying cycle', *Poult. Sci.*, 81, 805–8.

Wardecka, B., Olszewski, R., Jaszczak, K., Zieba, G., Pierzchala, M. and Wicinska, K. (2002), 'Relationship between microsatellite marker alleles on chromosome 1–5 originating from the Rhode Island Red and Geen-legged Partrigenous breeds and egg production and quality traits in F2 mapping population', *J. Appl. Genet.*, 43(3), 319–29.

Whitworth, K. M., Rowland, R. R. R., Ewen, C. L., Trible, B. R., Kerrigan, M. A., Cino-Ozuna, A. D., Samuel, M. S., Kightner, J. E., McLaren, D. G., Mileham, A. J., Wells, K. D. and Prather, R. S. (2015), 'Gene-edited pigs are protected from porcine reproductive and respiratory syndrome virus', *Nat. Biotechnol.* doi:10.1038/nbt.3434.

Wilkanowska, A., Mazurowski, A., Mroczkowski, S. and Kokoszynski, D. (2014), 'Prolactin (PRL) and Prolactin Receptor (PRLR) genes and their role in poultry production traits', *Folia Biol.* 62, 1–8.

Wolc, A., Lisowski, M., Hill, W. G. and White, I. M. S. (2011a), 'Inheritance of variation in production traits of laying hens', *Br. Poult. Sci.*, 52, 537–40.

Wolc, A., Stricker, C., Arango, J., Settar, P., Fulton, J. E., O'Sullivan, N. P., Habier, D., Fernando, R. L., Garrick, D. J., Lamont, S. J. and Dekkers, J. C. M. (2011b), 'Breeding value prediction for production traits in layers using pedigree and marker based methods', *Genet. Sel. Evol.*, 43, 5.

Wolc, A., Arango, J., Settar, P., Fulton, J. E., O'Sullivan, N. P., Preisinger, R., Habier, D., Fernando, R. L., Garrick, D. J., Hill, W. G. and Dekkers, J. C. M. (2012), 'Genome-wide association analysis and genetic architecture of egg weight and egg uniformity in layer chickens', *Anim. Genet.* 43(S1), 87–96.

Wolc, A., Arango, J., Jankowski, T., Dunn, I., Settar, P., Fulton, J. E., O'Sullivan, N. P., Preisinger, R., Fernando, R. L., Garrick, D. J. and Dekkers, J. C. M. (2014), 'Genome wide association study for egg production and quality in layer chickens', *J. Anim. Breed. Genet.*, 131, 173–82.

Wolc, A., Zhao, H. H., Arango, J., Settar, P., Fulton, J. E., O'Sullivan, N. P., Preisinger, R., Stricker, C., Habier, D., Fernando, R. L., Garrick, D. J., Lamont, S. J. and Dekkers, J. C. M. (2015), 'Response and inbreeding from a genomic selection experiment in layer chickens', *Genet. Sel. Evol.*, 47, 59.

Wolc, A., Kranis, A., Arango, J., Settar, P., Fulton, J. E., O'Sullivan, N. P., Avendano, S., Watson, K. A., Hickey, J. M., de los Campos, G., Fernando, R. L., Garrick, D. J. and Dekkers, J. C. M. (2016), 'Implementation of genomic selection in the poultry industry', Anim. Frontiers.

Wragg, D., Mwacharo, J. M., Alcalde, J. A., Hocking, P. M. and Hanotte, O. (2012), 'Analysis of genome-wide structure, diversity and fine mapping of Mendelian traits in traditional and village chickens', *Heredity*, 109(1), 6–18.

Wright, D., Kerje, S., Lundström, K., Babol, J., Schütz, K., Jensen, P. and Andersson, L. (2006), 'Quantitative trait loci analysis of egg and meat production traits in a red junglefowlxWhite Leghorn cross', *Anim. Genet.*, 37(6), 529–34.

Xu, H. P., Shen, X., Zhou, M., Luo, C. L., Kang, L., Liang, Y., Zheng, H., Nie, Q. H., Zhang, D. X. and Zhang, X. Q. (2010), 'The dopamine D2 receptor gene polymorphisms associated with chicken broodiness', *Poult. Sci.*, 89, 428–38.

Yao, J. F., Chen, Z. X., Xu, G. Y., Wang, X. L., Ning, Z. H., Zheng, J. X., Qu, L. J. and Yang, N. (2010), 'Low-density lipoprotein receptor-related protein 8 gene association with egg traits in dwarf chickens', *Poult. Sci.*, 89(5), 883–6.

Yi, G., Shen, M., Yuan, J., Sun, C., Duan, Z., Qu, L., Dou, T., Ma M., Lu, J., Guo, J., Chen, S., Qu, L., Wang, K. and Yang, N. (2015), 'Genome-wide association study dissects genetic architecture underlying longitudinal egg weights in chickens', *BMC Genomics*, 16, 746.

Youngren, O., Chaiseha, Y., Al-Zailaie, K., Whiting, S., Kang, S. W. and El Halawani, M. (2002), 'Regulation of prolactin secretion by dopamine at the level of the hypothalamus in the turkey', *Neuroendocrinology*, 75, 185–92.

Yuan, J., Sun, C., Dou, T., Yi, G., Qu, L., Qu, L., Wang, K. and Yang, N. (2015), 'Identification of promising mutants associated with egg production traits revealed by Genome-Wide Association Study', *PloS ONE*, 10, e0140615.

Zhang, G. X., Fan, Q. C., Wang, J. Y., Zhang, T., Xue, Q. and Shi, H. Q. (2015), 'Genome-wide association study on reproductive traits in Jinghai Yellow Chicken', *Anim. Reprod. Sci.*, 163, 30–4.

Zhang, L. C., Ning, Z. H., Xu, G. Y., Hou, Z. C. and Yang, N. (2005), 'Heritabilities and genetic and phenotypic correlations of egg quality traits in brown-egg dwarf layers', *Poult. Sci.*, 84(8), 1209–13.

Zhang, Y., Hou, Z. C., Chen, Z. X., Zheng, J. X., Chen, S. R., Qu, L. J., Li, J. Y., Xu, G. Y. and Yang, N. (2011), 'Low-density lipoprotein receptor-related protein 2 gene is associated with egg-quality traits in dwarf layers', *Poult. Sci.*, 90(12), 2718–22.

Index

Lightning Source UK Ltd.
Milton Keynes UK
UKOW06n1410280317
297705UK00001B/2/P

9 781786 760760